CRC Series in
Materials Science and Technology

Series Editor
Brian Ralph

Control of Microstructures and Properties in Steel Arc Welds
Lars-Erik Svensson

The Extraction and Refining of Metals
Colin Bodsworth

The Quantitative Description of the Microstructure of Materials
K.J. Kurzydłowski and Brian Ralph

Grain Growth and Control of Microstructure and Texture in Polycrystalline Materials
Vladimir Novikov

Corrosion Science and Technology
D. E. J. Talbot and J. D. R. Talbot

Image Analysis Applications in Materials Engineering
Leszek Wojnar

Surface Engineering
of
Metals
Principles, Equipment, Technologies

Tadeusz Burakowski
Radom University of Technology
Radom, Poland
  and
Institute of Precision Mechanics
Warsaw, Poland

Tadeusz Wierzchoń
Warsaw University of Technology
Warsaw, Poland

Translated from Polish by Dr. Witold K. Liliental

CRC Press
Taylor & Francis Group
Boca Raton London New York

CRC Press is an imprint of the
Taylor & Francis Group, an **informa** business

CRC Press
Taylor & Francis Group
6000 Broken Sound Parkway NW, Suite 300
Boca Raton, FL 33487-2742

First issued in paperback 2019

© 1999 by Taylor & Francis Group, LLC
CRC Press is an imprint of Taylor & Francis Group, an Informa business

No claim to original U.S. Government works

ISBN-13: 978-0-8493-8225-3 (hbk)
ISBN-13: 978-0-367-40012-5 (pbk)

Library of Congress Cataloging-in-Publication Data

Burakowski, Tadeusz.
      Surface engineering of metals : principles, equipment,
technologies / Tadeusz Burakowski, Tadeusz Wierzchoń .
           p.   cm. -- (Materials science and technology)
      Includes bibliographical references and index.
      ISBN 0-8493-8225-4 (alk. paper)
      1. Metals--Surfaces. 2. Surfaces (Technology) I. Wierzchoń ,
Tadeusz. II. Title. III. Series: Materials science and technology
(Boca Raton, Fla.)
TS653.B87 1998
620.1′6—dc21

                                                                                98-42176
                                                                                CIP

Tadeusz Burakowski: part I; part II – chapter 1, 2, 3, 4, 6
Tadeusz Wierzchoń :   part II – chapter 5

Library of Congress Card Number 98-42176

**Visit the Taylor & Francis Web site at
http://www.taylorandfrancis.com**

**and the CRC Press Web site at
http://www.crcpress.com**

Preface

Surface engineering is a new field of science and technology. Although the specific topical groups included in its domain have been known and practically applied in other areas, it is only in the last few years that surface engineering has been recognized as an individual discipline of applied science. This book is the first in the world to provide a complex treatment of problems related to surface engineering.

The material of this book has been treated in two parts, so designed as to allow extension in future editions.

Part I, devoted to general fundamentals of surface engineering, contains a history of its development and a distinction is suggested between superficial layers and coatings. Further, but foremost, the basic potential and usable properties of superficial layers and coatings are discussed, with an explanation of their concept, interaction with other properties and the significance of these properties for the proper selection and functioning of surface layers. This part is enriched by a general description of different types of coatings.

Part II contains an original classification of production methods of surface layers. This part presents the latest technologies in this field, characterized by directional or beam interaction of particles or of the heating medium with the treated surface. Due to its modest length, the book does not discuss older methods which are well known and widely used.

This edition is a revised version of the first Polish edition of the book entitled "Surface Engineering of Metals - Principles, Equipment, Technologies", published by Wydawnictwa Naukowo-Techniczne (Science-Technological Publications), Warsaw, 1995.

The authors express their gratitude to all who in any way contributed to the presentation of the broad array of problems of surface engineering in this form. In particular, our thanks go to professors: J. Kaczmarek, K.J. Kurzydłowski, R. Marczak, B. Ralph, J. Senatorski, J. Tacikowski and W. Włosiński, as well as doctors S. Janowski, K. Miernik and J. Walkowicz for their discussion regarding the book and constructive suggestions.

Special words of thanks are due to Dr. A. Mazurkiewicz, Director of the Institute of Terotechnology in Radom, for his invaluable help in the preparation of the work for print, and to the publisher, Wydawnictwa Naukowo-Techniczne, for making information material from the first edition available. The authors thank Mr. A. Kirsz for expert technical assistance for providing camera-ready text.

The authors

Table of Contents

Part II. *The newest techniques of producing surface layers*

1. Formation of technological surface layers 235

General Fundamentals of Surface Engineering

chapter one

The concept of surface engineering

1.1 The term "surface engineering"

The word *engineering* stems from the French language (*s'ingenier* - to contemplate, rack one's brains, strain oneself, exert oneself) and in the past had one meaning, while presently it has several meanings, all fairly close. In the past it was a skill; presently it is mainly a science relating to the design of shape or properties of materials and their manufacturing processes.

Originally engineering encompassed the art of building fortifications, strongholds and other elements of defense systems. In 18th - 19th century Europe we see the beginnings of differentiation between military and civilian engineering. In more modern times the concept of engineering embraced the art of design and construction of all types of structures (with the exception of buildings) and various engineering branches were distinguished: civil, hydro-, maritime, sanitary, forestry.

After World War II, the influence of Anglo-Saxon countries caused the spread in Europe of the US - born concept of social engineering. Quite recently, a new branch of science, termed environmental engineering, came into existence.

It was also during this last century, especially after World War II, that the term engineering was broadened to encompass some areas of human knowledge, more particularly those connected with applied research, e.g. the science of unit operations used in the chemical and related industries and the subsequent development of chemical equipment (chemical engineering), or the applied science drawing on the theoretical achievements of genetics in the breeding of animals, cultivation of plants and in medicine (genetic engineering).

Created and in use are such concepts as: biomedical engineering, electrical engineering, reliability engineering, programming engineering, communications engineering, aerospace engineering, process engineering, mechanical, ion beam, corrosion and other types of engineering.

The early 70s saw the importation from the US to Europe of the concept of *material engineering*, created in the 60s and embracing the "scientific discipline dealing with the investigation of the structure of materials, as well as improvement and the obtaining of new materials with predicted and reproducible properties." (Scientific and Technical Lexicon, WNT, Warsaw 1984).

Departments, chairs, institutes and even entire faculties of material engineering have sprung up.

It follows from the above definition that materials engineering deals with the investigation of the structure of and the design of different materials, including composites. It does not follow, on the other hand, although it cannot be excluded, that materials engineering deals specifically with problems of enhancement or modification of surface properties of materials.

It is probably due to this that the new term *surface engineering*[1] was coined for the first time in England in the 70s. In the early 70s the Surface Engineering Society, affiliated with the Welding Institute in Abington, was inaugurated. At first, it focused mainly on various aspects of welding and thermal spraying and gradually it broadened its scope of interest. Next, the Wolfson Institute for Surface Engineering was created at the University of Birmingham, initially concerned mainly with problems stemming from surface diffusion treatments and their connection with vacuum technology, gradually broadening the range of activity to other methods of formation of surface layers.

The year 1985 saw the first edition of the quarterly "Surface Engineering", published by the Wolfson Institute for Surface Engineering jointly with the Surface Engineering Society. As of 1987 another quarterly of a scientific-research and technical nature was published under the same title, as the combination of two periodicals: "Surfacing Journal International" and "Surface Engineering". This quarterly deals with thermal spraying technologies, layer formation by PVD and CVD, electron and laser beam hardening, ion implantation, shot peening, surface alloying by conventional and plasma processes and generally with technologies of surface layer formation and with some coating technologies. Problems of coatings, especially paint, plating and other types, are dealt with by other periodicals (e.g., "Surface and Coatings Technology", "Coatings", "Metalloberfläche" and "Metal Finishing").

In October 1986, at the V International Congress of Heat Treatment of Materials in Budapest, the name of the International Federation for Heat Treatment of Materials, by then in existence for over 10 years, was changed to International Federation for Heat Treatment of Materials and Surface Engineering. For obvious reasons, both the Federation as well as Congresses convened under its auspices prefer mostly problems connected with heat treatment and, to a lesser degree, other problems connected with surface engineering.

Over the past most recent years, many international conferences, meetings and discussions devoted to surface engineering and its connections with other fields of science and technology were organized.

---

[1] This term was later translated into French (*l'ingenierie de surfaces*), Russian (*inzhinerya poverkhnosti*), and German (*Oberflächeningenierie*) but to this day used in these languages only sporadically.

1.2 Scope of topics forming the concept of surface engineering

Surface Engineering is almost as old as structural materials used by man. From the beginnings of time until the early 70s of our century, mankind has worked on the development of surface engineering, although not aware of the concept. The term of surface engineering, in use in the world for over ten years, remained undefined and its topical scope is still the subject of discussions, especially on the aspect of definitions.

In various ways, attempts have been made to define and to conduct a broader discussion of selected problems of surface engineering, especially those viewed through the techniques of formation covered by this scope [1, 4]. Various book and handbook type publications presented different, chronologically older technologies, within the scope of surface engineering. There was a lack of publications dealing with the newest methods of manufacturing.

Earlier, generally the concept of surface enginnering was understood as solely different techniques of forming superficial layers prior to the beginning of service. Nothing was said about the formation of superficial layers during service, about research and propertiers or about modeling of these properties for concrete examples of application. Even newer literature does not present a modern approach to the overall concept of surface engineering [5, 6].

Today, such narrow understanding of surface engineering does not suffice. In fact, this would be a far-reaching simplification. For this reason, it was broadened during the years 1993-1995 to include problems of utilization of superficial layers, as well as problems of their design [3, 4].

Based on research conducted since the 80s, as well as available scientific and technical literature, the following topical scope and a definition of surface engineering are proposed:

Surface engineering is a discipline of science, encompassing:

1) manufacturing processes of surface layers, thus, in accordance with the accepted terminology - superficial layers and coatings, produced for both technological and end use purposes,

2) connected phenomena,

3) performance effects obtained by them.

Surface engineering encompasses all scientific and technical problems connected with the manufacture of surface layers prior to end use or service (technological layers) or during service (service-generated layers), on or under the surface (superficial layers) or on a substrate (coatings), with properties differing from those of the material which may be introduced to the surface of the core in the form of gas, liquid or solid (Fig. 1.1). It also includes research of connected phenomena and of potential and usable properties of surface layers, as well as problems connected with layer design.

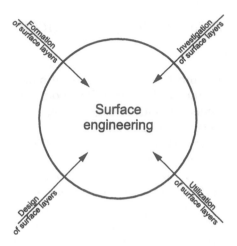

Fig. 1.1 Schematic representation of the area of activity of surface engineering.

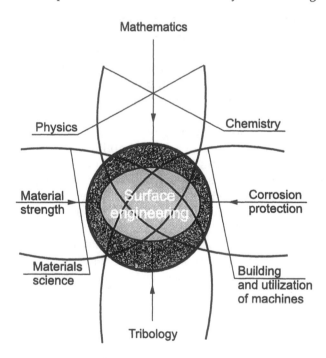

Fig. 1.2 Scientific and technical activity adding up to create surface engineering.

Thus, *surface engineering encompasses the total field of research and technical activity aimed at the design, manufacture, investigation and utilization of surface layers, both technological and for end use, with properties better than those of the core, such as mainly anti-corrosion, anti-fatigue, anti-wear and decorative. Other applications include properties such as optical, thermophysical, electrical, magnetic, adhesive, ablation, passivation, inhibition, catalytic, biocompatibility, diffusion and others.*

In the meaning as defined above, surface engineering has a lot in common with fundamental and applied (technical) science.

Surface engineering draws inspiration from (Fig. 1.2):

1) Fundamental sciences: physics, chemistry, partially mathematics and constitutes their application to material surface;

2) Applied (technical) sciences:

– sciences dealing with materials science and material engineering, with special emphasis on heat treatment,

– construction and use of machines, with special emphasis on material strength, primarily fatigue, tribology and corrosion protection,

– electrical engineering, electronics, optics, thermokinetics, the science of magnetism, etc.

The object of material science and material engineering - the material constitutes the fundamental substance, the surface properties of which are improved, enhanced and controlled by surface engineering. The knowledge of material substrate or core structure is the basic condition of producing layers on it. Methods of *formation* (producing) surface layers are included in the area of machine building, as manufacturing methods.

The properties of surface layers produced are evaluated by methods used in surface engineering, as well as in *investigation* and use of machines. These methods are used predominantly in areas such as: tribology, corrosion protection, material strength, etc.

Some methods of *designing* of surface layer properties, used in surface engineering, are also derived from - besides mathematics - material engineering and machine building. This pertains primarily to material strength and tribology.

The *utilization* of surface layers or their production during the course of service belongs to the area of machine service and takes into account, first and foremost, problems of tribology and corrosion protection.

References

1. **Bell, T.**: Surface engineering, past, present and future. *Surface Engineering*, Vol. 6, No. 1, 1990, pp. 31-40.
2. **Burakowski, T.**: *Metal surface engineering - status and perspectives of development* (in Russian). Series: Scientific-technical progress in machine-building. Edition 20. Publications of International Center for Scientific and Technical Information - A.A. Blagonravov Institute for Machine Science Building Research of the Academy of Science of USSR, Moscow, 1990.
3. **Burakowski, T., Roliński, E., and Wierzchoń, T.**: *Metal surface engineering* (in Polish). Warsaw University of Technology Publications, Warsaw, 1992.

4. **Burakowski, T.**: A word about surface engineering (in Polish). *Metaloznawstwo, Obróbka Cieplna, Inżynieria Powierzchni (Metallurgy, Heat Treatment, Surface Engineering)*, No. 121-123, 1993, pp. 16-31.
5. **Tyrkiel, E.** (General Editor), **and Dearnley, P.** (Consulting Editor): *A guide to surface engineering terminology.* The Institute of Materials in Association with the IFHT, Bourne Press, Bournemouth (UK), 1995.
6. **Stafford, K.N., Smart, R. St. C., Sare, I., and Subramanian, Ch.**: *Surface engineering: processes and applications.* Technomic Publishing Co. Lancaster (USA) - Basel (Switzerland), 1995.

chapter two

Development of surface engineering

2.1 History of development of surface engineering

2.1.1 General laws of development

Material development of human civilization was made possible mainly owing to the material base of this development, i.e.,

1) utilization of the Earth's natural resources as a source of structural material

2) development of techniques of manufacture of material means from these resources, dependent primarily on the utilization of fire, and later, other sources of manufacture and utilization of thermal energy, in particular, of electrical heating.

The role of surface engineering in the process of manufacture of the material product is shown in Fig. 2.1.

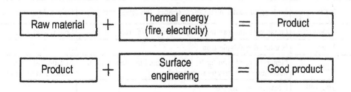

Fig. 2.1 Role of surface engineering in the process of manufacture of a material product.

2.1.2 History of development of metallic structural materials

The development of material history of mankind should undoubtedly be associated with the appearance of the first stone implements in Central-East Africa, about 1,700,00 B.C. (Table 2.1), the harnessing of fire by man about 1,400,00 years later (China) and finally, about 200,000 years after that - the skill of making fire. Several tens of thousands of years B.C. man already used improved stone and bone utensils, like knives, awls, engravers, saws and drills) and several thousand years ago mastered the skill of mining for flint-stone. The **era of the cleaved stone**, which had lasted for over a million years, ended with the appearance of the first tools and ornaments made from metals, although stone implements were still in use. Chro-

Surface engineering of metals

nologically, the most important role was that of copper, tin, gold, silver, lead and, finally, iron, as well as antimony, mercury, nickel, chromium, manganese, tungsten, aluminum and other metals.

Table 2.1
Chronological development of the most important
achievements of materials engineering

Era	Years	Achievement	Place
1	2	3	4
		B.C.	
Stone	1,700000--1,000000	Incidental use by "homo habilis" (skilled man) of found objects (mainly stones for tools)	Central-East Africa
	600000	Production of first stone tools - (fist hammers) by "homo erectus" (erect man)	China, Indonesia
	300000	Use of fire of natural origin (lightening)	China, Indonesia, South Africa
	350000-200000	Use by "homo sapiens" (thinking man) of knives, stone scrapers, wooden bludgeons and spears	Northern and Central Europe
	100000	Skill of making and lighting fire and use of fire obtained by burning wood, dried excrements and grass. Use of spears singed in fire	Eurasia, Africa
	12000	Beginnings of open-pit mining of flint-stone	Poland, Hungary
Bronze	5000	Finding on the surface of the earth of native metals, e.g. gold, silver, copper and iron from meteorites	
	5000	Processing of native copper	Egypt
	4000	Melting of copper from ores, first metal tools. Beginnings of foundry practice	Mesopotamia, Egypt
	3400	Beginnings of mining and processing of iron	Egypt
	3200-3000	Beginnings of mining and processing of iron	Mesopotamia, Egypt
	2994	Production of first copper needles	Egypt
	2000	Melting of easier metal ores: copper, tin, lead, iron	Orient, Mediterranean Basin
Iron	1900	Ornaments and natural weaponry made of iron	South-West Africa, Armenia
	1400	Hardened weldable iron	Armenia
	1000	General use of iron tools; iron obtained in dough form becomes basic structural material	Middle East, south-central Europe, Mediterranean Basin
	ca. 700-400	Obtaining of pig iron	China
	412	Utilization of iron beams in building construction	Greece (Athens)
	400	Cast iron	Europe
	480-220	Production of steel	China
	90	Iron posts of 18 m height and 17 ton weight	India (Delhi)
		A.D.	
	ca. 100	Utilization of coal as fuel	Rome
	ca. 400	Utilization of gas as fuel	Greece
	1000	Smith's forges	Stiria
	until 1200	Charcoal smelting furnaces, combined with forging to obtain steel	
	1200	Tall furnace (greater charcoal smelter) for production of steel	
	ca. 1300	Furnaces for repeatable melting of steel in charcoal flames	
	ca. 1450	Beginnings of general use of cast iron for manufacture of everyday products	Germany
	ca. 1500	Charcoal-fired blast furnace. Pig iron	

1	2	3	4
	ca. 1678	Cast iron becomes basic material for machine construction	Europe, China
	1709	Coke-fired blast furnace	Great Britain
	1721	Metallic zinc obtained for the first time by J.F. Henckel	Europe
	ca. 1750	Crucible furnace. Melted steel	
	1751	Discovery of nickel (A.F. Cronstedt)	Sweden
	1783	Obtaining of tungsten from ore (d'Elhuyar brothers)	France
	1784	Forgeable steel from a flame furnace, so-called puddle iron (H.Cort)	Great Britain (Lancaster)
	ca. 1800	Tool steel for cold work, bearing steel (Stribeck)	
	1811	German firm F. Krupp begins production of cast steel	Germany (Essen)
	1821	Bethier produces chromium alloy steel	France
	1828	Obtaining of pure aluminum from clay (F. Wöhler)	Germany
	1850	Commercial production of nickel alloy steel (Wolf)	Germany (Schweinfurt)
	1855-1856	H. Bessemer designs converter. Melted steel	Great Britain
	1857	Tool steel with additions of chromium and tungsten (R.F. Mushet)	Scotland
	1858	Oxland produces tungsten alloy steel	Germany
	1862	First synthetic material produced by A. Parks	Great Britain
	1865	Alloyed steel with chromium	Germany
	1883	Manganese steel (R.A. Hadfield)	United States
	1886	Industrial-scale electrolytic production of aluminum (P.L. Heroult)	France, China
	1889	Nickel alloy steel produced by Riley	United States
	1899-1900	Tungsten-containing high speed steel (F.W. Taylor, M. White)	United States
	1906	Development of alloy named duralumin (A. Wilm)	Germany
	1912	Production of nickel-chromium alloy steel by F. Krupp	Germany (Essen)
	1913	Chromium-based stainless steels (H. Brearly)	Great Britain
	1922	Sintered carbides	
	1950	Sintered metal ceramics	
	1965	Metal composites	

The bronze era began with the finding by man of pieces of copper, 4000 to 5000 B.C. Copper is the oldest metal known to mankind and although its content in the earth's crust is small (estimated to be ca. 0.01% by weight), it has played an unusually significant role in the history of man's evolution [1]. Initially it was used in the form of native copper (forged objects), later in the form of alloys with other metals: bronzes (alloys of copper with tin and possibly other components) and brasses (alloys of copper with zinc and possibly other components). Copper and its alloys, particularly the bronzes (harder than copper and at the same time easier to melt), formed the material basis for the manufacture of the first implements and ornaments. Bronze was used primarily to manufacture vessels for everyday use and for rituals, weaponry, lamps, mirrors, ornaments, instruments, for sculpture and for astronomical devices. Depending on the alloy composition, these objects were of dark brown, red, and even silver or green color.

Tin in its native form has been known since the beginnings of civilization, but it was only later that it became known in the form of a pure metal.

Ancient civilization, extending from the confluence of the Euphrates and Tigris (today's Iraq) to northern Africa (Egypt) and southern Europe (Greece, Italy), brought with it the skill of obtaining and use of several other metals, besides copper and its alloys, namely - gold, silver and lead. The remaining two metals known in ancient times - mercury and antimony - were probably found later, during the golden age of Greece and later of Rome [2].

Lead was discovered earlier as a by-product of melting zinc out of zinc-lead ores for the production of brass. In its purer form it found use in the manufacture of coffins, barrel girdles, wire and cannon balls - until the moment of mastering the production of cast-iron balls. Roman aqueducts were made of lead pipes [1].

The iron age basically began ca. 1000 B.C. although iron of meteorite origin had been known 3 - 4 thousand years earlier. Iron makes up about 5% of the earth's mass. However, during the bronze age, it found only marginal use, on account of the difficulty connected with its obtaining and processing. The melting point of iron, particularly that of its alloys, including those most popular, i.e., with carbon, was much higher than the melting point of copper alloys. Such temperatures could not, at first, be generated artificially by man. Attempts at iron processing date back to the middle of the 4th century B.C. but success came only during the 1st and 2nd century B.C. The oldest written relics and excavations show proof that the production or, at least, the use of iron was not unknown to almost all the peoples of the ancient world already in the beginnings of history. Wrought iron objects, mainly of meteorite origin, were used sporadically by the 4th and 3rd century B.C. in Egypt and western Asia. The European continent acquired the skill of iron processing first in the Aegean Basin, ca. 1000 B.C. About the 4th century B.C. iron began to slowly to oust bronze and zinc bronze.

The development of technology of iron processing proceeded primarily in the direction of improvement of smelting furnaces, and the production of alloys of iron and carbon. Later, other alloying elements were added, mainly: tungsten, chromium, aluminum, nickel. Still later came also the utilization of other manufacturing and processing technologies, e.g., forging, casting, and machining of wrought and cast steel, both carbon and alloyed.

Toward the end of the 19th century, hard tool steels, both carbon and alloyed, came to be used, followed by high speed steels and finally by sintered carbides, metalo-ceramic and ceramic sintered materials. During the second half of the 20th century, metal composites were developed. It should be remembered that the very first composite, universally used in building construction since ancient times, was a mixture of hay with clay (ca. 3000 B.C. in Mesopotamia). Superalloys, high strength alloys of Ti and Cr, microalloyed steels, duplex steels and metal glazes have been developed.

In the development of human civilization, metallic materials have always played a significant role, usually a predominant one. Over thousands of years, the percentage share of metallic materials grew up to about 1960. Approx. 10,000 years B.C. metallic materials constituted only several percent, while ceramic materials as much as 40%, non-metallic composites ca. 10%, and polymers (wood, fibers, hides and glues) ca. 45%. By 1960, the share of metallic materials rose to almost 80%. It is estimated that by 2020, the share of metallic materials may decline to approximately 50%.

Fig. 2.2 shows the development of the most important metallic substrate materials, from the point of view of their usable properties.

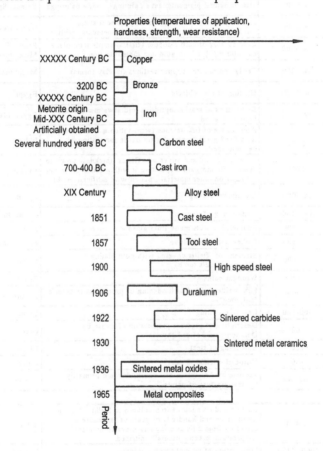

Fig. 2.2 Historical development of the most important metallic substrate materials.

2.1.3 *History of development of the technology of surface improvement of structural materials*

Products of the ironmaking and metal industries may be and sometimes are used without any surface improvement. Usually, however, various manufacturing and processing technologies have been applied from ancient times to this day (Table 2.2) to improve the service life of the material by several to several thousand percent in comparison with the raw substrate.

Table 2.2
Chronological development of the most important
achievements of surface engineering

Era	Years	Achievement	Place
1	2	3	4
		B.C.	
Stone Age	ca 35000	Use of natural pigments: cave paintings, body paintings	France, Spain
	ca 7000	Artificially smoothened (ground, polished) stone implements. Making holes in stone implements (drilling)	South-western Asia
Bronze	ca 6000	Cold forging of native copper (first attempts took place ca 8000 B.C.), silver, gold and meteorite iron	Persia, Egypt, Mesopotamia
	ca 4000	Fairly complicated copper castings (in clay forms)	Mesopotamia
	ca 3500	Glazing of clay objects	Egypt
	ca 3000	Beginnings of machining: drills and saws made of stone, bone and wood	Egypt, Mesopotamia
	2874	Appearance of first faience (semi-vitreous China ware with fired glaze, containing tin oxide for sealing of ceramic shell)	India
	2708	Forging of gold and silver into flakes (thickness down to even 0.001 mm) and depositing it with a brush on waxed wood or on other materials, plated with artificial marble	Egypt
	ca 2000	Investment casting	Egypt
Iron	2000-1500	Carburizing of iron	Egypt, India
	ca 1600	Enameled objects made from glass and ceramic	Syria, Egypt
	ca 1500	Lathes with string drive	Mesopotamia, Egypt
	ca 1400	Statuette of Osiris covered with gold plating (Tutenkhamon's tomb)	Egypt
	700	Wrought coins	Egypt
	ca 700	First written record of hardening of projectiles (Homer's "Odyssey")	Mediterranean Basin
	700-500	Cast iron casting	China
	200	Description of pigment preparation (Treatise by Democrite of Bolos: "Physica")	Greece
	120	Hardening of steel	China
	100	Nitrocarburizing of steel	China
	100 -80	Treatise by Anaxylaos of Larissa, devoted primarily to the manufacture of paints	Greece
		A.D.	
	ca 414	Casting and erection of iron column in Delhi which, as a result of several hundreds of years of exposure to ammonia from animal urine and elevated ambient temperature, became naturally nitrided	India
	1000- 1100	Propagation of wrought iron products	Europe
	ca 1100	German monk Theophilus accurately described pack carburizing in his work: "Schedule diversarum artium"	Germany
	1400	Manufacture of cast iron castings	France
	1500	The implementation of a vertical drill (with horse gear drive) to machine holes	Europe
	1620	Beginnings of manufacture of tin-plated steel sheet	Saxony
	1722	French physicist Rene Antoine Fechrault de Reamur gives accurate (close to modern-day) procedure for pack carburizing of steel in his work: " The Art of Changing Wrought Iron into Steel".	France
	1742	Frenchman Malouin discovers that zinc-plated steel sheet does not rust	France

1	2	3	4
	1784	Englishman H. Cortl conducts first modern process of steel rolling (from steel ingot to finished product)	England
	1786	W. Watson develops specification for producing of zinc-plated steel sheet. Production of hot-dip galvanized steel	Great Britain
	1786	Hot dip gold-plating	
	1789	Discovery by L. Galvani of the so-called "animal electricity"	Italy
	1790	J. Keir discovers possibility of passivating iron	Great Britain (West Bromwich)
	1802	First hydrogen-oxygen welding torch	United States
	1803	Englishman W. Cruikshank deposits copper and gold by electrolysis	Great Britain
	1805	L.G. Brugnatelli accomplishes electrolytic gold plating	Italy
	1805	L. Stone invents gas burner (town gas and oxygen)	Great Britain
	1805	Hobson and Silvester discover that zinc heated to 100-150°C can be easily rolled into sheet	Great Britain (Sheffield)
	1840	Sheet electro-tinning (Frenchmen: Roseleur and Bucher)	France
	1848	Burnishing of shafts and journals for the railroad industry. Tradesmen use burnishing ealier	Great Britain
	1856	W.H. Perkin discovers first synthetic pigment (blue)	Great Britain
	1856	Chemist Pantocek conducts industrial-scale deposition of coatings on glass. Coatings are metallic, tinted by bismuth oxides, and exhibit rainbow colors. (Earlier, colored coatings were deposited on glass by tradesmen in Egypt and Greece)	Hungary
	1857	M. Faraday deposits metal vapors on a glass substrate in vacuum	Great Britain
	1862	Russian N. Bernados and Pole Olszewski obtain patent for arc welding by a carbon electrode	Poland, Russia
	1871	W. Kruk defines and describes a method for pack carburizing of steel	Poland
	1878	J.W. Gibbs introduces the concept of interface	United States
	1883	T.A. Edison accidentally discovers the phenomenon of thermal emission of electrons from a lit bulb filament in vacuum	United States
	1900	Two Frenchmen: Picard and E. Fouche achieve first lighting of oxy-acetylene torch	France
	1905	Frenchman E. Fouche introduces flame welding	France
	1908	Industrial application of the Sol-gel method of metal plating. (Tin, zinc or lead in the form of a suspension is deposited on the metal surface with a brush and dried with a soldering burner)	Germany (Berlin)
	1909	M.U. Schoop develops a pot spray gun	Switzerland (Zurich)
	1913	M.U. Schoop and others develop wire flame spraying (WFS)	Switzerland (Zurich)
	1914	M.U. Schoop and others develop an arc spray gun (AC)	Switzerland (Zurich)
	1916	First attempts at burnishing	Germany
	1916	A. Einstein introduces the concept of forced emission	Germany
	1921	M.U. Schoop obtains patent for flame-fired powder spray gun	Switzerland (Zurich)
	1924	First industrial implementation of gas nitriding (A. Fry)	Germany
	1928	R. Mailänder determines the effect of type of polishing on fatigue strength of steel	Germany
	1928	I. Langmuir introduces the concept of plasma	United States
	1930	B. Berghaus obtains patent for glow discharge nitriding of steel	Germany
	1938	First practical utilization of the thermal effect of an electron beam on metal	
	1943	Development of basics of shot-peening by J. O. Almen	Germany
	1944-1945	Glow-discharge nitriding of gun barrels	Germany

1	2	3	4
	1940-1948	Electro-discharge coatings	USSR
	1948-1950	Dynamic development of electron beam technology	
	1950	Investigations of fundamentals of ion implantation	
	1953-1954	Implementation of forced emission to amplify microwaves and the construction of the first maser by Ch. H. Townes and J. Weber - USA and N.G. Basov and A.M. Prokhorov - USSR	United States, USSR
	1954	W. Shockley obtains patent for implantation of dopants to semiconductors	United States
	1955-1956	R.M. Poorman and others conduct first detonation guns spraying of metal substrate with powdered substance	United States
	1956	D.N. Garkunov discovers the phenomenon of wear-free friction	USSR
	1956-1957	Beginning of practical implementation of PVD	United States
	1960	A.R. Stetson and C.A. Hauck conduct controlled atmosphere plasma spraying (CAPD)	United States
	1960	Invention of the laser by T.H. Maiman	United States
	1964	C.K.N. Patel constructs first molecular CO_2 laser	United States
	1972	Production of industrial furnaces for ion nitriding in Europe	Germany
	1972	H.A. Köhler and others construct the first excimer gas laser	United States
	1973	E. Mühlberger conducts first vacuum plasma spraying (VPS)	
	1982	J. Browning conducts high velocity oxy-fuel spraying (HVOFS)	United States

The development of structural metallic materials is accompanied, although not quite keeping pace but slightly lagging behind, by the striving toward ensuring of best properties of the surface:
  - resistance to oxidation and other forms of corrosion, including high temperature corrosion and corrosion in environments of different aggressiveness,
  - resistance to sliding, abrasive and erosion wear,
  - raising static and dynamic (fatigue) strength,
  - giving the surface special physical properties, e.g., improving electrical conductivity,
  - facilitating the carrying out of subsequent technological operations.
This direction was manifest in the implementation of [3]:

1) various physical, chemical, thermal and electrical phenomena - both single and complex - in order to impart the required properties to the surface of the structural material,

2) various materials and their compositions, in order to give the surface other properties than those of the substrate, by coating of the surface by various means (immersion, spray, sputtering) with:
  - metals and alloys,
  - non-metals (e.g., C, N, B),
  - intermetallic compounds,
  - silicates (metals, ceramic and glass),
  - paint products (paints and varnishes),

– plastics,
– oils, greases, wax, paraffin, gum, indiarubber, tar, bitumens.

Chronologically, the earliest technologies for surface improvement were those which shaped the form of the metal object: by changing shape, usually the surface properties were improved. Improvement of the surface by reducing its roughness went hand in hand with giving the object the required shape. The first such tools were artificially smoothened stone implements known and used in Asia since 7000 B.C.

Regarding metals, among the earliest technologies connected with surface engineering is **cold forging**. The beginnings of forgings (smithies) date back to 5000 B.C. when man could not yet smelt metals from ores but already knew how to cold forge native copper, silver, gold and iron from meteorites, using stones as tools [4].

A specific version of cold forging was the obtaining of very thin foil from gold and cladding with it objects made from other metals. The first object found clad with a gold layer (gold-plated) was a wooden sculpture of an animal's head, dating back to the 3rd century B.C. The same gold plating was present on the statuette of Osiris, made from bronze, found in the tomb of Tutankhamon, dating back to the 16th century B.C. [4].

On a broader scale, and predominantly regarding bronze objects, metal forging was known in Egypt in the 3rd century B.C. Besides the earliest recorded free forging, die forging was known already in ancient times, initially used mostly for coins. First such coins appeared ca. 7th century B.C. The oldest known evidence of forging of iron were iron coins from Sparta, dating back to ca. 5th century B.C. [4].

An especially significant role in the development of forging was played by the development of ironmaking which, in a way, forced the general use of forging as a technology. In the 10th and 11th centuries A.D. forged metal objects were widely known throughout Europe. A strong development of the artistic forging trade came about in the 16th and 17th centuries A.D. [5].

Besides forging, other methods of cold as well as **hot plastic forming** developed. Significantly later times brought the development of rolling, drawing and press forming. Today, plastic forming, as a process in which shaping or division of metal, bringing about changes in its physico-chemical properties, structure, surface roughness or the formation of residual stresses is brought about by plastic deformation, constitutes one of the basic types of technologies in the manufacturing and processing industries. Since the middle 19th century, volume hot plastic forming has been broadened by surface forming, accomplished first by journeyman methods of hammering, to be replaced later by mechanized treatment. In the early 20th century surface treatment was supplemented by surface forming, known as **burnishing**, which is a combination of sorts of plastic forming with machining.

A second, almost equally as old forming technology is **casting**, known as a trade beginning in antiquity. The Sumerians, inhabitants of Mesopotamia, knew the art of obtaining even fairly complicated copper castings as early as the 4th century B.C. The oldest known Egyptian bronze castings date back to the 3rd century B.C. In approximately 2000 B.C. investment casting was known in Egypt. Casting was used in ancient Greece and Rome. In what is now Poland, beginnings of casting date back to the Neolithic era, more than 2000 years B.C. In ancient times, beside sculptures, coins and jewelry, technical equipment and machines were manufactured by the casting method. In medieval times, casting was used for bells, cannon, cannon balls, etc. [6].

In China, iron casting was probably known starting in the 7th - 5th century B.C. and during the 13th and 14th centuries A.D., its level was higher than that in contemporary Europe. It may be that the skill of casting iron reached Europe only toward the end of the 14th century when it began developing with the manufacture of cast iron cannon balls. Strong development of iron casting in Europe came in the 16th century with such products as gun barrels, water piping and furnace wall plates. A major step forward was the beginning of iron casting in molds, in 1708 by the Englishman A. Darby, as well as the construction in 1792 by the Englishman J. Wilkinson of the first coke fired cupola. In the 19th century, casting of iron was also adopted by the building construction industry. In 1824, American J. Laing designed the first equipment for continuous casting, in 1838 pressure casting was introduced and in 1890 this pressure casting was used for zinc alloys. In 1851, J. Meyer, in Germany, produced the first steel casting. At the end of the 19th century the casting of aluminum alloy products was begun and by the 1930s, casting of magnesium and zinc products was implemented. The technology of precision casting and shell molds was mastered in the days of World War II. Today, foundry engineering constitutes one of the most basic branches of the metal industry.

The third technology, almost as old as the former two and closely connected with surface engineering, is **machining**. Prototypes of the first lathes were the string friction drill and rock saws, known ca. 5000 B.C. Among the earliest machining tools used by man in the primitive form up until ca. 3000 B.C. were grindstone (sandstone), files, chisels, knives, drills - initially made of bone and flintstone, later of bronze, iron, cast iron and finally steel. The first primitive machine tools with string drive were invented: drills, lathes (ca. 1500 B.C.). In approximately 1500 A.D. the application of a vertical drill for machining gun barrel bores with horse-gear drive took place. When fairly good machine tools with mechanical drive were designed (initially for woodwork, including the smoothing of bores of cylinders), the end of the 18th century saw the popular use of mechanical tools. Until 1900 these were made from plain carbon steel; the years 1900-1906 mark the beginning of the use of high speed steels. Since 1926 sintered carbide tools and, since 1932, sintered metal oxide tools are in use.

From machining evolved the already mentioned strengthening of the surface by burnishing, used not semi-unwittingly beginning during the 19th century, and applied intentionally for the first time in 1916 in Germany. An offshoot of electro-erosive methods was **spark-discharge coating** (1940s).

Currently, machining constitutes about 80% of the overall operations used in the production of machine components [7].

The first known technology belonging to surface engineering was **heat treatment**, more precisely, surface heat treatment, both non-diffusion (hardening) and diffusion (carburizing, nitriding). As was the case with most inventions, it was mainly tied to the war effort when iron weaponry began to be used. The first attempts at enhancing of iron by the introduction into its surface of elements which hardened it took place in Egypt and in India ca. 2000 - 1500 B.C. Probably, it was carburizing of iron surface. The first description of hardening projectiles must be attributed to Homer (Odyssey, ca. 700 B.C.). In approximately 120 B.C. hardening of steel was used in China and about 20 years later, soy bean grains were used for carburizing or, more strictly speaking, carbonitriding of steel [8]. These beans, rich in carbon and nitrogen, were used to saturate steel, heated to red heat. In a sense, heat treatment of steel is tied to the production of Damascus steel, distinguished for its high hardness and elasticity without hardening. Developed initially in India during the early Middle Ages (4th - 11th centuries A.D.), perfected by the Arabs and propagated by them throughout Europe ca. 1400, this method consisted of welding together spliced rods or wires of steel with a different carbon content, ranging from 1.2 to 1.8%, by their annealing and multiple forging. Similar to Damascus steel, another such material was made by Romans and by the Japanese. In later times, Damascus-type steel from Persia was highly valued.

The first relatively precise description of pack carburizing was given by the German monk Theophilius ca. 1100 A.D. in his work: "Schedule diversarum artium" [8] and not, as was until recently supposed, in 1772 by R.A. Fechrault de Reamur in "L'art de convertir le fer forge en acier et l'art d'adoucir le fer fondu" [9]. In 1871 W. Kruk from Poland rendered a precise definition and description of pack carburizing [10].

It would be worth mentioning that nitriding, used on an industrial scale in Germany since 1924, also has its prehistory. In 415 A.D. a forged iron column was erected in Delhi which, surprisingly, did not corrode with passing years. Investigations conducted 15 centuries later showed that the surface of the column is covered with a thin layer of iron nitrides, ensuring the iron perfect anti-corrosion protection. This column, however, was never intentionally nitrided by anyone. It is now supposed that the cause of the enhanced corrosion resistance of the column can be attributed to high concentration of ammonia in the surrounding air, originating from the vapors of animal urine, coupled with prolonged effect of the subtropical climate of India [11].

Presently, several tens of different surface heat treatment processes are in use, both involving diffusion and without diffusion, as well as several versions of the same process. The latest diffusion technologies, since the end of World War II, are carried out in conditions of glow discharge.

It was relatively early that **painting technologies** began to develop but it must be emphasized that for the first several tens of thousands of years they were used mainly for decorative purposes. The first cave paintings and the use of natural pigments for painting bodies were known as early as 35000 B.C. From ancient times come treatises about the preparation of natural pigments (Democrites of Bolos, ca. 200 B.C.) and the manufacture of paints (Anaxylaos of Larissa, ca. 100 - 80 B.C.). In 1856 A.D. an Englishman, W.H. Perkin, obtained synthetic pigments. Generally, up until the end of the 19th century, painting products were used mainly for artistic and decorative purposes. Since the turn of the century, they came to be used in the form of coatings for both protective and decorative purposes. It can be accepted as a fact that up to about 1930, paint formula principles had not undergone any basic changes. To the earth-based paints of various versions, known since ancient times, and water-based (e.g., lime, casein, tempera, aquarelle, silicate) new types were added, based on resins and natural oils [12]. Some time later came such new materials as nitrocellulose and alkyd resins, while by the late 1930s - PVC. Between 1950 and 1960 there came about a rapid development of new polymers: condensation, polymerizing, addition, etc., for the paint and varnish industry. Since 1960 a great stride has been made in the popularization of new painting techniques: pneumatic, hydrodynamic and electrostatic, with the use of dry, wet and electrophoretic paints [15]. Presently, increased use is made of ecologically friendly water-soluble paints and recycling of paint products.

Discovery by L. Galvani in 1789 of the so-called "animal electricity" is accepted as the beginning of development of those technologies within surface engineering which make use of the flow of current through an electrolyte in order to deposit an element contained in it on the surface of a metal or non-metal. Later research by the Italian A. Volta (1801) and the Russian B.S. Jacobi (1838), as well as by M. Faraday and H. Davy, laid the groundwork for **electroplating**. Its fast development came during the years 1940 - 1956. Presently, electroplating, or to use a more correct term, galvanostegy, is one of the fundamental technologies of surface engineering [13].

Deposition of coatings by the method of **thermal spraying** is tied in chiefly to the development of welding, more strictly speaking, to the development of heat sources for softening or remelting of the sprayed material. The first natural source of heat for welding was the flame and the oldest flame obtained artificially was the forge flame, which for many thousands of years had also been the basic source of heat for heat treatment. In the early 20th century a hydrogen-oxygen welding torch was used (1900), applied practically about 1905. Since the beginning of the 20th century, the electric arc became the most important source of heat, later to be overtaken by the plasma burner [14]. The beginnings of thermal spraying may be assumed to have taken place at the turn of the century when A. Schoop from Switzerland [15-17] atomized molten metal by a high velocity stream of gas and placed a metal sheet in the way of the stream. Later, he and his associates developed spray guns: pot, flame, wire and powder, as well as arc. Intensive development of thermal spraying began in the second half of the 20th century by the practical utilization of plasma, controlled atmosphere, vacuum and supersonic spraying [16, 17]. In 1955, R.M. Poorman accomplished the first utilization of the energy of detonation of explosive material to deposit a coating on a metal substrate. Already in 1786, thus substantially earlier than **spray metallizing, dip metallizing** came to be used [16].

Toward the end of the 19th century, increasingly acute problems began to appear, related to ensuring of essential product life, its tribological and corrosion resistance, decorative value and other special properties.

Table 2.3
Power densities introduced to the load

Type of technology	Method of heating	Power density [W/cm^2]	
		Possible to achieve	Most frequently used in practice
No beam technologies	Glow	$10 \div 10^2$	$0.2 \div 0.7$
	Indirect resistance- controlled atmosphere- fluidized bed	$0.5 \div 10$ $2 \cdot (10 \div 10^2)$	$0.5 \div 1.5$ $3 \cdot 10^1$
	Direct resistance	$10^2 \div 10^5$	$2 \div 6$
	Radiant	$1.0 \div 3 \cdot 10^2$	$5 \div 10$
	Electrode	up to 10^2	$5 \div 5 \cdot 10$
	Welding torch	$5 \cdot 10^2 \div 10^4$	$10^2 \div 10^3$
	Induction	up to $2 \cdot 10^4$	10^3
	Arc	$10^2 \div 10^4$	10^5
	Plasmotron	$5 \cdot 10^5$	$1.0 \div 6 \cdot 10^2$
Beam technologies	Ion	$6 \cdot 10^2$	
	Electron: - low energy - high energy	up tp 10^4 up to 10^{12}	$10 \div 10^2$ $10^3 \div 10^9$
	Laser: - continuous - millisecond impulse - nanosecond impulse	10^8 10^9 $4 \cdot 10^{15} \div 10^{20}$	$10^3 \div 10^6$ $10^3 \div 10^8$ $10^3 \div 10^{10}$
For comparison	- solar constant - solar (no condensation) - solar(condensed by lens)	0.1367 0.1 $5 \cdot 10^3$ and higher	 10^2

Note: Power densities introduced to load, utilized for technological purposes, are smaller by several times to several orders of magnitude than power densities possible to achieve.

During the 20th century, strictly speaking, during its first 50 years, practical utilization began of the interaction of the electron beam with materials. The basics of shot peening were developed and glow discharge in gases at partial pressure were implemented. The fundamentals of ion implantation in semiconductors and metals from the gas phase were developed. These works were intensified, especially after World War II and during the cold war. During the 1950s and 1960s, forced emission was utilized to amplify microwaves, the laser was built and implemented. Ion implantation was practically utilized, along with methods of and chemical vapor deposition from the gas phase (so-called PVD and CVD methods). Finally, besides plasma, detonation gun spraying came to be used.

Special mention should be made of the fact that the 1960s and later years constitute a period of rapid introduction and development of methods, techniques and technologies using a concentrated or, at least, a directed beam of high power density, solar energy, infrared radiation, plasma, ion beam and coherent photon beam (Table 2.3). In the majority of cases these newest methods of surface engineering are based on the latest discoveries in science and technology. They utilize the skills of mastering, creating and controlling beams of ions, photons and electrons which are all "hi-tech", latest, highly specialized and high efficiency, although high cost, techniques of surface enhancement.

2.2 Surface engineering today

2.2.1 General areas of activity of surface engineering -

During the past fifteen or so years and still to this day, surface engineering has been undergoing very dynamic development.

Every year, there are some fifteen scientific conferences held, dedicated to surface engineering or its particular fields. Each year, several books are published on the subject, mainly in the form of conference materials. Various scientific and technical journals publish many specialized works in the field of broadly understood surface engineering. These amount to several hundred or more annually.

Periodicals dedicated to surface engineering have appeared, and handbooks, reference books and monographies, dedicated to various problems covered by surface engineering, are being published. Various scientific organizations, some of an international status, dealing with aspects of surface engineering, are being launched. Surface engineering has been recognized as a scientific and technical discipline.

In substance, one observes an ever greater integration of object shaping techniques with those that impart special properties to their surfaces. An obvious broadening is noticed of the various areas covered by surface engineering, i.e., formation, design, investigation and utilization of surface layers, along with their progressing integration. Most advanced is the field of methods of manufacture of surface layers, while the connected field of property testing lags behind somewhat. An increasing number of reports are published, related to the area of utilization of surface layers. This is research conducted by the broad base of tribologists and machine users. Clearly least advanced is research in the field of design of surface layers.

Formation (manufacture) of surface layers. In the field of production techniques, surface engineering is involved with the constitution of surface layers, usually in the form of material, which from the point of view of properties is basically a composite [1, 2, 11, 18]. In this involvement, surface engineering takes into account the material of the core (or the substrate) and the interaction of the environment, both chemical and physical. In considering the concept of surface layers and coatings, the following distinctions should be made:

– **Technological layers** - are produced as the result of application of various methods, either independently or jointly. Depending on the set of effects utilized to this end, methods of producing of surface layers may be divided into 6 groups: mechanical, thermo-mechanical, electro-chemical, chemical and physical. In each group different methods are utilized to produce surface layers of determined thickness and designation [19].

– **Service-generated layers** - are produced as the result of the utilization of technological layers in conditions either natural or artificial. Utilization causes these surface layers to have properties that differ from those of the initial, technological layers [19].

Just as it is possible to affect the properties of technological layers in the course of their manufacture, it is also possible to affect those of service-generated layers or to create such properties during service itself.

Manufacturing of surface layers is traditionally the oldest but, at the same time, fastest growing field of surface engineering. Even today, this field is sometimes directly identified with the concept of surface engineering itself.

Designing of surface layers. This field of activity of surface engineering involves such design of surface layers which will allow them to meet service requirements. This area is, as yet, weakly developed. To this day, designing of surface layers is most often reduced to the utilization of "methods of those who have done it in the past," in other words, to reproduce the structure of layers already known, enriched by latest technological and service know-how. The design of a process, such as to obtain a predetermined structure and properties of surface layers, the correlation of technological properties with usable service properties, and the final decision regarding a manufacturing process which ensures the obtaining of such properties are practiced only in exceptional cases. More often, although still seldom practiced, is mathematical modeling of surface layer properties for cases of already known and practiced manufacturing techniques.

Investigation of surface layers. This field of surface engineering involves experimental research of the structure and properties of surface layers, relative to various parameters, both technological and connected with service conditions, and the acquisition of knowledge about related effects and applicable rules. The results of this research are integrated with the particular manufacturing processes and their parameters and constitute a database of technological know-how serving the design of new surface layers or their composition. The accomplishment of this research requires the implementation of newest methods of investigation, including physical, chemical, biological, corrosion, strength, tribology, etc.

Service utilization of surface layers. This area comprises two problem groups:

– **service testing** of behavior of surface layers in different working conditions (different external hazards). Usually, tests cover the change in behavior with progressing time of service. Because investigation of layer properties during service encounters numerous difficulties, layers are usually tested after certain predetermined periods or after completed service. This so-called post-service testing is carried out by methods close to those used in investigations of surface layers, taking into account duration of service, and broadened by specific tribological, strength and other tests. Investigation of the structure and properties of surface layers during service requires special physico-chemical methods and is not, to this date, well developed;

– **production of service-generated layers** during service, due to interaction with the material, by design, of substrates from the environment, e.g., originating only from that environment (atmosphere or technological medium) or from a different rubbing layer and a lubricating medium, under conditions of forced pressure, temperature, velocity, etc. [17, 18].

2.2.2 Significance of surface engineering

The development of surface engineering has been dynamic due primarily to the fact that this is a discipline of science and technology which meets the expectations of modern technical science: energy and material efficiency, as well as environmental friendliness. Besides the fact that it allows the investigator to live a passionate scientific adventure in the field of shaping the properties of matter, it is very solidly set in practical reality. Everyone has daily contact with products of surface engineering because all objects have a surface with given decorative or utilitarian value. Thanks to surface engineering we gain [18]:

– the possibility of producing tools, machine components and whole appliances from materials with lower properties, usually cheaper, and giving their surfaces improved service characteristics (usable properties). This is conducive to a reduction of mass and energy consumption necessary to manufacture them, retaining same strength characteristics and usually better tribological, decorative and numerous other properties;

– improvement of reliability of work of tools, machine components and appliances and reduction of failures. Poor design and improper service conditions are the cause of 15% of down time, while improper selection and poor manufacture of surface layers are responsible for as much as 85% of failures;

– diminishing of energy losses to overcome resistance caused by friction, due to mass reduction of moving machine components and appliances, and due to enhancement of tribological properties of the rubbing surfaces. Usually, 15 to 25% of the supplied power is spent on overcoming friction resistance and in some branches of industry, e.g., textile, as much as 85% of the supplied energy is lost in this way;

– reduction of frequency of replacing used tools and machine parts, as well as frequency of maintenance overhauls;

– reduction by 15 to 35% of losses due to corrosion, which is of great significance when it is realized that the impact of corrosion on economy may even reach 5% of gross national product;

– reduction in energy consumption by the industry due to the fact that methods used in surface engineering are usually energy efficient, and high energy techniques are used only in the treatment of selected sites of machine components or tools, without the need to heat the entire mass of the material. Also, the time of application of such methods to the treated material is extremely short, usually seconds or even less;

– minimization of environmental pollution, primarily due to reduction of energy consumption by burdensome branches of the industry and low rate of energy consumption by methods used in surface engineering, besides low amounts of waste, effluent, smoke, dust and industrial gases. Moreover, the small amounts of solid waste, after treatment, may be recycled; dust may be separated from gases and also recycled. These gases usually contain relatively small amounts of components which are indirectly harmful by intensification of the greenhouse effect (CO_2, NH_3, freon, N_2O, O_3), or which constitute a source of acid rain (SO_2, NO_x, volatile hydrocarbons), replete the ozone layer (chlorocarbonates, NH_4, NO_x) or, finally, directly harmful to human and animal organism, as well as plants (SO_2, NO_x, lead oxides, heavy metal vapours).

2.3 Directions of development of surface engineering

In the near and foreseeable future, surface engineering will be in a continued state of intensive development, proportional to the level of general development of a given country. Surface engineering belongs to a group of technologies based on latest discoveries and inventions and it is expected that it will remain in the forefront of technical science. The general directions of development will constitute a synthesis of the particular domains of science and technology, which together form surface engineering.

2.3.1 Perfection and combination of methods of manufacturing of surface layers

The development of manufacturing methods of surface layers will largely depend on the method adopted and its significance to the development of technology, stemming from the benefits of its application and the degree of practical utility.

General trends in the development of manufacturing methods of technological surface layers may be summed up by the following points:

1. Utilization of the effect of *synergy* by the application of:
- techniques allowing the development of sandwich layers, produced by the same method but from different substrates;
- duplex, triplex and multiplex techniques, in order to obtain surface layers with improved usable properties and longer service life, e.g., application of metal-paint coatings (thermal spray + pneumatic or electrostatic painting) with a life of 25 to 40 years without need to renovate, in the place of only paint coatings with a life of maximum years; the application of nitriding of prior hardened substrate, followed by deposition of titanium nitride coating; combination of burnishing with heat and thermo-chemical methods of surface hardening; combined application of nitriding and implantation of nitrogen ions or boriding with implantation of boron ions; application of vacuum deposition of coatings by PVD and CVD methods with simultaneous ion implantation; two-flux implantation; plasma heating combined with simultaneous nitriding.

2. Reduction of energy consumption of surface layer production methods and elimination of high energy consuming methods, e.g., recuperation of heat in organic coating drying rooms; utilization of new methods of electric heating of aluminum, alloy and other diffusion baths; elimination of salt baths and their replacement by fluidized beds, atmospheres and vacuum; the application of high energy (but low energy consumption) beam, methods and techniques (laser, electron, ion, plasma); application of different methods of burnishing to replace heat and thermo-chemical treatment.

3. Reduction of the share of material and raw material consuming methods of surface layer production, e.g., replacement of pneumatic spraying of paints by electrostatic deposition of liquid or powdered paints, or the application of glow discharge diffusion methods to replace salt bath and gas treatments.

4. Increasingly accurate preparation of substrate to accept the coating, taking into account its chemical activation, as well as the application of increasingly productive methods of cleaning and washing, including ultrasonic) and rinsing, as well as deposition of intermediate passivating and transition layers.

5. Application of ecologically friendly technology, resulting in less pollution of the natural environment, i.e. emitting a decreased amount of greenhouse dust and gases, gases that deplete the ozone layer, promote acid rain or are directly harmful to human health, animals and plants. This is manifest in the trend to use powdered paints in the place of liquid paints and varnishes, aqueous solvents in the place of organic (chiefly xylene, toluene and hydrogen chloride); application of anti-corrosion coatings which self-stratify in the process of drying, forming a primer and a surface layer; total elimination of freon as a washing medium; elimination of salt baths for hardening and their replacement by polymers. Another natural trend will be that of neutralizing wastewater, effluents and dusts.

6. Concentration of techniques of surface layer production at sites of production of blanks, i.e. mainly in steel mills, in order to eliminate the unnecessary transportation and to utilize recuperation of surplus heat. As an example, the growing trend to coat blanks already in steel mills, especially cold-rolled steel and profile bars, round bars and wires by organic, hot dip, electrolytic and thermal spray coatings may be estimated at 5 to 15% for Western European countries. In the majority of Central and Eastern European countries, continuous coating lines in steel mills occur only sporadically. The focus here, unfortunately, is on dispersed paint, enamel and zinc plating shops where, at a higher cost and by methods which pollute the environment, components made from mill blanks are coated. In Poland, for example, there are approximately 1000 paint shops and an equal number of zinc plating shops.

7. Mechanization and automation and even robotization of surface layer production methods, especially organic and thermally sprayed, on account of their being burdensome and harmful to the operator.

8. Growth in the application of microprocessor and computer control of not only single systems, but of entire cells and production lines. In the future one can expect the creation of entire departments which are computer controlled, especially those which combine the functions of blank production (e.g., of cold-rolled sheet) and their corrosion protection (e.g., organic or electrolytic coatings), production of blanks and their hardening, production of tools and coating them with anti-wear coatings, etc.

9. Increasing application of recycling, either in the form of utilization of wastes from technological processes as substrates (e.g., copper scale as abrasive) or return of materials used in the process of surface layer production for reuse after processing, e.g., paint wastes and electroplating deposits.

2.3.2 Design of surface layers, based on mathematical modeling

Inasmuch as in the field of broadly understood design activity, humankind has great achievements, in the field of designing surface layers or, in a stricter sense, giving them such predetermined properties as to allow them to fulfill their function in the best possible way, these achievements are not very significant. The best results are obtained when utilizing mathematics, correlating process parameters of surface layer production with their service properties and even with strength characteristics of objects on which they are developed. To this day, we have not learned to design in the same way we design gears, frameworks or cranes. Today, to this end, we still use traditional methods or develop a surface layer with various properties which is later tested in different conditions, first in the laboratory, next in the industry, to finally determine its range of applications. It is only in very rare cases that the opposite order is practiced, i.e., for given applications (expressed by numerical values); technological parameters are designed for an optimum surface layer production process [18].

Knowledgeable, mathematical design of surface layer properties for strictly determined needs, along with their practical verification, constitutes a very important, although very difficult and, as yet, quite distant problem to be solved by surface engineering.

It seems that chronologically this trend will be put into practice by:

– development of physical models, based on experimental data for the particular processes of surface layer production;

– development of partial mathematical models (for particular processes) which would combine selected technological and service parameters;

– mathematical modeling of surface layers and their practical verification;

– designing (mathematical determination of correlations between physico-chemical parameters of production and required service parameters) of different types of surface layers for selected working conditions;

– striving to arrive at a general model for the design of surface layers (will this be accomplished?);

– mathematical optimization of models of surface layer design.

2.3.3 Micro and nanometric testing

Testing of surface layers, consisting of determination of various physical and chemical properties, will utilize the same investigative methods which are used on a broad scale in material testing and in material science, supplemented by specialized methods for testing the properties of surfaces. It is foreseen that some investigative methods may be combined, such as:

– those typically used in tribology, strength of materials and corrosion protection with strictly material methods;

– subtle nanometric methods, used to investigate atomic and crystal layer structure and in experimental physics with technical testing in the micromillimeter scale.

2.3.4 Rational application of surface layers

Rational application of surface layers requires a very good knowledge of their characteristics, both potential and, especially, service. The chief tasks here will be

– reduction of energy and material consumption during service of components and appliances operating under conditions of tribological, fatigue or corrosion hazard. This implies the application of such surface layers and their utilization under such conditions which would minimize the rate of energy consumption and where losses attributable to corrosion would be minimal and, at the same time, rubbing surfaces of mating components would wear least. The preference here would be wear-free friction;

– diagnostic analysis of the state of utilized surface layers (wear, stresses, unit pressures, etc.) of working and mating components in such a way that the collected data are helpful in preventing failures and ensuring safe work;

– selection of surface layers for service conditions, directed toward an as-far-as-possible even wear of mating components.

References

1. **Burakowski, T.:** Heat treatment in China (in Polish).*Metaloznawstwo i Obróbka Cieplna (Metallurgy and Heat Treatment)*, No. 71-72, 1984, pp. 3-9.
2. **Wesołowski, K.:** *Metallurgy and heat treatment with exercises* (in Polish). Edition 6, PWSZ, Warsaw 1962.
3. **Bell, T.:** Surface engineering, past, present and future. *Surface Engineering*, Vol. 6, No. 1, 1990, pp. 31-40.
4. *Chronik der Technik.* Chronik-Verlag, Dortmund 1992.
5. *DK Science Encyclopedia.* Dorling Kindersley Ltd., London 1995
6. **Burakowski, T., Miernik, K., and Walkowicz, J.:** Production technology of thin tribological coatings with the use of plasma (in Polish). Proc.: *VII Conference on Utilization and Reliability of Technical Equipment.* Porąbka-Kozubnik, Poland, 26-28 October, 1993, *Tribologia*, No. 3/4, pp. 21-42.
7. *Encyclopedia of nature and technology* (in Polish). Edition II, Publ. Wiedza Powszechna, Warsaw 1967.
8. **Kaczmarek, J.:** *Principles of machining by cutting, abrasion and erosion.* Pregrinus Ltd., Stevenage, United Kingdom, 1976.
9. **Kosieradzki, P.:** *Heat treatment of metals* (in Polish). Edition I. PWT, Warsaw 1954.
10. **Kowal, S., and Witek, W.:** *Carburizing of steel* (in Polish). PWT, Warsaw 1957.
11. **Burakowski, T.:** Development, present state and perspectives of nitriding and derivative processes (in Polish). Proc.: Monotheme Conference: *Nitriding and Related Processes*, Rzeszów, Poland, 26 June,1980, pp. 5-23.
12. *Modern Painting Methods* (in Polish). WNT, Warsaw 1977.
13. *Electroplating. Fundamental Problems* (in Polish). WNT, Warsaw 1963.
14. *Welding, Vol. I. Engineering Handbook* (in Polish). WNT, Warsaw 1983.
15. **Linnik, V.A., and Peskev, P.Yu.:** *Contemporary technology of thermal gas deposition of coatings* (in Russian). Publ. Masinostroyenye, Moscow 1985.
16. **Pawlowski, L.:** *The science and engineering of thermal spray coatings.* John Wiley & Sons, Chichester 1995.
17. **Kirner, K.:** Geschicht des Termischen Spritzens Entwicklung zu den verschiedenen High-Tech-Verfahren. Proc.: 3 Kolloquium on *Hochgeschwindigkeits-Flammspritzen*, 10-11 Nov. 1994, Ingolstadt. *Tagungs Unterlagen.*
18. **Burakowski, T.:** Surface engineering yesterday, today and tomorrow (in Polish). *Inżynieria Powierzchni (Surface Engineering)*, No. 1, 1996, pp. 3-10.
19. **Burakowski, T., and Marczak, R.:** The service-generated surface layer and its investigation (in Polish). *Zagadnienia Eksploatacji Maszyn (Problems of Utilization of Machines)*, Polish Academy of Sciences - Committeee for Machine Building, Book 3 (103), Vol. 30, 1995, pp. 327-337.

chapter three

The solid surface

3.1 The significance of the surface

The surface of living organisms limits them and protects them from the environment. Similarly, in technology, the surface limits structural materials, separates them from the surrounding medium or the environment, but, at the same time, establishes contact with surrounding medium. In surface engineering, the fundamental object of research, design, enhancement (during the manufacturing process) and, finally, of wear (during service) is the solid surface.

The **solid surface** - of a tool, of a part of a machine, of an element or even of a finished product - has for many years been the object of physical and chemical processes which impart to it required service properties, better than those of the core (or substrate). When the usable properties of the objects are related to its chemical resistance, wear resistance, partially to fatigue strength, thermal or electrical conduction - proper modification of the surface of cheaper and less resistant materials can successfully replace the use of those materials which feature in their entire volume the right properties, such as anti-corrosive, anti-wear or decorative, typical of costly materials (e.g., gold), or difficult to machine (e.g., austenitic stainless steels, sintered carbides).

According to A. A. Griffith, as quoted by L. Szulc [1], the image of the real structure of a solid, including metals, which are of special interest to us is "a set of interruptions of macro and microscopic continuity, consisting of crevices, porosity and irregularities in structure, laminar or mosaic in character, or caused by the inclusion of foreign matter". If we further accept his premise that faults and irregularities of structure originate at the surface or occur chiefly in its direct vicinity, it is difficult to underestimate the significance of the surface in the process of technological shaping of properties and their utilization during service.

The surface of a solid is usually characterized by a structure and properties which differ from that of the core of the material. This difference stems predominantly from the following:
- a distinct energy condition, causing a state of elevated energy and enhanced adsorption activity [2],
- combination of mechanical, thermal, electrical, physical and chemical effects at the surface during processing of the object,
- cyclic or continuous: mechanical, thermal, chemical or physical action of the environment of the object on its surface during service.

The surface exerts a fundamental influence on the usable properties of objects and solids. Several physico-chemical effects, such as chemical catalysis, corrosion, wear (abrasive, adhesive, combined abrasive-adhesive, erosion, clotting, cavitation, fatigue, oxidation or flaking), adhesion, adsorption (physical and chemical), flotation, diffusion and passivation all depend on and occur at the material surface or with its participation [3].

The surface of a solid constitutes a specific research, technological and design problem.

The concept of the surface is perceptible and understandable by intuition; it is, however, quite difficult to define and understand in a precise manner. Usually, the definition of surface is not clear-cut. In fact, this concept has been defined in a number of ways, depending on the discipline of science or technology for the purpose of which it is used.

3.2 *The surface - geometrical concept*

In the mathematical sense, and more strictly, geometrical, *the surface is a two-dimensional geometrical figure*, e.g., a sphere or a cylinder, and constitutes one of the basic concepts of geometry. In elementary geometry, surfaces are described as certain *sets of points or straight lines with certain properties, i.e., loci of points with a given characteristic.*

It is evident from such definitions that the mathematical concept of the surface is purely theoretical and not material.

3.3 *The surface - mechanical concept*

Closer to reality is the concept of the surface used in applied mechanics and related technical sciences.

The **surface** is defined here as the *edge (or limit)* of material bodies. This is a very general concept, dependent on the scale of the effect considered - molecular, micro and macro.

The **material surface** is defined as a *continuous material system in the form of a surface, comprising material points.*

Finer differentiation in the description of surface concepts is attributed to a team of Polish scientists, under the leadership of Prof. Jan Kaczmarek, who in the 1950s laid down some scientific foundations for such processes as machining, abrasion and erosion [4]. Since the first of these has been the oldest and the most widely used of processes forming the geometry of objects - it was in this field that the most significant first achievements were accomplished. These are contained in many published documents, of which of greatest significance was the Polish Standard PN-73/M-04250 [5], and they formulate the following new definitions:

Nominal surface - *the surface as described by a blueprint or technical documentation, with the omission of roughness, waviness and shape errors.* This, of course, is the theoretical surface; earlier called the geometrical surface.

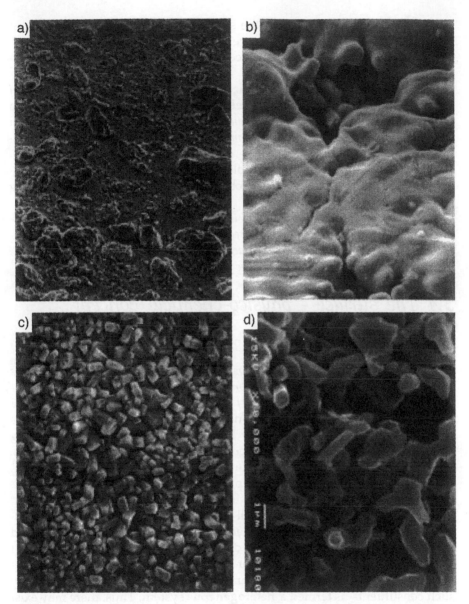

Fig. 3.1 Images of observed surfaces: a) chromium-aluminium coating on austenitic 300 series steel, 300 ×; b) Discalloy sintered P/M, 1000 ×; c) phosphate coating with gold vapor deposit, 2000 ×; d) surface of oxynitride layer, obtained by the ONC method, 1000 ×.

True or **real** **surface** - *the surface limiting the object (its solid shape), separating it from the environment.* The concept of the real surface, correct in the general sense, depends, however, on the scale of the phenomena considered. For the "macro" scale the concept is adequate to the definition. But for the "micro" scale, especially down to the scale of the atom, where there is a necessity to take into account the subtle interaction of the environment with the object, this concept is more difficult to be precisely understood. The concept of the observed surface substantially facilitates this understanding.

The **observed (measured) surface** - *an approximated image of the real surface of an object, obtained as a result of observation* (e.g. with the aid of a scanning electron microscope) *or measured within the bounds if precision achievable by observation or measurement* (by a given method of measurement) (Fig. 3.1).

The above concepts may be supplemented by general descriptions connected with the processing of the object, proper not only for machining but for all types of processing.

The **surface being processed** - *the surface which constitutes the boundary of the processed object in the area subjected to processing.*

The **processed surface** - *the surface which constitutes the boundary of the processed object in the area where the processing was carried out.*

3.4 The surface - physico-chemical concept

3.4.1 The phase

The phase is a homogenous part of a system with same physical properties in its entire mass and with same chemical composition, separated by an interface surface (phase boundary from another part [phase] of that system). For example, a gas or a gas mixture constitutes one phase, similarly to a homogenous liquid or a solution. On the other hand, a mixture of two non-mixing constituents, as e.g., oil and water, constitutes two inhomogenous phases. The system, ice - water - water vapour, is a homogenous tri-phase system. The phases of the system may be separated from one another by mechanical means, e.g., by decanting, filtration, centrifuging, sifting.

Constituents of the system are all those substances from which the systems are built. Further, an independent constituent is such that its type and quantity, if known, are sufficient to determine the chemical composition of each phase of the system. A phase or a system, composed of only one independent constituent is called **homogenous**, and one that is composed of many such constituents is termed a **heterogeneous** system. Thus, there may exist the following [6]:

- monophase and one-constituent system (e.g., pure water), or uniform and homogenous system,
- monophase and multi-constituent system (e.g., solution of sugar in water) or uniform and heterogeneous system,
- multi-phase and one-constituent system (e.g., ice and water) or non-uniform and homogenous system,
- multi-phase and multi-constituent system (e.g., solution of alcohol in water and their vapor) or non-uniform and heterogeneous system.

A transition within the considered system from one crystallographic form to another (e.g., α-iron to γ-iron) or the creation of a new chemical bond (e.g., of an intermetallic compound during solidification of a metal alloy) is always accompanied by the creation of a new phase.

In all dispersed systems (e.g., in emulsions) one of the phases is made up of the entire mass of the dispersing medium (e.g., water) while the second phase is the entire mass (all droplets) of the dispersed substance (e.g., oil).

A homogenous metal alloy constitutes only one phase; although it contains many grains of varying shapes and sizes, which can be separated from one another, all grains have the same chemical composition, therefore they constitute only a portion of the system. In the case of melting a metal alloy the system initially comprises two phases: the liquid and the solid crystals within it. These crystals can be separated mechanically from the liquid (e.g., with the use of a sieve). A fully melted alloy does not, as a rule, contain any additions which could be separated out of it by mechanical means. It has the same chemical composition throughout its mass, is homogenous and comprises one phase only. This stems from the ability of the majority of metals to dissolve in one another in the liquid phase in any given proportions. The only exception is the alloy of lead and iron [7].

A non-homogeneous metal alloy is multi-phased. The particular alloy phases usually differ quite significantly in properties, and their number, type, as well as properties depend on the chemical composition of the alloy.

3.4.2 Interphase surface - a physical surface

The dividing surface between phases (phase boundary) constitutes and *interface*, termed also *inter-phase boundary* or *boundary surface* (in fluid mechanics); in the case of the liquid-gas system the interface is termed level (as in water level).

The concept of interface was first introduced in 1878 by J. W. Gibbs who considered the simplest theoretical mono-constituent, biphase system. In this system one can distinguish three areas (Fig. 3.2): a homogeneous zone of phase A, a homogenous zone of phase B and a heterogeneous interface zone C. Zone C does not constitute a third phase in the strict sense of the phase rule. It can be treated as a fictitious individual portion of the system which is in equilibrium with the extraneous interactions between the phases A and B, i.e., temperature, pressure, chemical potentials, concentration, specific mass, etc. Gibbs treated the interface as a **physical inter-phase surface**. He moreover proposed the concept of a mathematical two-dimensional boundary surface, characterized by a directed force of surface tension, connected with pressure. Such a surface embodies the physical interface. This zone is characterized by an anisotropy of pressure, connected with the heterogeneity of the interface zone across its thickness.

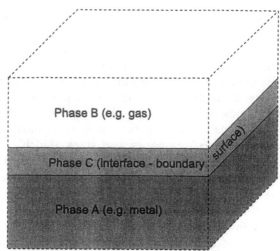

Fig. 3.2 Schematic representation of an interface, i.e., surface separating two phases.

Gibbs' proposition has been accepted in physico-chemical science.

In the physico-chemical concept, the surface is constituted by a boundary between two (or more) phases. In this book, it is the boundary between the chief object of our interest, i.e., the solid, in the form of a metal or its alloy (treated generally as one phase although it may also constitute a multiphase system) and the phase surrounding it, e.g., gas (most frequent case), or liquid or even solid (usually in the form of small particles dispersed in a gas). **For this reason the meaning of the surface in the physical sense is a material, three-dimensional object, although the third dimension - its thickness - is very small.**

This thickness is very difficult to determine. It can, however, be determined experimentally that the overwhelming portion of the surface energy of the solid is concentrated in a layer equal in thickness to only several diameters of molecules forming that solid.

Some authors [2] have partially equated the concept of the physical surface with that of the observed surface.

3.4.3 Surface energy

From the thermodynamical description of the surface by Gibbs stem two very significant statements, both valid for a system in the state of thermodynamic equilibrium:

1) interaction of interface with the phases surrounding it: between the inter-phase surface and the neighboring phases there may occur an exchange of mass and entropy. This means that that the interface formed under the influence of homogeneous phases at the same time affects the state of these phases and behaves actively with respect to them, and has an effect on the formation of the phase with which it is surrounded (e.g.,

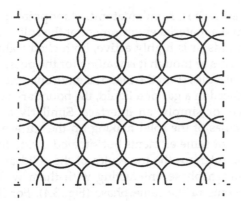

Fig. 3.3 Pattern representation of "hanging bonds" of a surface with atomic (covalent) bonds. (From *Hebda, M., and Wachal, A.*[8]. With permission.)

water at the interface with a solid occurs in a different from normal allotropic form which does not expand upon solidification);

2) equalization by the interface of the differences between the adjoining; the heterogeneity of the interface is expressed by the occurrence of internal electrical and mechanical fields causing a compensation of the gradients of chemical potential in the direction of maintaining a constant value of the electrochemical potential in the entire system (which constitutes a condition of thermodynamic equilibrium).

The clarification of the above statements facilitates an atomistic description of the simplified system, in the form of an ideal (i.e., no defect structure) solid crystal (e.g., of a metal), situated within ideal vacuum. The inside of that crystal can be identified with that of a system of atom structures, surrounded by potential fields of their interactions. The distribution of these fields determines the situation of valence electrons which are responsible for chemical bonds between atoms. Such bonds are, of course, mutually compensated within the crystal but there is no compensation on the crystal surface. The surface layer of atoms has unsaturated chemical bonds in the direction perpendicular to it and these may be termed "hanging bonds" (Fig. 3.3) [8]. The potential energy of valence electrons of surface atoms, therefore of surface atoms themselves, is different from that of atoms from within the crystal.

The described surface is not an inter-phase surface because while the solid constitutes a phase, ideal vacuum cannot be termed a phase. (Technical vacuum may be a phase - the lower the degree of vacuum, the more its properties approach those of a real phase). The said surface, however, does constitute a boundary surface which may, in the general sense, be identified with a real surface.

Boundary surfaces exist only when component elements of the solid - its atoms, ions or molecules - are mutually bonded by strong forces of adhesion. If these forces are small, the dissolution of one substance in the other takes place [8].

The described surface is, from an atomic standpoint, pure, not contaminated by atoms from a material environment of the solid. Since it has unsaturated chemical bonds, it is highly active, both chemically and physically. Such a surface behaves as though it is waiting for the possibility of attaching atoms, ions and molecules from the material environment.

If the solid is placed in a gas or a liquid, the boundary surface becomes an inter-phase surface, or simply, an interface. Similarly as in the case of a vacuum, the molecules of the solid making up the surface are in conditions different from those of same elements but situated inside the solid. From the external side of the interface the molecules of the solid are in contact with molecules of a foreign phase, interacting with them with forces different from those of elements of the same phase (Fig. 3.4). For liquids and gases, such forces are substantially weaker than those acting from the side of their own phase. As a result, the molecules of the solid have a portion of the forces not compensated and the surface is richer in energy than the inside of the solid [8]. Naturally, the molecules of the solid are subjected to interaction from molecules directly adjoining them (strongest forces), as well as from those situated deeper. The further away from the surface, the weaker the interaction with molecules at the surface (Fig. 3.5) [8].

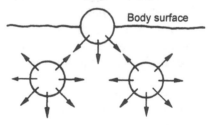

Fig. 3.4 Representation of forces acting on particles situated inside the solid and at its surface. (From *Hebda, M., and Wachal, A.* [8]. With permission.)

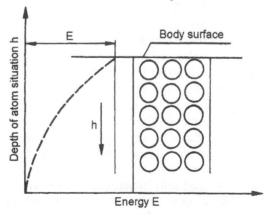

Fig. 3.5 Share of energy by atoms of the subsurface layer in the total energy of the surface. (From *Hebda, M., and Wachal, A.* [8]. With permission.)

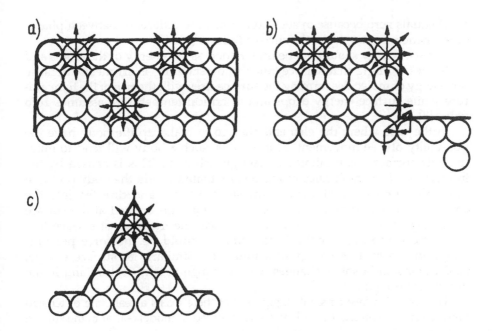

Fig. 3.6 Schematic representation of field of forces on surface of different shapes: a) plane surface; b) edge; c) corner. (From *Hebda, M., and Wachal, A.* [8]. With permission.)

Atoms at the surfaces of solids have a very limited freedom of movement. Saturation of forces of adhesion between them depends on other atoms in their vicinity. The smaller their number in direct proximity of a given atom, the lower the degree of saturation of adhesion forces and hence, the higher the surface energy (Fig. 3.6) [8].

Grain edges of crystals are richer in energy than a flat surface which manifests itself by the greater reactivity of atoms situated at grain boundaries, leading to e.g., the development of intergranular corrosion. More active, both chemically and physically, is a rough surface, in comparison with a smooth one. Similar effects are caused by structural defects, residual stresses, microcracks, porosity, scratches and crevices.

A measure of undersaturation of adhesion forces between molecules of a solid - both inside and on the surface - is surface energy. It is an inseparable property of the surface and its magnitude and distribution depend on the type of chemical bonds, i.e., on the type of the body. It is mainly a property of solids, regardless of the degree of structural ordering.

Surface energy is the difference between the total energy of all atoms or surface molecules and the energy which they would have if they were situated inside the solid. A measure of surface energy is the work which must be carried out to displace atoms from inside the solid to its surface. Surface energy in the critical condition (i.e. at critical temperature and

---

[1] The critical state - a state of a biphase (one component) system in which the physical properties of both phases existing in equilibrium are identical.

sure) equals zero because in such conditions the difference between phases fades, hence an atrophy of the surface follows [8].

Surface energy should not be understood as the energy of the atoms and molecules forming that surface. Such understanding is erroneous because the energy of molecules forming the surface rises with the rise of the temperature while surface energy drops and at critical temperature assumes zero value [8].

In the case where the elements that go to make up the body have the possibility of free movement, as in liquids, such a body will tend to minimize its surface, i.e. minimize its energy-rich zone. This is caused by the interaction of the molecules of the body situated inside the body on those molecules which are situated in the surface layer, and directed into the core of the body from the surface. The tension thus created at the surface of the liquid is called *surface tension*. Hence, the measure of surface tension - from the mathematical standpoint - could be the force per unit length or the surface energy of a unit area. Similarly to surface energy, surface tension in solids changes with a change in temperature and in the critical state equals zero [3].

The term "surface tension" suggests that there exists a real state of tension between surface molecules and even - as assumed in models - that in the surface zone there exists something in the form of a flexible membrane [3].

3.4.4 Surface phenomena

The occurrence of surplus free energy of particles making up the surface, i.e., of surface energy, their greater activity and changed orientation, as well as structural and chemical differences between the surface, the underlying matrix and the surrounding medium, cause that the physical surface is the site of several characteristic phenomena. Generally, these are connected with the spontaneous tendency to reduce the surface energy, proportional to the surface area on which they occur.

Of special significance are surface phenomena occurring in highly dispersed (colloidal) systems. These are the generation of colloidal systems by condensation or dispersion, joining of droplets or tiny blisters in emulsions, mists and foams (coalescence), the coagulation of the dispersed phase and its generation due to the presence of three-dimensional structures (chains and nets). These phenomena also affect the thermodynamic equilibrium of phases in well developed surfaces [6].

The solid is a material object, rigid and reacting with resistance to stresses. It can be said that under the influence of applied forces, the solid undergoes some elastic deformation and that its shape is determined more by its "past history", i.e., by the method of its preparation, than by the forces of surface tension. The surface of crystalline bodies differs from that of liquid in that the components of its structure have only limited freedom of movement. It is assumed that at ambient temperature, surface molecules are simply imprisoned in the crystal lattice and have no freedom of movement. The growth of their mobility is caused by extraneous factors, e.g., rise of temperature. When heating up a solid to melting point, the mobility of surface atoms dramatically rises, followed by enhanced diffusion of these atoms in the direction of the inside; finally, there is some movement toward the surface, caused by evaporation [6]. At temperatures where some atom mobility occurs, there is a tendency to equalize energy in those zones in which it achieves high values, i.e., in places with enhanced curvature, crystal corners, microcrevices, etc. By way of example, if a silver or copper sphere is placed on a flat surface, made of the same material, at a temperature close to melting point, the gap between the sphere and the flat surface will become filled. Thus, in practice, the surfaces of solids are sufficiently "plastic" to be able to "flow", albeit very slowly, in certain conditions. The mobility of surface atoms at temperatures close to the melting point is utilized in such technological processes as sintering or diffusion welding [9].

In the liquid - gas system, such as water and water vapour, at room temperature, for each 1 cm^2 of water surface, $3 \cdot 10^{21}$ new molecules reach the surface during each 1 s but the same number departs from it. Thus, it is a very turbulent state. The time of dwell of one molecule at the surface is of the order of a microsecond. In the said system there also occurs an exchange of molecules between the surface zone and the adjoining layers of the liquid. The diffusion coefficient of the majority of liquids is of the order of 10^{-5} cm^2/s. A molecule reaches the depth of 10 nm in a time of approximately 10^{-6} s [6]. It follows that the exchange of molecules between the surface and the adjoining zone of volume phase is very rapid. Thus the apparently "still" water, and, more generally, liquid, is in a state of turbulent movement at the molecular level [3].

On the other hand, in the case of a metal of low volatility, such as tungsten (with a high melting point: 2400°C), whose vapour pressure at ambient temperature is estimated at approximately 10^{-43} hPa, the number of atoms colliding with the surface is approximately 10^{-20} per cm^2·s, while the average dwell time of an atom at the surface is approximately 10^{37} s. Even for metals with higher volatility (with relatively low boiling points) these times at room temperature are very long. Thus, in reality the molecules of a solid at its surface are quite immobile when considering changes at the surface during evaporation and condensation [3].

At temperatures above 0.75 of the melting point (temperatures at which sintering and diffusion welding processes are carried out) dwell times of atoms at the surface may be very short. For example, copper at 725°C has a vapour pressure of the order of 10^{-6} Pa. It follows that the dwell time of atoms at the surface is of the order of 1 s. The general picture of the phenomenon is similar when diffusion rate is considered. In the case of copper at 725°C the coefficient of self-diffusion in the volume phase is approximately 10^{-11} cm^2/s. The time needed to move an atom to a depth of 10 nm is 0.1 s. At room temperature this time would be 10^{27} s.

From the examples quoted here it stems unequivocally that the movement of atoms at the surface of the solid depends on temperature and that for solids at room temperature the picture of the surface zone is quite different from that of the surface of a liquid where a very turbul+ent movement of molecules crossing the interface takes place. And it is because of the fact that surface molecules of solids are practically immobile in normal conditions, the surface energy and other physical properties of the surface depend to a large extent on the "history" of the given substance. For instance, a fresh fracture surface (a cleaved surface of the crystal) of a brittle substance will have a different surface energy than a surface prepared by grinding, polishing or by thermochemical treatment [6].

At the solid surface, besides the already mentioned surface mobility of atoms, there also occur effects of cohesion, adhesion, wetting, activated and chemical adsorption and propagation of the formed surface layer across the absorbing surface. These are accompanied by two-dimensional migration of atoms and particles, i.e., two-dimensional diffusion, friction, corrosion, nucleation of new phases, condensation, and crystallization, capillary and electro-capillary effects, electro-kinetic, temperature and thermoelectronic emission and many others [6].

Among the group of surface phenomena are those which occur within the multi-phase solid at interfaces (phase boundaries), formed as the result of defects of the crystalline lattice, during deformation (slip planes) and chipping of solids, causing the exposure of new surfaces, nucleation of new phases, etc. The dimensions and properties of interfaces, themselves dependent on the type of particles and their surface structure, affect thermal and mass exchange processes, i.e., the transport of substance from one phase to another by diffusion. Other such processes include: dissolution, evaporation, condensation, crystallization, multi-phase chemical processes, such as intercrystaline and stress corrosion, multi-phase catalysis and others [6].

The knowledge of surface phenomena and purposeful exertion of influence on them enables the shaping of properties of surface layers.

References

1. **Szulc, L.:** *Structure and physico-chemical properties of treated metal surfaces* (in Polish). Special edition by Warsaw University of Technology, Warsaw, September 1965.
2. **Kolman, R.:** *Mechanical strain-hardening of machine part surfaces* (in Polish). WNT, Warsaw 1965.
3. **Burakowski, T., Roliński, E., and Wierzchoń, T.:** Metal surface engineering (in Polish). Warsaw University of Technology Publications, Warsaw 1992.
4. **Kaczmarek, J.:** *Fundamentals of machining, abrasive and erosion treatment* (in Polish). WNT, Warsaw 1970.
5. PN-73/M-04250. Polish Standard Specification. The Surface Layer. Terminology and Definitions.
6. **Adamson, A.W.:** *Physical chemistry of surface.* Interscience Publishers, Inc., New York, Los Angeles 1960.
7. **Domke, W.:** *Werkstoffkunde und Werkstoffprüfung.* W. Girardet Buchverlag, GmbH, Düsseldorf 1986.
8. **Hebda, M., and Wachal, A.:** *Tribology* (in Polish). WNT, Warsaw 1980.
9. **Iżycki, B., Maliszewski, J., Piwowar, S., and Wierzchoń, T.:** *Diffusion welding by pressure* (in Polish). WNT, Warsaw 1974.

chapter four

Surface layers

The physical surface - as stated before - is considered as a heterogeneous zone between two adjoining phases. Atoms (component particles) of the surface are distributed quite differently to those inside the solid body. They are therefore subjected to entirely different energy conditions than these same atoms situated within the solid. Due to a less dense distribution at the surface they have fewer directly neighboring atoms. Therefore, the external atoms (those at the surface) have a higher potential energy than those inside the solid. In the case of an interface between a solid and a gas, the surface is acted upon from the gas phase side by forces substantially smaller. As a result, some of the forces acting on the surface particles are not compensated and the surface is energetically richer than the inside. For that reason, the energy required to remove an atom from the surface is significantly smaller than that required to remove an atom from any location in the bulk of the crystal. The asymmetry of the field of forces acting on the atom (or particle) - or element of the surface - affects the value of surface tension which has a tendency to pull the surface particles (or atoms) into the bulk [1].

As has been stated before, the atomically pure surface of a solid is very active both physically and chemically. Besides surface energy and surface tension, at the surface of solids with metallic bonds there occurs electrical voltage with a very high gradient, reaching tens of millions V per cm [1].

Each contact of the surface of a solid with a material body, e.g., gas or liquid, release processes conducive to lowering of tension and to saturation of the surface with molecules of gas, liquid and solids which are situated in the vicinity of the interface. These processes which are accompanied by an accumulation of accidental substances have been given the name of **sorption**. Surface sorption is usually termed **adsorption**.

The rate of attachment of substances (adsorption, sorption) by solids to the atomically pure surface and the force of their bonding with the solid are both substantially greater than in the case of a surface with earlier adsorbed foreign atoms.

Processes of attaching may occur:

- **spontaneously** - in such cases accidental substances are attached, e.g., molecules of water vapour or oxygen from the surface, particles of a lubricant, worn-off particles of metal and then a **natural surface** is formed;

– **artificially** - as a result of intentional action, during the execution of a technological process of enhancing properties (by the creation of new surfaces) of objects embraced by the range of surface engineering. Such enhancement leads to changes in the microstructure, chemical composition, residual stresses, etc., resulting in the creation of a **technological surface**. It is also possible to artificially enhance the technological surface during service in which case **service-generated** (usable) **surface** is formed.

In both cases a new surface is created, with properties different to those of the original surface. A sudden limitation of the atomic lattice at the time of creation of the new surface causes the formation of numerous structural defects on this newly created surface. These are formed as a result of displacements of atoms from their ideal positions which cause, among other effects, the creation of dislocations, stacking faults, etc. Each of these structural imperfections has its own free energy which exceeds the total surface energy of the solid [1].

The new technological or usable surface may be a new phase, several new phases, or it may also be a different material. In all cases, however, it constitutes a zone which differs by its state of energy from the rest of the material (substrate, core).

A characteristic feature of the physical surface, besides the energy barrier, the surface tension, different character of chemical bonding from that in the bulk of the solid, as well as great physical and chemical activity, is the heterogeneous structure and hence the anisotropy of properties in directions vertical and parallel to the surface.

If we assume the real surface to be a physically pure metal surface and we subject it to the action of a gas medium, the gas will have an effect on the metal, just as the metal will have an effect on the gas. As a result, on both sides of the idealized physically pure surface, an interface zone will be created, in the form of a system of layers in a direction normal to the surface. These layers will be basically parallel to the physically pure surface and their structure will be non-uniform in both the parallel as well as the normal direction to the surface. Such layers may be called **surface zone layers**.

The layer of deformed (by the production process) metal or alloy - physically (by heat, force, diffusion of foreign atoms), chemically (e.g., by oxidation) and structurally situated below the physically pure surface may be called the **subsurface layer** (or **layers**). Since the situation of this layer (or layers), relative to the core of the object, is on the side of the real surface, the term applied is **superficial layer**.

Layers of adsorbed gas, water vapour, sweat, lubricant, and solid particles (dust, material debris), situated above the physically pure surface, may be termed **supersurface layers**. During technological processes of manufacturing and (although very rarely) during service, these layers form the source of nucleation of a new phase, leading to a new layer, or are removed (to activate the real surface) before being deposited on the almost physically pure surface of a layer of new material, different from that of the core. Since some 30 - 40 years ago this layer has been referred to as **coating**.

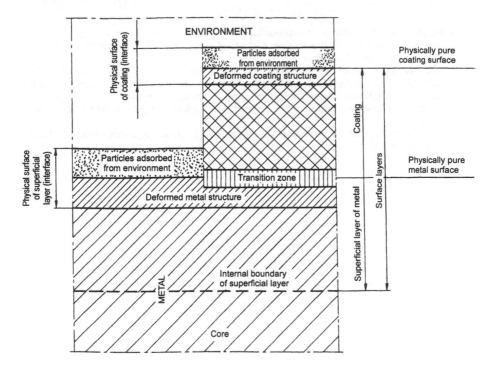

Fig. 4.1 Schematic representation of surface layers.

This book, going in the footsteps of publications [1, 3 - 5], has recognized a distinction between concepts of superficial layer and coating and assigned a common term of **surface layer** to both (Fig. 4.1).

In a narrower, stricter sense of the word, a physical surface is an inter-phase zone (interface) between a solid and gas (liquid); in a broader sense it includes the superfical layer, and in an even broader sense, the coating. Thus, surface layers constitute, in a broad meaning, a physical surface.

Since a coating is manufactured from a material different than that of the core, in reality it is a different material deposited on the core of another one. Therefore, the coating has its own physical surface.

References

1. **Burakowski, T., Roliński, E., and Wierzchoń, T.:** *Metal surface engineering* (in Polish). War-saw University of Technology Publications, Warsaw, Poland, 1992.
2. **Adamson, A.:** *Physical chemistry of surface.* Interscience Publishers, Inc., New York, Los Angeles, 1960.

3. **Burakowski, T.**: Methods of manufacture of superficial layers - metal surface engineering (in Polish). Proc.: Conference on *Methods of manufacture of superficial layers*, Rzeszów, Poland, 9-10 June, 1988, pp. 5-27.
4. **Burakowski, T.**: *Metal surface engineering - status and perspectives of development* (in Russian). Series: Scientific-technological progress in machine-building, Edition 20. Publications of International Center for Scientific and Technical Information - A.A. Blagonravov Institute for Machine Building Research of the Academy of Science of USSR, Moscow 1990.
5. **Burakowski, T.**: Metal surface engineering (in Polish). *Normalizacja (Standarization)*, No. 12, 1990, pp. 17-25.

chapter five

The superficial layer

5.1 Development of concepts regarding the superficial layer

Since the close of the previous century, intensive research has been conducted to learn about the structure and properties of the surface obtained by machining, later to be broadened to include abrasive and even later to erosion treatment. In significantly earlier times not much was known about the fact that as a result of any reduction machining, shaping the final object, specific properties are given to the surface. It was only known that by machining one can obtain a smoother or a rougher surface, which was for a long time thought to be the only property of the surface.

Research of the real surface intensified after the discovery, at the turn of the century, of a layer of destroyed crystals at the surface by G. Beilby [1]. This layer later came to be known as Beilby layer or superficial layer. After the end of World War II it was recognized that the structure of superficial layers of objects plays a major role in shaping their usable properties during service.

In Poland research of properties of superficial layers was conducted by scientists involved with various treatments by reduction - the most commonly used of all mechanical technologies of yesterday and today. These pioneering works did not encompass general investigations of physical surfaces, created as a result of various technological processes (with use of external force), but naturally were directed at the broadly understood machining treatment with the possible inclusion of other smoothing treatments. For this reason, even for many years after World War II, the term **surface treatment** was understood either only or mainly as machining, abrasive or erosion treatment with one addition - selected from the entire family of electrochemical treatments - electropolishing [2-4].

Today, an approach like this is an evident anachronism; nevertheless, research in the field of machining was the topic of pioneering technological works which made possible closer insight and widening of knowledge about the physical surface. Several new concepts and definitions were developed, of which the most important are the following.

By **superficial layer** of a material we should understand a set of material points, contained between its external surface and an apparent surface which is the limit of changes of properties of the subsurface layer, these changes having come about as a result of external forces: pressure, temperature, chemical and electrical factors, bombardment by particles charged and electrically neutral, etc. The remaining portion of the material beyond the superficial layer is the **core**.

From the point of view of standardization, the superficial layer is defined as follows [5]:

Superficial layer[1] - a layer of material, limited by the real surface of an object, which includes this surface and that portion of the material inward from the real surface which exhibits changed physical, and sometimes chemical properties in comparison with properties of the material core.

The above definition basically includes the **technical superficial layer** which constitutes the domain of shaping, research and technical utilization and which is the subject of technical science. The introduction seems also possible of another concept: that of a **physical superficial layer**, with a thickness of several angstroms (thus several orders of magnitude thinner than that of the technical superficial layer), within which the change in physical and chemical properties is greatest. Further in this book we will consider mainly the technical superficial layer.

It should be emphasized that the properties of the superficial layer, i.e., its structure and properties - imparted to it as a result of treating of the object, depend predominantly on the type of this treatment - on the technology of shaping of the object (both by material reduction and without reduction) as well as on the technology of giving it special physico-chemical properties. The obtained service properties are usually not constant with time:

– they may change spontaneously, in a natural manner, without the participation of extraneous technical factors, e.g., as a result of natural aging, stress-relieving;

– they may change non-spontaneously, under the influence of extraneous technical factors (forces) occurring during the service of the object, e.g., as a result of loading, wear, chemical action of the environment.

The structure of the superficial layer, constituting a "set of elements of the real surface of the object and structure of its superficial layer, such as geometrical elements of the surface and physical properties of the material, e.g., grain size," evidently also depends on the type of treatment.

Researchers emphasize the zoned structure of the superficial layer, which is a derivative of the type of treatment, and define the concept of the superficial layer zone, its internal limit and thickness.

The superficial layer zone is the portion of the superficial layer, the volume of which is determined by the existence of a given property. The individual layers may permeate one another, with a transition from one to the other, or they may occupy together the same part of space.

The internal limit of the superficial layer is the theoretical surface determined by points corresponding to boundary values of that property of the superficial layer for which the thickness of that layer, thus defined,

---

[1] *Superficial layer (Eng), couche superficielle (Fr), poverkhnostniy sloy (Rus), Oberflächenschicht (Ger).*

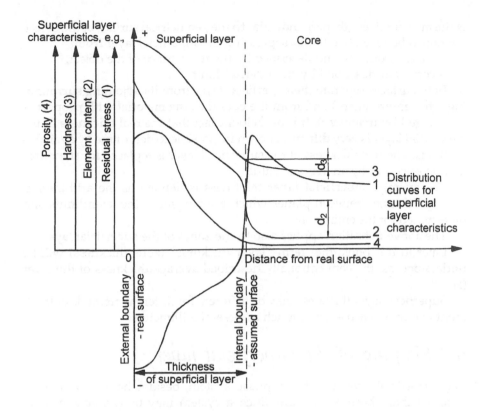

Fig. 5.1 Method of determination of internal boundary of superficial layer.

is greatest (Fig. 5.1). The determination of the internal limit of the superficial layer is very difficult and in most cases is only an approximation, an assumed value. This is because the internal limit, as a two-dimensional plane, is practically non-existent. For that reason, the internal limit of the superficial layer is assumed as an arbitrarily determined depth of penetration of changes of one or more properties of the layer, called boundary value or assumed value. Such properties of the superficial layer, also referred to as parameters of the superficial layer, could be e.g., microhardness, residual stresses, type of structure, distribution or diffused or implanted element. In the standardized sense, the depth of penetration of the considered property is defined by:

The boundary value of a property - the value differing from the value of that property in the core by a conventional differential **d** (Fig. 5.1). This value **d** is, naturally, different for different properties, and, in many instances, for the same property but defined by different researchers or different schools of thought, e.g., conventionally assumed hardened case depth or conventionally assumed depth of diffusion.

The concepts of internal limit of the superficial layer together with the earlier discussed real surface of an object are associated with yet another concept of great importance - that of the **superficial layer thickness.** This

is the measured length, perpendicular to the geometrical surface of the object, contained between two limits, separating the superficial layer from:
- the environment and described by the real surface of the object,
- core, and described by the internal limit.

Both surfaces separate the superficial layer from its adjoining surroundings (the environment) and from the core (i.e., the material with properties unchanged by treatment). It is an evident fact that the real thickness of the superficial layer is very difficult or even impossible to determine by available methods. The real thickness of the superficial layer is replaced for standardization purposes by an average value:

The **average superficial layer thickness** is defined as the arithmetical average distance between planes which are tangent to the layer limits and determined for the entire zone.

The above definition applies also to the zones of the superficial layer.

Later in this book, the term "layer thickness" (zone thickness) will be understood as the conventionally described average thickness of the layer (zone).

Superficial layer thickness may be different for different methods of treatment and may reach even as much as several millimeters.

5.2 Shaping of the superficial layer

The superficial layer is a heterophase system, besides having properties different than those of the core. Such a system may be characterized by the application of the generalized equation of the I and II principle of thermodynamics [6]:

$$dG = dH - TdS = -SdT + Vdp + sds + S\,\mu_i dn_i + jdq \qquad (5.1)$$

In the above equation:

dG is the change in Gibbs' energy (free enthalpy), i.e., that portion of internal energy of the physical system which in isothermal - isobaric conditions may be transformed into work;

dH	- change in enthalpy,
TdS	- change in free energy,
$-SdT$	- change in heat energy,
Vdp	- change in mechanical energy,
σds	- change in surface energy,
$\Sigma\,\mu_i dn_i$	- change in chemical energy,
φdq	- change in electrical energy,

where: T - absolute temperature; S - entropy, $S = dQ/T$ (Q - heat absorbed by system at temperature T); for all reversible processes $S = $ const, for irreversible processes the entropy of the system rises; H - enthalpy (heat capacity): sum of internal energy U (contained in the thermodynamic system and connected with the movement of masses, charges and atoms, as well as their interaction) and their product pV; V - volume; p - pressure; σ - surface tension; s - surface; μ - chemical potential; n - number of moles; φ - electrical potential; q - electrical charge.

The system remains in equilibrium when the sum of the components of the equation (5.1) is constant. In the case where the system is subjected to an extraneous force which changes the value of any of the forms of energy, the values of the remaining forms of energy also change. This change depends on local structural conditions, as a result of which appropriate transformations of energy take place in appropriate order [6]. For example, as a result of mechanical action on the already formed superficial layer, a refinement of its structure takes place which will cause a change in its chemical reactivity and the occurrence of electrical, thermal and other effects. Similarly, as a result of thermal action, there will come about a change of the original structure to a different one, accompanied by changes in chemical potential, surface tension, resistivity, volume, residual stresses, etc.

The superficial layer acquires its specific properties during the formation of a new surface or as a result of changes of properties of the already formed surface for other properties. The obtaining of these properties is accomplished by subjecting the material to the treatment operation.

The **treatment operation** can serve either **exclusively** as the means of imparting properties to the surface which enhance its service value (e.g. strengthening and improvement of properties determining its durability and strength, as a result of burnishing, surface hardening, carburizing, nitriding) or as imparting these properties as a **side effect**, while shaping the object (e.g. by casting or machining).

The superficial layer may be given various usable properties as a result of subjecting the material to various treatment operations or by changing the parameters of the same operation. Nevertheless, the formation of a new physical surface or a new superficial layer requires the delivery of a certain amount of energy, which may differ in form and value for different treatment operations or - more broadly speaking - for different technological processes and their parameters [7].

Fig. 5.2 shows a diagram of the effect of the superficial layer shaping process on the usable service properties of the object.

The technological process is characterized by the type of successive treatment operations. The chief aim of each such operation is the delivery to the treated object of appropriate energy. Strictly speaking, especially in the case of local treatment (spot or point) or the energy comparison of various treatments, this is the energy introduced to a unit area in unit time, i.e., appropriate power density. Each operation is characterized by a range of power densities which stem from the changes in process parameters required for its accomplishment. Fig. 5.2 characterizes selected operations of surface treatment by power density delivered to the treated object, identified with the power density needed to create a new surface. The given values are approximate and in principle do not apply to rough

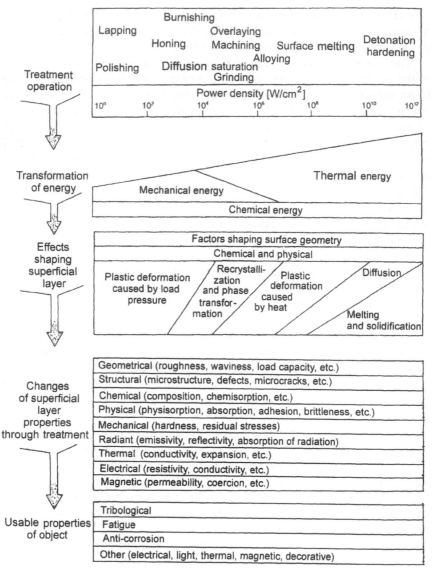

Fig. 5.2 Diagram showing the effect of the technological process of shaping of the superficial layer on the usable properties of the object.

machining or forming technologies (like cold forming, casting, compacting with sintering, etc.). There is a noteworthy regularity in that with a rise in power density, the contribution of mechanical energy drops while thermal energy rises significantly.

The energy introduced into the treated object, be it mechanical, chemical, thermal, a combination of these or of all energies together, may be transformed into a different form of energy. As a result, in the superficial layer different types of phenomena may occur: chemical (change of chemical composition), physical (diffusion, permanent deformation of mechanical and thermal origin, melting or solidification), structural (recrystallization, phase transformations causing a change of properties, relative to those of the core, e.g., microgeometry of the surface, microstructure, microhardness, brittleness, adhesion, adsorption, residual stresses, etc.).

Changes in the properties of the superficial layer relative to core properties, obtained as a result of the treatment operation, are of a **potential character**. In simple words, the superficial layer has such and such properties. There may be changes several times in the desired direction by subjecting already obtained surface layers to successive treatment operations. Always, the **final** properties of superficial layers, obtained as a result of the considered treatment operation, also called **initial**, feature **initial** properties for the next operation, etc. And, each time, these properties are of a potential character.

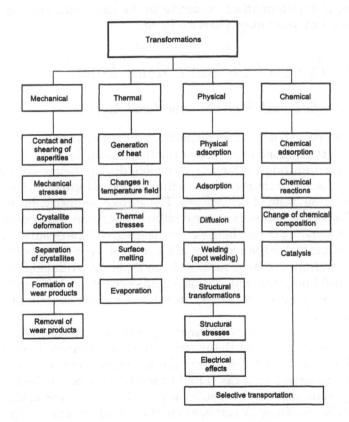

Fig. 5.3 Diagram of basic transformations taking place in the superficial layer as the result of friction.

These properties acquire a **usable** or **functional** character only after exposure of the superficial layer (usually formed on a tool, a machine component or other object) to a definite external technical hazard (single or multiple) like contact with another system (e.g., air, lubricant, material) or to the effect of an external system (e.g., external mechanical or thermal loading).

A change of potential solid geometry and physico-chemical properties of the superficial layer causes changes of usable properties of the object. These changes usually, although not always, go in the direction of the enhancement of those properties which are connected with the physical surface of the object, beginning or occurring at the surface, such as: tribological, fatigue, anti-corrosive, decorative, electrical, thermal, optical and other.

During service, the technological superficial layer will be subjected to similar external hazards as during the formation process. These may occur locally in certain zones or in the entire layer. As a result, the initial properties of the technological superficial layer undergo changes. These usually, but not always, go in the direction of deterioration. Fig. 5.3. shows a diagram of transformations occurring in the superficial layer of a metal alloy, caused by wear during service [3, 8].

5.3 Structure of the superficial layer

5.3.1 Simplified models of the superficial layer

Surfaces of solids always reflect methods of their formation, e.g., crystallization (during melting and casting), deformation (during forging), deformation and heat effect (during machining), diffusion (during thermo-chemical processes), etc. Each surface, regardless of the method of formation, is characterized by a certain state of unevenness. The surface geometry, including the height of asperities, depends on the treatment operation. Surfaces machined by turning, grinding or polishing exhibit clear signs of the treatment (a representation of the tool shape and path) in the form of repetitive asperities. Even the most carefully prepared surfaces have asperities with heights ranging from 0.01 to 0.1 μm, those roughly machined - even above 1000 μm (e.g., after rough turning of ductile materials, the thickness of the superficial layer exceeds 1000 μm [3].

Regardless of the type of treatment operation or type of treated material, the nascent atoms of the metal surface are characterized by high chemical activity which influences the interaction between the surface and the environment, be it gas, liquid or solid. This leads to the adsorption of foreign substances which causes a drop in the surface energy of the superficial layer. Thus, in common conditions of storage or service, the unprotected surfaces of metals are usually covered by a layer of oxides, adsorbed organic compounds, dust or gases [9].

Moreover, at different depths of the superficial layer there occur different effects, stemming from the method of formation. These effects are reinforced (occurring usually not homogeneously but at point sites) or they are partially dampened by changes caused by service conditions, e.g., during wear (Fig. 5.3), or caused by physico-chemical phenomena. As a result, the surface layer has a different structure from that in the initial condition. It is harder, but more brittle, may exhibit lowered cohesion, lower resistance to variable stresses, etc.

The formed superficial layer always has a structure and properties which depend on the core material (chemical composition and physico-chemical properties) but also on the type and conditions of the treatment operation. Since for any given core material the first factor may vary widely and there may be different treatment operations, and since the second factor may also vary within a very broad range, it is difficult to develop a general regularity of the superficial layer structure, leading to its generalized model. It is, however, possible to develop models for the various treatment operations.

Further, the only models discussed will be those of the superficial layer formed as the result of machining. This type of treatment is the most popularly used form of shaping technology and, in most cases, constitutes an initial basis for other surface treatments. Usually, after machining there follow other treatment operations, or after other treatments there follows a final machining operation.

The models of the superficial layer after machining differ among themselves chiefly by the degree of detail, or by taking into account different number of phenomena. What they have in common is the laminar structure: a multi-layer superficial layer usually consists of many sublayers, with different structure and properties but with often intangible transition from one sublayer to the next. Such a sublayer is called a **zone**.

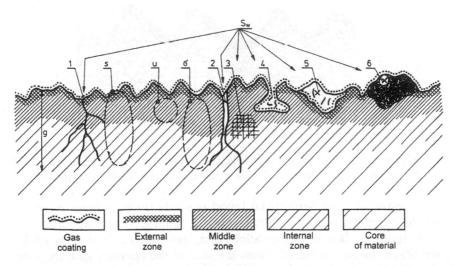

Fig. 5.4 Diagram showing a set of factors characterizing the superficial layer of a solid, against the background of a 3-zone model of the superficial layer [3]: g - thickness of superficial layer; s - structure of surfical layer; u - strengthening of the middle zone; s - residual stresses in the superficial layer; S_f - flaws of the superficial layer; 1 - microfracture; 2- fissure; 3 - microshrinkage; 4 - porosity; 5 - gap; 6 - inclusion. (From *Kolman, R.* [3]. With permission.)

The 3-zone model. The simplest 3-zone model is that proposed by R. Kolman[1] [3] in which, besides the structure of the superficial layer, different factors characterizing the layer and those occurring in other models but in different mutual proportions are given (Fig. 5.4).

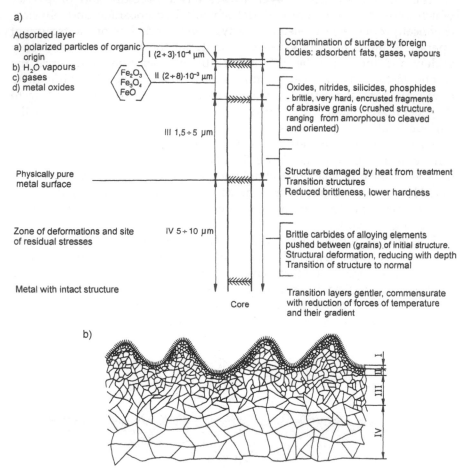

a)

Adsorbed layer
a) polarized particles of organic origin
b) H_2O vapours
c) gases
d) metal oxides

Fe_2O_3
Fe_3O_4
FeO

I $(2 \div 3) \cdot 10^{-4}$ µm

II $(2 \div 8) \cdot 10^{-3}$ µm

III $1,5 \div 5$ µm

Physically pure metal surface

Zone of deformations and site of residual stresses

IV $5 \div 10$ µm

Metal with intact structure

Core

Contamination of surface by foreign bodies: adsorbent fats, gases, vapours

Oxides, nitrides, silicides, phosphides - brittle, very hard, encrusted fragments of abrasive granis (crushed structure, ranging from amorphous to cleaved and oriented)

Structure damaged by heat from treatment
Transition structures
Reduced brittleness, lower hardness

Brittle carbides of alloying elements pushed between (grains) of initial structure. Structural deformation, reducing with depth Transition of structure to normal

Transition layers gentler, commensurate with reduction of forces of temperature and their gradient

b)

Fig. 5.5 Schematic of the 4-zone superficial layer model: a) per Szulc (From *Szulc, L.* [4]. With permission); b) per Okoniewski. (From *Okoniewski, S.* [10]. With permission.)

The **external** zone (superficial layer) is made up by a layer of foreign particles (particles of tool material or of the mating friction material, coolant or lubricant liquid, perspiration, dirt, dust, etc.) mixed with spalled particles of the core material. From the side of the environment, the external zone is covered by a layer of adsorbed gases: oxygen, nitrogen and water vapour. The thickness of the external zone is 0.001 to 0.02 µm while that of the gas layer approximately $(2 \text{ to } 3) \times 10^{-4}$ µm [3].

---

[1] Original terminology, proposed by the authors of the discussed models, is used.

The **middle** zone (layer adjacent to surface) consists of strongly deformed grains of the core material and in many cases may be significantly textured. Its thickness ranges from 0.5 to 500 μm. Main usable properties of the physical surface depend on the structure of this zone [3].

The **internal** zone (subsurface layer) consists of grains which are not permanently deformed but has a different structure than that of the core, e.g., as a result of transformations induced by heat. Residual stresses reach this zone. The transition of the internal zone to the core is difficult to observe. The thickness of this zone may reach several thousand micrometers [3].

The 4-zone model. A fairly good description of the 4-zone model has been given by L. Szulc [4] (Fig. 5.5a). It differs from the model proposed by Kolman mainly by the following factors:

– introduction of a physically pure surface,
– division of external zone into two sub-zones,
– a different division of the middle and internal zones,
– absence of zone names.

A somewhat differently presented description [10] of the 4-zone model is shown in Fig. 5.5b:

– layer I is constituted by particles of oxygen, nitrogen and water vapor, adhering to the surface of all metals which are surrounded by air. The thickness of the layer does not exceed several Lngstroms;

– layer II (superficial layer) is constituted by a mixture of fine dusts of spalled material of the tool, of the superficial layer and of the coolant-lubricant liquid. The thickness of this layer depends mainly on the method of treatment and is usually from several thousandths to several hundredths of a micrometer;

– layer III (subsurface or middle layer) is made up exclusively of grains of the treated material which, due to the action of the machining tool, became deformed. The thickness of this zone depends more on the type of treatment than on the method of its actual execution. As an example, after polishing, this thickness is approximately 5 μm, while after rough machining, it may be from 40 to 100 μm;

– layer IV (internal layer) consists of crystals of undeformed material. (Nothing was said in [10] whether these crystals had been earlier subjected to any type of influence, other than deformation or whether they constitute the material of the core).

The 5-zone model. This model, currently termed "simplified," was, chronologically, one of the earliest. It is embraced by different sources (Fig. 5.6) [5].

The **subsurface** zone consists of the part of the superficial layer adhering directly to the real surface. It is built up of ions, adsorbed or chemically bonded to the core, and originating from the environment or from elements in contact with the object.

The **directional** zone, lying under the subsurface zone, constitutes a portion of the zone of deformation, with a clearly defined orientation of material grains.

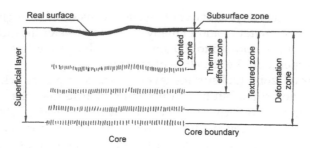

Fig. 5.6 Schematic of simplified 5-zone model of a superficial layer obtained as result of machining.

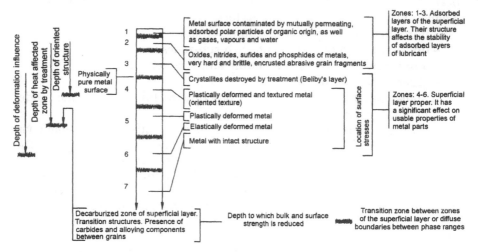

Fig. 5.7 Schematic of the 6-zone model of the superficial layer. (From *Nowicki, B.* et al. [2]. With permission.)

The **heat affected** zone constitutes a portion of the zone of deformation in which significant changes took place as the result of heat processes, e.g., changes of grain size, phase transformations and physical changes.

The **preferred orientation** zone constitutes a portion of the zone of deformation with a preferred crystal or grain orientation, from the point of view of the crystal lattice.

The **zone of deformation** makes up the overwhelming majority of the superficial layer where permanent deformation took place, and consists of the three above described zones.

The 6-zone model. An earlier, 5-zone model has been broadened by the Institute of Machining [5] by the addition of an amorphous layer (so-called Beilby's layer), forming a part of the subsurface zone (Fig. 5.7).

A somewhat different, more detailed approach was used in the development of the model proposed by [2]. It basically constitutes the prototype of the most detailed, 8-zone model.

5.3.2 The developed model of the superficial layer

This model presents a description that comes closest to the real structure of the superficial layer after machining (Fig. 5.8). It constitutes a developed combination of the earlier models, especially of the 5- and 6-zone ones and consists of 8 zones.

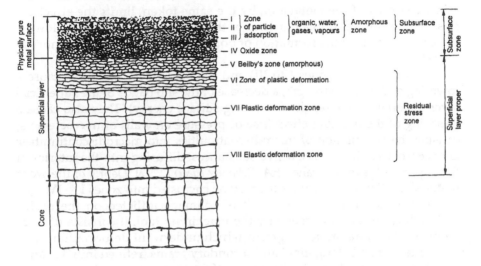

Fig. 5.8 Schematic of the 8-zone model of the superficial layer.

Zone I is created as the result of adsorption by the metal surface of polarized particles of organic origin (grease. lubricants, perspiration, etc.).

Zone II is created as the result of adsorption of water particles (usually from vapours).

Zone III is created as the result of adsorption of gases (nitrogen, sulfur and phosphorus vapours).

Zones I to III blend one into another and are created by the adsorption of particles of a dipole structure. In the case of a pair of rubbing surfaces, these layers cause the existence of technically dry friction between non-lubricated surfaces of machine components. The presence, at the friction surface, of adsorbed layers prevents dry friction (in the strict sense of the term) for which the friction coefficient would be several times higher than for the technically dry friction condition [12, 13]. The structure of the discussed zones is amorphous and their joint thickness is $(2 \text{ to } 3) \times 10^{-4}$ µm. On the surface of NiTi alloys with shape memory, the industrial atmosphere causes the formation of a thin layer containing significant amounts of carbon, oxygen and sodium, as well as small amounts of chlorine and sulfur [14].

Zone IV constitutes a layer of oxides of the core metal, created as the result of a chemical reaction between oxygen and the core material. The thickness of this layer depends on the chemical properties of the core metal and the rate of diffusion of oxygen through the oxide layer. This zone forms a layer protecting the core from corrosion. It secures the core against further diffusion of oxygen and other aggressive substances, e.g. sulfur or phosphorus [2, 12]. Sulfides, phosphides, nitrides and all hard and brittle compounds can build themselves into the oxide layer. Moreover, the oxide layer saturates the surface forces of the metals and - in the case of friction - limits the participation of adhesion forces in the fusion of rubbing elements. At the same time, however, it limits the value of the surface energy of the metal and, by the same token, limits the effectiveness of bonding of dipole particles of the lubricant. This, in turn, lowers the effectiveness of the formation of boundary layers (boundary lubrication) [8].

Adsorbed layers (zones I to III) and the oxide layer (zone IV) cause a lowering of the yield strength, a decrease of the slip bands of the superficial layer material (as the result of lowering surface energy). They also cause a lessening of the bursting effect (rise of pressure on walls of microcrevices, causing the coagulation of microdiscontinuities into macrodiscontinuities: crevices, cracks) by surface-active substances, diffusing into the interior of the superficial layer (so-called *P.A. Rebinder effect*) [2, 12]. Debris of the worn material, particles of dirt and dust may be encrusted into zones I to IV.

Zone V is formed as the result of damage inflicted on the grains (crystallites) of the core metal by the machining tool. It consists of fragments of the initial material grains which underwent deformation (usually flattening and elongation) and secondary grains (refined initial grains) which were created due to the action of forces accompanying the treatment (e.g., machining), or, finally, due to friction and wear of the material, partially during machining but mostly during service. This introduces defects to the crystal lattice and causes the structure of that zone to be virtually amorphous, containing fragments of grains. This is Beilby layer or amorphous layer [12, 13]. The thickness of this layer varies from 0.1 to 1 µm [14, 15].

Zone VI comprises the zone of permanently deformed metal, characterized by its significant fibrous structure and sometimes by preferred orientation. This orientation preference occurs as the result of unidirectional forces permanently deforming the superficial layer material. Physical, chemical and especially strength characteristics of the textured and deformed grains are different in different directions [9]. The thickness of this zone usually does not exceed several tenths of a millimeter [16].

Zone VII comprises the zone of material which is only permanently deformed but with no preferred orientation. Permanent deformation of the metal in the superficial layer may come about by slip or by twinning. During slip, there occur displacements of thin atomic layers relative to other (formation of slip bands) in a defined direction and along closest packed crystallographic planes [12]. The number of planes and directions depends on the structure of the crystal lattice. Distances between slip lines range from 0.0001 to 0.0004 mm. During twinning (mirror-like reflection) one part of the grain

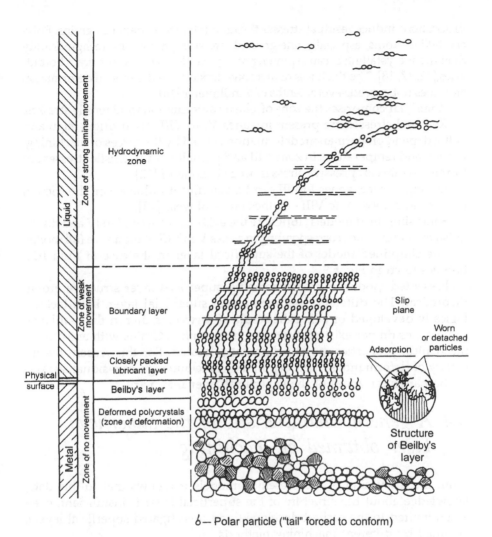

б— Polar particle ("tail" forced to conform)

Fig. 5.9 Simplified schematic of the superficial layer in service conditions - for fluid friction; model of boundary layer formation per Marelin. (From *Hebda, M., and Wachal, A.* [8]. With permission.)

is symmetrically displaced. Each permanent deformation is accompanied by **strengthening** of the metal, dependent on the rate of deformation (grows with a rise in deformation rate) and inversely dependent on temperature. The effect of strengthening is explained by the fact that the permanent deformation, once begun, faces, as it progresses, an ever greater resistance. Therefore, to sustain the continuation of deformation, greater forces must be used to overcome resistance offered to dislocations during their movement [12]. It is this retardation of the movement of dislocations, as a result of the growth of their density from 10^6 to 10^8 cm^{-2} to 10^{10} to 10^{12} cm^{-2} which occurs during permanent deformation that is the main cause of strengthening. In the deformed zone, dislocations induce residual stresses through interaction between grains. Poly-crystalline atoms, especially fine-grained, are strengthened more significantly than monocrystalline because pinning of slip bands occurs across grain boundaries [12, 17, 18]. The thickness of this zone depends on the time of deformation and does not exceed several tenths of a millimeter [16].

Zone VIII comprises the zone of elastic deformations and tensile stresses.

Residual stresses are present in zones VI to VIII. Their origin, sign and value depend on permanent deformation caused by the forces of machining, friction and temperature. In zone VIII and partially in VII, tensile stresses are formed, while compressive stresses occur in zone VI [12].

During service, zones I to III exert a significant influence on the friction process, and zones V to VIII - on the course of wear [13].

Some simplified models [16] ignore the existence of zones I to IV and distinguish only four zones corresponding to zones V to VIII of the above full model.

The simplified model of the superficial layer in the case of fluid friction is shown in Fig. 5.9.

Presented models of post-machining superficial layer structure are all theoretical. The difference between a real superficial layer, both technologically developed or being the result of service, is that in the real layer, some zones do not exist or blend into one another. Grains with anisotropic mechanical properties may be differently located. For this reason, the intensity of deformations, their depth, the value and sign of residual stresses in the particular zones of the superficial layer may vary.

5.4 A general characteristic of the superficial layer obtained by machining

From the discussion of the presented models it follows that to this date, knowledge about the structure of the superficial layer is insufficient, even when limited to the earliest developed and investigated superficial layers, obtained by different machining methods.

The removal of the external layer of material from an object causes direct contact between the freshly exposed surface with the environment surrounding it, in the majority of cases with air or coolant. Upon such exposure, an immediate reaction occurs between the physically pure metal surface with chemically active components of the environment - oxygen, sulfur, nitrogen and various other compounds. Consequently, a thin layer of approximately 2 μm, consisting of oxides, sulfides, nitrides and other compounds, is formed at the surface of the metal, strongly adhering to it and reflecting micro-asperities and discontinuities of the physically pure surface. This thin layer even assumes some of the properties of the base metal [13].

This thin layer absorbs contaminations from the environment in the form of ionized foreign matter - dust, grease, and condensed vapors which cover the layer from the external side [13].

The structure and thickness of metal layers depend on the plastic-elastic properties of that metal, on machining conditions, on the type of the cutting tool and method of treatment (dry or in the presence of a coolant) and on its intensity, i.e. cutting variables. Under the influence of loads present during the machining operation, cutting forces and friction forces cause deformations of the superficial layer, but the directions in which these two forces act are mutually opposite. In zone VIII (see Fig. 5.8) they cause the formation of elastic deformations and tensile stresses. After the pass of the tool, the return of stresses in zone VIII is countered by resistance of plastically (permanently) deformed zones VI and VII. This, in turn, causes the formation of tensile residual stresses in zones VIII and VII, and compressive stresses in zone VI. The distribution of residual stresses depends on the forces of friction and decohesion[1] of the chip material - they raise the temperature of the treated surface. During turning, the temperature may rise up to approximately 300 to 600°C and during grinding and polishing, up to 400 to 800°C with local surges up to 1200°C, lasting 1×10^{-3} to 3×10^{-3} s [13]. After the passing of the cutting edge, during cooling down of the treated surface, a change in the sign of residual stresses takes

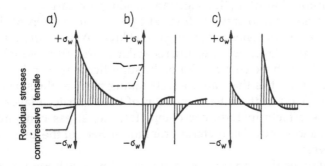

Fig. 5.10 Changes in distribution of residual stresses in zones VI to VIII according to the 8-zone model, during the action of heat created by friction during machining: a) at the moment of machining; b) during cooling; c) after completed cooling. (From *Szulc, S., and Stefko, A.* [13]. With permission.)

place in zones VI to III. Heat originating from friction usually induces tensile stresses in the treated surface (Fig. 5.10) [12, 13].

Rough machining causes the formation of a thicker superficial layer with greater deformation and residual stresses than in the case of final ma-

---

[1] *Decohesion* - the destruction of intermolecular cohesion between solids. The opposite of decohesion is *cohesion* (from the Latin: *cohaesio* - make contact) - the mutual attraction by particles of the same material, caused by the action of intermolecular forces, countering any changes in the state of aggregation, e.g., separation of particles of a solid. In contrast to adhesion where intermolecular forces bind two different materials, cohesion occurs within and between molecules of the same material. Cohesion forces are strongest in solids, especially in metals. In liquids they are significantly smaller and in gases - very small.

chining. The latter to some degree removes the defects of rough machining and oscillation superfinish eliminates the weakening of the surface after preceding operations, raising the bearing surface and causing a lowering of deformations and damage in the superficial layer during subsequent service.

During polishing a thin layer is formed which does not exhibit crystalline structure. Its thickness does not exceed several Ångstroms. This layer was discovered in 1900 by G. Beilby. It is formed with the participation of big amounts of heat evolved during polishing. Local high temperatures cause partial melting and smearing of the surface, thereby creating a surface with a roughness not very different from molecular. After polishing, the superficial layer is usually characterized by good anti-corrosion properties and favors the adhesion of coatings, especially those developed by electro-plating [16].

During service of the superficial layer exposed to friction, the following effects occur [12, 13]:

– overlapping of tangential stresses in points of real contact between the mating pair,

– contact stresses in macrozones present during rolling friction,

– action of friction forces, causing grain elongation (fibrous structure, preferred orientation) in layers adhering to the friction surface,

– unwedging (Rebinder effect) of forces of viscosity by particles with high adsorption characteristics (e.g., fatty acids) permeating into valleys between asperities and to crevices formed by the deformation of the crystal lattice in the external zones of the superficial layer. As a result of discontinuity the superficial layer acquires microcapillary properties which facilitate communication between the interior of the superficial layer and the observed layer. These properties that cause effects taking place at the surface are dependent on effects in the interior portion of the superficial layer and vice-versa. These effects include corrosion, fatigue, stress concentration at microflaws, and removal of material from the surface of the layer, as a result of friction wear.

In many cases the superficial layer formed by machining is ennobled; usually it is strengthened by other surface treatment operations, significantly changing its properties. During service, both thickness, as well as properties of the superficial layer, may undergo changes as a result of oxidation, friction, corrosion and fatigue. In the event of mating with other surfaces in conditions of friction, wear of the superficial layer progresses with time. This wear leads not only to a thinning down of the layer but also to changes in the surface geometry and physico-chemical properties. Moreover, it often leads to the gradual displacement of the superficial layer in the direction of the core. In the case of modification of the superficial layer after machining by heat treatment (e.g., hardening), thermochemical treatment (e.g., nitriding) or implantation (e.g., of nitrogen ions), the depth of migration of the nitrogen ion layer exceeds its initial thickness by a factor of several tens, which is due mainly to the rise of temperature during friction.

5.5 Physical description of the superficial layer

The superficial layer formed on the metal core assumes many of the core's properties.

The most characteristic property of metallic substances is their **non-homogeneity** - chemical, physical and structural, existing between the various microzones of the metal alloy. Two types of non-homogeneity can be distinguished, mutually correlated and mutually dependent. They cause the differentiation of:

– spatial distribution of various elements of microstructure, i.e., **non-homogeneity of structure**,

– content of components in the various microzones, i.e., **chemical non-homogeneity**. This stems from non-uniform distribution of the alloy's components and from incomplete filling of the volume.

These non-homogeneities are caused by numerous factors, beginning with the metallurgical process of alloy smelting and adding of alloying additives or mechanical alloying of powder metallurgy products and ending with the process of shaping (casting, forging) and giving the surface special properties (e.g., through heat treatment). In the process of shaping and structure stabilization there occurs enrichment or depletion of some component of the grain or grain boundary. The concentration of the particular components in the grains or zones depends on chemical affinity and thermodynamic conditions[19].

The grain structure of metal alloys causes the necessity of a precise definition of the volume size within which non-homogeneity is discussed or properties are averaged for the discussed volume. It has been agreed to assume that:

– a **macro non-homogeneity** is a zone with a volume greater than that of the grain; macro non-homogeneity is a sum of micro non-homogeneities,

– a **micro non-homogeneity** is the non-ordering of the chemical composition within a volume not exceeding that of one grain,

– **submicro non-homogeneity** is the non-ordering of the chemical composition in the crystal lattice, caused by defects of the crystalline structure of metals and alloys, changing both the degree of filling of the volume taken up by the crystal (e.g., dislocations, vacancies, interstitial atoms), as well as elementary composition of the crystal (e.g., presence of foreign atoms). Because of the minute affect of the defect, the affect of single structural defects on the properties of the entire metal is insignificant, although it is clearly enhanced at sites of defect concentrations, which change point defects into linear and spatial defects [20].

Relative to the core, the superficial layer exhibits a very high degree of submicro, micro and macro non-homogeneities.

Considering zones approximating the diameter of particular atoms one can see that there is a variation of content of particular elements in the superficial layer and that their spatial distribution is similar to laminar.

At the next step of averaging, when not single atoms but chemical compounds formed by them are considered, the micro non-homogeneity of particle location and the variation of their electric fields are revealed. This, jointly, forms the secondary molecular roughness. The primary molecular roughness, initiated by the treatment operation of material shaping, is later (during surface modification treatment) modified by the joining of particles into agglomerations of varied size and with varied degree of internal ordering [20]. At this level of averaging, different properties of particle agglomerations are revealed, due to the imperfections of internal structure, e.g., different crystal structure (as in twinning), foreign atoms, vacancies, dislocation exits. Such imperfections cause non-homogeneity of energy of the superficial layer by forming active centers, i.e., zones of higher energy potential, which are sites of enhanced absorption, catalytic, corrosion and emission capability [20]. The subject of the next degree of averaging are concentrations of agglomerations with varied spatial location. On the surfaces of one type of particle agglomerations there occur faces, edges and apexes of crystals, as well as a step-like structure; jointly they form a three-dimensional submicroroughness of the superficial layer. These concentrations are divided from one another by boundaries which are also defects of the crystal structure of metals. The spatial distribution of particle agglomerations forms the phase structure of the superficial layer, a decisive factor in the shaping of mechanical, thermal and electrical properties [20].

In the next stage of averaging we may consider the average microstructures and properties of the particular, differing zones of the superficial layer. These zones are initially situated parallel to the surface asperities, and deeper - to the nominal surface.

In the final averaging step we do not differentiate zones of the superficial layer but rather consider its structure as a whole, relative to that of the core [20]. At this stage of averaging it may be assumed that the superficial layer exhibits an abnormal structure in comparison with that of the core which takes up a volume greater by a factor of several to several hundred and even more. As a result, the superficial layer exhibits other properties than the core, usually - but not always - better. In many practical cases it would be beneficial if the core exhibited the same properties as those of the superficial layer. This, however, is impossible.

5.6 Strengthening and weakening of the superficial layer

The superficial layer obtained as the result of a treatment operation, carried out with known parameters, is usually subjected to one or several subsequent operations (e.g., rough turning - finish turning - carburizing - hardening - grinding - implantation). Thus, new superficial layers will be formed, with different properties from those of the initial superficial layer. Similarly, during service, as was mentioned earlier, the superficial layer may change its dimensions and properties, described by service conditions.

The direction in which these parameter changes may go may vary. As for surface strength, the superficial layer may, during both its formation or during service, be enhanced or lowered, relative to the initial state.

Strengthening - the raising of the surface strength, consisting of changes of mechanical properties under the influence of cold plastic deformation, e.g., by technological burnishing or friction during service. The strengthening is expressed by a change in physical, chemical and, most of all, mechanical properties: a rise in hardness, fatigue strength, yield strength and a drop in impact strength and elongation. Due to permanent (plastic) deformation, a portion of the superficial layer may exhibit preferred orientation. The depth of this preferred orientation depends on the value of forces causing the deformation, the duration of their action and plastic-elastic properties of the material. As the result of preferred orientation the superficial layer exhibits an anisotropy of mechanical properties and an enhanced resistance to wear. The strengthening is accompanied by a lowering of electrical conductivity, permeability and magnetic susceptibility, coupled with a rise of the magnetic hysteresis.

Weakening - is the lowering of surface strength, manifest at the same time by a drop in fatigue strength. The causes of this weakening are the following [16]:

– drop in hardness of the superficial layer (so-called de-strengthening) either below its initial value or below core hardness which may occur, e.g., after heat treatment (technological tempering and annealing) or during service (tempering due to temperature developed by friction);

– lowering of intercrystalline cohesion of the superficial layer (so-called loosening) which is accompanied by a lowering of hardness and surface strength, due to exceeding of boundary deformation, chemical action (corrosion) and physical action (Rebinder effect), the occurrence of dry friction with its high surface temperatures or as the result of refinement of the primary grains of the material in the superficial layer, mainly of its amorphous zone;

– non-homogeneity, both structural (defects) and physico-chemical, and random location of the superficial layer zones.

5.7 Potential properties of the superficial layer

The superficial layer exhibits a great heterogeneity of the physico-chemical state. The existence and thickness of its particular zones, as well as the total thickness of the entire superficial layer, depend on the type of treatment operation, the chemical composition of the core material, its mechanical properties, especially plasticity, on the value of forces involved in the treatment and the amount of heat dissipated, on the chemical reactions involved and on the exchange of components between the superficial layer and the surrounding system by either diffusion or sublimation, on the method of exposing the surface and on the intensity (parameters) of the treatment operation, etc. Moreover, the imbalance of intermolecu-

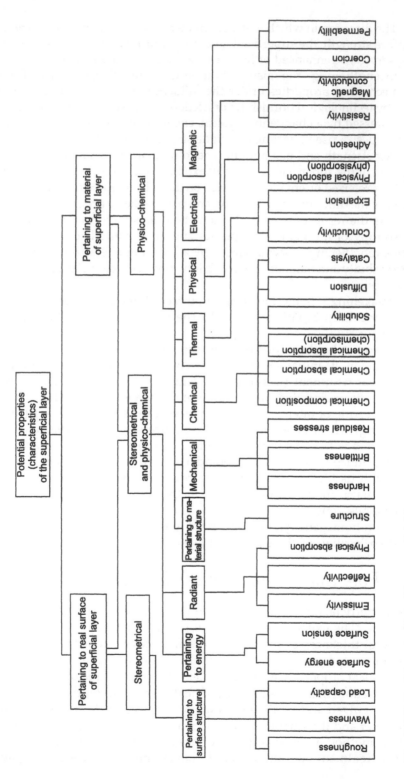

Fig. 5.11 Potential properties of the superficial layer.

lar forces at the interface between the superficial layer and the environment is the cause of stresses at the material surface. The multitude of simultaneously acting factors and phenomena taking place do not allow an unequivocal description of their effect on the properties of the superficial layer in a general manner [13].

The properties of the superficial layer may be expressed by a series of parameters that are mutually interdependent. A method of linking these parameters is usually complex and either very difficult or impossible to describe mathematically. If the correlation between treatment parameters is known for a given operation and a strictly defined given material, the superficial layer could be designed in advance in such a way as to meet all the requirements, depending on the conditions it is to be subjected to during service.

The qualitative correlation between the basic parameters is usually known, but between auxiliary parameters - usually to a much lesser degree. For example, a change in three-dimensional structure of the surface changes the contact area between mating surfaces, affecting texture, strengthening, residual stresses, microhardness and fatigue strength. On the other hand, a quantitative correlation between these parameters, at least between some of them, is - unfortunately - known only very seldom. Currently, there are only a few countries of the world where research is conducted to determine the mutual correlation between several parameters which could enable the design of a superficial layer.

The condition of the superficial layer at the time of its formation consists of a set of its potential properties which can be described by parameters: three-dimensional (stereometrical), stereo-physico-chemical and physico-chemical (Fig. 5.11).

The condition of the superficial layer is represented by properties, as described by parameters, of a surface ready to be exposed to various extraneous hazards.

Further, a broader discussion will be accorded to only those parameters which:

– are most significant and whose value strongly depends on the type of treatment operation and which clearly differentiate properties of superficial layer and core,

– are not clearly understandable on the basis of terminology alone,

– have not been discussed up to this point.

5.7.1 Geometrical parameters of the superficial layer

5.7.1.1 Three-dimensional structure of the surface

All real surfaces of solids always exhibit a deviation from ideal smoothness - they are rough to a lesser or greater degree. The overall set of deviations from the ideally smooth surface, reflecting the condition of the real surface, is described by the **three-dimensional structure of the surface**.

Viewed as a measure, it constitutes a three-dimensional set of geometrical elements of the real surface, determined by shape, size and distribution of asperities which are usually produced by the treatment operation (e.g., casting, forging, rolling, machining) or by wear. The structure may be **anisotropic** - with clearly oriented peaks and valleys, most often corresponding to machining traces or grain orientation (we call this oriented structure) or **isotropic** - which exhibits no orientation.

The three-dimensional structure of the surface of any material of known composition depends on the treatment operation. The latter is a kinematic-geometrical representation of the following:

– during the technological process of surface formation: of the tool movement along the treated surface, as well as of the tool roughness (casting moulds, stamp, cutting tool),

– during service: of the movement and roughness of the mating surface (countersurface).

The traces of surface interaction constitute the basic element of the three-dimensional structure. They often undergo changes under the influence of other physico-chemical effects accompanying the treatment operation or taking place during service.

In the case of machining, this is the representation of the movement and geometry of the cutting tool. Machining is accompanied by processes such as elastic and plastic deformation, formation of chip segments, friction between tool and chip and the treated surface, thermal effects, and vibration and decohesion of the treated material [3, 14, 18].

The three-dimensional structure of the superficial layer has a very significant effect on the service properties, such as wear resistance (forces of friction and grinding-in time), rigidity of contact connections, fatigue strength, thermal conductivity, emissivity, flow resistance, tightness, etc.).

The three-dimensional structure of the surface is characterized by several tens of parameters, of which the most important ones are discussed below.

5.7.1.2 Surface roughness

The three-dimensional structure of the surface is made up of surface **asperities**, or peaks and valleys which are usually traces of treatment or wear (Fig. 5.12).

Surface asperities directly participate in the interaction of the treated surface with the liquid or gas environment surrounding the object or with the asperities of the mating surface in contact with them. They are the agents passing on the results of this interaction (e.g., heat, force, diffusion, etc.) to the interior of the superficial layer. These asperities are described by parameters of roughness and waviness, as well as flaws in the geometrical structure of the surface. Therefore, they should be described in all three dimensions. However, practical difficulty with their measurements causes that the problem is reduced to a two-dimensional plane on which a roughness profile is traced.

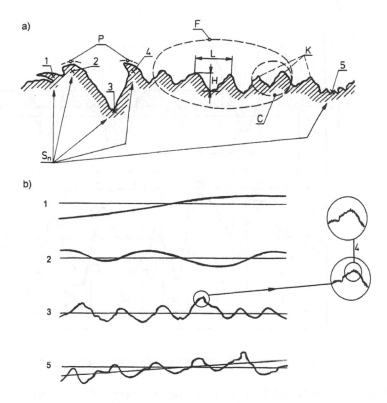

Fig. 5.12 Diagram showing: a) profile of asperities on surface of a solid: H - peak to valley height; L - peak to peak distance; F - waviness; C - roughness; P - adhesion; K - orientation of asperities; S_n - asperity flaws; 1 - flaking; 2 - folding; 3 - scratch; 4 - burr; 5 - pit (From *Kolman, R.* [3]. With permission.); b) elements and resultant unevenness of surface: 1 - shape flaw; 2 - waviness; 3 - roughness; 4 - submicroroughness; 5 - resultant structure of real surface.

The observed (measured) profile is constituted by the line of intersection of the observed surface with a defined orientation relative to the nominal surface (defined by the technical drawing or by technical documentation with the omission of roughness, waviness or other shape errors (Fig. 5.13).

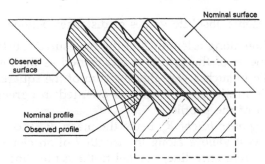

Fig. 5.13 Nominal and observed profile.

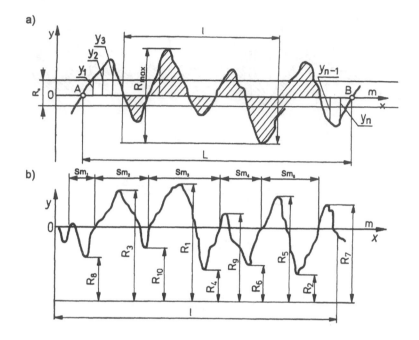

Fig. 5.14 Auxiliary diagrams for determination of values of a) R_a and b) R_z.

Surface roughness is a mode of unevenness with usually small distances between peaks and valleys, less than in the case of waviness. Roughness is defined as a set of asperities of the real surface, conventionally described as deviations of the measured profile from a reference line within the limits of a length along which waviness is not taken into account [5].

The following surface roughness parameters by ISO 4287/1-1984 (E/F/R) are defined (Fig. 5.14):

– *mean arithmetical deviation R_a of the profile from the center line average (CLA)*

$$R_a = \frac{1}{n} \sum_{i=1}^{i=n} |y_i|$$

(5.2)

where $|y_i|$ represents absolute values of distances between profile points and the center line along a length L of the measured surface (Fig. 14a).

The center line m of the profile is understood as the line dividing the roughness profile in such a way that the sum of the squares of deviations from that line is minimum; this line is oriented in agreement with the general direction of the profile;

– *10 point roughness height R_z*, being the mean distance of five highest peaks to five lowest valleys along the length l of an elementary interval, measured from a reference line, parallel to the center line (Fig. 5.14b).

$$R_z = \frac{(R_1 + R_2 + \ldots + R_9) - (R_2 + R_4 + \ldots + R_{10})}{5}$$ (5.3)

– *highest peak-to-valley height* R_{max}, or maximum distance between two lines, both parallel to the reference line, of which one passes through the highest peak and the other through the lowest valley of the profile, within limits of the elementary length *l*;

– *roughness spacing S* of the surface, being the mean spacing between peaks of roughness, measured along the elementary distance *l*.

Depending on the values of R_z and R_a, roughness is calibrated by 14 classes. The asperity height varies within a range of Angstroms to milli-meters ($R_z = 0.05$ to 320 µm; $R_a = 0.01$ to 80 µm). Asperities with heights less than 0.01 µm, termed *submicroasperities* or *subroughness*, may themselves constitute roughness or be superimposed on roughness [15]. Smaller than geometrical roughness, *molecular roughness* is the set of surface uneven-ness, dependent only on the size of molecules in the superficial layer of a solid.

The size and shape of roughness also affect such service properties of the surface as: wear resistance, adaptability to bear static and variable loading, corrosion resistance, emissivity, flow resistance, fatigue strength (drops with a rise of roughness) [3, 13, 21]. By its deepest valleys, rough-ness acts in a manner similar to a flaw - it constitutes a sudden reduction in the active cross-section of the machine component - leading to a stress concentration in that area to the value at which microcracks are initiated [15]. In the valleys of a rough surface corrosion sources are created with-out the presence of variable loading.

Surface roughness significantly affects flow resistance, which is very important in all cases where liquid movement must be assured relative to the surface of solids, e.g. in shipbuilding and pipelines. For instance, the unevenness of cladding of ship hulks grows during service (its value before overhaul may exceed that of any newly placed cladding by several times) and its increase by approx. 10 µm calls for an increase in the ship's power by about 1% [15]. Besides causing a rise in flow resistance, a rise of roughness also causes a rise in the noise of movements in a liquid envi-ronment. This facilitates sonar detection of submerged bodies, e.g., sub-marines.

Roughness and subroughness of the surface very significantly affects its radiant properties; by intensifying emissivity it reduces heat reflection and raises the diffusivity of the reflection [22, 23].

Surface waviness also constitutes a form of unevenness, with wide spacings between crests, in comparison with wave height [3]. It is de-scribed as a set of random or close to periodically repeatable unevenness, characterized by the fact that spacings are significantly larger than those in roughness and that heights are significantly, at least 40 times (but not exceeding 1000 times) smaller than the average spacing between peaks.

Most often, waviness is caused by vibrations during machining and non-uniformity of the machining process. In industrial practice waviness is inspected very seldom, since the height of the waves is small in comparison with allowable deviations.

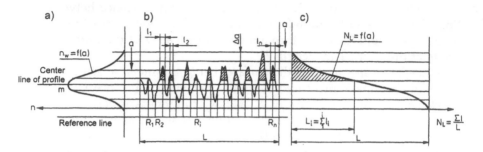

Fig. 5.15 Parameters describing the geometrical microstructure of the surface: a) profile ordinate density curve $n_w = f(a)$; b) profile of microroughness; c) linear bearing curve N_L.

Waviness, just like roughness, is the site of residual stresses in the superficial layer, possibly leading to the formation of cracks [3]. Waviness exerts a varied effect on the wear of mating surfaces, depending on wear conditions. For example, in frictional wear it causes a rise in the amount of worn material (accelerates wear), while in adhesive wear it renders adhesion difficult and prevents seizure during work under heavy surface loading (reduces wear) [3]. In conditions of boundary friction, very high surface loads occur at wave peak sites, causing fast wear of initial roughness and the tendency to adhesion and galling [15].

As opposed to **technological (initial) roughness**, the effect of which is fast reduced as a result of the formation of **service-induced (secondary) roughness**, the effect of waviness on service properties usually occurs throughout the entire service life of the machine, not only during the initial grinding-in. Waviness, similarly to shape errors, usually is subjected to elastic stresses. Moreover, waviness of machine components causes, during service, reduction of tightness of fittings, a rise in flow resistance and the induction of vibrations [15].

A derivative of the three-dimensional structure of the surface is a concept introduced specially for description of service, strictly speaking, tribological, a concept which characterizes the degree of adherence of a given surface to the mating surface [3]. Ideal, 100% adherence of real surfaces is impossible; moreover, in real conditions it depends on the loading force. For that reason, the concept of surface **contact capacity** is described as the area of contact of the real surface with a standard countersurface, loaded by a known force, or as the length of linear contact of an observed profile with a line (parallel to a reference line, at a fixed distance from it [Fig. 5.15c]). The parameter describing this surface contact capacity is the linear contact fraction N_L of the profile and its surface area analogy [5]. Both parameters, especially the first, are strongly correlated with service properties of the surface, e.g., pitting resistance, wear resistance, and rate of grinding-in of mating surfaces [21].

5.7.1.3 Structural flaws of the three-dimensional surface

The three-dimensional structure of the surface is to a greater or lesser extent contaminated by discontinuities of orientation or of geometrical character. These are called **flaws**.

Over 30 different types of surface flaws are distinguished [15]. They are formed by:

– interaction with other bodies: furrows (grooves), scratches, buckling, rounded corners, chip remnants, bands;

– stresses and flaws of the machined material: cracks, tears (crevices, gaps), delaminations (decohesion), blistering, nipples, shoulders, swellings, flaking and inclusions;

– corrosion: pitting, corrosion spots;

– erosion: pits, craters.

The most often occurring types of flaw of the three-dimensional structure are

Defects - flaws formed during the treatment operation or during service, as the result of mechanical damage (e.g., indentations, wear scars, splinters), chemical or erosive damage (e.g., pits, spots) or hidden material flaws (e.g., blisters, cold shuts) revealed during machining by exposing them to the surface.

Grooves - surface damage formed by the movement of an element of the mating surface or an element of the cutting tool, forced into the material surface. A characteristic of the groove is the plastic stacking of material along its side faces (raising of sides above the surface) and ahead of the grooving tool, as well as rounding of the groove bottom [21].

Scratches - surface damage caused by the same mechanism as grooves, but without rounding of the bottom.

Cracks - surface flaws caused by exceeding the material's strength, as the result of a point concentration of surface stresses. The cross-section of a scratch is characterized by high slenderness.

Pores - surface flaws, exposed by surface treatment and constituted by empty spaces inside the material in the form of crevices, canals or blisters. Pores formed by design (e.g., in order to enhance lubricity after filling with other material) are not considered surface flaws.

Besides the flaws enumerated above, the surface structure is also disarranged by: nicks, dents, seams, kneads, folds, flakes and other defects.

The flaws of the three-dimensional structure discussed above are most often created as the result of shortcomings or irregularities of the treatment process or in service and always constitute an undesired anomaly. They are always stress-raisers and with appropriate combinations of extraneous loading may become sources of fatigue cracking [3]. Nicks on the internal surface of a hollow turbine shaft in one engine of a Polish Airlines IL-62 jet became the source (3 such sources were detected) of fatigue cracks, leading to the tragic crash in 1982. In the presence of a corrosive environment, surface flaws may become sources of corrosion. If rising above the surface, they intensify the wear process.

When selecting the right three-dimensional surface structure to meet service requirements, a general rule prevails that with the rise of loading level, relative movement velocity and accuracy - the allowable roughness and waviness, as well as the size and amount of surface flaws must be reduced. A deviation from the above rule occurs in the case of boundary friction conditions, where friction resistance and wear depend on roughness height [15].

5.7.2 Stereometric-physico-chemical parameters of
   the superficial layer

These parameters depend simultaneously, although to varying degrees, on the three-dimensional surface structure and on the type of material of the external zone of the superficial layer and they characterize a set of geometric-material properties of the real surface and of the adherent thin subsurface layer. They describe mainly properties related to:
   – energy, predominantly the earlier discussed surface energy and surface tension,
   – radiation, predominantly emissivity (or the corresponding in value radiation absorption rate) and reflection.

5.7.2.1 Emissivity
Emissivity constitutes the main parameter, characterizing the quality of the superficial layer as a thermal radiator. Similar to other radiation parameters, two types of emissivity are distinguished, i.e. total emissivity ε_T and monochromatic emissivity ε_λ [22-25].

Total emissivity ε_T (Fig. 5.16a) describes what portion of radiant energy M is emitted by a unit surface, relative to a unit surface M_{bb} of a hypothetical blackbody in same temperature conditions,

$$\varepsilon_T = \frac{M}{M_{bb}} = \frac{\int\limits_0^\infty m_\lambda d\lambda}{\int\limits_0^\infty m_{\lambda,bb} d\lambda} \qquad (5.4)$$

where: m_λ - monochromatic density of radiant energy of tested body; $m_{\lambda,bb}$ - monochromatic density of radiant energy of blackbody.

Monochromatic emissivity ε_λ (Fig. 5.16a) describes the appropriate ratios of monochromatic radiant power densities of the tested body and blackbody at the same temperature for any chosen radiation wavelength λ, as long as it is the same for both bodies.

$$\varepsilon_{\lambda 1} = \frac{m_{\lambda 1}}{m_{\lambda 1,bb}}; \qquad \varepsilon_{\lambda 2} = \frac{m_{\lambda 2}}{m_{\lambda 2,bb}}; \qquad \varepsilon_{\lambda n} = \frac{m_{\lambda n}}{m_{\lambda n,bb}} \qquad (5.5)$$

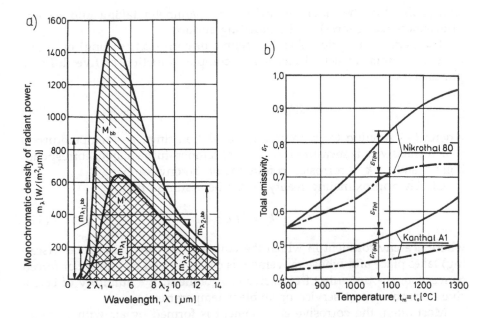

Fig. 5.16 Emissivity: a) diagram explaining the concepts of radiant power at temperature T (Numerical values are for $T = 650$ K); b) effect of surface roughness (n) and degree of oxidation of material (m) on total emissivity ε_T for two different resistance heating alloys, under different temperature conditions (t_m - measurement temperature, t_o - oxidation temperature.)

Monochromatic emissivity changes with the wavelength of the emitted radiation.

For the same temperature conditions, total emissivity is the same as the total absorption rate ($\varepsilon_T = A_T$) and monochromatic emissivity is equal to the monochromatic absorption rate ($\varepsilon_\lambda = A_\lambda$) [24].

Emissivity depends on many factors.

Many authors have attempted to assign a dependence of ε_T and ε_λ solely on resistivity, totally neglecting the three-dimensional state of the surface or describing it in broad general terms as "smooth". Despite this simplification, they obtained only approximate correlations, true only for a certain range of temperatures or wavelengths. Besides, these correlations did not pertain to metal alloys but pure metals and physically pure surfaces, i.e. to metals surrounded by deep vacuum. For this reason, they are of little use in surface engineering [23].

Generally, emissivity can be represented as a function of three basic parameters [22]

$$\varepsilon = F\left[\varphi(m), f(k_s), \Psi(n)\right] \tag{5.6}$$

where: m - parameter dependent on the type of material emitting radiation, including its resistivity; k_s - parameter describing the degree of corrosion (scaling) of the radiating surface; n - parameter taking into account unevenness (roughness) of the radiating surface.

The parameter k_s describes the degree of corrosion (scaling) only in a general manner. In actual fact it is a complex function of five different parameters:

$$k_s = f(t_o, \tau_o, t_m, v, k_w)$$ (5.7)

where: t_o - oxidation temperature, τ_o - oxidation time, t_m - temperature at which emissivity is measured, v_t - rate of temperature change, correspondingly t_o and t_m; k_w - corrosion conditions of the environment.

Closely approximating reality, it can be assumed that:

$$k_w = \xi(u, p, h)$$ (5.8)

where: u - environment in which the temperature corrosion process (scaling) takes place and whose movement is characterized by natural or forced convection; p - pressure of corrosive environment, h - humidity of corrosive environment at service or ambient temperature.

Most often, the corrosive environment is formed by air with pressure close to atmospheric; less frequently by a technologically created atmosphere with controlled composition.

Emissivity is, therefore, an implicit function of nine variables, the majority of which are mutually interrelated.

$$\varepsilon = F \{ \varphi(m), f[t_o, \tau_o, t_m, v_t, \xi(u, p, h)], \psi(n) \}$$ (5.9)

For a given material, the correlation (5.9) is simplified by the value m but only in the case when the structure and phase composition of the radiating surface do not vary. In the opposite situation, occurring practically in more than 90% of cases, the chemical composition and structure of the radiating surface do change. This leads to the conclusion that for a given initial tested material, emissivity is a function of nine variables. As the scaling progresses, the unevenness of the radiating surface changes.

Attempts at separating the effects of the particular parameters on the value of emissivity face major difficulties. The theoretical resolution of the function (5.9) is impossible, but it can be done empirically by investigating emissivity vs. each of the particular parameters, keeping the remaining parameters constant.

From among the nine variables, the biggest effect on emissivity is exhibited by:

– surface roughness, to a lesser degree surface flaws and to a minute degree waviness,

– physico-chemical condition of the emitting material (non-oxidized material, material covered by a film of oxides or other chemical compounds).

It should be borne in mind that metals and their alloys, existing in a given corrosive environment at low temperatures, usually corrode only insignificantly. With a rise of temperature the intensity of corrosion increases, especially of oxidation corrosion. Beginning from certain temperatures the material is covered totally by a thick layer of corrosion products, in most cases by an oxide layer. Thus, initially only the clean surface of the metal radiates; next, the clean surface and the gradually growing corrosive layer which finally becomes so thick that it takes over the radiation completely (Fig. 3.15, part II). Besides, with the degree of corrosion, the roughness of the surface changes, along with the chemical and phase composition of the emitting material (metal - metal oxides or other corrosion products).

Thus, the total emissivity of the metallic material, working in an atmosphere composed of air, is expressed by:

$$\varepsilon_T = \varepsilon_{T(init)} + \Delta\varepsilon_T \qquad (5.10)$$

where: $\varepsilon_{T(init)}$ is the initial emissivity of the non-oxidized material, dependent only on the type of emitting material and its smoothness; $\Delta\varepsilon_T$ - is the increment of emissivity, resultant from surface oxidation and:

$$\Delta\varepsilon_T = \varepsilon_{T(n)} + \varepsilon_{T(m)} \qquad (5.11)$$

where: $\varepsilon_{T(n)}$ is the rise in emissivity, stemming from the rise in surface unevenness, being the result of oxidation; $\varepsilon_{T(m)}$ is the rise in emissivity stemming from the change in chemical and phase composition of the emitting material (metal - metal oxides).

For the case where the character of the unevenness profile can be simplified to a series of wedge-shaped cavities, the cavities as shown on the profile plot are represented by triangular-shaped teeth, regardless of the direction in which the profile is measured. The value determined is the mean emissivity of the cavity material - a component of emissivity, modified by changes of unevenness of the oxide scale. It can be expressed by the formula:

$$\varepsilon_{T(n)} = \varepsilon_{T(init)} \frac{W}{1+(W-1)\varepsilon_{T(init)}} ; \qquad (5.12)$$

where W - relative cavitivity, expressed by the formula:

$$W = \frac{n_{w(t)}}{n_{w(init)}} ; \qquad (5.13)$$

where: $n_{w(t)}$ - cavitivity at temperature t; $n_{w(init)}$ - initial cavitivity.

Also [24]:

$$n_w = \sqrt{1+4\left(\frac{R_z}{S}\right)^2}$$ (5.14)

where: R_z - height of unevenness; S - surface roughness spacing.

Fig. 5.16b shows the effect of the component $\varepsilon_{T(n)}$ on total emissivity ε_T. The effect of the second component $\varepsilon_{T(int)}$ is extremely difficult to determine. It can be approximated, knowing the value of $\varepsilon_{T(init)}$ by measuring, at given temperatures, the total emissivity ε_T, determining $\varepsilon_{T(n)}$ from the expression (5.12), making use of (5.11).

Generally, the emissivity of every material increases with a rise of the unevenness of the surface, the degree of corrosiveness of the environment (e.g., a rise of its humidity) and the degree of corrosion (most often the degree of oxidation). All of these factors intensify with a rise of temperature. Greatest emissivity is exhibited by surfaces which are rough, matte, dark, oxidized and corroded.

Emissivity constitutes a significant parameter for the description of temperature exchange, especially in industrial and residential thermotechnical installations, where the exchange takes place mainly by radiation [22-24].

5.7.2.2 Reflectivity
Reflectivity R depends on the same parameters as emissivity. Since for non-transparent bodies,

$$\varepsilon_T + R = 1$$ (5.15)

reflectivity also depends on surface roughness and the degree of its oxidation, but as opposed to emissivity, it is reduced with a rise of surface unevenness and degree of oxidation. Highest reflectivity for heat radiation is exhibited by surfaces which are smooth (polished), shiny and bright [24, 25].

5.7.3 Physico-chemical parameters of the superficial layer

5.7.3.1 General characteristic
Physico-chemical properties of the superficial layer vary from one zone to the next within the superficial layer. The existence, location and thickness of these zones are the resultant of core material properties and extraneous effects. They are, therefore, strongly dependent on the type of treatment operation but independent or dependent only to a minor degree on surface geometry. For example, after cold working there usually occurs a clear-cut grain orientation which does not occur in the superficial layer shaped by electro-discharge machining. On the other hand, E.D.M. leaves a heat affected zone with thermo-chemical changes which is not found after machin-

ing with a cutting tool. Similarly, after burnishing there is a distinct zone of deformed grains with a high degree of strain hardening, without changes to chemical composition, whereas after ion implantation, a small degree of strain hardening occurs, along with a high degree of structure refinement, the existence of a heat-affected zone and significant changes of the chemical composition.

Physico-chemical properties of the superficial layer are synthetically determined by superficial layer parameters. These parameters characterize a set of properties of the superficial layer material, mainly in the entire volume of the layer. Properties are usually considered on the cross-section of the superficial layer; sometimes through their measurement along the layer depth, the so-called **distribution profile** is generated: of values, concentrations of the particular diffusing or alloying components, implanted ions, residual stresses, etc. In both groups, the significant parameter is the depth of the superficial layer and the thickness of its particular zones.

The most important physico-chemical property is the metallographic microstructure on which other properties of the superficial layer depend. These are

– mechanical (hardness, plasticity, residual stresses, fatigue strength, wear resistance),

– chemical (absorption, chemisorption, resistance to chemical corrosion),

– electrochemical (resistance to electrochemical corrosion),

– thermo-physical (conductivity, expansion, physisorption, adhesion),

– electrical (resistivity, conductivity),

– magnetic (coercion, permeability).

5.7.3.2 Metallographic structure

Metallographic structure is defined as the internal structure of the superficial layer, including the distribution of constituent elements (crystals, grains, atóm arrangement in the crystal lattice) and the set of correlations between these elements, characteristic of the given system. We distinguish **macrostructure** - the structure visible with the unaided eye, and **microstructure** - which can be seen under a microscope. In surface engineering, similarly to metallurgy, microstructure is of basic significance [26].

Metals and the overwhelming majority of non-metals feature a crystalline structure, i.e., internal system of strictly determined distribution of atoms, ions or molecules in the elementary cells of the crystal. The structure depends on the treatment operation to a greater degree than any other parameter of the superficial layer. The microstructure of the particular zones of the superficial layer is characterized predominantly by the type, amount, shape and distribution of the solid phases and other microstructure constituents. The superficial layer may have the following type of microstructure:

– **primary** - formed during the transition of the liquid metal phase to solid, i.e., during the solidification of the metal or alloy,

– **secondary** - formed from the primary structure after recrystallization in the solid state, as the result of phase transformations or cold working.

Usually, the primary structure occurs rarely in the superficial layer, e.g., after remelting (by laser, electron beam or plasma) to a certain depth of the primary superficial layer. The secondary structure almost always occurs but it should be remembered that structural transformations may occur several times either during the technological operation or during service.

An orientation of structural elements of the superficial layer which, in the statistical sense, is preferred, is referred to as **texture** and pertains to:
– grain boundaries, so-called grain boundary texture,
– spatial crystal lattice, so-called crystalline or crystallographic texture.

In the superficial layer the main type of texture found is the **deformation texture**, caused by oriented action of forces during cold working. More rare are **casting texture**, caused by oriented outflow of heat, e.g., during partial melting of the superficial layer, and **recrystallization texture**, dependent on the annealing temperature, as well as on deformation texture, chemical composition, etc. [27].

Texture is generally an undesired effect, causing anisotropy of some properties of the superficial layer, e.g., magnetic, physical, chemical and mechanical.

Structure significantly affects the properties of the superficial layer. By selection of the treatment operation it may be shaped in a manner that is most appropriate for the given material - the material of the initial superficial layer may be mechanically strengthened, remelted, alloyed by diffusion or by remelting or implanted by ions. During service, especially in conditions of dry or boundary friction, at peaks of surface asperities in contact with the mating surface, temperatures are developed which exceed those of structural transformations. The result is tempering of formerly hardened sites, local diffusion of alloying elements; there may also occur burns similar to grinding burns, constituting so-called stress-raisers which are considered a **structural flaw** [3]. Structural stress-raisers are also constituted by sudden transitions of structure from one zone to that of another, with differences in the specific volume of structural components.

The zone structure of the superficial layer very strongly depends on the chemical composition of zones, usually determined along the cross-section of the superficial layer. Less frequently, the distribution of elements is determined on the real surface of the layer and in cross-sections, parallel to the nominal surface of the object.

Locally, the structure may be contaminated by **material flaws**: microcracks, crevices, shrinkage porosity, inclusions, pores, indentations, tears, etc. In service conditions, structural flaws may expand or become the nuclei for new flaws, e.g., stress-raisers. Material flaws cause lowering of strength, especially of the fatigue limit.

5.7.3.3 Hardness

Hardness is the property of all solids, by which the solid offers resistance to plastic deformation or cracking by another, harder solid, exerting a local, strong force on its surface. In a narrowed-down sense, it is the resistance of the material to plastic deformation by concentrated forces, acting on a small surface area.

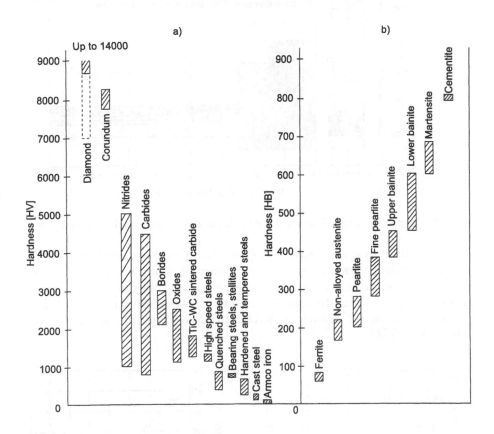

Fig. 5.17 Hardnesses of different materials (a) and steel microstructures (b).

The definitions presented here do not, however, render the physical meaning of hardness. In other words, hardness cannot be defined by other physical quantities. For this reason, hardness has been defined as a purely experimental concept (in a similar manner to some other technological properties, as press-formability, castability, forgeability) and is expressed by different measuring methods. Hardness is a conventionally defined characteristic, allowing the comparison of the resistance of different materials to surface damage (Fig. 5.17, 5.18). It is at the same time the most frequently measured mechanical property. Hardness is of special importance to tooling and strongly loaded surfaces of machine components.

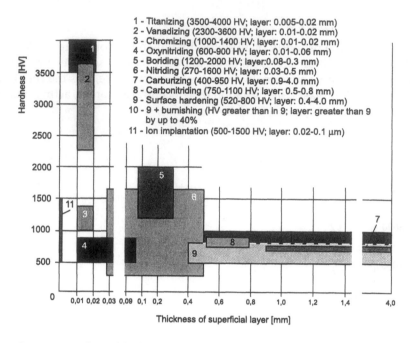

Fig. 5.18 Orientation values of thickness and hardness of some superficial diffusion layers.

In addition to surface hardness (the measurement of which was introduced to industry as late as the 20th century) it is important to know the hardness of structural elements of the particular zones of the superficial layer, e.g., grains and structural components, especially on cross-sections. This last parameter, known as **microhardness**, came into use only after World War II [28]. ·

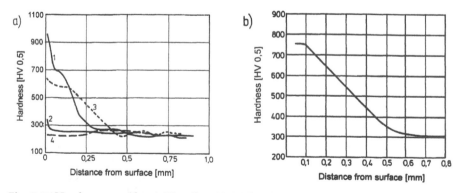

Fig. 5.19 Hardness profile: a) Nitralloy 135M, hardened and tempered to 30±2 HRC; 1 - glow discharge nitrided at 520°C for 9 h; 2 - implanted by nitrogen ions with energy of 100 keV and ion dose of $2 \cdot 10^{17}$ ions of N_2^+ per cm^2; 3 - electron beam hardened with power density of 2230 kW/cm^2 and exposure time $0.74 \cdot 10^{-4}$ s; 4 – laser hardened with power density of 1.4 kW/cm^2 and exposure time of 0.13 s; b) 18HGT grade steel, gas nitrided at 530°C for 36 h.

Hardness (microhardness) is one of the most basic, universally accepted properties of materials, especially of metals and their alloys, easily measured by various methods, and connected with many other properties of the superficial layer, e.g., wear resistance, strength, residual stresses, plasticity. Usually, the higher the stress loading to which the part is subjected, the higher should be the hardness of the surface. Unfortunately, a rise in hardness is often connected with a rise in brittleness.

Hardness depends on the type of material and its structure which, in turn, depends on treatment, especially strain-hardening, heat and thermochemical treatment (Fig. 5.19). The hardness of crystalline bodies depends on the limit of elasticity under compressive loading and on the modulus of elasticity. The microhardness of superficial layer zones may change during service, especially during wear, as the result of microstructural changes caused by surface tempering, secondary hardening (grinding burns), the breakdown of residual austenite and other factors [21,32].

5.7.3.4 Brittleness

Brittleness is a material property, consisting of permanent partition of material under the influence of internal or external forces. The partition begins at the tip of the propagating crack and is formed without the presence of any significant plastic deformation. Brittleness depends on the type of material, its phase composition, structure, etc. and on external factors such as stress distribution, method of loading, temperature, chemical composition of the environment and others. Usually, brittleness occurs in solids within certain temperature ranges [26]. The majority of materials exhibit brittleness at ambient temperature (so-called cold shortness); others, as e.g., unalloyed open-poured steel, exhibit greater brittleness at elevated temperatures (so-called hot shortness). Metals may exhibit different types of brittleness, e.g., the already mentioned cold shortness and hot shortness, hydrogen embrittlement (caused by excessive diffusion of hydrogen into the metal), pickling embrittlement or embrittlement caused by electroplating of metal objects, temper embrittlement, blue brittleness, etc.

In the case of superficial layers and coatings, brittleness is an undesirable effect, e.g., brittleness of superficial layers after diffusion, caused by excessive concentration of saturating element, like nitrogen. Often, although not always, brittleness is connected with hardness: the higher the hardness, the greater the brittleness of the layer.

A property opposite to brittleness is **ductility** - the susceptibility of metals to permanent plastic deformation without the formation of cracks. Ductility is one of the basic characteristics of the metallic state. Often the term "ductility" is used as a synonym of plasticity but it means a qualitative, non-measurable characteristic, strongly dependent on structure, processes occurring at the atomic level and on the type of slip.

Usually it is desired that hard but not brittle layers be formed over a ductile core [26, 27].

5.7.3.5 Residual stresses

Types of residual stresses. In all materials subjected to extraneous effects - be they mechanical, thermal, chemical or a combination of any or all of them - there occur non-uniform volume changes, both reversible and irreversible, causing the formation of stresses. Stresses describe the state of internal forces and moments of forces, brought about by the interaction, in a given locality, of two parts of the material, situated on either side of an apparent cross-section, the forces in question acting on a unit area of the cross-section.

After the removal of external effects, reversible changes (elastic deformations) undergo atrophy, along with stresses caused by them. However, some irreversible changes (plastic deformation) remain in the material, along with stresses caused by them which are referred to as residual stresses [33].

Residual stresses, in earlier times referred to as rest or final stresses, are those which are in mutual equilibrium within a certain zone of the material and which remain after the removal of external loading. Depending on the zone where this equilibration occurs, the following types are distinguished:

1) according to the classification by E. Orowan [34], two types of residual stresses include:

– **macrostresses** - formed as the result of any external loading, and balanced out in the entire volume of the body. They are regarded as the result of the joint, average interaction of microstresses. A definition of this type assumes the material to be homogenous, i.e., having isotropic properties;

– **microstresses** - formed as the result of heterogeneity of the material (blocks of grains, single grains), which usually generate a non-homogenous stress field, often connected with texture and therefore exhibiting preferred orientation (so-called stress-texture) [29];

2) according to the classification by N.N. Davidenkov [34, 35] three types of residual stresses are distinguished (Fig. 5.20):

– **stresses of the Ist kind**, termed **macrostresses** (*body stresses*), caused by the mutual interaction of macroscopic-size zones of the material, balancing out within volumes of the same order of magnitude as the object, within the limits of the entire superficial layer, in zones of dimensions approximating those of the superficial layer or in major zones of the superficial layer (e.g., in a zone with a very big number of grains). They are formed when external effects in the form of, e.g., mechanical loading causes non-uniform plastic deformation or as the result of thermal effects, causing non-uniform expansion of neighboring macrozones. For this reason, they were once referred to as thermal stresses. The conservation of body continuity requires the formation, between such macrozones, of mutual interaction, tensile or compressive, which we call macrostresses [33]. Macrostresses are caused directly by non-uniform plastic deformation, temperature changes, changes in the material structure or a combination

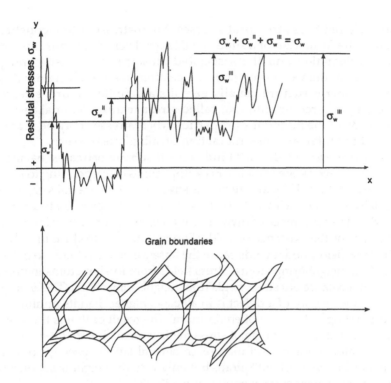

Fig. 5.20 Schematic representation of residual stresses, σ_w of the first kind (σ_w^I), second kind (σ_w^{II}) and third kind (σ_w^{III}) along the surface of a grain block.

of the above. They may be regarded as the result of the total average interaction of microstresses, assuming homogeneity (practically never existent in the micro scale) of the material microstructure, i.e. isotropic properties. These stresses cause changes of dimensions, deformation or cracking of the object or its superficial layer;

– **stresses of the IInd kind**, termed **microstresses**, formed as the result of non-homogeneity of metallic structure, consisting of grains and grain blocks. These stresses balance out within zones of dimensions comparable with those of grains, thus within limits of a small number of grains or even within parts of grains of the superficial layer. In materials not exhibiting texture, grains and grain blocks are randomly oriented and may be characterized by an anisotropy of elastic and plastic properties or of the thermal expansion coefficient while their boundaries may be sites of dislocation pile-ups and other defects. When the grains belong to different phases, greater differences in properties occur. As the result of different changes due to transmittal of mechanical, chemical or thermal effects through grains and blocks, microstresses are formed at their boundaries, balancing out within microzones of grains or within a small number of grains [33]. Microstresses are formed between separate components of the microstructure and act upon various microstructure elements which already are subjected to different microstresses. Therefore, they create a non-homogenous stress field. For this reason they were sometimes referred to as structural stresses. Microstresses often constitute the result of the formation of a superficial layer. Their chief source is different crystal orientation and the associated anisotropy of elastic and plastic properties of the various crystals. Since after treatment (mainly deformation) the microstructure usually exhibits a definite texture, stresses also exhibit a preferred orientation, called **stress texture**. Its final result is the anisotropy of the material's properties. Microstresses may be regarded as the result of total, average interaction of submicrostresses;

– **microstresses of the IIIrd kind**, termed **submicrostresses**, balancing out within the space of one crystal, thus within zones corresponding to the crystal lattice parameters. They are treated as stresses of the material's crystal lattice, especially in zones with defects. In such zones the proper structure is disrupted by the occurrence of own or foreign atoms in improper interstitial and nodal sites or the existence of voids. Foreign atoms introduce into the lattice their stress fields, nodal voids cause the absence of stress fields to balance the fields from neighboring atoms. Stress fields from foreign atoms in nodal sites also do not balance out stress fields from neighboring atoms. The energy of the lattice in the vicinity of a defect is in all cases higher than its minimum value corresponding to the state of equilibrium. The result of that is the stress field around the defect. The range of stress fields is small due to the small range of action of atomic forces and may reach several lattice spacings. Stress fields around defects interact with atoms but only with the neighboring ones, upsetting them from their state of equilibrium [33, 35-37].

If an atom of gas, e.g., hydrogen, is introduced by diffusion into the crystal lattice of steel, it generates around it compressive residual stresses of the IIIrd kind. Next, as the result of desorption of gas molecules in the internal discontinuities of microstructure, very high pressures are generated in such sites, giving rise to compressive residual stresses of the IInd kind. After the saturation of the superficial layer with this element it is usual that a gradient of its concentration will occur (and along with it a gradient of properties). The final result will be that residual stresses of the Ist kind will be generated between layers or between the superficial layer and the core [38].

In the superficial layer there exist three kinds of residual stresses; they are manifest predominantly as macrostresses. Micro and submicrostresses affect the limit of elasticity of the material but have only a small influence on its strength. They are added to stresses caused by external effects and for that reason they determine the moment of exceeding of the material's strength, manifest by the formation of microcracks. Submicrostresses may be the cause of high hardness and strength of metal alloys [33]. Independently of the kind of stresses, the result of their action is the same - they always induce defects and elastic deformations of the crystal lattice. Further on in this book the term "residual stresses" should be understood as residual stresses of the Ist kind.

Each surface treatment in which the limit of elasticity is exceeded by any element of the superficial layer or core structure leaves behind a trail in the form of residual stresses, especially those of the I[st] kind. In the majority of finished machine parts and structures there exist residual stresses left behind by treatment or assembly operations.

Residual stresses are characterized by their sign ("-" compressive and "+" tensile), their value, distribution, gradient and depth of penetration.

Factors causing the formation of residual stresses. Such factors can usually be classified as being of three kinds:

– **mechanical,** stemming from non-uniform plastic deformation of superficial layers during mechanical cold work. They are accompanied by non-uniformly distributed and interconnected processes of force action, reorientation, refinement, expansion or contraction of structural components. Macrodeformations give rise to reorientation of structural components in layers situated closer to the real surface relative to deeper situated zones. Microdeformations, on the other hand, reveal themselves within the volumes of separate components, due to their refinement into fragments and blocks and to mutual elastic-plastic interaction of neighboring grains. Resulting from that is local increase or decrease in material density, enhanced by the movement of dislocations, their distribution and kind [37]. Plastic deformation due to cold work causes changes in material density (a rise in volume of approximately 0.3 to 0.8 [21]), conducive to the rise of compressive stresses. Plastic stretching of the superficial layer by forces of friction and by machining chips also causes the formation of compressive stresses. Residual stresses caused by mechanical factors are sometimes termed **mechanical residual stresses;**

– **thermal,** caused by thermal expansion of the material and stemming from non-uniform heating or cooling of various layers of the material (macrodeformations) or of its particular fragments (microdeformations). During heating, especially if it is non-uniform, there occurs non-uniform thermal expansion causing plastic deformation which prevails all the way up to melting point. In the liquid state, the volume of all metals (with the exception of bismuth and antimony) is smaller than in the solid state. Fig. 5.21 shows a diagram of the formation of residual stresses using water quenching of 100 mm dia. heated steel bar as an example [39, 40]. Upon heating, surface temperature is usually slightly lower than that of the core. With progress of cooling time, the difference between surface temperature (curve S) and core temperature (curve C), in other words - the temperature gradient - rises. The material of the superficial layer and of layers situated deeper diminishes in volume with the progress of the cooling process, shrinking (linear changes of approximately 0.5%), causing the formation of tensile stresses (curve 1). At the same time compression of the still hot core, gives rise to compressive stresses there (curve 3). The temperature gradient between surface and core rises until it reaches point M. The maximum temperature difference (approximately 600 K) corresponds to maximum tensile stresses at the surface and maximum compressive

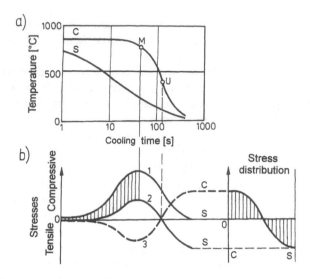

Fig. 5.21 Diagram showing the mechanism of formation of residual stresses during water quenching of a steel bar of 100 mm diameter, heated to above 700°C [39, 40]: a) cooling curves for bar surface S and core C; b) residual stress curve: 1 - when residual stresses do not exceed yield strength; 2 - when residual stresses exceed yield strength; 3 - stresses in core. (From [40]. With permission.)

stresses in the core. If the tensile stresses at the surface and compressive stresses in the core do not exceed yield strength, their variations with time proceed according to curve 1. However, at elevated temperatures plastic deformations usually occur, leading to a reduction of residual stresses, and their course is modified to that shown by curve 2. The shaded area between curves 1 and 2 represents plastic deformations which occur until the moment represented by point M is reached. From that moment on, the temperature gradient begins to diminish and so do the stresses, until core temperature is reached (point U), when stresses drop to zero value. Further shrinkage of the core causes the formation of tensile stresses in the superficial layer, which was earlier expanded due to plastic deformations. Variations of stresses vs. time in the cooling core (curve 3) constitutes a mirror image of stresses in the superficial layer. From the moment ambient temperature is reached by the entire cross-section of the bar, tensile stresses remain in the core while compressive stresses persist in the superficial layer [40]. The value of thermal residual stresses rises under the influence of all factors which increase the temperature gradient and drops with a rise of yield strength (which increases with temperature, shifting the curve upward) [36, 39, 40]. Stresses caused by thermal factors are termed **thermal residual stresses**;

   – **structural**, caused by phase transformations of structural components, occurring at different temperatures. In the case of steel, martensite has the biggest specific volume; next come bainite, troostite, sorbite, pearlite and austenite. The process of transformations, causing volumetric changes (these changes being of the order of 0.5 to 1.5% but may reach even 3%) is accompanied by a simultaneous process of stress formation. The stresses are compressive if the specific volume is increased and tensile if decreased. In turn, all volumetric changes within the volume of a given component are accompanied by changes in neighboring zones [37]. Greatest residual stresses are formed during hardening, caused by the transformation of austenite to martensite which proceeds at a very high linear rate (in ferrous alloys the rate of growth of martensite nuclei is approximately 33% that of the speed of sound in a crystal). Martensitic transformation in the heated material occurs as the result of quenching at a known rate of heat extraction, highest at the surface, causing a volumetric increase in the superficial layer. When the carbon content in martensite is 1%, volume increase of martensite relative to austenite is approximately 4%. In the slower cooled core, martensitic transformation is retarded. The core is subjected to stretching, causing compressive stresses at the surface. Next, the onset of martensitic transformation in the core causes the stretching of the outer layers which were hardened earlier and, in consequence, the compression of the core. Changes in specific volume which are due to structural transformations are greater than those brought about by thermal expansion. Stresses caused by these factors are termed **structural residual stresses**.

Other examples of external forces causing the formation of residual stresses with varied value and range of action may be, besides pressure (mechanical stresses) and temperature (thermal and structural stresses), chemical interaction (e.g., formation of chemical compounds by atoms introduced through diffusion and substrate atoms) and physico-chemical (e.g., implantation with the formation of chemical compounds). Through the change of chemical composition, such interaction causes changes in the specific volume of the material or in the coefficient of thermal expansion. As an example, the saturation of iron and its alloys with nitrogen increases volume and decreases the thermal expansion coefficient of the saturated layer relative to that of the core which causes compressive stresses to be set up in the layer and tensile stresses in the core.

Usually, residual stresses are formed as the result of joint interaction of several forces (causes) and their separation is usually difficult. For example, during hardening, when the effects of thermal and structural stress formation overlap, structural stresses tend to either raise or diminish thermal stresses, depending on the size and shape of the element's cross-section plane, rate of heat extraction and steel hardenability. Tying in the above to point U in Fig. 5.21 [40] the following can be stated:

– structural stresses raise thermal stresses if they are formed in the core before and in the superficial layer after reaching point U and vice versa;

– structural stresses across the entire cross-section or after passing through point U counteract thermal stresses;

– greatest compressive stresses in the superficial layer and tensile in the core are formed when transformation in the core occurs before and in the superficial layer after passing through point U.

When, after removing the external forces, residual stresses prove to be only slightly less than the material's strength, the material may deform, warp, suffer delamination or exfoliation. If they prove to be greater, the material will crack.

Residual stresses are superimposed on operating stresses, induced by external forces (see Fig. 5.44).

– They can be added to them, resulting in the material being destroyed already under operating stresses, lower than material strength, sometimes under quite small loads. Residual stresses can also cause the material to crack spontaneously [37]; it is said that residual stresses reduce material strength. In the superficial layer, these are usually tensile stresses.

– They may be subtracted from operating stresses, resulting in destruction of the material only when operating stresses exceed the material's strength; it is then said that residual stresses raise material strength. In the superficial layer these are usually compressive stresses.

Residual stresses are formed in the superficial layer and in the core. Usually, the value of residual stresses is greatest in the superficial layer and, the greatest stress gradients are located there, especially at the interface between the superficial layer and core (Fig. 5.22).

Residual stresses in the superficial layer usually occur in zones of texture, plastic deformation, and elastic deformation, but it is in the textured zone that they assume their highest values. Their distribution and value depend on the type of material and its three-dimensional and metallographic structure, on strength and thermal characteristics, on external factors (e.g., rate of heat extraction) and on the associated strain-hardening of the superficial layer, as well as on wear resistance.

General functional expression of residual stresses. In the broadest sense, residual stresses σ_w may be expressed by an implicit function of the most important, mutually interacting parameters in the form below:

$$\sigma_w = f(m, t, k, o) \qquad\qquad (5.16)$$

where: $m = f_1 (c, w, f, ch, s)$ - is the function of the primary material (core, superficial layer, coating), described mainly by its properties: c - thermal (especially: thermal conductivity, thermal expansion, specific heat); w - mechanical (especially strength: Young's modulus, Poisson ratio); f - physical (e.g., ion implantation); ch - chemical (especially: chemical composition, formation of chemical compounds of diffusing atoms with substrate atoms); s - structural (especially: roughness and valley bottom radius) and metallographic (especially grain type, size and orientation, defects); t - technology of formation of superficial layer or coating (type, number, sequence and parameters of treatment operations; temperature, temperature variation rate, temperature gradient, pressure, loading, feed rate, energy, element concentration, etc.); k - shape and size of component in which residual stresses are measured; o - interaction of core with superficial layer or coating.

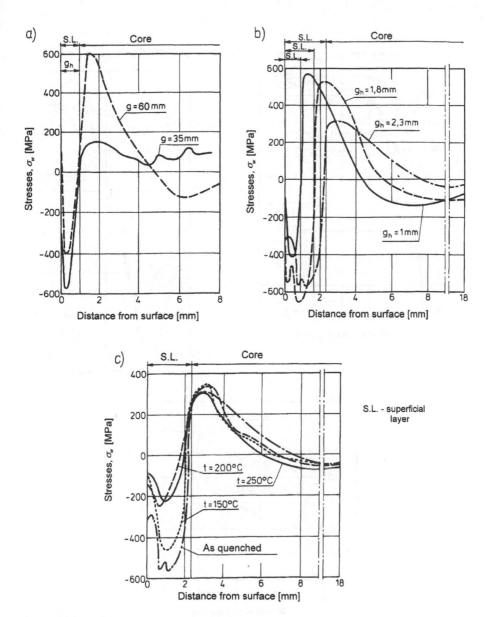

Fig. 5.22 Distribution of axially oriented residual stresses in induction hardened 1045 steel: steel bars of different thicknesses g, all hardened to a depth of 1 mm; b) hardened to different depths g_h; c) hardened to depth $g_h = 2.3$ mm and tempered at different temperatures t. (From *Janowski, S.* [41]. With permission.)

To this day, no mathematical expression has been formulated which would correlate the particular elements of the set (5.16). On the other hand, between ten and twenty different measurement methods have been developed, allowing the determination of residual stresses for strictly defined conditions.

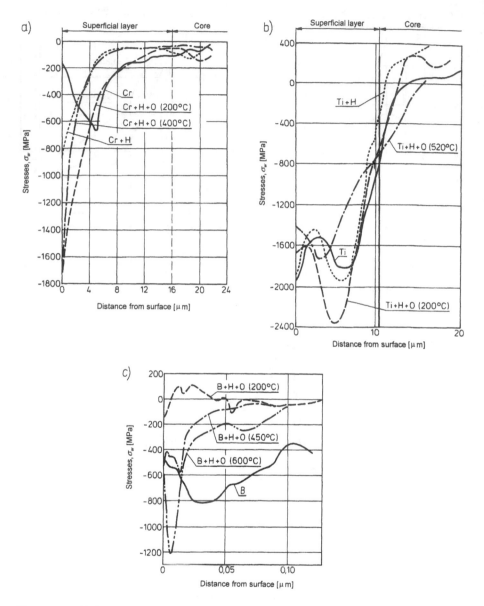

Fig. 5.23 Distribution of residual stresses, resulting from: a) diffusion chromizing of D2 grade steel; b) TiC coating of D2 steel; c - boriding of 1045 steel; designations: B - boriding; Cr - chromizing; Ti - TiC treatment; H - hardening; T - tempering. (From Janowski, S. [41]. With permission.)

In the absolute sense, a given value of residual stresses when all other parameters are equal depends heavily on the method of measurement. Numerical values of residual stresses, obtained by different measurement methods, may differ by several to several tens percent. In certain cases differ-

differences exceeding 100% and even results with opposite signs may be obtained [41, 42].

Residual stresses in a superficial layer directly affect the layer's cohesion but their action may also be of an indirect nature - by forcing the migration of atoms with small diameters (e.g., hydrogen, carbon, nitrogen, boron) through the crystal lattice of the host material. The force exerted by stress gradient on an atom in an interstitial position is, admittedly, not big in comparison with the force exerted by a concentration or temperature gradient. However, local stresses may cause migrations of interstitial atoms to sites preferred by geometry or thermodynamics (vacancy clusters, dislocation lines, grain boundaries and stacking faults) causing significant local stresses, favoring the initiation of cracks [38].

When knowingly shaping the properties of the superficial layer, it is endeavored to obtain, as the final result, compressive residual stresses in the superficial layer, while in the core - tensile residual stresses with a small gradient. Compressive stresses in the superficial layer may even attain a value equal to approximately 50% of the material's ultimate strength [37].

The value of compressive residual stresses obtained as the result of surface diffusion treatments may even reach 2400 MPa (Fig. 5.23) [41]. As an example, the value of compressive stresses in nitrided layers on low alloy nitriding steels and on high alloy structural steels may reach 900 MPa [38].

In the case of mechanical strain hardening, the depth of penetration of stresses is usually greater than the depth of hardening even by several tens percent. With a rise of stress value at the surface, the depth of their penetration diminishes [37]. The value of residual stresses rises when mechanical strain hardening is coupled with heat treatment of thermo-chemical treatment (Fig. 5.24).

Generally, with a rise in the strength of the mechanically strain-hardened material and in the strain-hardening parameters (mainly, the loading force), residual stresses in the superficial layer increase. Their value, depth of penetration and character of distribution may all be controlled by treatment operation parameters. In almost all cases the formation of compressive stresses in the superficial layer causes a rise of fatigue strength (with tensile stresses the effect is opposite) and hardness, wear resistance and corrosion resistance. A greater degree of plastic deformation causes an increase in residual stresses and in fatigue strength.

Regardless of the root cause of formation of residual stresses, their value and distribution affect strength properties, especially fatigue strength, resistance to dynamic loading and to brittle cracking (see Section 5.8.1), as well as tribological properties, especially contact fatigue (see Section 5.8.2) [42].

A particularly significant effect of residual stresses on mechanical properties, especially fatigue, is revealed in the case of superficial layers containing technological or structural flaws, surrounded by stress concentrations.

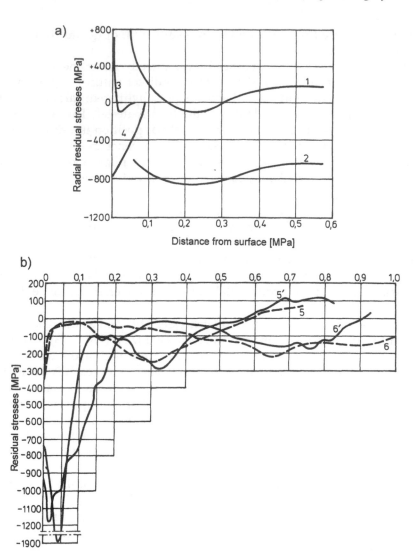

Fig. 5.24 Distribution of residual stresses in a) steels: *1* - 5140 grade, hardened, tempered and lathe machined; *2* - same, hardened, tempered and centrifugally shot-peened; *3* - T1 high speed grade, hardened and ground; *4* - same, hardened and pneumatically shot-peened (From *Tubielewicz, K.* [37]. With permission.); and b) in 20HNM grade, carburized to depth of 0.6 mm (curve *5*) and 1.3 mm (curve *6*), hardened (curve *5'*) and shot-peened (curve *6'*). (From *Nakonieczny, A., and Szyrle, W.* [76]. With permission.)

Residual stresses may be caused by treatment operations or service (e.g. through friction). In the first case, **treatment** or **technological residual stresses** are formed, while in the latter, **service residual stresses**. Service conditions may cause the relaxation and redistribution of technological residual stresses.

In surface shaping treatment processes the following types of technological residual stresses are formed:

– **quenching stresses**, caused by volumetric changes due to predominantly phase transformations but also to heating and cooling,

– **casting stresses**, caused by solidification and cooling,

– **welding stresses**, caused by phase transformations and thermal expansion.

In all superficial layer shaping treatment operations, the character and value of technological residual stresses change during the technological process (see Fig. 5.10) and from process to process [13] in the following manner:

– at first, the superficial layer contains only primary (initial) residual stresses, created during the previous treatment operation (in the steel-making process, forging, casting, cold forming or heat treatment) and being the net result of a superimposition of effects which had occurred prior to the considered operation;

– under the influence of the treatment operation considered, technological residual stresses are created which, when added to initial stresses, become resultant stresses;

– resultant stresses of the considered treatment operation constitute, at the same time, the initial stresses for subsequent treatment operation.

Technological residual stresses do not constitute a value which is constant in time or for any location. Under the influence of external forces occurring during storage or service, technological stresses become service stresses and their value and distribution change, due to processes of relaxation and redistribution (Fig. 5.25).

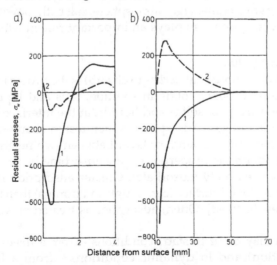

Fig. 5.25 Redistribution and relaxation of residual stresses during service: a) in 1045 steel, induction hardened and subjected to fatigue testing. (From *Janowski. S.* [42]. With permission.); and b) structural steel, subjected to wear testing. (From *Svecev, V.D.* [43]. With permission.); 1 - before test; 2 - after test.

5.7.3.6 Absorption

Absorption (from Latin: *absorptio* - imbibition) is a physico-chemical process of permeation of mass, consisting of the taking up of a constituent, usually a gas mixture called **absorbate**, by a liquid or a solid (called **absorbent**) and uniform dissolution of the former in the entire mass of the latter. This is a **volumetric process**, i.e., the entire volume of the absorbent uniformly takes up the absorbate. The effect of volumetric absorption is often accompanied by diffusion of the absorbate. In a simplified manner, absorption is treated as dissolution in a liquid (for that reason, the amount of equilibrium absorption is described by solubility) or - in a more general way - as the permeation of one phase into another in a diffusion process. Absorption is often accompanied by chemical reactions, e.g., in pack carburizing of steel, carbon from the carburizing powder pack reacts with oxygen contained in pores of the carburizing mixture, forming carbon monoxide CO which breaks down at the steel surface, due to its catalytic action: $2CO \rightarrow CO_2 + C$, giving off atoms of nascent carbon, capable of diffusing into the steel. In gaseous carburizing, some atoms are obtained from the breakdown of hydrocarbons [39]. The effect of absorption is widely used in the chemical and related industries in order to separate a harmful or a valuable component out of a gas mixture or to combine the gas with an absorbent to obtain a compound, an extraction of a substance dissolved in a liquid (e.g., in water) by another liquid which does not mix with the solvent, etc. In surface engineering, absorption of gases by metals and alloys is utilized chiefly in order to saturate the superficial layer by the diffusing element. The course of absorption is, in this case, dependent on the difference of chemical potentials in metals and alloys on the one hand, and the surrounding environment (gas atmosphere, salt bath, powder pack, paste) on the other. Absorption also plays an important role in tribology.

5.7.3.7 Adsorption

Adsorption (from Latin: *ad* - at, *sorbe* - to absorb) is the process of attraction of substances (gases, vapors, solids in solution, ions and liquids) and their collection at the surface of solids and liquids, at the interface between solid and gas or liquid and gas. Adsorption is manifest in changes of concentration of a substance in the boundary layer between two neighboring phases and depends both on the properties of the adsorbing body (adsorbent), as well as the adsorbed body (adsorbate). Greater adsorption is exhibited by bodies with a developed surface (e.g., rough and porous) than by bodies with smooth surfaces [9, 40-43]. Often, adsorption is treated as **surface adsorption**.

Adsorption may occur in static conditions - from a fixed volume phase (static adsorption) and in dynamic conditions - from a flux of gas or solution (dynamic adsorption).

A molecule from the volume phase, e.g., gas, having reached the surface of the solid or liquid adsorbent is maintained there (or adsorbed) by

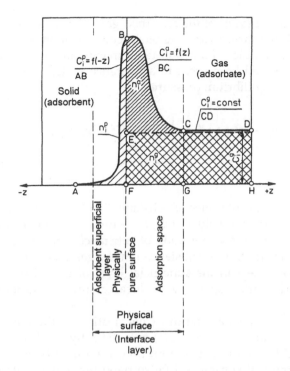

Fig. 5.26 Adsorption at solid/gas; solid line - profile of substance concentration (i) vs. distance from physically pure solid surface; dashed line - profile of substance concentration vs. distance from solid surface in reference system; surface concentration excess n_i is represented by the shaded area. (From *Ościk, J.* [45]. With permission.)

surface forces for a certain time, dependent on the character of the adsorbate and adsorbent, on temperature and pressure, and finally leaves that surface or is desorbed. Commensurate with the saturation of the surface, the rate of adsorption decreases while the rate of desorption increases. When both rates are equal, desorption equilibrium is set.

Molecules of the adsorbate at the surface of the adsorbent form **adsorption layers**.

We distinguish positive adsorption when the concentration of the substance is greater in the superficial layer than in the deeper phase, and negative adsorption when the concentration in the superficial layer is less than in the deeper phase.

In most cases, positive adsorption of gases, vapors and dissolved substances occurs at solid surfaces. The molecules of a very volatile phase (adsorbate) are then subjected to spontaneous densification in the thin layer at the surface of the very condensed phase (adsorbent).

Fig. 5.26 shows the profile of gas concentration at the interface with a solid, vs. distance z from the physically pure surface. The area covered between points BC and E expresses the surface excess (in concentration) of the adsorbed gas substance, relative to the reference concentration of the gas phase.

The surface excess n_i of the adsorbed gas substance i (or volumetric excess), which is the surface (or volumetric) concentration, expresses the excess in the number of moles of that substance in comparison with the number of moles which would be present in a reference system without adsorption, given the same equilibrium pressure

$$n_i = (n_i^a - n_i^g) + n_i^p = \int (C_i^a - C_i^g)dV_1 + \int C_i^p dV_2 \qquad (5.17)$$

adsorption	surface adsorbent
space	layer

where n_i^a - number of moles of substance i in field *FBDH*; n_i^g - number of moles of substance i in field *FEDH*; n_i^p - number of moles of substance in field *ABF*; C_i^a - local concentration of substance i in the adsorption space; C_i^g - local concentration of substance i in the gas phase; C_i^p - local concentration of substance i in the superficial layer of the adsorbent; V_1 - local volume of adsorption space; V_2 - local volume of superficial layer of adsorbent.

Due to the very small depth of permeation of the adsorbate into the adsorbent, the quantity n_i^p (or C_i^p) is sufficiently small to be neglected in expression (5.17). With this assumption, the quantity n_i corresponds to the total amount of substance i (adsorbate) remaining within the field of adsorbent forces.

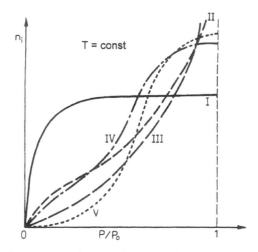

Fig. 5.27 Types of adsorption isotherms of gases and vapors, according to Brunauer; n_i - total amount of adsorbed substance i; p - pressure; p_o - pressure of saturated gas. Type I - typical curve for chemical adsorption, less frequent for physical adsorption; types II to V - various curves for physical adsorption; the most frequent is type II, least frequent - type V.

The amount of a substance adsorbed by the superficial layer depends on its pressure and on temperature. For a gas mixture, the partial pressure of the given substance is taken into consideration. At fixed pressure (p = *const*), the amount of adsorbed substance (gas, vapours) is only a function of temperature and usually decreases with its rise. For constant temperature (T = *const*) the amount of adsorbed substance, expressed by the so-called adsorption isotherms, depends only on pressure and increases with its rise (Fig. 5.27) [45, 46, 48].

Naturally, the amount of adsorbed substance depends on the material of the superficial layer (adsorbent) and the type (structure) of the adsorbate, as well as on conditions of adsorption (p, T), increasing with an increase of the adsorbent surface. The higher the molecular mass of the adsorbate and the higher the condensation temperature, the easier it is adsorbed. Usually, gases and vapors are adsorbed in amounts which grow with the temperature of the boiling point. For example, the volume of ammonia (113.4 cm^2/g) adsorbed by the surface of charcoal at room temperature is close to 40 times greater than that of hydrogen. Adsorbed to an even greater degree than gases are vapours of substances which are in the liquid state at room temperature, e.g., gasoline, ether, alcohol, etc. When the surface is reached by molecules of a substance which is adsorbed stronger than the considered molecules, the adsorption of the latter is reduced.

The process of surface binding of the adsorbate may be divided into three groups, mainly from the point of view of forces acting between the adsorbent and the adsorbate.

1. **Physical adsorption** (also termed: *molecular, surface, specific* or *physisorption*) consists of densification of a substance at the surface of the adsorbent under the influence of intermolecular forces of attraction, so-called Van der Waals forces. The character of these forces is the same as in intermolecular interaction in gases, liquids and in solids. These are forces induced by resonant vibrations of electrons in molecules coming into close proximity (so-called electro-kinetic or dispersion forces) and electrostatic forces associated with the presence of electrical dipoles in molecules of the adsorbate (so-called polar molecules), quadrupoles or, generally, multipoles, caused by a non-uniform distribution of electron density in molecules. In the case of an apolar adsorbent, it is mainly the action of forces of dispersive attraction; in the case of a polar adsorbent, the multipoles of the adsorbate molecules are additionally attracted by an electrostatic field which enhances the adsorption of these molecules. This is especially true if the surface contains ions of the same sign or dipoles of same orientation.

Adsorbed molecules cause a reduction in surface energy, as a result of which a certain amount of energy, called **heat of adsorption**, is exchanged with the environment. It assumes a value of the order of heat of evaporation of the adsorbate and usually is contained within the limits of 40 kJ/mole. Physical adsorption is thus an exothermic process.

Physical adsorption usually occurs instantaneously if not hampered by side effects (e.g., slow diffusion of the adsorbate to the surface or its slow permeation into pores within the adsorbent). It is a dynamic and reversible process which means that molecules of the adsorbate are not permanently connected to the surface of the adsorbent but are in a state of constant exchange with molecules of the gas phase. During adsorption equilibrium, the number of molecules settling down on the surface is equal to the number of molecules passing to the gas phase in the same time. As a result, the number of molecules at the surface remains constant. Adsorbed molecules of the adsorbate maintain their individual characteristics.

The energy of the superficial layer plays a significant role in the phenomenon of adsorption. Good adsorption properties will be featured by an adsorbent with high surface energy (e.g., resulting from an induced state of stress), as well as with a high surface to mass ratio. It is therefore obvious that with a rise of the surface, e.g., due to refinement of molecules forming it, the active surface also rises and so does the intensity of adsorption [9].

The effectiveness of physical adsorption increases with the lowering of temperature approaching the temperature of condensation of the adsorbed gas. On the other hand, a rise of temperature causes a decrease in the intensity of adsorption, unless this temperature rise causes effects of chemical activation and the associated presence of stronger chemical bonds. Further, with a rise of pressure, the amount of adsorbed substance rises out of proportion to the former and the higher the former, the slower the latter (see Fig. 5.27, curves II to V). Starting from a certain boundary pressure, sometimes difficult to determine, further rise in pressure does not affect the amount of the adsorbed substance. The adsorbent appears as if it were saturated (see Fig. 5.27, curve I). This case occurs seldom in physical adsorption but takes place mainly in chemical adsorption. The final mass of the adsorbed adsorbate is less in the case of chemical than in physical adsorption. In all cases (curves I to V) the effect of pressure on the amount of adsorbed substance is particularly big in the zone of low temperatures and pressures.

Physical adsorption - as was noted - is a reversible process and the adsorbate may be removed, e.g., by lowering the pressure. It is reclaimed in a condition that is chemically unchanged. Taking into account the very small value of the activation energy, of the order of 4 kJ/mole, physical adsorption is a process that is very fast even at very low temperatures [39], in particular on smooth surfaces. The thickness of physically adsorbed layers corresponds to several molecule diameters of the adsorbate [40].

2. Condensation adsorption (also called *capillary*) consists of such a high densification of gases and vapors of the adsorbent that after covering the surface with a monomolecular layer they undergo condensation to the liquid state. This type of adsorption takes a somewhat longer time than the physical, it is partially reversible, i.e., the desorption curve differs from that of adsorption (this is the so-called *sorption hysteresis*). It may be treated as a version of physical adsorption. Its course is plotted by isotherms of the type depicted by curves IV and V in Fig. 5.27. Maximum adsorption occurs when pressure p is lower than the pressure of saturated vapor p_o.

3. Chemical adsorption (*chemisorption*) is also often called **activated adsorption** because it calls for a much higher activation energy than physical adsorption and is of the order of 20 to 80 kJ/mole. Forces binding molecules of the adsorbate with surface molecules of the adsorbent are significantly greater but with a shorter range of effectiveness. These are forces of chemical bonds. For that reason the value of heat of chemical adsorption is significantly higher than the heat of physical adsorption. It is of the order of 30 to several hundred kJ/mole, thus of the same order as the heat of chemical reaction. It is usually an irreversible process. Gas, once chemically adsorbed, is very difficult to remove. If it undergoes desorption it usually changes its chemical state. For example, oxygen adsorbed on the surface of charcoal at room temperature is so strongly bound that it is released in the form of carbon dioxide. Chemical adsorption proceeds slowly, especially at low temperatures, and its rate rises with temperature, similarly to the rate of chemical reactions. Kinetics indicates the presence of energy of thermal activation. Chemical adsorption is limited to a monomolecular superficial layer. Additional amounts of gases or vapours may be adsorbed physically in the second and subsequent layers over the monomolecular, chemisorbed first layer. There is no sharp dividing line between physical and chemical adsorption, although extreme case may be unequivocally distinguished. This constitutes proof that usually chemical adsorption is the next phase of physical adsorption which cannot take place in the presence of additional energy, enabling a closer approach of atoms (molecules) of gases and vapours to those of the surface. Thus, considering the phenomenon of adsorption of nitrogen in iron it has been determined that at temperatures up to 200°C nitrogen is adsorbed physically and above 200°C chemically [39].

New bonds created as the result of chemical adsorption at the surface of a metal are always to some degree polarized, due to the difference in electronegativeness between atoms forming them. This causes an insignificant increase or decrease in the concentration of conducting electrons in the metal which may be detected by a measurement of changes in electrical conductivity. Physical adsorption does not bring about such electrical effects [40].

Chemical adsorption, to a degree greater than physical, depends on surface condition, i.e., on its structure and method of preparation. It should be remembered that the entire surface is not homogenous as regards energy. For this reason, the concept of **active centers** has been introduced in which adsorption takes place (see Section 5.7.3.11). The role of active centers - characterized by higher surface energy - is taken by areas with high free energy, particularly all defects of the crystal structure, atoms situated at edges and nodes of crystals. They exhibit highest adsorption energy.

During chemical adsorption, when additional energy appears, enabling an even closer approach of gas atoms to those of the surface, a chemical reaction may take place where a surface atom that joins with an atom (molecule) of the gas is "extracted" from the substrate structure and creates a new chemical compound [45]. In those cases, surface chemical bonds of the adsorbate with the adsorbent are created.

A chemisorbed molecule at the surface may undergo deformation, chemical bonds may be relaxed or even totally severed, with the formation of free atoms and radicals which takes place in the process of gas nitriding in an ammonia atmosphere: $2NH_3 \rightarrow 2N + 3H_2$ [9].

When a molecule of gaseous adsorbate undergoes dissociation into component atoms or radicals which, in turn, undergo adsorption, a process of this type is called **dissociation chemisorption** [46]. Dissociation chemisorption of gases on transition metals is a non-activated process and, consequently, it is determined by thermodynamics and not by its kinetics. There are, however, exceptions, e.g., a small amount of activation energy is necessary in the case of chemisorption of nitrogen on the surface of steel. Transition metals are particularly active in chemisorption [41].

A chemisorbed molecule is more chemically active than the non-adsorbed molecule. For example, the nascent nitrogen released during the dissociation of NH_3, whose lifespan is 1 to 1.5 s, undergoes chemisorption at the metal surface and later diffusion during the nitriding process [44]. The heat of binding of atomic nitrogen is close to twice that of molecular nitrogen. At the surface of tungsten its value is 646.4 kJ/mole.

Chemical adsorption may be treated as a chemical reaction between molecules of the adsorbate with atoms of the superficial layer of the metal [32]. Energy of chemisorption bonding has a value close to that of the energy of chemical bonding in free molecules. For example, the heat of chemisorption of carbon monoxide on the surface of transition metals is 170 to 350 kJ/mole [9].

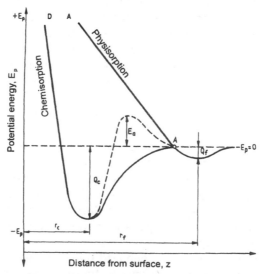

Fig. 5.28 Potential energy vs. distance of adsorbate molecule from metal surface.

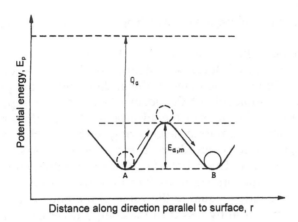

Fig. 5.29 Potential energy curve in plane perpendicular to ideal metal surface; $E_{a,m}$ - activation energy of migration of adsorbed particle from site A to unoccupied adjacent site B; $E_m \ll Q_a$, where Q_a - heat of adsorption.

Chemical adsorption often takes place on the surface of catalysts (see Section 5.7.3.11).

All adsorption phenomena are always accompanied by the release of heat (4 to 800 kJ/mole of adsorbed substance), depending on the type of adsorption.

Fig. 5.28 shows the dependence of potential energy E_p on the distance z between molecules of the substance and the metal surface in cases of physical and chemical adsorption. The zero level ($E_p = 0$) corresponds to the energy of a molecule at infinite distance from the surface ($z = \infty$). The physisorption curve is characterized by a small amount of heat Q_f and big equilibrium distance z_f, while the chemisorption curve is quite the opposite: by a big amount of heat Q_c and a small equilibrium distance z_c ($Q_f < Q_c, z_f > z_c$), in agreement with the close range of chemical forces. The intersection of the chemisorption curve with $E_p = 0$ (point A) indicates that chemisorption is a non-activated process.

The case for which the transition of a molecule to the adsorbed state requires activation energy E_a is shown by the dashed curve [41].

As has already been mentioned, the surface of a solid is non-homogenous in energy, geometry and structure. For theoretical considerations it can be assumed that it constitutes a geometrical plane in the form of a plane regular lattice, filled by metal ions, reflecting the spatial structure of the metal. The potential energy at the metal surface varies approximately like a sine curve. The energy of bonding of an adsorbed molecule attains maximum values at sites corresponding to the minimum of the sine wave. These are the so-called **adsorption sites** [41]. The number of such minima per unit surface may be determined, knowing the geometry of the metal lattice. The height of the potential barrier, limiting the mobility of adsorbed molecules (molecules, atoms and radicals) between adsorption sites, is small in comparison with bonding energy of the molecule at any given site. As the result, molecules are mobile already at temperatures at which adsorption does not yet occur (Fig. 5.29).

The case where the displacing molecules dwell for a long time at adsorption sites is termed **localized adsorption**. Molecules present at adsorption sites vibrate in the plane of the surface and in the plane perpendicular to it. If a molecule of the adsorbate takes over a portion of the thermal energy of vibrations of the adsorbent lattice, sufficient to surpass the potential barrier separating neighboring minima of the oscillating potential energy, migration of an adsorbate molecule from its occupied site to a free site may take place. This process is called **surface diffusion** and consists of activated jumps from one site to another, the activation energy of migration $E_{a,m}$ being approximately equal to the value of the potential barrier. At the same time it is significantly less than the energy necessary for desorption. Surface mobility of molecules constitutes a dominant factor in the fast attainment of adsorption equilibrium on a non-homogenous surface. An increase in surface mobility of adsorbate molecules is favored by heating of the system [9, 11].

Adsorption of a liquid on solid surfaces can be basically reduced to the case of wetting of the surface, leading in some cases to adhesion.

In the case of adsorption from solutions the situation is complicated by the fact that besides molecules of the dissolved substance the solvent is also adsorbed and both types of molecules interact with one another. Since there are more solvent molecules, they attain a numerical dominance at the surface of the adsorbent even when they are adsorbed somewhat slower than molecules of the dissolved substance. This leads to a significant drop in the concentration of the latter in comparison with the amount adsorbed from the gas phase. On account of the lower rate of diffusion in the liquid phase, the rate of adsorption also decreases. The ability to be adsorbed by the dissolved substance depends to a high degree on the same ability of solvent molecules. It can be shown qualitatively that the better the solubility of a given substance in a solvent, which indicates increased forces of interaction, the more difficult it will be for it to be adsorbed at the surface of a solid. Stronger interaction causes molecules to be more strongly displaced from the surface of the adsorbent and pulled into the solution. The amount of adsorbed substance will, therefore, depend on the value of interphase tension, occurring at the interface between solid and liquid. The better the substance is adsorbed, the higher the tension.

The diminishing of surface tension at the interface indicates a rise in the wettability of the given solid by the liquid. If the solvent is water, adsorbents can be, therefore, divided as follows:

1) **hydrophobic** (water-repellent) - poorly wetted by water. At the interface of between the apolar adsorbent with a polar liquid, significant surface tension is formed, and for that reason water wets the given adsorbent poorly. On the other hand, substances dissolved in it will be well adsorbed;

2) **hydrophilic** - well wetted by water. At the interface, very good adsorption of water is observed, while substances dissolved in it are

adsorbed very poorly. In this case we are dealing with the effect of so-called **negative adsorption**, i.e., molecules of the solvent displace other molecules, adsorbed earlier at that surface.

A quantitative measure of the above-described properties is the so-called **surface activity** which characterizes variations of surface tension at interfaces, as influenced by the concentration of the given substance. Substances which lower interphase tension are called **surfactants** or simply **active**, while those which raise it or have no effect are called **surface - inactive** or simply **inactive**. In the case of aqueous solutions, the first of the above groups includes organic acids, alcohols, aldehydes, ketones and generally compounds with long carbon chains. Examples of the second group include predominantly electrolytes, sugars, proteins, glycerine and urea [8].

Thus, adsorption from solutions to the boundary interphase surface affects surfactants. Inactive substances "flee" from that interface into the solution in a process of negative adsorption. Surfactants have found application - taking their properties into account - in highly refined lubricants, in the flotation process, in chromatography and in electroplating [8, 58].

Adsorption at the surface of crystalline substances usually lowers their strength properties.

Adsorption is utilized in processes of machining, by the introduction of adsorbable components to lubricants and coolants; in service, by using lubricants containing adsorbable components for machine lubrication in order to prevent dry friction [45, 46]; in thermo-chemical processes and in plating metal substrates.

The phenomenon of adsorption is also utilized for enhancing the formation of vacuum.

5.7.3.8 Solubility

Solubility is the ability of a substance in the solid, liquid or gaseous state to form, together with other substances, mixtures which are homogenous from a physical and chemical point of view. A measure of solubility is the amount of a substance being dissolved in a given amount of substance of the solvent at a given temperature and under a given pressure. Besides these extraneous conditions (temperature and pressure), solubility depends on the state of aggregation of the dissolved substance and of the solvent.

Solubility of gases in liquids rises proportionally to a rise in pressure and drops with a rise of temperature; usually it is higher when the dissolved substance reacts chemically with the solvent. The solubility of liquids in liquids may occur in any proportions, in limited proportions, or may not occur at all and it may both increase or decrease with temperature. Solubility of solids in liquids usually rises with temperature and is pressure dependent only to a minor degree.

The solubility of gases in metals is the ability to form liquid or solid solutions of the gas in a metal, in accordance with equilibrium conditions.

Physical solubility consists of the formation of interstitial solutions, while chemical solubility means the formation of a special type of chemical compounds. Physical solubility rises with temperature and pressure, according to the following expression [42]:

$$s = K_1 \sqrt{p} e^{-\frac{b}{T}} \tag{5.18}$$

where: s - solubility of gas in the metal; K_1 - constant, dependent on pressure p and temperature T; b - constant, characteristic of given metal.

Solubility of gas in metal can also be expressed as

$$\frac{dC}{dt} = \alpha (C_S - C) \tag{5.19}$$

where: C - concentration of gas in metal after time t; C_S- concentration C in condition of saturation; α - coefficient, dependent on physical conditions and on the surface-to-volume ratio of the liquid metal.

If p = const.

$$C = C_S (1 - e^{-\alpha t}) \tag{5.20}$$

The solubility in a metal of monogases, e.g. nitrogen, changes by a leap after exceeding temperatures of allotropic transformations and melting point. Crystallization of a metal with normally applied cooling rates renders impossible the diffusion of the gas out of the metal, and its excess, due to the drop in solubility with falling temperature, may cause the formation of blisters. Gases dissolved in the metal have a strong, not always benign effect on metal alloy properties (e.g., hydrogen in steel = hydrogen embrittlement) [42].

Mutual solubility of metals in the liquid state is due to slow solidification of liquid solutions of these metals or to appropriate heat treatment of the alloys and is aided by similarity of size and shape of component particles. A significant effect on solubility is exhibited by impurities (dopants). With total solid solubility of metals forming a given system, continuous solid solutions are obtained. These are of big technical significance. As a rule, however, mutual solubility of metals in the solid state is limited and there are cases where it does not occur at all. Systems with saturated solutions (so-called boundary solutions) with variable solubility, usually decreasing with a drop in temperature, are heat treatable (e.g., by solution annealing) [26].

5.7.3.9 Diffusion

Types and mechanisms of diffusion. Diffusion (from Latin: *diffundere* - to pour out, propagate) in the most general case consists of relative changes in the locations of atoms or particles in a stationary system, driven by thermal excitation [49]. When two bodies in any state come in contact, atoms of one of these bodies permeate into the other one, due to random thermally driven movements. In a stricter sense, diffusion is the transportation of particles of one substance relative to particles of another substance within the same phase (gaseous, liquid or solid), driven by concentration gradient, chemical potential gradient, temperature gradient (thermodiffusion) and electrochemical gradient (electrodiffusion) [49-57].

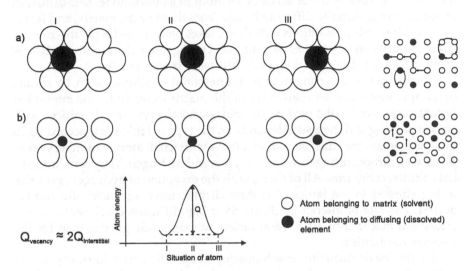

Fig. 5.30 Diagrams showing mechanisms of diffusion: a) vacancy; b) interstitial; *I* - atom in initial position; *II* - atom in activated position; *III* - atom in final position.

We distinguish diffusion in solids, liquids and gases. Diffusion in solids may be subdivided as follows [47-50]:
  – lattice type: occurring in crystals containing vacancies and dislocations,
  – dislocation type: pipe diffusion,
  – surface type: occurring across a free surface of a crystal.
  Diffusion accompanies almost all microstructural phenomena occurring in metals and alloys during their heating, soaking, during tranformations in the solid state (i.e., during heat and thermo-chemical treatment), during solidification and cooling. It may be recognized as a thermally activated movement of atoms (of the same type as the host metal or other) in the spatial lattice of the metal or alloy, and generally oriented in the direction of concentration equalization.
  Depending on the kind of diffusing atoms, the following types of diffusion are distinguished:
  – **self-diffusion**, when mutual intermixing of atoms of the same kind takes place,
  – **chemical diffusion** (heterodiffusion), when displacement of atoms of different kinds takes place. Such atoms form interstitial solutions (interstitial

mechanism) or intermetallic compounds (reactive diffusion).A condition for chemical diffusion to take place in a solid is solid solubility of the saturating element in the material of the matrix.

In a solid crystalline material there occur two basic diffusion mechanisms (Fig. 5.30) [26, 27, 49, 51–53, 55–57]:

– **vacancy** - occurring mainly in substitution-type solutions, when the displacement of atoms takes place by way of vacancies, i.e. point defects of the lattice, created by the absence of an atom in a lattice node. Self-diffusion of atoms, insignificantly differing in size from those of the matrix, and forming substitution-type solutions with them, takes place according to this mechanism. In diffusion in iron these are atoms of manganese, chromium, molybdenum and nickel. The rate of vacancy diffusion is very small;

– **interstitial** - occurring mainly in interstitial solutions, when, by means of jumps, atoms smaller than those of the matrix move from one interstitial (interatomic) void to the next. Such voids occur always, even in lattices with closest packing and their size depends on the type of lattice. It is according to this mechanism that diffusion of elements with small atomic numbers takes place, e.g., carbon, nitrogen, boron, oxygen and hydrogen. They form interstitial solutions with iron. All of them, with the exception of hydrogen, give rise to big stresses in the lattice. For their displacement vacancies are not required. The rate of interstitial diffusion is big: diffusion coefficients for the interstitial mechanism are several orders of magnitude greater than for the vacancy mechanism.

Besides basic diffusion mechanisms, in specific conditions, especially in the case where a tendency exists to form intermetallic compounds or specific lattice defects (dislocations), special mechanisms may occur [49, 52, 53]:

– **reactive diffusion** - consisting of the formation, at phase boundaries in the matrix, as the result of reaction between the guest element and the matrix or precipitations and the formation of new phases with different lattice structures. These new phases are intermetallic compounds, with a thickness of $g = P\tau^{0.5}$, where P is a constant, exponentially dependent on temperature T. Reactive diffusion plays a special role in thermo-chemical treatments like nitriding, boriding, chromizing, etc. Atoms adsorbed at the surface of the metal or alloy by surface defects penetrate into the lattice and, in appropriate conditions, may be displaced across distances of many grain diameters (up to several mm), forming compounds with the host metal. These compounds may be nitrides, carbides, borides, silicides, etc. For example, nitriding brings about the formation of Fe_2N and Fe_4N nitrides, while boriding - of FeB and Fe_2B borides. Chromizing causes the formation of chromium carbides. These compounds usually have a good effect on mechanical (tribological and fatigue) properties of the superficial layer. It is usually endeavored to form rather monophase diffusion layers, on account of the residual stresses present in them;

– **dislocation diffusion** (diffusion through dislocations) - occurring in the case of linear defects in crystals (present even in the annealed state in

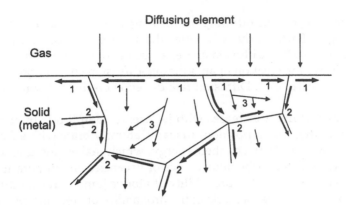

Fig. 5.31 Diagram showing diffusion paths: *1* - easiest diffusion - along surface; *2* - more difficult diffusion - along grain boundaries; *3* - most difficult diffusion - across grains (inside grains).

an amount of 10^{16} in 1 cm^3), which constitute passages of easy diffusion at lower temperatures. Diffusion is facilitated more by edge-type than by screw-type dislocations. Dislocation diffusion plays a significant role in thermo-mechanical burnishing where plastic deformations raise the density of dislocations. Heating causes them to atrophy but aids diffusion (recrystallization, homogenization and recovery occur);

– **grain boundary diffusion** - occurring in polycrystalline materials, along surface defects such as grain boundaries (Fig. 5.31). A grain boundary is a flat channel of width equaling approximately 2 atom diameters along which diffusion proceeds up to 10^6 times faster than across the crystal. Acting in a manner similar to that of the grain boundary is the dislocation line. Both grain boundaries and dislocations are passages of easy diffusion at lower temperatures. The more grain boundaries, the easier the diffusion process. Fine-grained polycrystalline materials are, therefore, more susceptible to diffusion than coarse-grained. Besides, not all types of grain boundaries exhibit same action. The most effective are wide angle boundaries with random orientation, less effective are small angle boundaries and least effective are twin or special boundaries;

– **diffusion along interfaces (surface diffusion)** - occurring with high intensity both across interfaces between the solid phase and gas or liquid, as well as inside multi-phase bodies. Boundaries between phases are structurally defective areas, thus they constitute paths of easy diffusion. Effects of diffusion along interphase boundaries depend on surface tension between the particular phases: when tension values are close there is a tendency to form globular phase particles; when a new phase with a lower surface tension than that of already existent phases is created, due to diffusion, there is a tendency to penetrate the new phase along interphase boundaries of the primary phases [49].

As far as surface treatment is concerned, dislocations and grain boundaries exert a benign effect on the diffusion of components in low-temperature technologies, like nitriding, sulfurizing, as well as aided CVD and PVD technologies. Particularly in the latter, the occurrence of dislocation diffusion and grain boundary diffusion causes a rise of adhesion of the coating to the substrate and the creation of an adhesive-diffusion connection.

Laws of diffusion. Diffusion rate depends on the following factors [26, 49–57]:

– **Temperature**. It increases with its rise: the amplitude of atomic vibrations about their mean positions (usually, atoms vibrate with a frequency of approximately 10^{13} s^{-1}), i.e., their energy, which makes possible the generation of lattice defects (the dependence of vacancies on temperature is exponential) and increases the probability of atomic jumps causing diffusion;

– **Time**. It increases with time; the probability of atom jumps rises;

– **Type of bodies** participating in diffusion and conditions which enhance or impede diffusion: concentration, pressure, stresses, atom size, valence, type of lattice and its defects, etc.

Quantitatively, the rate of diffusion is determined by laws formulated in 1858 by A. Fick. Their form is analogous to that of J.B.J. Fourier's equations, describing conduction flow of heat.

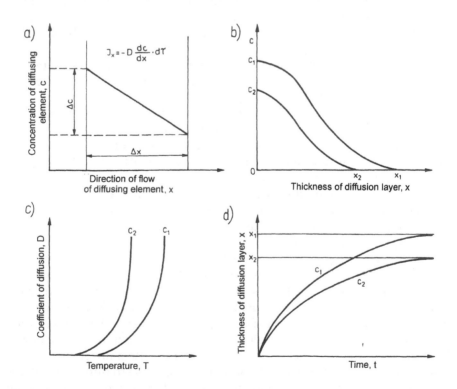

Fig. 5.32 Four diffusion-related correlations: a) diffusion process; b) thickness of diffusion layer vs. concentration of diffusing element; c) diffusion coefficient vs. temperature for different concentrations of diffusing element; d) thickness of diffusion layer vs. time for different concentrations of diffusing element.

Fick's first law describes the dependence of a flux of diffusion on the concentration gradient of the diffusing component $\partial c/\partial x$ (Fig. 5.32) [49–57].

$$J = -D\frac{\partial c}{\partial x} \qquad (5.21)$$

where: J - diffusion flux, i.e. amount of matter flowing during unit time in the direction x through a unit surface, normal to the flux; c - concentration of the diffusing component; D - coefficient of diffusion [m^2/s].

The value of D is determined by the equation developed by S.A. Arrhenius:

$$D = D_0 e^{-\frac{Q}{RT}} \qquad (5.22)$$

where: D_0 - constant, dependent on crystalline structure, referred to as frequency factor or the Arrhenius constant; Q - diffusion activation energy, independent of temperature (see Fig. 5.30), reduced to 1 mole and equal to minimum energy needed to bring an atom out from its nodal location; for the process of self-diffusion $Q \approx 150 T_m$ [J], T_m being melting point temperature; in chemical diffusion, the activation energy decreases with a rise of valence of diffusing atoms and increases with a rise of lattice density and size of diffusing atoms; R - universal gas constant (8.345 J/mole); T - absolute temperature.

The sign "–" in equation (5.21) means that the flux of diffusion is directed opposite to the concentration gradient.

The factor $e^{-Q/RT}$ in equation (5.22) represents the probability that in the vacancy mechanism of diffusion, at a given temperature, a vacancy will occur beside a diffusing atom and that the atom will attain the energy necessary to jump over. In the interstitial mechanism this factor expresses only the probability of attaining the energy sufficient for the atom to jump from one interstitial vacancy to the next neighboring one, without the necessity of the existence of vacancies. Since the value of Q is, in this case, approximately twice less, the result is faster diffusion of interstitial atoms [49].

A measure of diffusion rate is given by the coefficient D which indicates, for a unit concentration gradient, the rate of equalization of the concentration gradient. For solids in the temperature range of 20 to 1500°C the value of D varies from 10^{-20} to 10^{-4} cm^2/s [49].

Fick's second law describes the correlation between the concentration gradient and the rate of its change at a given point, i.e. the dependence of concentration distribution of the diffusing element on time [49-57], and is expressed by the equation:

$$\frac{\partial c}{\partial \tau} = \frac{\partial}{\partial x}(D\frac{\partial c}{\partial x}) \qquad (5.23)$$

where τ - time.

From expression (5.22) it follows that the path of diffusion $a = 2(\tau D)^{1/2}$ [40].

Phenomenological diffusion equations. Complex phenomena of diffusion may be considered with the help of thermodynamic descriptions, derived from the action of forces connected with heterogeneity of concentration of system components. The most general form of the thermodynamic equation for diffusion of the first component is the following [49]:

$$J_1 = -M_{11}\frac{d\mu_1}{dx} - M_{12}\frac{d\mu_2}{dx} - \ldots - M_{1n}\frac{d\mu_n}{dx} - M_{1T}\frac{dT}{dx} - M_{1p}\frac{dp}{dx} - M_{1\Phi}\frac{d\Phi}{dx} \quad (5.24)$$

where: J_1 - diffusion flux of first component; m_1, m_2, m_n - chemical potential of first, second, n-th component; M_{11}, M_{12}, M_{1n} - kinetic constants describing, in sequence, the diffusion of first, second and n-th component, M_{1T}, M_{1p}, $M_{1\Phi}$ - kinetic constants, describing, in sequence, the diffusion of the first component in the presence of a gradient of temperature T, pressure p or other potential Φ (e.g., concentration).

Phenomenological equation (5.24) may be written in an analogous manner for each component of the thermodynamic system.

If diffusion occurs in isothermal ($T = $ const), isobaric ($p = $ const) or isopotential ($\Phi = $ const) conditions, the corresponding components of the phenomenological equations (5.24) will equal 0 and the flux of diffusion will depend only on the gradients of chemical potentials [49].

It follows from these equations that for multi-component systems, the driving force of diffusion is not concentration gradients but chemical potential gradients. In multi-component systems, the components mutually interact to affect chemical potentials. It may, therefore, happen that the concentration gradient $dc/dx = 0$ but the gradient of chemical potential $dm/dx \neq 0$. The phenomenon of **uphill diffusion** will then take place, i.e., from places of lower to places of higher concentration. This occurs, e.g., during diffusion chromizing of alloyed tool steels, containing tungsten and vanadium, resulting in a higher concentration of these elements in the superficial layer than in the core. This, in turn, has a benign effect on the properties of the core and of the diffusion layer [47].

A second conclusion from the phenomenological equation is the possibility of diffusion, driven only by a temperature gradient, i.e., **thermodiffusion** (also known under the name of *Soret's effect*). Besides, what has been experimentally proven, atoms may diffuse with the temperature gradient (e.g., carbon, nitrogen or zinc in iron) or against it (e.g., hydrogen in iron). The result of thermodiffusion is self-diffusion of vacancies, with the temperature gradient (in platinum) or against it (in gold). The phenomenon of thermodiffusion may be of major significance in cases of these surface treatments where big temperature gradients occur, with the time of their duration defined, as in e.g., thermal spraying or pad welding. When the time of existence of a temperature gradient is very short, thermodiffusion is negligible (e.g., in pulse laser or electron beam heating) [26, 49, 52, 55, 57].

A third conclusion from the phenomenological equation (5.24) is the possibility of diffusion, driven by a pressure gradient or - to put in other words - under the influence of a stress gradient. The direction in which atoms diffuse will cause a lessening of residual stresses: interstitial atoms will diffuse from compressed zones to those subjected to tensile stresses, while vacancies will be displaced in the opposite direction. This effect is the cause of internal friction. In the case of a solution composed of atoms of different sizes, the bigger atoms will migrate to the zone under tensile stresses, while the smaller atoms to the zone with compression. These effects may occur during burnishing, thermal (and explosive) spray, ion implantation and induction hardening [49, 52].

Finally, the phenomenological equation gives rise to the conclusion that there is a possibility of diffusion, forced by a variable magnetic field, in other words, by a gradient of the magnetic potential, or by the flow of a strong electrical current (electrical potential gradient). In the first case, the magnetostrictive effect causes volumetric changes of the material and these, in turn, force the movement of interstitial atoms. In the second case, during self-diffusion, atoms have a tendency to migrate in the direction of the anode, while vacancies migrate toward the cathode. Interstitial atoms in the majority of metals migrate in the direction of the cathode. These processes have been observed in induction hardening, without, however, any significant effect on hardening results [49, 52].

Diffusion plays the following roles:

– primary, in the case of medium and high temperature treatments of long duration (temperatures above 500°C; times, several hours and longer). Examples: thermo-chemical treatments, CVD, dip metallizing.

– secondary, in the case of short processes at low temperatures (e.g., PVD) or at high temperatures (e.g., thermal spraying, pad welding);

– ternary, or almost none, in the case of treatments carried out at ambient temperature and slightly above (e.g., burnishing, electroplating).

In all cases, a rise of temperature significantly intensifies diffusion processes and extension of time causes an increase in the amount of the diffused element. Intensification of diffusion is also aided by the existence of residual stresses and, to a lesser extent, by electric and magnetic fields. In all cases, of capital importance are element concentration and chemical potential.

In surface engineering, the most important role is that played by the diffusion of particles of gases and metals or non-metals into metal alloys, resulting in the formation of chromized, borided, silicized, sulfurized and other layers (including combinations).

5.7.3.10 Adhesion

The concept of adhesion. Adhesion (from Latin: *adhesio* - cling together) is a phenomenon of permanent and strong joining of superficial layers of two different (solid or liquid) bodies (phases) brought into mutual contact. A specific case of adhesion is **cohesion** which occurs when the bodies in contact are of the same material. Adhesion may be caused by adsorption.

Adhesion is caused by the presence of attraction forces (e.g. Van der Waals, ion and metallic bonds) between particles of touching bodies. A boundary case of adhesion is the occurrence of chemisorption bonding at the interface with the formation of a superficial layer, constituting a chemical compound [58].

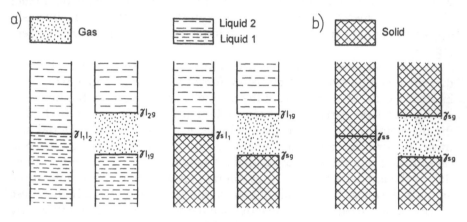

Fig. 5.33 Diagrams showing: a) adhesion of two different bodies; b) cohesion of two identical bodies; γ - surface stresses at interfaces.

The strength of adhesion is described by the value of force (or amount of work) necessary to separate adhering bodies, applied to a unit contact surface [8]. The work of adhesion (adhesive separation) W_a for a reversible and isothermal process is characterized by a decrement of free energy f_a per unit surface of adhesive contact, equal to the difference in surface tensions of the two surfaces (Fig. 5.33). In the case of adhesion of liquid (1) to liquid (2) we have

$$W_a = -f_a = \gamma_{l1l2} - (\gamma_{l1g} + \gamma_{l2g}) < 0, \tag{5.25}$$

where: γ_{l1l2} - surface tension between liquids (1) and (2); γ_{l1g} - surface tension between liquid (1) and gas; γ_{l2g} - surface tension between liquid (2) and gas.

Wettability. In the case of adhesion of liquid (1) to a solid, equation (5.25) cannot be used to calculate the work of adhesion W_a because surface energies between solid and liquid γ_{sl} and between solid and gas γ_{sg} cannot be measured directly. The difference in surface energies of an unwetted and a wetted surface, i.e., the so-called wetting tension $\beta = \gamma_{sg} - \gamma_{sl}$ is expressed by the cosine of the **boundary angle** Θ (also known as the **wetting angle**) corresponding to the state of equilibrium:

$$\beta = \gamma_{sg} - \gamma_{sll} = \gamma_{llg} \cos\Theta \qquad (5.26)$$

The surface tensions γ_{sg} and γ_{sl} are not, as a rule, equal. For that reason, when a liquid is in a vessel, the horizontal level of the liquid near the walls is curved, rising upward or dropping downward along the vessel wall, depending on which of the surface tensions is greater.

If $\gamma_{sg} > \gamma_{sl}$ the interface forms a concave meniscus (Fig. 5.34a) and the surface tension force g_{sl} is tangent to the curved surface of the liquid. The

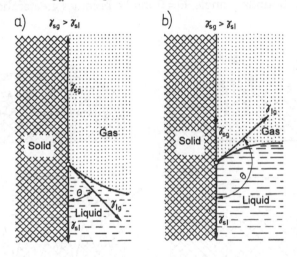

Fig. 5.34 Schematic representation of interfaces forming a concave or convex meniscus: a) wetting liquid (concave meniscus); b) non-wetting liquid (convex meniscus).

vertical component of that force is equal to $\gamma_{l1g} \cos\Theta$ and the state of equilibrium is reached when the following condition is met:

$$\gamma_{sg} = \gamma_{sl} + \gamma_{lg} \cos\Theta \qquad (5.27)$$

The above equation is known as the **wetting equation**.

If $\gamma_{sg} < \gamma_{sl}$ the interface will have a convex meniscus because the liquid will drop at the side wall (Fig. 5.34b). The condition of equilibrium remains the same since in both cases the boundary (wetting) angle is [9, 59]

$$\cos\Theta = \frac{\gamma_{sg} - \gamma_{sl}}{\gamma_{lg}} \qquad (5.27a)$$

If the wetting angle Θ is equal to zero, full wetting occurs. If $\Theta = 180°$, the case of absolute non-wettability occurs [9, 58]. In accordance with the correlation expressed by (5.27a), full wettability of a solid by a liquid takes place when $\gamma_{sg} - \gamma_{sl} \geq g_{lg}$ (for an angle equal to zero).

The wetting angle is an angle formed between the wetted surface of the solid and the tangent to the curvature of the meniscus of the wetting liquid, at the point of contact between the liquid and the solid (Fig. 5.35). A knowledge of the value of the wetting angle is of significant importance in flotation processes, lubrication and in the production of laundry agents [58].

From an equation equivalent to (5.25) it follows that $W_a = \gamma_{sg} (1 + \cos\Theta)$ for the full range of values: $0 \le \Theta \le 180§$. When total wettability occurs ($\Theta = 0$), $W_a = 2\gamma_{sg} = W_p$ which means that the work of adhesion W_a equals the work of decohesion W_d. When the liquid totally wets the surface of the solid, the boundary angle $\Theta = 0$ and W_a becomes greater than $2\gamma_{sg}$ (when $\gamma_{sg} - \gamma_{sl} > \gamma_{lg}$).

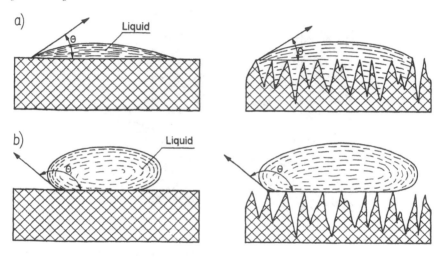

Fig. 5.35 Wetting angle of a solid: a) for a wetting fluid; b) for a non-wetting fluid.

Physical description of adhesion. A case that is important in practical application, especially in tribology, is that of adhesion occurring between two solids. The equation (5.25) no longer holds true since adhesive separation is accompanied by some irreversible processes. Examples of such are: dispersion of energy, due to non-elasticity of permanent deformation (flow) of material prior to the separation and the transformation of energy of plastic deformation into heat with sudden relaxation at the moment of detachment. The work of adhesion also depends on the rate of separation and usually increases rapidly with its growth which is not addressed in equation (5.25). This phenomenon is explained by the already mentioned irreversibility of mechanical processes (dispersion of deformation energy before the detachment) and the occurrence of electrical effects, due to the formation of a double electrical layer at the contact between two different bodies, causing attraction of opposite-charged surfaces, aiding the action of interparticle forces. The superposition of these effects causes an ambiguity in the theoretical determination of the value of adhesion between solids and, hence, the necessity of reverting to experimental methods [58].

When adhesive joining of solids is effected it is justifiable to aim for high values of surface energy. However, high energy surfaces easily absorb vapors, gases and contaminations, conducive to a drop in free energy. In order to obtain a good adhesive connection it is, therefore, necessary to remove the adsorbed layers by e.g., creating a vacuum or elevating the temperature. The process of adhesion of solids is also intensified by other forms of surface activation, e.g., by the introduction of energy in the form of ultrasonic vibrations, radiation (microwave and corpuscular), by defecting the superficial layer (e.g., deformation, treatment at cyclically varied temperatures, oxidation and immediate reduction, pulsed pressure) [59].

When two different bodies adhere, the value of adhesion force is particularly big in the case of full contact across the entire surface of the two bodies:

– at the interface of two liquid phases (e.g., water - mercury);

– upon significant plastic deformation of contacting bodies (e.g., cold-welding of metals), conducive to total contact and strengthening of the structure of the adhesive connection (seam);

– upon the introduction of a liquid on to the surface of a solid in conditions of total wettability (e.g., glueing, welding, painting), conducive to (after solidification) obtaining of exceptionally durable adhesive connection (seam);

– upon the formation on a solid surface of a second solid as a new phase, due to the creation and growth of two-dimensional crystallization nuclei(e.g., electroplating of metals, vacuum deposition of solid particles on metal surfaces by PVD methods);

– with dry friction, occurring particularly intensively in the case of metals with same or similar chemical composition, conducive to the creation of adhesive spot welding and causing adhesive wear [16, 72].

In the case of strong adhesion of solids, with time and due to diffusion, there comes about a progressively stronger bond. This so-called "intergrowth" or "in-growth" may lead to an atrophy of the interface as the result of an unlimited solubility in the solid state, i.e., the transformation of two joined phases into a single phase. This process occurs mainly in the adhesion of same materials.

Adhesion occurs often in everyday life. Dust particles are attached to walls by adhesion, chalk adheres to the classroom board, glue joins the glued material, etc.

5.7.3.11 Catalysis

Concept and types of catalysis. Catalysis (from Greek: *katalisis* - decomposition) is a term introduced in 1836 by J.J. Berzelius and used to describe a phenomenon consisting of acceleration and deceleration of certain chemical reactions by substances called **catalysts**. (In stricter terms, the effect consists of variations in the rate at which a chemical reaction achieves the state of equilibrium.) Catalysts participate in the chemical reaction (but do not participate in the stoichiometric equation), themselves neither being used up nor appearing among the reaction products. Their amount and chemical composition do not undergo any changes during the reaction). Usually, to effect a significant change in the rate of a reaction, only a small amount of catalyst is sufficient, relative to the amount of reacting substances. The following types of catalyses are distinguished:

– **positive catalysis** - when the (positive) catalyst accelerates the rate of a reaction; this is the most frequently used type of catalysis;

– **negative catalysis** - when the (negative) catalyst, in this case called **inhibitor**, decelerates the rate of reaction (as well as the stability and selectivity of the catalyst); this is the type of catalysis used less often, mainly to slow down corrosion processes by the application of various corrosion inhibitors;

– **autocatalysis** - when the product of the reaction (or one of the intermediate products) exerts a catalytic effect, which is usually positive. In such a case, the reaction rate rises with the accumulation of that product.

All catalytic reactions are, from the point of view of thermodynamics, spontaneous, i.e., they are accompanied by a drop in free energy. A catalyst does not change the state of chemical equilibrium but only the time of achieving that state. The same catalyst usually changes the rate of a reaction both from left to right, as well as from right to left. Catalysts act selectively, changing the rate not of every reaction but only of one of those thermodynamically possible within a given system.

Catalysts may have the form of solids (such catalysts are called contacts), liquids and gases. Examples of good solid catalysts are platinum, palladium and oxides of certain metals. Catalysts of numerous biochemical processes (digestion, oxidation of sugars in the bloodstream, fermentation) are called enzymes.

To date, there is no satisfactory theory which would explain the action of catalysts. The mechanism of catalysis may be interpreted as the formation by the catalyst with one or more substrates (in this case - initial substances) of a non-stable intermediate bond. This bond suffers immediate dissolution which indirectly or directly leads to the formation of the final products of the reaction and to a regeneration of the catalyst (its return to the initial form). A reaction with the formation of the intermediate bond is faster than without it, i.e. without the catalyst. From the point of view of kinetics, the catalytic reaction is one with a lower activation energy.

Depending on the physical state of the catalyst, we distinguish the following types of catalyses [60, 62]:

– **homogenous** - in which the catalyst occurs in the same phase (solid, liquid or gaseous) as the reacting substances;

– **heterogeneous** - in which the catalyst occurs in a different state of aggregation than reacting substances; most often, the catalyst is a solid and comes in contact with reacting substances only through its surface.

Heterogeneous catalysis. In the case of the system occurring most frequently in surface engineering: i.e. alloy - gas, in which the catalyst is separated from the substrates by an interface, heterogeneous catalysis takes place [62]. It is connected by an unbreakable bond with the formation of an adsorption layer, its structure and with the character of interaction between surface and metal. In the system: solid - gas, the chemical reaction occurs at the interface. The phase containing the substrates, i.e., the gas phase, is simply a "reservoir" of particles which are subject to transformation, as well as those created by the reaction. Classical heterogeneous catalysis is based on reactions caused by the action of the solid's field of forces on substrate particles. The range of action of the forces is limited to distances comparable to an atomic or particle diameter, thus, of the order of tenths of a nanometer [8].

The mechanism of heterogeneous catalysis is complex. It is, however, an indisputable fact that in this case, a significant role is played by the adsorption of substrate particles at the surface of the catalyst, by chemisorption, to use a stricter term.

The reaction of heterogeneous catalysis comprises five stages: diffusion of substrates to the catalyst, adsorption, chemical transformations at the surface, desorption, and diffusion of reaction products from the catalyst surface [60, 62].

Due to diffusion, substrate particles approach the catalyst surface and become adsorbed by it. However, not every process of adsorption is conducive to catalysis but only such which is accompanied by the creation of a chemical bond between the substrate and the surface, in other words, by chemisorption. This process is accompanied by the coming close of particles of reacting substances and the simultaneous rise of their chemical activity, under the influence of forces exerted by surface atoms of the catalyst. In the next stage, the newly created products break away from the catalyst and finally, by diffusion, permeate into the core of the other phase. Thus, compounds created at the catalyst surface in the case of heterophase catalytic reactions are intermediate compounds [9].

In order for catalysis to occur, a condition must be met. This condition is that the binding energy of the adsorption compound be contained within certain limits, i.e., that it be neither too small nor too great, because the formation of an excessively stable bond between substrate or product with the surface renders further reaction difficult. The rate of completion of a catalytic process depends on its conditions and is determined by the rate of the slowest of the above-mentioned stages of the process.

The change of energy in the second, third and fourth stage is illustrated by Fig. 5.36. Adsorption of substrates is connected to activation energy, corresponding to an increment in enthalpy along the length 1 - a'. The stage of adsorption ends at the point marked 3. Next, an active complex a is formed which is adsorbed at the surface of the catalyst (stage: 3 - a"). In the next phase (a" - 4) the products of reaction are adsorbed and these finally break away from the catalyst surface. The interval 4 - a'''

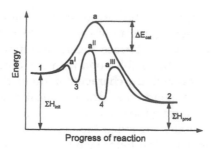

Fig. 5.36 Schematic representation of a process with heterogeneous catalysis (Curve *1 - a' - 3 - a" - 4 - a''' - 2*) and of a process without a catalyst (curve *1 - a - 2*); ΔE_{cat} - decrease of activation energy due to the action of the catalyst; *1* - energy of substrates; *2* - energy of end products; *3* - chemically adsorbed substrates; *a* - adsorbed complex; *4* - adsorbed products.

corresponds to desorption. In heterophase reactions the accelerating action of the catalyst is, therefore, also connected with the formation of intermediate compounds which causes a drop in activation energy [61]. The process of formation of intermediate compounds (catalysis) at the catalyst surface occurs at **active centers** of that surface, i.e. sites characterized by highest adsorption energy [58, 61].

In 1925, H.S. Taylor demonstrated that only a very small portion of the catalyst surface is catalytically active. He named this portion "active centers", thinking that their role is played by atoms located at geometrically prominent sites of the surface, e.g., at corners or edges of crystals. Presently, active centers are associated rather with the presence, at the surface of a solid, of various disturbances in its lattice, e.g., dislocations, foreign atoms, atoms with incorrect electrical charges or atoms incorrectly situated [60, 62].

The field of forces of active centers weakens bonds in particles of adsorbed substances which, in turn, raises substrate activity. In order to increase the number of active centers it is necessary to increase the catalyst surface. Of significance is not only the number of active centers but also the way in which they are distributed on that surface. The more the distribution of atoms in adsorbed substrate particles approximates the distribution of active centers, the more active the catalyst [43].

Heterogeneous catalysis always occurs at the surface of a metal. For that reason, catalytic activity is influenced by the size and character of the surface, chemical composition of superficial layer, its structure, inclusions of foreign substances, in short, all those factors which affect the energy of the superficial layer. It should also be stressed that transition metals are particularly active both in catalysis and in chemisorption. This activity is associated with the development of the *d* electron subshell and with the participation of unpaired electrons of the same subshell in the formation of strong chemisorptive bonds. A hypothesis has been put forward that the unpaired electrons *d* are necessary for the loose attachment of the adsorbed particle to the surface in the initial phase, after which the which the particle later goes into a state of permanent bonding to the surface. The intermediate phase enables the drop of activation energy for adsorption to a comfortably low value. This value, however, remains high in the case of metals not containing unpaired electrons d. For that reason, metals which have valence electrons only on the s or p shell belong to those, characterized by weak chemisorption. Among such metals, also, are alloy components of steel, such as Fe, Cr, Mo, Ni, Ti, Co. In this way, the steel surface affects heterogeneous reactions occurring during thermo-chemical treatment [9].

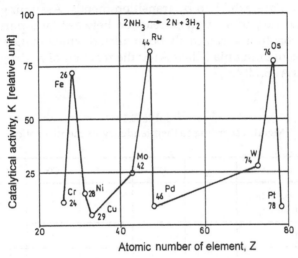

Fig. 5.37 Dependence of catalytic activation k of some elements on the atomic number Z of the element, for the reaction of ammonia dissociation at 800°C, under a pressure of 0.1 MPa. (From *Karapetjanc, M.Ch.* [61]. With permission.)

Fig. 5.37 shows examples of the correlation between the atomic number Z of an element and the catalytic activity of certain metals for the reaction of ammonia decomposition at a temperature of 800°C and pressure of 0.1 MPa. Catalytic activity was determined by investigating the rate of decomposition occurring during the contact of substrates with a known mass of catalyst in given conditions of pressure and temperature, in other words, as a certain empirical measure, enabling a comparison of catalysts. As can be deduced from Fig. 5.37, the highest catalytic activity was exhibited by Fe, Ru and Os. A practical conclusion follows that iron contained in steel catalyzes the decomposition of ammonia during the process of gas nitriding in an atmosphere of NH_3; on the other hand, it is difficult to nitride e.g., nickel and its alloys in this way [9].

Catalytic interaction of alloying elements of the steel matrix also takes place in the process of formation of titanium carbide layers in an atmosphere of $TiCl_4 + H_2 + CH_4$ during the initial stages of layer formation on high chromium tool steels (the effect of chromium)[63], as well as in the process of formation of duplex titanium nitride layers on top of nitrided layers (the effect of the nitrided surface) [64, 65]. The condition for a catalytic reaction of particles at the metal surface is their prior chemisorption. When two particles react, at least one of them, but most often both, are chemisorbed. Thus, chemisorption constitutes a basic stage, preparing the particle for reaction.

Generally, metals may be categorized into various groups, regardless of the number of gases which may be chemisorbed by them. This division is shown in Table 5.1 [62]. It should be emphasized that the categorization is only qualitative. It follows unequivocally from the table that properties of chemisorption are exhibited by transition metals. Assuming the premise that the condition of a catalytic reaction between two particles is their prior chemisorption, it can easily be predicted which metals will catalyze the synthesis of ammonia (Class A) or the reaction of hydrogen with oxygen (Classes A, B_1, B_2).

Table 5.1
Metals according to their tendency to chemisorption

Metals	N_2	CO_2	H_2	CO	C_2H_4	O_2
Ti, Zr, Hf, V, Nb, Ta, Cr, Mo, W, Fe, Ru, Os, Ni, Co	+	+	+	+	+	+
Rh, Pd, Pt, Ir, Mn	−	−	+	+	+	+
Al, Cu	−	−	−	+	+	+
Li, Na, K	−	−	−	−	+	+
Mg, Ag, Zn, Cd, In, Si, Sn, Ge, Pb, Sb, Bi	−	−	−	−	−	+

It should be noted that one of the better known and most undesirable properties of heterogeneous catalysts is their tendency to become deactivated or poisoned by so-called toxins. These may enter the substrates as contaminations and act in a momentary or permanent manner, depending on whether their action stops or not after their expulsion from the system. A toxin may also be a by-product; an example of that is the formation of hydrogen chloride and its chemisorption during the formation of titanium carbide in a CVD process, carried out in an atmosphere of vapours of $TiCl_4 + H_2 + CH_4$ [66]. Metallic catalysts are particularly sensitive to toxins, especially to compounds of sulfur and nitrogen, containing free pairs of electrons which form strong coordinate bonds with the metal surface. A toxin is, therefore, a substance which is adsorbed more than the substrates and in that way renders their access to the reactive surface impossible [62].

In surface engineering, catalysis facilitates the carrying out of certain diffusion processes, as well as the application of some PVD and CVD technologies. Its most significant role, however, is in thermal spraying (catalytic action of exhaust gases). Of great ecological significance are catalytic coatings, sprayed onto surfaces of fuel-fired heating equipment. Exhaust gases or metal and ceramic catalysts, introduced into the gas stream, flow around them, including exhaust gases from vehicles with combustion engines, in order to reduce the content of nitrous oxides [67] and carbon monoxide [68] which are harmful to the environment. For those applications catalysts are made of e.g. oxides of manganese, vanadium, titanium, and aluminum oxides or their mixtures, with strongly developed surfaces (porous and rough coatings).

5.8 Practically usable properties of the superficial layer

The superficial layer is always produced with a clearly defined aim; it is always designated to be exposed to appropriate external hazards, both chemical and physical. Practically usable properties of the superficial layer, beneficial to the service of the part in conditions of one type of hazard, may prove to be less beneficial in conditions of a different type of hazard. For example, anti-corrosion superficial layers usually impair fatigue strength [69].

Practically usable properties of the superficial layer are, therefore, the result of matching its potential with external hazards (Fig. 5.38). Service properties change in the process of service, with time of use of the product. In only very few cases can potential properties be equivalent to usable properties of the superficial layer.

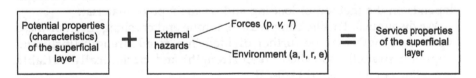

Fig. 5.38 Usable properties of the superficial layer (p - unit pressure; v - velocity; T - temperature; a - atmosphere; l - lubricant; r - radiation; e - electromagnetic field).

Appropriate combinations of parameters (characteristics) of the superficial layer may be best for appropriate combinations of external hazards. The latter may be mechanical stresses (including variable), friction, chemical interaction with the environment (oxidizing, reducing, inert), physical, by electric current or magnetic field or a combination of any of them.

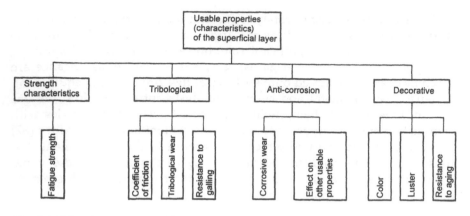

Fig. 5.39 Most important usable properties of the superficial layer.

Among the most important usable properties of the superficial layer are the following (Fig. 5.39): strength, tribological, anti-corrosion, decorative and some others.

5.8.1 Strength properties

5.8.1.1 General characteristics

In a general sense, strength is the resistance to destructive action of mechanical factors, such as various types of loading. In a strict sense it means the value of resistance put up by the forces of cohesion of a solid to external loading or the ability to withstand this loading and determines the boundary value of stresses which, when exceeded, cause fracturing of the solid (machine part or any structural component). The strength of a material is usually described as a load per unit area of cross-section. Strength depends on the method of loading and on the type of material.

We distinguish tensile, compressive, torque and crushing. Each of these types of strength may be further divided into **fixed** strength (including long-term strength, so-called creep strength) and **periodically variable** (including fatigue strength) [70-73].

By changing the properties of the material of the superficial layer, average properties of the material in its entire volume are changed. The degree to which a change of superficial layer properties affects average bulk properties of the material is proportional to the ratio of the cross-section of the superficial layer to that of the core.

The greatest influence of the superficial layer is not on **static** strength but on **dynamic** strength, especially in conditions of multiple periodically variable loading. This causes the creation of a mutually interacting set of effects, and successively developing properties which significantly reduce material strength and often lead to its failure, described as **material fatigue** (first time by J.V. Poncelet in 1939).

Destruction of material as a result of fatigue exhibits a different character than the one caused by static loading and occurs without major plastic deformations even in ductile metals and alloys. A fatigue fracture develops gradually. As the number of changes of loading increases in metals and alloys, successively, there come about local plastic deformation in particular grains, formation of slip bands, division of grains into blocks and submicroscopic fissures. Local microcracks propagate and join up with one another. The process of cracking is initiated in a site of strong concentration of stresses. Next, the crack develops gradually up to the moment when the remaining portion of the cross-section becomes too weak to support the load and suffers catastrophic decohesion. On the fracture surface thus formed, which in the macroscopic scale bears the character of a brittle fracture, it is clearly possible to distinguish the fatigue zones of the developing fracture. It has a characteristic, so-called *"beachmark" or "seashell" appearance* [70].

Stresses giving rise to material fatigue may change their values in a periodic or irregular manner which usually occurs in service conditions where loads are statistically random.

5.8.1.2 Fatigue strength

General notions. Variable loads causing variable stresses may be randomly distributed in time. Often, there are loads consisting of identical, repeatable values and frequencies of occurrence in fixed time intervals - periods (Fig. 5.40a). The simplest model case is the harmonically variable load. An example of that is the machine shaft which, while

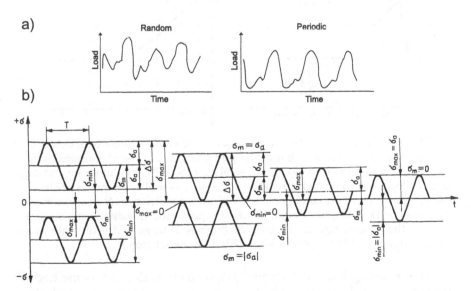

Fig. 5.40 Spectra of fatigue loading (a); and sine-wave forms of variable stresses (b). (From *Kocańda, S., and Szala, J.* [70]. With permission.)

under a fixed value of bending and torque moment, is subjected to sine-wave variable loads [70].

Different Standards (e.g., international - ISO R 373-1964 [74], British - BS 3518-63, Polish - PN-71/H-04325) distinguish: maximum cyclic stresses σ_{max}, minimum cyclic stresses σ_{min}, cyclic stress amplitude σ_a, mean stress σ_m and period of stress variation T. They are interconnected by the following correlations (Fig. 5.40b):

$$\sigma_m = \frac{\sigma_{max} + \sigma_{min}}{2} \tag{5.28}$$

$$\sigma_a = \frac{\sigma_{max} - \sigma_{min}}{2} \tag{5.29}$$

Moreover, the quoted Standard distinguishes the stress range

$$\Delta\sigma = 2\,\sigma_a = \sigma_{max} - \sigma_{min} \tag{5.30}$$

and the stress ratio

$$R = \frac{\sigma_{min}}{\sigma_{max}} \tag{5.31}$$

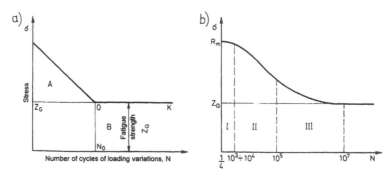

Fig. 5.41 Fatigue curves according to F. Wöhler: a) incomplete diagram for description of fatigue strength (A - zone of limited fatigue strength; B - zone of unlimited fatigue strength, e.g. for normalized 1045 grade steel $Z_G = 280$ MPa; for aluminum and magnesium alloys, and some high strength alloy steels, as well as all metals at elevated temperatures in corrosive conditions, line OK is not parallel to the N axis; b) complete curve with marked fatigue strength zones: I - quasistatic (quasistatic cracking); II - low cycle (low cycle fatigue); III high cycle (high cycle fatigue); R_m - ultimate tensile strength. (From *Kocańda, S., and Szala, J.* [70]. With permission.)

Fatigue strength Z_G, also termed *fatigue limit*, is defined as the highest stress σ_{max}, variable in time, which can be sustained by the material. Practically, it is the highest stress under which a specimen or a tested element does not suffer destruction after attaining a conventionally assumed boundary number of cycles $N_{o'}$ as determined by the so-called *F. Wöhler fatigue curves* (Fig. 5.41).

Fatigue strength is lower than static strength. For example, in mild carbon steel, subjected to symmetrically alternating tensile-compressive loads, fatigue strength is approximately 0.4 to 0.6 that of static ultimate tensile strength R_m. Practically, in the case of steel, fatigue strength is determined for 10^7 cycles, while in the case of non-ferrous metals - for $(5$ to $10) \cdot 10^7$ cycles.

Effect of properties of superficial layer on fatigue strength. Fatigue strength is affected, besides type and value of loading and its variations with time, by the following factors:

- selected technological properties of the superficial layer, mainly spatial geometrical, mechanical, structural and chemical,

- sudden changes of these properties, brought on by different effects and causing local disturbances in the distribution of stresses, both residual and load-induced, their concentration in the neighborhood of such changes and - in turn - rise of local stress values. Such changes are, therefore, stress-raisers and act as flaws - geometrical, structural, corrosion, etc. They cause a lowering of fatigue strength relative to that of a flawless material,

- technological environment to which the material is exposed during service, especially its corrosiveness and temperature.

The strongest effect on fatigue strength is exerted by surface **roughness**. Asperity valleys form geometrical flaws which lower fatigue strength. By reducing surface roughness (e.g., the average arithmetical deviation of the roughness profile from the center line R_a from 2.5 to 0.16 µm), it is possible to obtain improvement of fatigue strength by several tens percent (e.g., by 25 to 100%) [15, 31].

The smoother the surface, the less and the smaller the geometrical flaws and microflaws on it and the more difficult the conditions necessary to initiate fatigue microcracks. Of equally significant importance is recess radius: the greater the radius, the higher the fatigue limit.

Roughness of the superficial layer affects fatigue strength to a degree proportional to the static strengths of the superficial layer and core. Since this is usually associated with a lesser number of material flaws, the role of profile recesses as geometrical flaws rises [31].

The effect of surface roughness on fatigue strength is taken into account in engineering calculations by the introduction of stress-concentration coefficients, correlated by a mathematical function to roughness parameters. For tensile and compressive stresses, the formula to use is given below [31, 69]:

$$\alpha = 1 + 2\sqrt{\gamma \frac{R_z}{r}} \qquad (5.32)$$

where: α - coefficient of stress concentration; γ - load coefficient, dependent on the ratio of mean roughness deviation S_m from asperity height R_z across 10 points (Fig. 5.42); r - mean radius of recess curvature of roughness profile.

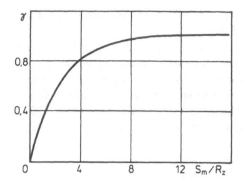

Fig. 5.42 Dependence of load coefficient γ on the S_m/R_z ratio. (From *Zielecki, W.* [31]. With permission.)

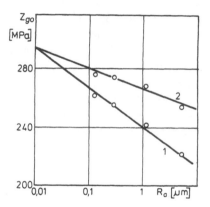

Fig. 5.43 The effect of R_a roughness and orientation of the geometrical structure of the surface on Z_{go} fatigue limit for the XH56BMKJu* alloy: *1* - structure perpendicular to the direction of external load; *2* - structure parallel to direction of external load. (From *Zielecki, W.* [31]. With permission.)

Fatigue strength is also significantly affected by the orientation of the three-dimension surface structure, resulting from machining. Machining traces usually form a series of multiple microflaws. When this orientation is perpendicular to the loading force, fatigue strength is 10 to 50% lower than that of an object with structure oriented parallel to the direction of the load. The effect of surface structure orientation on fatigue strength increases with a rise in surface roughness (Fig. 5.43) [69].

Among various metallic materials, particular susceptibility to a change of Z_g upon machining is exhibited by titanium alloys: the fatigue limit of roughly turned specimens may be lower than that of polished specimens by as much as 30% [70].

Scratches, crack networks, blisters, pits, grinding burns and local de-carburization (so-called soft spots) on the superficial layer all significantly reduce fatigue strength [70].

Cold plastic deformation causing strain hardening of superficial layer grains favorably affect fatigue strength. As the result of fragmentation of the substructure an increase in dislocation density and their uniform distribution, the formation of compressive residual stresses and hardening of the superficial layer, fatigue strength rises significantly: for example, after burnishing, by 25 to 90% for parts without stress-raisers and by 150 to 200% for parts with stress-raisers [72, 75, 76]. Relative increment of fatigue strength is, in the case of uniform cold work, directly proportional to the relative increment hardness caused by burnishing, as in the expression:

$$\frac{Z_1}{Z_0} = \frac{H_1}{H_0} \tag{5.33}$$

where: Z - fatigue limit prior to (Z_0) and after (Z_1) burnishing; H - hardness prior to (H_0) and after (H_0) burnishing.

The above condition is usually not met by brittle materials with high hardness.

Fatigue strength Z is correlated to yield strength R_e by the expression [21]:

$$Z = c\,R_e \tag{5.34}$$

where: c - coefficient (for steels, $c = 0.3$ to 0.7)

Because yield strength is proportional to hardness, the dependence of fatigue strength on hardness is almost ideally proportional, particularly for materials which are not brittle [21].

Grain size and material structure also affect fatigue strength. Fine-grained steels are characterized by better fatigue properties than coarse-grained steels. Steels with a higher carbon content and high tensile strength also exhibit higher fatigue strength. The most favorable fatigue properties are exhibited by steels with a martensitic structure [69, 72]. The effect of material structure on fatigue strength is usually associated with residual stresses.

Similarly to surface structure orientation caused by machining, anisotropy (grain orientation) of mechanical properties of the superficial layer and core, caused mainly by the type of deformation, has an effect on fatigue strength. For smooth, symmetrically round objects, the latter may be higher by as much as 50% in the direction parallel to grain flow than in the direction perpendicular to it [72].

All structural discontinuities in the material of the superficial layer in the form of cracks, foreign inclusions, blisters, pores, cold shuts, etc., forming structural flaws, have a negative effect on fatigue strength.

Residual stresses have a significant effect on fatigue strength. These stresses superimpose themselves (i.e., add algebraically) on those stemming from external loading and cause either a rise or a drop of the fatigue limit of an object, depending on whether the net result will be an increase

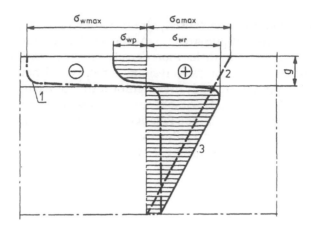

Fig. 5.44 Distribution of stresses in a bar with a hardened superficial layer of thickness g, subjected to bending: *1* - distribution of residual stresses σ_w; *2* - distribution of nominal stresses caused by external loading σ_a; *3* - distribution of resultant stresses (σ_{ws} - highest compressive stresses at surface, σ_{wc} - highest tensile stresses in core). (From *Kocańda, S., and Szala, J.* [70]. With permission.)

or decrease of the sum total of stresses at concentration sites (Fig. 5.44) [31]. Tensile residual stresses cause the decrease of highest compressive stresses and their shift inward, to a depth dependent on the thickness of the hardened layer, often below it. It is below that layer, then, and not at the surface (which is weakened by flaws with both geometrical and structural-like inclusions or decarburization) that the fatigue failure source is located. It consists of a concentration of stresses, usually at a site coinciding with the highest stress gradient, the initiation of microcracks and material damage. At that moment, surface flaws cease to be dangerous from the point of view of fatigue strength. If the hardened layer thickness is increased beyond its optimum value the situation is reversed and the surface again becomes the fatigue failure source. A thick hardened layer, especially in elements with sharp notches, defeats the purpose because usually in such cases the fatigue failure source is located at the bottom of the notch [70].

Fig. 5.45 shows the effect of residual stresses on the distribution of net (i.e. residual together with external) stresses in conditions of tension and bending, all within the range of elastic deformations.

In the case of stretching of unnotched specimens (Fig. 5.45a), residual stresses cause a rise of sum total maximum tensile stresses, creating conditions conducive to a drop in strength. From conditions of equilibrium it follows that the occurrence of compressive residual stresses in one portion of the cross-section is of no special significance, since they have to be balanced by tensile stresses in the remaining portion of the cross-section [42].

On the other hand, when stretching notched specimens, the interaction of residual stresses brings about a local concentration of stresses caused by external loading (Fig. 5.45b). Compressive tensile stresses in the

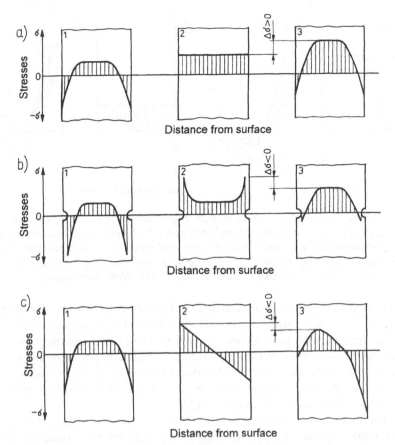

Fig. 5.45 Effect of residual stresses on the total stress distribution in tensile and compressive loading: a) unnotched tensile specimen; b) notched tensile specimen; c) bending test specimen; 1 - distribution of residual stresses; 2 - distribution of stresses caused by external loading of specimen; distribution of sum total stresses; $\Delta\sigma$ - increase of tensile stresses. With cyclic variation of loading, the distribution of stresses will vary in a cyclic mode from state 1 to state 3. (From *Janowski, S.* [42]. With permission.)

vicinity of a notch may effectively lower the maximum value of total tensile stresses, which helps to raise strength [42].

In the case of bending, a similar phenomenon may occur (Fig. 5.45c). A drop in the value of maximum tensile stresses is accompanied by a rise of total compressive stresses. Strength, however, is associated predominantly with tensile stresses [42].

Fig. 5.46 shows the effect of residual stresses on total stresses after locally exceeding the fatigue limit which causes a redistribution of stresses as well as their partial relaxation. In the subsurface layer a rise of stresses will then occur in areas which are not plastically deformed (Fig. 5.46d, curve 2). When the loading force is removed from the specimen a residual stress distribution remains which differs from the initial one (Fig. 5.46d, curve 3) [42].

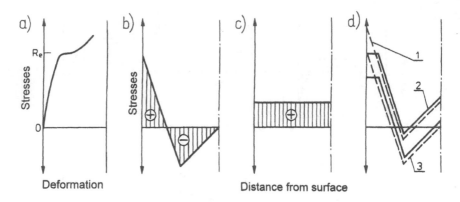

Fig. 5.46 Effect of residual stresses on sum total stresses after locally exceeding the yield strength limit: a) fragment of tensile test curve; b) distribution of residual stresses; c) stresses caused by external forces; d) sum total stresses; *1* - sum residual stresses and stresses caused by internal forces; *2* - real state of stresses; *3* - state of residual stresses after removal of load. (From *Janowski, S.* [42]. With permission.)

The result of addition of residual and loading stresses is a change in the value of the mean stress σ_m, and what follows of the coefficient of loading stability. Tensile residual stresses, by causing a rise of the mean cyclic stress, bring about a lowering of fatigue strength, while compressive stresses are conducive to its increase. By the lowering of the mean cyclic stress it is possible to raise the cycle amplitude more than twice in ductile materials and almost three times in brittle materials [42].

The effect of residual stresses on fatigue strength is, in most cases, determined experimentally. It usually depends on the character of the external load. For an asymmetrical cycle, when the middle of the cycle occurs on the side of compressive stresses, fatigue strength is higher than if the middle of the cycle occurs on the side of tensile stresses (Fig. 5.47). In the case when stresses caused by external loads vary according to a symmetrical cycle, the presence of residual stresses changes the cycle to an asymmetrical one. If the surface layer contains compressive stresses, the middle of the cycle passes to the side of compressive stresses and the fatigue limit is raised [21].

At high service temperatures of metal objects, residual stresses, regardless of their sign and value, do not have a significant effect on fatigue properties. The action of elevated temperature is conducive to an atrophy of the fatigue limit. Besides, the decrease of the fatigue limit with temperature is usually irregular and depends mainly on temperature ranges in which structural transformations occur. At depressed temperatures the fatigue strength of metals usually rises [70].

The effect of residual stresses on fatigue strength is also insignificant in those cases where the sum of residual stresses and those caused by external loads exceeds the yield point of the material (see Fig. 5.46). In both cases residual stresses are rapidly relaxed in the superficial layer [31, 69].

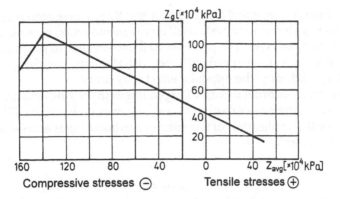

Fig. 5.47 Fatigue strength Z_g of nitrided EI355 grade steel; Z_{avg} - mean stress in a cycle. (From Łunarski, J. [21]. With permission.)

Corrosive environments such as moist air, tap water, sea water, and electrolyte solutions all lower fatigue strength. The Wöhler curve does not feature an interval that is parallel to the N axis; therefore, there does not exist a fatigue limit. Instead, it is said that there occurs a limited fatigue strength for a given number of cycles N. With a rise of corrosive activity of the environment, fatigue strength decreases, as it does with a rise in the time of its activity (rise in number of cycles). An especially negative effect on fatigue strength in corrosive media is that of flaws, including corrosion flaws (corrosion sources). On the other hand, compressive residual stresses in the superficial layer effectively improve fatigue properties of metals in a corrosive environment [70].

Surfactants which do not have a corrosive effect on metals or affect them to only a minute degree, e.g. oils with purifying additives or surfactant solutions, upon entering micropores in the surface, have a de-wedging effect (P.A. Rebinder effect - see Fig. 5.60), causing fatigue strength to drop significantly but with a clearly marked fatigue limit [70].

As a result of various types of external effects occurring during treatment operations, fatigue strength varies according to the formula:

$$Z_{po} = \frac{Z_{wo}}{\beta} \tag{5.35}$$

where: Z_{po} - post operation fatigue limit (of unnotched or notched specimen); Z_{wo} - fatigue limit without any treatment operation (of unnotched or notched specimen); β - coefficient of effect of treatment operation on fatigue limit; for $\beta <1$ a positive effect is noted; for $\beta >1$, a negative effect (Table 5.2).

In summing up data from literature [31, 69-78] it can be generally stated that:

– the majority of hardening treatments significantly improve fatigue properties, particularly those of machine components with notches (Table 5.2) [70]. Their effect may be varied and is associated with changes of microstructure, chemical composition and three-dimensional geometry of the surface but always manifests itself by the introduction of residual stresses into the superficial layer. The presence of these stresses (in particular of compressive stresses in the superficial layer) usually improves fatigue strength, although sporadically, the opposite effect may occur;

– the highest improvement of fatigue strength may be achieved through an increase in surface smoothness, on an average by 50 to 70%. That is obtained by the application of smoothing treatments (superfinishing, grinding, polishing, etc.);

Table 5.2
Values of coefficient β for different treatment operations (From *Kocańda, S., and Szala, J.* [70]. With permission.)

Type of operation	Type of specimen finish	Specimen diameter [mm]	Coefficient β
Roll hardening	notched	7-20	0.71-0.83
		30-40	0.80-0.91
	unnotched	7-20	0.45-0.67
		30-40	0.55-0.77
Shot peening	notched	7-20	0.77-0.91
		30-40	0.91-0.93
	unnotched	7-20	0.40-0.70
		30-40	0.57-0.90
Carburizing and hardening (case depth: 0.2-0.6 mm)	notched	8-15	0.47-0.83
		30-40	0.67-0.90
	unnotched	8-15	0.40-0.67
		30-40	0.50-0.83
Nitriding (case depth: 0.1-0.4 mm) HV = 730-970	notched	8-15	0.80-0.87
		30-40	0.87-0.90
	unnotched	8-15	0.33-0.52
		30-40	0.50-0.77
Cyaniding (case depth = 0.2 mm)	notched	10	0.55
Induction hardening	notched	7-20	0.63-0.77
		30-40	0.67-0.83
	unnotched	7-20	0. 36-0.63
		30-40	0.40-0.67
Chrome plating (0.04-0.2 mm)	notched	8-18	1.1-1.5
Nickel plating (0.03-0.1 mm)	notched	8-18	1.0-1.5

– a lesser degree of improvement of fatigue strength, 40 to 45% on an average, is achieved by changes in the chemical composition of the superficial layer (diffusion alloying with carbon, nitrogen, chromium, boron, etc. or their combinations) [78];

– an even smaller degree of improvement is achieved by mechanical strengthening (by static or dynamic pressure) of the superficial layer - on an average: 20 to 30%;

– surface hardening (flame, immersion, induction, by electron beam and laser) improves fatigue strength less than saturation and diffusion alloying [31]; it is noteworthy that hardened layers should not begin or end suddenly (i.e., with sharp interfaces) but as far as possible, exhibit a gentle and continuous transition into the core structure, thus avoiding the presence of a structural flaw [70];

– electroplating, as a rule, reduces the fatigue limit; the strongest negative effect is that of chrome, nickel, cadmium, and iron plating (a chrome plated layer of 0.1 mm thickness on steel causes a drop of fatigue limit by 30 to 40%. This drop may increase with greater core strength. The effect of copper and zinc plating is less pronounced.

The values of the β - coefficient quoted here have a purely orientation character. It is obvious that the various methods of improvement of fatigue strength do not exclude one another but that their effects may be combined.

5.8.2 Tribological properties

5.8.2.1 Types of basic tribological properties
Tribological properties[1] comprise those properties which constitute conditions of mutual interaction of surface and the environment of bodies in frictional contact.

Since, under the same loading force P, the force of friction T depends on the friction coefficient f, according to the formula:

$$T = fP \tag{5.36}$$

and the effect of interaction of bodies rubbed against each other is tribological wear, it will be understandable that tribological properties describe: the friction coefficient and wear intensity, as well as, to some degree, resistance to seizure (galling).These depend on conditions in which the process of friction occurs, as well as on the potential properties of rubbing surfaces, i.e., type of friction, method of lubrication and nature of wear.

5.8.2.2 Types of friction
Friction occurs universally in nature and in technology. It is essential for the movement of living beings and vehicles, makes work possible to be carried out and constitutes the basis of many technical devices, such as brakes,

---

[1] *Tribology* - the science of friction and related phenomena (from the Greek: *tribo* - rub + *logos* - science).

clutches and belt transmissions. At the same time, in many cases friction is an undesired effect, causing significant energy losses to overcome the frictional resistance, e.g., in bearings and bushings of machines and other vehicles. Moreover, friction causes the wear of machine components, often contributes to damage of working technical devices. Approximately 80 to 90% of machine components work in conditions of friction [31].

Friction is a physical phenomenon, always involving mutual displacement of particles of matter in different states of aggregation. It may be termed:

– **external**, when it involves the relative displacement of surfaces of two solids in contact with each other. This type of friction is characterized by mutual interaction of bodies on their touching surfaces, manifest by resistance to relative displacement in a direction tangent to the surface of contact;

– **internal**, when it involves the relative displacement of particles of the same body: of a liquid separating the surfaces of solids (e.g., in fluid friction) or particles of a solid (e.g., in deformation).

The object of interest to surface engineering is, of course, external friction.

From the point of view of wear intensity, friction may be divided as below:

– **wear**, occurring in the majority of practical cases, associated with a smaller or greater degree of destruction of the rubbing surfaces, accompanied by high intensity of wear;

– **non-wear**, occurring only in special conditions and associated with the spontaneous constitution, during service, of new rubbing surfaces, as the result of so-called selective translocation of material. In these cases the intensity of wear is several orders of magnitude smaller than in wear type friction.

Depending on the relative velocity of rubbing materials, friction may be divided thus:

– **static** - this is the friction between two bodies in mutual contact which do not change their relative positions, expressing a force which must be overcome to initiate relative movement of these bodies;

– **kinetic** - this is the friction between two bodies in relative movement, expressing force which must be overcome to maintain the movement of these bodies.

From the point of view of the presence of a lubricant between rubbing surfaces, the following types of kinetic friction may be distinguished (Fig. 5.48):

– **physically dry friction** - when no other bodies (solid, liquid or gaseous) are present between the rubbing surfaces while the rubbing surfaces themselves are not coated by any adsorbed chemical compounds; in practice, this type of friction occurs extremely seldom in vacuum;

– **technically dry friction** - when the rubbing surfaces may be coated with oxides and layers of adsorbed gases or vapors. Presently, science

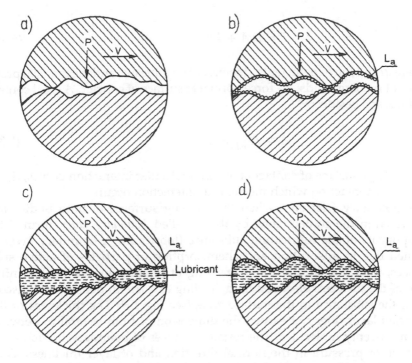

Fig. 5.48 Types of kinetic friction, exemplified by sliding wear occurring in pairs: a) physically dry friction; b) technically dry - boundary friction; c) mixed friction; d) fluid friction; L_a - adsorbed layers (zones I, II, II - organic particles - see Fig. 5.8) of dust, water vapor and gases.

accepts a dualistic character of the origin of friction forces, based on Kragielski's molecular-mechanical theory, according to which the total force of dry friction [9]:

$$T = T_{adhesion} + T_{cohesion} \qquad (5.37)$$

where: $T_{adhesion}$ - component of friction force, originating from the overcoming of adhesion forces which are caused by molecular interaction in the contact zone; it may create local connections, especially at asperity peaks, subsequently destroyed by the cohesion component; $T_{cohesion}$ - component of friction force, originating from the overcoming of cohesion forces, caused by elastic and plastic mechanical deformation, especially by shearing of surface asperities, formation of grooves, severance of material continuity at sites of strong adhesive connections (see Fig. 5.55). This component causes mechanical damage to interacting surfaces.

The force of technically dry friction may also be expressed by the formula [38]:

$$T = \alpha A + \beta P \tag{5.38}$$

where: α and β are coefficients, dependent on adhesive and mechanical properties of rubbing solids; P - force normal to surface of contact, A - real contact surface, and besides:

$$A = A_{mol} + A_{mech} \tag{5.39}$$

where: A_{mol} - surface of contact on which molecular interaction occurs, A_{mech} - surface of contact on which mechanical interaction occurs;

– **boundary friction** - when the rubbing surfaces of both bodies are separated in the contact zone by the so-called *boundary layer,* formed by components of the lubricating substance (lubricant). A boundary layer is formed due to adsorption and chemisorption of particles of active substances, their densification and ordering of polarization under the influence of the surface's electrical field, leading to a rise in density and viscosity of the substance in the surface zone (see Fig. 5.9) The thickness of the lubricant layer does not exceed the dimension of from several to between ten and twenty molecules. The resistance forces of boundary friction are a function of pressure in the normal direction and of layer thickness. During relative movement the lubricant layer may quite easily be broken by surface asperities. The removal of the boundary layer is succeeded by the onset of dry friction, leading to intensive wear of superficial layers and even to scuffing. Boundary lubrication should be stable, which is of great significance in unstable conditions of machine work, during startup and shutdown and at times of failure of the lubricating system [38];

– **fluid friction** - when between the surfaces of rubbing bodies there is a continuous, unbroken layer of liquid or gaseous substance, under pressure which balances out forces of normal load pressure from both bodies to such an extent that surface asperities do not come into mutual contact. External friction of the mating surfaces is thus replaced by internal friction. The only mechanism of damage of superficial layers is by the formation of chemical compounds on their surfaces, e.g., by oxidation, and subsequently by their removal. For most lubricants this is a very slow process. Resistance forces in fluid friction depend only on the thickness of the lubricant layer and on its viscosity (which itself varies with temperature and pressure);

– **mixed friction** - which is an intermediate case between dry and fluid friction. In this case, in the zone of contact between the rubbing pair there occur phenomena which are characteristic of at least two of the above mentioned types of friction. This is prevalent in frictional nodes of machines, especially with low relative velocities, high unit loads and unstable conditions.

The coefficient of friction varies, depending on conditions of friction (Fig. 5.49). It assumes lowest values in fluid and mixed friction [80].

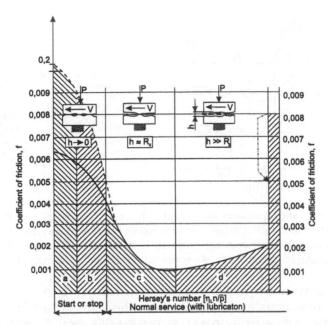

Fig. 5.49 Dependence of friction coefficient on conditions of sliding wear, characterized by the dimensionless Hersey number: a) physically dry friction, b) technically dry or boundary friction, c) mixed friction, d) fluid friction; h - thickness of lubricant film; R_z - surface asperity height; $\bar{p}$ - average unit loading force; v - sliding speed; n - rotational speed; p - mean unit pressure; η_L - dynamic viscosity.

Depending on the method of relative movement of the rubbing pair, friction can be divided into sliding, rolling and mixed: rolling-sliding friction [79].

5.8.2.3 *Sliding friction*

Sliding friction is an example of kinetic friction in which relative velocities of bodies at sites covered by the contact zone are varied and are situated in a plane tangent to the contact surface. It occurs when there is an absence of lubrication (dry friction) or in the presence of any type of lubrication and is practically utilized in bushings, gears, guide rails, piston systems, clutches and brakes.

A modification of sliding friction is **torque** or **butt friction** of two bodies with a flat contact surface, in relative rotational movement about an axis which is perpendicular to the contact (sliding) surface [79]. This type of friction occurs in bush bearings loaded parallel to the axis of rotation.

If two surfaces of solids come into mutual contact or slide against each other, the real contact between them occurs only in a limited number of small zones (Fig. 5.50)

The ratio of real surface contact area A_r (increasing with a rise of load value) to nominal surface A_n is usually contained within the range of 1/1000 to 1/10 000. Local pressures at contacting peaks of surface asperi-

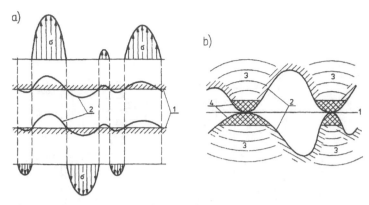

Fig. 5.50 Schematic representation of contact between two rubbing surfaces: a) distribution of stresses resulting from contact between rough surface and nominal surface; b) real contact between two rough surfaces; *1* - nominal surface; *2* - surface asperities; *3* - zones of elastic deformation; *4* - zones of permanent deformation.

ties may even under small loads be sufficiently big to cause significant plastic deformations, e.g., in metal alloys. This leads to the formation of connections between asperity peaks, increasing this area of contact with progressing deformation of the material. Such connections create resistance to bodies in relative motion. When the tangent force increases, there occurs shearing of the connections formed during sliding [81]. The force of friction needed to shear all such connections is directly proportional to the shear strength of the material at such sites [8]:

$$T_{cohesion} = F_s = A_{mech}\tau_s = \frac{P}{p}\tau_s = \frac{\tau_s}{p}P = fP \qquad (5.40)$$

where: F_s - force of friction causing shearing of connections; A_{mech} - real surface of contact of the formed mechanical connections; P - loading force, τ_s - material shear strength; p - pressure causing plastic deformation of material; f - coefficient of sliding friction.

When two materials with two different hardnesses slide on one another the coefficient of friction depends mainly on shear strength and on yield strength of the softer material.

In conditions of sliding wear there may occur additional effects, such as squeezing out of the contact surface of the softer material by the protruding asperities of the harder material and closing of the surface in both materials. The peaks of asperities of the harder material may scratch the softer material, in particular when asperity heights of the former are big and if that material contains particles with sharp edges. During sliding wear of a soft material against a hard material there is no scratching of the hard material, but small fragments of the soft material may adhere to the hard one [9].

If effects associated with scratching of the surface and blocking its unevenness may be neglected, the force of sliding friction F_s depends on the real contact area A_r and on the material shear strength τ_s. For that reason, in order to ensure low friction it is necessary that both A_r and τ_s be as low as possible. This means that material with high hardness and low shear strength would be the most appropriate. Such a combination is, however, practically impossible because materials with high hardness usually have high shear strength.

If, on the other hand, a hard material coated with a thin layer of soft material is used, shear strength will be determined by the softer material of the superficial layer, while resistance to deformation will depend on the hard material of the substrate. In such a case the real contact area A_r will remain practically constant, even for high loads, while friction will be low. This principle is used in the design of modern bimetallic bushings, especially those working under high pressures and at elevated temperatures [9].

Sliding wear occurs mainly in bushings and in all types of sliding connections with rotational and progressive motion or their combination.

5.8.2.4 Rolling friction

Rolling friction is also a case of kinetic friction but occurs when one body rolls on the surface of another or, in a stricter sense, when there is no sliding of one body against the other in the zone of contact and when the zone of contact is displaced with a relative velocity, while the time of contact is very short. For ideally rigid bodies the zone of contact is a momentary axis of rotation in the form of a point or a line [8, 12].

In practice, rolling wear does not occur in its pure form. In real conditions of deformable materials, the phenomenon of pure rolling occurs only in the case when both mating bodies are made from the same material and have the same diameter and length. Surface roughness of bodies should be as low as possible. In such conditions, with the absence of a lubricant, there is only a hysteresis of the elastically deformed material of both bodies in the contact zone [81-83].

In real conditions, if the curvatures of mating surfaces are different, deformations in the contact zone are accompanied by microslip. If, however, the velocities of both mating surfaces are the same (i.e., their relative velocity is equal to zero), their relative movement is called slipless rolling. Rolling friction occurs at contact sites (point or linear) of the rolling bodies in such a way that material deformed under normal pressure forms zone contact across a certain area, in other words, surface contact. The rolling element elastically deforms the material of the race which offers resistance. When, during the rolling motion of two bodies their circumferential velocities are different, friction with slip occurs [82].

Rolling friction occurs in rolling bearings in which the loaded rolling element, usually in the form of a ball, cylinder, cone or barrel, rolls against a surface (e.g., in ball, cylinder, roller, pin, barrel or cone bearings) called a race.

Friction in these bearings is accomplished only in the presence of a lubricant. Besides exceptional cases in which lubrication is by solid lubricants
(e.g., graphite, molybdenum disulfide), two types of friction occur: *fluid* and
mixed. The first occurs when an elastohydrodynamic film of the lubricant
totally separates the mating surfaces while the second type constitutes a
combination of fluid, dry and boundary friction. The thickness of lubricant
films in rolling bearings is 0.1 to 1 µm and is comparable with the roughness
value [81].

The coefficient of rolling friction is directly and proportionally dependent
on forces of resistance to movement, determined empirically for different friction conditions and different geometry of the rubbing pair. It is also inversely
proportional to the loading force; for technical pairs it varies within the range
of 0.05 to 0.0005 [12].

The working temperature of rolling bearings is usually within
40 to 130°C [82].

Friction forces of resistance in rolling motion are incomparably smaller
than those in the case of sliding motion. In roller bearings, friction resistance forces are composed of rolling resistance and sliding resistance. Rolling resistance, which is the basic component, is caused mainly by losses
of elastic energy in different zones of superficial layers of rolling bodies
which are alternately loaded and relaxed during motion [9], as well as by
internal friction in the lubricant. Rolling resistance depends on the value
of the normally oriented force, the geometry of contact and the rigidity of
materials of the rubbing pair [38]. Sliding resistance is caused by slips and
microslips due to deformations, geometry of contact and relative movement of the rolling elements [82].

Rolling friction with slip is a case of mixed kinetic friction of bodies, the
relative movement of which causes the simultaneous occurrence of rolling
and sliding friction in the contact zone [79]. The amount of slip depends on
the velocity of friction.

5.8.2.5 *The role of surface in the friction process*

Friction is a very complex process, difficult to present in one simple theory.
The force of friction depends on the loading force which presses the rubbing surfaces together, on the type of friction and on the coefficient of
friction, which all depend, in turn, on the type of superficial layers of the
rubbing pair (potential properties of the superficial layers) and on the
type and properties of the substance present between the mating surfaces.
Moreover, it depends on velocity, temperature and duration of the friction
process.

Metal surfaces exposed to air are always covered by thin layers of
oxides or adsorbed gases which to a significant extent affect adhesion and
friction between mating surfaces. As long as these layers are present at
the metal surface, the coefficient of friction is low and only very seldom
attains the value of 1 to 1.5, characteristic of friction of pure metallic surfaces,
obtained in high vacuum after heating the metals. With pure metallic sur-

faces adhesion is so strong that the mutual bonding of both surfaces occurs at sites of contact of asperities by forces of metallic bonds which, in turn, causes the creation of adhesive microfusion. The fast development of adhesive bonding of surfaces of the rubbing pair causes leaps in the rise of friction resistance forces and in intensity of wear, conducive to deep destruction of the surface and often to a stoppage of the relative motion of the bodies. Such a process is known as **seizure (galling)** [9].

Similar effects have been observed in other materials. For example, diamond has a very low coefficient of static friction (0.05) in air, but in vacuum this coefficient reaches the value of 0.5. Graphite has a low coefficient of friction not only on account of its laminar structure, but also on account of adsorbed superficial layers such as water and gases. In vacuum, graphite elements are subject to seizure and degreasing, but with access of air, and in particular moisture, its coefficient of friction and wear drop significantly. Depending on the moisture in air, the coefficient of friction of graphite on graphite may vary from 0.06 to 1.0.

The coefficient of friction on snow or ice is only 0.03 because due to local very high pressure the temperature of water-ice phase transformation is lowered and a layer of water is created. At low temperatures (-40°C and lower) the layer of water is not formed and the coefficient of friction rises to a value normal for two sliding solid surfaces, i.e., 0.7 to 1.2 [9].

5.8.2.6 Thermal effects of friction

During relative motion of sliding surfaces, a significant amount of heat is dissipated, causing a rise of temperature even when loads and sliding velocities are relatively low.

Heat created as the result of friction is uniformly distributed in the zone of rubbing materials but is localized at peaks of asperities. This leads to rises in temperature all the way up to the melting point of the material and, consequently, to local structural transformations and changes in residual stresses of the material or to the formation of local microwelds. The surface temperature depends on the loading force and sliding velocity, but also on thermal conductivity and coefficient of friction [9]. The dissipation of heat is a self-accelerating process since the rise of friction and adhesion at local hot sites of microwelds causes a rise in the rate of heat dissipation. This, in turn, contributes to an increase in the number of such microwelds and leads to galling or even to bonding of the two mating surfaces (friction welding) [84].

The phenomenon of heat dissipation during friction can play a favorable role in the obtaining of very smooth surfaces in such processes as polishing with abrasive medium. This process consists of smoothing of unevenness by wearing down material from peaks of asperities and transferring it to grooves. As the result of friction between tiny particles of the abrasive powder and the polished surface, hot spots are formed within the contact zone at which temperature rises to melting point. Melted or softened metal is smeared on the surface and upon rapid resolidification forms a characteristic amorphic layer, known as the *Beilby layer*. This layer is built up of exceptionally small crystallites, formed as the result of a sudden recrystallization of the metal [9].

5.8.2.7 Lubrication

Lubrication consists of the introduction of a lubricant between surfaces in relative motion, in order to reduce friction (by reducing the coefficient of friction) through the elimination of dry friction and its replacement by other types friction. Lubrication allows a reduction of wear rate and of damage to the rubbing surfaces, enhancement of vibration damping, intensification of heat conduction from the rubbing surfaces, removal of wear products, counteraction to corrosion and, in some instances, the formation of a usable superficial layer.

From the point of view of reduction of friction, the optimum type of lubrication is such which allows the formation, between the rubbing surfaces, of a stable, continuous (unbroken) layer of the lubricant in the form of a cushion (exclusive to hydrostatic lubrication) or a wedge (in cases of hydrodynamic lubrication and appropriate design of rubbing surfaces) [12, 83].

This is not possible with every design of the friction node and in all service conditions. For this reason, in practice there occur different types of friction, described by different laws, on account of the function of the lubricant.

Extreme types of friction, on account of the nature of lubrication, is liquid and boundary friction. The main discriminant is the relative thickness of the lubricant layer R, connecting the absolute thickness of lubricant layer to the R_a roughness parameter of the rubbing surfaces (Fig. 5.51).

Lubrication with fluid friction, also known as **hydrodynamic lubrication**, involves the total separation of sliding surfaces by a thick layer of liquid lubricant (lubricating film) in such a manner ($5 \leq R \leq 100$) that no direct contact exists between the rubbing surfaces (Fig. 5.52a). Internal friction of surfaces of rubbing bodies has here been replaced by internal friction of particles of the lubricant. Fluid friction is governed by laws of hydrodynamics, while the friction coefficient:

$$f = \eta \frac{\upsilon}{P} \qquad (5.41)$$

where: η - viscosity of lubricant; υ - relative velocity of rubbing surfaces; P - loading force.

Most often, the lubricant is in the form of liquid oils, but also gases and thick (plastic) lubricants may be used, because with high unit loading usually occurring in a rubbing pair, lubricants may behave like liquid. The coefficient of fluid friction is very low ($f \geq 0.001$).

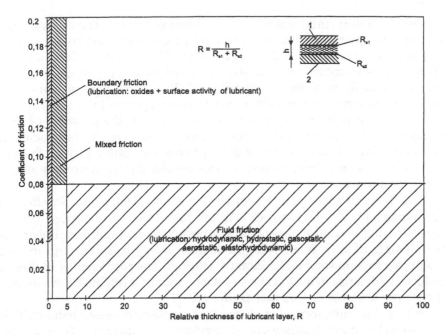

Fig. 5.51 Friction zones from the point of view of lubrication.

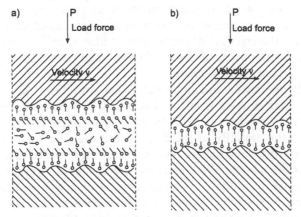

Fig. 5.52 Lubrication of solid surfaces: a) fluid; b) boundary.

Lubrication with boundary friction occurs when the rubbing surfaces are not separated by a layer of lubricant but are covered with an adsorption layer of that medium, with a thickness of only several particles ($1 \geq R$), comparable to the height of surface asperities (Fig. 5.52b). Boundary lubrication is significantly influenced by the surface oxide layer, formed as the result of surface oxidation of the atom layer at the rubbing surface which adsorbs gas atoms, water particles and also polarized particles from the environment. The joint layer is named boundary lubrication film.

Between the above two extreme types of lubrication there exists a middle zone - **lubrication with mixed friction** - in which there occur simultaneously boundary, fluid and elastohydrodynamic friction. The last one takes into account elastic deformations of the rubbing surfaces and changes in the viscosity of the lubricant with rising pressure. The rubbing surfaces are partially in direct contact, but partially separated by a layer of the lubricant which is usually of a thick composition ($1 \leq R \leq 5$). This type of lubrication occurs most often in the majority of machine components [8].

Lubrication with boundary friction takes place in the presence of relatively small amounts of polar organic compounds such as fatty acids in the lubricating oil. The reaction of polar groups, e.g., carboxyl, with the metallic surface causes the formation of a monomolecular layer, strongly adherent to the surface.

This layer attracts particles which are more distant and, in effect, a subsequent layer is formed, etc. As the result of the presence of these layers the number and surface of metallic connections are reduced. The crushing strength of such layers is high but shear strength is low, with the net result that friction wear is significantly reduced. The coefficient of friction is reduced with a rise in the thickness of the boundary layer, but below 100 nm it becomes unstable and friction again rises.

A boundary film, formed as the result of sorption processes, exhibits high mechanical strength. It is, however, dependent on temperature. For example, at a temperature close to that of melting of soaps (organic acid - metal) there occurs a desorption of organic particles connected with the substrate. The ordered state undergoes a transition to the disordered state which significantly lowers the strength of the boundary film, manifest by its resistance to pressure and other forces. Only the chemisorbed particles, e.g., those of MoS_2 solid lubricant, are not desorbed.

At high pressures and high velocities sometimes additional friction heat is dissipated. As a result, at high surface temperatures a breakdown of the lubricating film occurs. In such cases, special agents are added which raise the stability of the film on the metallic surface. These agents are able to withstand higher temperatures and exhibit satisfactory shear strength, besides being able to sustain higher loads. Among such additives are organic compounds containing groups of active radicals and sulfur, chlorine or phosphorus. These compounds react with the surface of the metal, forming unlimited layers of sulfides, chlorides or phosphates. A lubricant with such additives ensures the reproducibility of the thin superficial layer if it is destroyed in the course of friction.

Lubrication with boundary friction may be attained with the application of solid lubricants such as graphite, molybdenum disulfide (MoS_2) or tungsten disulfide (WS_2). Solid lubricants may be present in a colloidal solution of oil or resin or the self-contained form. These materials form thin layers at the metal surface and ensure high shear strength, as well as high temperature [9].

All of the above-mentioned lubricants have a laminar structure. Atoms located in a flat layer are bound by covalent bonds which are strong. Between the layers much weaker Van der Waals intermolecular bonds

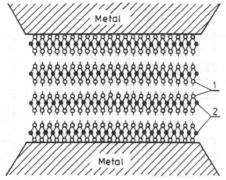

Fig. 5.53 Schematic representation of layers of molybdenum disulfide during friction: 1 - sulfur; 2 - molybdenum. (From *Janecki, J., and Hebda M.* [12]. With permission.)

exist. For that reason they are characterized by clearly defined slip and cleavage planes which run along the layers, giving them good lubricating properties. For instance, crystalloids of the widely used MoS_2 have many easy slip planes; practically each plane of elementary crystal connections (molybdenum atom surrounded from both sides by sulfur atoms) offers very little resistance to deformation, when acted upon by tangential forces (Fig. 5.53) [12].

5.8.2.8 Tribological wear and its various versions

The result of interaction of the rubbing elements is tribological wear, understood as the process of destruction and removal of material from the surface of solids, due to friction, and manifest by a continuous change of dimensions and shapes of the rubbing elements [31]. The causes of wear are in most cases of a mechanical character, less often mechanical, combined with the chemical interaction of the surrounding medium. The basic causes of wear are [85]

– elastic and plastic deformation of peaks of asperities and their work-hardening,

– the formation at the rubbing surface of oxide layers, preventing, on the one hand, galling and deep detachment of particles, but, at the same time, layers which are brittle, flaky and easy detachable; exposed surfaces may undergo secondary oxidation, etc.,

– building-in of fragments of the superficial layer of one rubbing material into the surface of the second material. During sliding wear this causes scratching of the surface and, with extended time (multiple formation of new asperities), destruction of surface,

– adhesive bonds between contacting elements of the surface, conducive to transportation of metal from one superficial layer to the other which accelerates wear,

– accumulation of hydrogen in the superficial layer of steel and cast iron elements which, depending on service conditions, may accelerate wear even by a factor of 10.

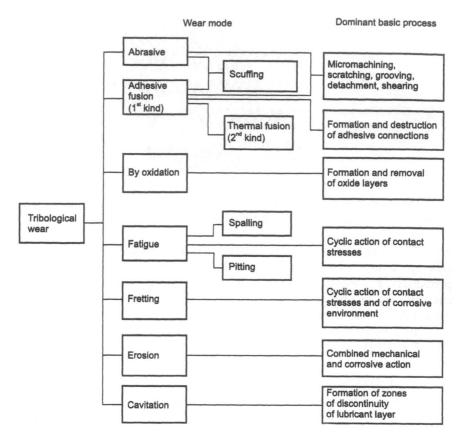

Fig. 5.54 Classification of tribological wear processes.

Many processes of tribological wear may be distinguished, of which only the basic types will be discussed here (Fig. 5.54).

In practice, pure wear processes, i.e., those existing in only one classical form, are not encountered. Usually, combined wear processes take place with one of them being the dominant which decides the amount of wear [21].

National Standards (e.g., Deutch - DIN 50323 [79], Polish - PN-91/M-04301) usually distinguish ten types of wear, covering both **elementary processes** (Table 5.3, Table 5.4) of destructive changes to the surface caused by friction in microzones of solid surfaces, and **technical processes**, observed in the work of machine parts.

The most frequently encountered type of wear is **abrasive wear**, which is responsible for 80 to 90% of all tribological wear. This is a process associated with bulk properties of the material and for that reason it is the best known, purely mechanical wear process. It involves the separation of small particles of the material of the superficial layer in conditions of friction, usually sliding, caused by the presence in the rubbing zones of elements which fulfill the role of an abrasive, harder than the material of the rub-

Table 5.3
Elementary wear processes and the accompanying types of rubbing elements (From *Zwierzycki, W.* [20]. With permission.)

Type of interaction		Number of cycle N, leading to destruction	Name of type of wear
Character of interaction	Schematic diagram of contact and conditions		
mechanical	$(l/R < 5.4(c.R_e/E)^2$ $(l/R < 0.01$ steels, $(l/R < 0.0001$ for non-ferrous metals)	$N \to \infty$	elastic
	$l/R > 5.4(c.R_e/E)^2$ $l/R < 0.1$ dry friction, $l/R < 0.2 \div 0.5$ lubrication	$1 < N < \infty$	plastic, with flow of material around asperities
	$l/R > 0.5 (1-2t/Re)$ $l/R > 0.1$ dry friction, $l/R < 0.2 \div 0.5$ lubrication	$N \to 1$	micromachining
molecular	$t/R_e > 0.5$	$N \to \infty$	adhesive (without material transportation): fusion of boundary layers adsorbed layers
	$t/R_e < 0.5$	$N \to 1$	cohesive (with transportation material): fusion of same material

Table 5.4
Elementary processes of change of state and destruction of metals
by friction and wear (From *Janecki, J., and Hebda, M.* [12]. With permission.)

Changes					
of state				of structure	of chemical composition
material loss	displacement of material	discontinuities in material	buildups		
micromachining, detachment of asperities, brittle cleavage, flaking	grooving, polishing, indentation	surface scratches, deep cracks	smearing of antibody, oxide films, deposits	cold work, orientation of structure, phase transformations	new structural components at surface, oxides

bing surface. During the relative motion of the rubbing surfaces there occurs the driving in and displacement of particles of the abrasive into rubbing surfaces, causing micromachining, scratching, grooving, shearing and detachment of asperities, coupled with removal of the worn material, similarly to the process of abrasive treatment of metals, e.g. grinding. When the rubbing surfaces are in direct contact with the environment, the role of abrasive is played by dust, pollen, tiny fragments, oxides, etc. Also, asperity peaks of the rubbing surfaces may be the cause of abrasive wear [21]. Fig. 5.55 shows models of elementary processes of abrasive wear.

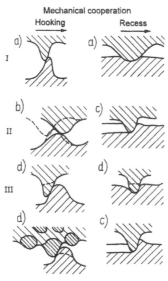

Fig. 5.55 Models of elementary abrasive wear processes: *I* - maintaining an untouched state of rubbing elements in contact (with elastic deformation); *II* - with shape alteration of rubbing elements (plastic deformation); *III* - with detachment of fragments of one of the elements in contact (by shearing off or extraction); a) forcing in; b) kneading; c) rifling; d) shearing. (From *Łuszczak, A., et al.* [83]. With permission.)

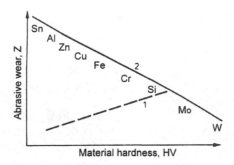

Fig. 5.56 Dependence of abrasive wear on hardness: *1* - for some materials; *2* - for the majority of materials.

The process of abrasive wear is dominant in conditions of dry friction. During fluid friction abrasive wear occurs only when the lubricant contains abrasive particles (usually these are products of wear and contaminants).

Abrasive wear depends on the type, structure and properties of mating materials. It is usually assumed that with a rise in hardness, there is a rise in the abrasive wear resistance of metals and alloys (Fig. 5.56). There are, however, exclusions from that rule:

– in some instances the softer metal wears less than the harder metal. This happens when hard abrasive particles embed themselves in the soft metal in a permanent way and act as abrasives with respect to the harder metal;

– if the surface is excessively hard, it may also be very brittle. Consequently, the material may crack around points of contact with abrasive grains and relatively large portions of the material may become detached from the surface. In such cases, wear and damage of the surface may be very significant.

In sliding connections of machines the main source of abrasive wear is constituted by hard particles of carbide or silicide inclusions at the steel surface, hardened wear products, very hard oxides, etc. [9].

During the process of sliding wear, if the relative velocities are small and loading pressures high, **adhesive wear** may occur, due to the creation of adhesive connections (so-called cold welds) at sites of real surface contact and their subsequent tearing off in motion. If this fusion is intensive, the material often exhibits greater strength in the fusion zone than the material of one of the rubbing elements. In such cases particles of the weaker material are torn out, leaving craters (Fig. 5.57).

A condition for the occurrence of fusion is the close proximity of mating surfaces, such that the distance between them is less than the range of action of intermolecular forces. A further condition is the absence of adsorbed or oxide layers with bonds of a non-metallic character and, for that reason, not exhibiting tendencies to create adhesive joints (so-called fusion) [21].

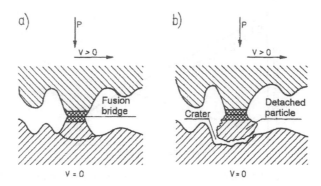

Fig. 5.57 Diagram explaining the mechanism of adhesive wear: a) adhesive fusion; b) deep extraction of adhesion, fused particle.

Adhesive wear occurs predominantly at peaks of asperities where contact of two metallic surfaces takes place. Its intensity is proportional to the affinity between the contacting metals. Steels exhibit both a tendency to fuse and to oxidize. An increase in lattice energy, attained by a rise in temperature, cold work and elastic deformation, causes an increased tendency to fuse. The introduction of foreign atoms to the metal lattice impedes fusion. Alloying elements which inhibit oxidation of steel also, as a rule, impede fusion [21]. Significant tendency to fuse with steel is exhibited by titanium, chromium, nickel, molybdenum, copper, aluminum, zinc and magnesium, while for tin, lead, bismuth and antimony this tendency is meager. Copper alloys containing tin, antimony or lead exhibit a lesser tendency to fuse than pure copper.

Adhesive wear occurs as the result of tearing of fused joints from one of the materials and the transportation of particles of that material in the form of buildups onto the rubbing surface of the other metal. The hardness of such buildups is usually harder than that of the host material and, besides, their shapes are sharp and irregular. The superficial layer is subjected to deep reaching plastic deformations; distinct deformations of grains are observed and even a loss of cohesion between them. The superficial layer becomes loose and its hardness decreases [12].

If the strength of the adhesive joint is less than that of materials comprising it, shear occurs of the joint. In such a case, both wear, as well as surface lesions, may be small. This takes place mostly when the wear process occurs within the superficial oxide layer zone [81].

If the adhesive joint has higher strength than that of materials forming it, shearing will take place in the weaker of the two materials. A consequence of this is severe wear and intensification of the process of surface damage. The force of friction, on the other hand, may rise only insignificantly. A surface with higher hardness is progressively covered by a thin layer of carried-over softer material. After some time, friction takes place between the same materials. Carrying over of material from the harder surface to the softer surface is also possible [9].

In the event of friction between same materials it is probable that adhesive joints will have greater strength than the host materials. The cause of this phenomenon is strain hardening of the joints during friction. In such an event shearing of the joints occurs, as a rule, deeper in the host material and is accompanied by severe wear of the surface. The wear may, however, turn out to be mild because carryover of material occurs in both directions [9].

A modification of adhesive wear (also called fusion of the I^{st} kind) is **thermal wear**, sometimes termed fusion of the IInd kind. It occurs with high relative velocities and high loading forces of mating surfaces in sliding friction. It is caused by insufficient lubrication or the breaking of the lubricant layer, heating of the rubbing surfaces and a change in their properties (a rise of ductility causing smearing of material).

With thermal wear welds are formed (thermo-adhesive macrofusions), followed by their tearing in the zones of contact. The rise in intensity of the phenomena described here may cause an avalanche process [21]. This may lead to serious mechanical damage of the mating surfaces and also to galling [8, 86, 87].

Mixed abrasive-adhesive wear (*scuffing*) occurs as the result of the joint influence of abrasive and adhesive wear [79]. It is a form of extremely rapid wear, caused by breaking the lubricating film under high load. Due to mutual interaction of asperities of both rubbing surfaces, it brings about the fusion and detachment of these asperities in microzones of contact [8].

Wear by oxidation[1] consists of adsorption of oxygen in zones of friction and its diffusion into plastically and elastically deformed layers of metal, and their mechanical removal due to abrasive wear and spalling of the metal surface. Wear by oxidation, sometimes called normal wear, is viewed as the only admissible process of wear.

During the oxidation process, several layers of oxides of different thickness are formed in a predetermined sequence, e.g., for unalloyed steels they are Fe_2O_3, Fe_3O_4, FeO, core. The hardness of oxidized zones on steel is higher than that of the core material or of the hardened core. In the case of oxidized aluminum, a severalfold rise in hardness occurs. This is one reason given in explaining the relatively mild wear by oxidation (Fig. 5.58) [21]. Wear by oxidation occurs when the rate of formation of the oxide layer is greater than the rate of its wear from the surface. It takes place in sliding and rolling friction. In conditions of rolling friction, wear by oxidation always accompanies fatigue wear. In sliding friction, wear by oxidation is usually dominant when conditions of boundary friction prevail and, in some cases, in conditions of dry friction. Intensive wear by oxidation often takes place in condition of fluid friction. It is most

---

[1] According to new opinions, wear by oxidation, as a hydrogen wear and another type of wear with the participation of chemical reactions, is assumed by the general term *tribochemical wear*.

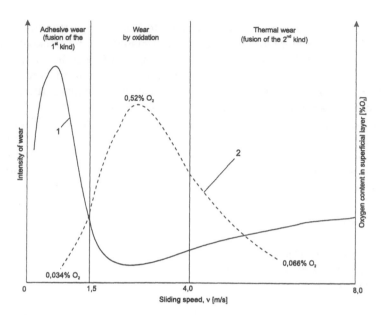

Fig. 5.58 Dependence of type and intensity of wear of plain carbon steels (1) on sliding speed with varying oxygen content in the superficial layer (2).

severe for rubbing pairs in which materials have different mechanical properties. This is the most characteristic type of wear of multi-component alloys, i.e. materials with a high yield strength and high hardness [9].

Fatigue wear takes place when adhesion at the surface of mutual contact of the rubbing materials does not play a major role in the friction process, while a long-term cycling interaction of (elastic-plastic) contact stresses occurs in the superficial layer. Wear is caused by fatigue of the superficial layer, often termed **surface** or **contact fatigue**. It occurs in cases of mating of two elastic bodies with a straight and a curved surface of friction or with curved surfaces, having point or linear contact where rolling friction or rolling with slip takes place [9, 12, 21].

Very often fatigue wear occurs in well-lubricated sliding pairs, as the result of multiple repetitions of collisions between surface asperities. Surface asperities may sometimes not form direct metal-metal contact but only be subjected to strong elastic deformations by the lubricant layer. In consequence, peaks of asperities suffer fatigue and become separated from the surface [9].

The occurrence of fatigue wear is equivalent to scrapping the rolling or sliding pair from service.

The mechanism of initiation of fatigue wear is the following: residual stresses and stresses caused by external loads are superimposed; at sites of stress concentrations, the greatest material effort is located at the so-called *Belayev point*. Maximum shear stress occurs at a depth of 0.39 of length of real contact. From this point it travels in the direction of the surface with the rise of the force of friction which affects the contact zone jointly with normal stress. This may be the root cause of the initiation of microcracks, expanding to macrocracks. The end effect is the detachment from the core of fragments of the superficial layer. Flaws and structural discontinuities of the superficial layer and surface itself may also constitute sources of fatigue cracks [21].

Two types of fatigue wear are distinguished:

– by **spalling** which occurs in conditions of dry friction or insufficient lubrication. It is manifest by local material loss in the form of flakes or scales. The intensity of wear by spalling is high and depends on the depth of plastic deformation of the superficial layer, the value of unit load at the contact site, number and frequency of cycles, as well as dimensions and mechanical properties of the rubbing elements;

– by **pitting** which occurs in conditions of lubricated wear. The lubricant protects the surface from metallic contact, thus from adhesive wear. Besides, it fulfills the role of a shock absorber for contact loads carried over from one surface to the other. In the initial period of friction the presence of the lubricant inhibits the migration of the point of maximum material effort by attenuating unit loads at contact points and causes the retardation of fatigue wear. After the appearance of fatigue microcracks the lubricant plays an unfavorable role (Fig. 5.59). It penetrates the microcrevices and by being forced into them by the mating surface it

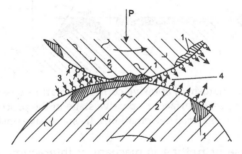

Fig. 5.59 Mechanism of pitting wear in the case of rolling friction: *1* - zones with reduced cohesion (extracted); *2* - microcracks; *3* - compressed particles of lubricant; *4* - stretched particles of lubricant, bound to the surface by forces of adsorption or chemisorption.

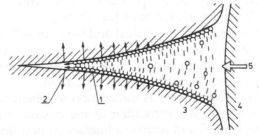

Fig. 5.60 The Rebinder effect - surface-active particles forcing wedge fissures: *1* - polar particles; *2* - pressure forcing wedge fissure; *3, 4* - mating elements; *5* - direction of pressure of element 4.

creates wedges and increases the size of the cracks (Fig. 5.60). When a network of cracks and crevices develops in the superficial layer, single particles partially lose cohesion with the matrix and are torn out by lubricant layers at the surface (adsorbed or chemisorbed by the surface), the latter subjected to periodic compression and stretching. The intensity of pitting wear is smaller than that of spalling wear by a factor of 2.4 [9, 12, 21, 83].

Spalling and, in particular, pitting are typical processes of fatigue wear of ball bearings and oil-lubricated gear transmissions [21].

Abrasive-corrosive wear (*fretting*, fatigue-friction wear) is the phenomenon of destruction of the superficial layer, consisting of the formation of local material losses in elements subjected to the action of vibrations (e.g., in fitted joints) or small slip in forward and reverse motion, due to the cyclic interaction of loads and intensive corrosive action of the environment (Fig. 5.61).

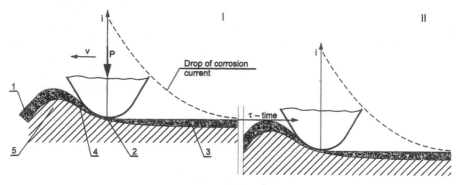

Fig. 5.61 Model of mechanism of abrasive-corrosive wear: *1* - passive layer formed at surface; *2* - freshly exposed surface; *3* - passive stage in the process of regeneration; *4* - site of destruction; *5* - material deformation; I, II - successive position of instrument.

The direct cause of fretting is mechanical interaction. Characteristic of this interaction are strong corrosion effects, accompanying all stages of destruction [46]. Products of abrasive-corrosive wear are usually made up of metal oxides with relatively high hardness, acting as an abrasive medium. Fretting is manifest by the presence of fine brown powder, collecting in the vicinity of the friction joint [21, 83, 88].

It is worth mentioning that polished and very smooth surfaces corrode slower than surfaces which are rough. The reason for that is the more developed area of the rough surface, able to accommodate more moisture which, in turn, accelerates corrosion and oxidation.

The concept of **erosive wear** is understood as phenomena of destruction of the superficial layer, consisting of the creation of local material losses, due to mechanical and corrosive interaction of a flux of particles of solids or liquids with high kinetic energy [79]. This is encountered predominantly in machines involving flows. A flux of particles of gas or liquid with particles of solids suspended in it may cause erosion of some

elements, especially on turbine and compressor blades. From the point of view of type of particles causing erosion, wear may be classified as:
 – erosive, in a stream of solid particles,
 – erosive, in a stream of liquid (hydroerosion),
 – erosive, in a stream of liquid containing solid particles (hydro-abrasive).

A specific type of erosive wear is **cavitational wear** (cavitation erosion) - the phenomenon of material losses due to mechanical interaction of a liquid in which cavitation occurs, i.e., the formation within the flowing liquid of discontinuity zones, filled with a heterogeneous mixture of gas and liquid.

Table 5.5 shows the dependence of basic types of tribological wear on the relative motion of mating elements.

Table 5.5
Elementary processes of change of state and destruction of metals
by friction and wear (From *Senatorski, J.* [80]. With permission.)

Type of relative motion of mating elements 1 and 2	Dominant mechanisms of wear			
	abrasive	by oxidation	adhesive	fatigue (pitting, spalling)
Sliding	●	◑	●	●
Rolling	◑	●	◑	●
Dynamic	●	●	○	○
Oscillation	Wear by fretting = combined effect of all types of friction			

● – Occurs with strong intensity
◑ – Occurs with mild intensity
○ – Practically does not occur

5.8.2.9 Factors affecting tribological wear
The strongest effect on tribological wear is that exhibited by the three-dimensional condition of the rubbing surfaces - both in dry and lubricated friction. Roughness has an especially unfavorable effect. With the presence of major

unevenness, asperity peaks catch one another and cause rapid abrasive, thermal and adhesive wear. The shape of the roughness profile (sharp peaks, big angles of inclination of asperity sides, the character of contact capacity curves of profile, mean roughness spacing) significantly affects tribological wear. A similar effect is that of orientation of unevenness of the rubbing surfaces (Fig. 5.62) [21, 31].

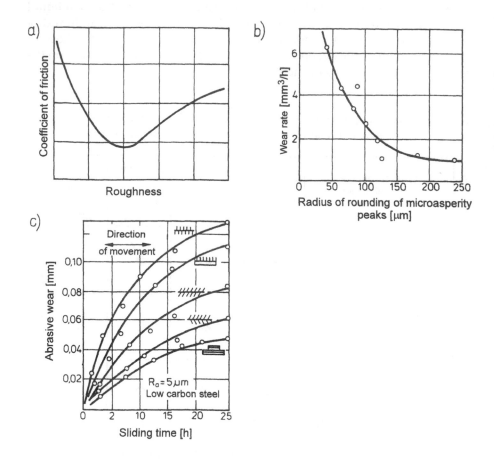

Fig. 5.62 Dependence of tribological parameters on roughness: a) coefficient of friction vs. roughness; b) intensity of wear vs. radius of asperity roundness; c) abrasive wear vs. time of wear test. (From *Zielecki, W.* [31]. With permission.)

Assuming conditions of friction with lubrication, with the same wettability of material by the lubricant and with the same unit loading, rough surfaces will wear more than smooth surfaces. Valleys between asperity peaks form microreservoirs for the lubricant. On the other hand, smooth surfaces do not exhibit good retention of lubricants which may be pushed out, leading to metallic contact, fusion and adhesive or thermal wear in large areas, and finally to rapid seizure (Fig. 5.63) [21].

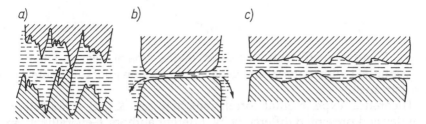

Fig. 5.63 Effect of roughness of rubbing surfaces on lubricating conditions: a) excessively high roughness; b) excessively low roughness; c) optimum roughness. (From Łunarski, J. [21]. With permission.)

From the point of view of wear, each pair is characterized by time of wearing-in, time of normal wear, also called service life, and time of accelerated wear (Fig. 5.64a).

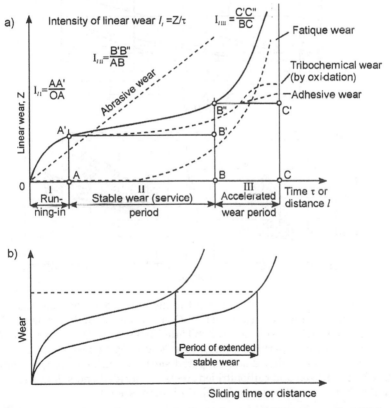

Fig. 5.64 Typical corelation between tribological wear and time (so-called *W. Lorenc's curve*): a) kinds of wear; b) extended service life due to reduction of running-in time.

The rate of wear is usually characterized by **intensity of linear wear** I_l, experimentally determined from plots of tribological wear by a method shown in Fig. 5.64a, or theoretically from the expression:

$$I_l = Kp^x \qquad\qquad (5.42)$$

where: K and x - coefficients, dependent on the type of wear and on properties connected with material and geometry of friction joints; p - contact pressure (stresses).

Theoretical-experimental correlations between K and x are relatively complex and presented differently by different authors. Generally, it can be stated that the more the technologically obtained roughness is closer to optimum (service roughness) for the given application, the milder is resultant wear, shorter wearing-in time and longer service life (Fig. 5.64b). Optimum surface roughness for the majority of joints is small. Usually the arithmetical average deviation of profile should be $R_a = 0.1$ to 1.25 μm [21].

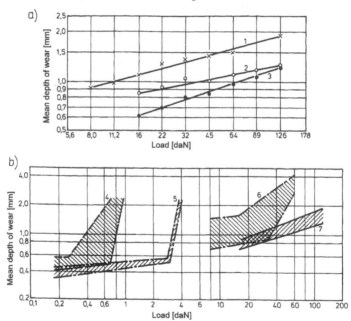

Fig. 5.65 Resistance to galling of different materials: a) pairs of materials: *1* - PA6 aluminum alloy - normalized 1045 steel; *2* - ZL300 cast iron - sulfonitrided 1045 steel; *3* - sulfonitrided ZL300 cast iron - sulfonitrided 1045 steel; (From *Senatorski, J.* [92]. With permission.); b) superficial layers on steels rubbing against hardened and tempered 1045 steel: *4* - hardened and tempered steels; *5* - pickled steels; *6* - induction hardened, carburized and carbonitrided steels; *7* - nitrided and sulfonitrided steels. (From *Rogalski, Z., and Senatorski, J.* [93]. With permission.)

Hardness significantly affects tribological wear - for most metals wear resistance rises linearly with hardness (see Fig. 5.56). Particularly good tribological wear resistance is exhibited by superficial layers obtained by heat and thermo-chemical treatment (Figs. 5.65 and 5.66), containing hard, finely dispersed carbides, nitrides and borides [20, 31, 83, 89-93].

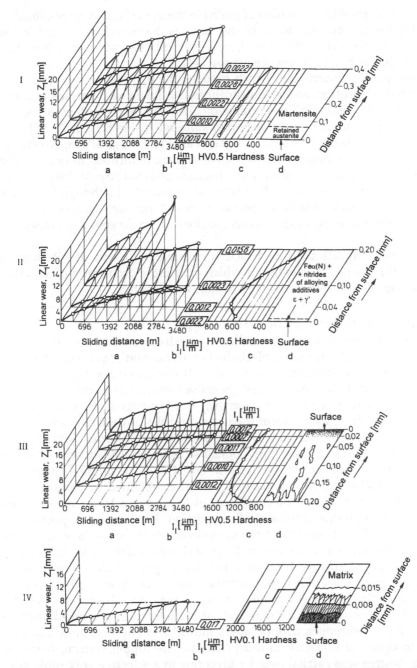

Fig. 5.66 Tribological characteristics of superficial layers obtained by different thermo-chemical treatment, correlated with their microstructure: I - carburized; II - nitrided; III - borided; IV - chromized; a - linear wear vs. sliding distance and depth under surface at which test begins; b - intensity of wear vs. distance from surface; c - hardness distribution in layer; d - layer microstructure. (From *Senatorski, J.* [80]. With permission.)

Crystallographic orientation of grains in the superficial layer obtained in a technological process should be the same as that which is present in the friction process. When that condition is met, tribological wear will be less intense than with a different orientation [94].

Cold plastic deformation which hardens the material also gives rise to enhanced adsorption and chemisorption activity by the surface and the increase in its susceptibility to diffusion. This favors the enhanced occurrence of oxidation and the formation of adhesive fusion [21, 31, 75, 95].

Residual stresses in the superficial layer affect tribological properties only slightly [31]. Since during friction intensive plastic deformations occur in the superficial layer, they cause a rapid relaxation of initial stresses [94]. During the wearing-in process, they affect tribological wear [96] although this effect is not unequivocal; usually compressive stresses increase resistance to abrasive wear, but in some cases a favorable effect of tensile stresses is observed [21].

The most significant set of factors causing reduction of tribological wear is the creation of conditions which enable the occurrence of selective carry-over of particles, conducive to non-wear friction.

5.8.2.10 Non-wear friction (selective carryover)

The phenomenon of selective carryover was discovered in the 1950s by D. N. Garkunov [85]. This takes place during lubrication of rubbing elements between which fluid friction does not occur. It was later termed the Garkunov effect or non-wear friction effect [97] because it characterizes minimum wear of rubbing elements and minimum energy lost to overcome friction forces of resistance. In its classical form it can be described in the following way: during friction of bronze on steel in the presence of glycerine, with a mean velocity lower than 6 m/s, high contact stresses (unit load over 40 MPa) and at temperatures of 40 to 60°C, the intensity of wear could be lowered as much as 1000 times in comparison with that obtained in conditions of boundary friction, while the coefficient of friction was lowered to a value comparable with that obtained in fluid friction. This was made possible as the result of the formation on both surfaces of a thin copper layer of quasi-crystalline structure, thickness of 1 μm and the number of vacancies two orders of magnitude greater than in normal copper, which to some degree exhibited properties of a liquid crystal. Due to lower strength of copper layers than that of the substrate material, it was only the copper that deformed plastically while the materials of the substrates deformed only insignificantly. Moreover, copper layers moved relative to each other, similarly to layers of lubricant in fluid friction. The copper coming from the bronze source was carried over by glycerine in a selective way (only copper) and deposited on the bronze and steel [20]. This may be explained in the following way: bronze is first mechanically rubbed onto the steel, while simultaneously, bronze components form microcells Pb-Cu or Zn-Cu. Removal of the less noble metal (Pb, Zn) from both rubbing surfaces continues until on both surfaces, the only material to remain in contact will be pure, defective copper. Moreover, there occur transformations connected with sorption of organic compounds of copper, formed in the glycerine layer and polymer layers [101].

The effect of selective carryover is based on phenomena stemming from the consequence of the electromotive series and is classified as one belonging to the so-called *self-organization of matter*, consisting of the lowering of entropy of the system by spontaneous restructuring of the system [6]. Self-organization phenomena are characteristic of living organisms. In nature, the ideal (sliding) friction node is the human or animal joint - ideal because it lasts a lifetime and efficient, practically not subject to wear or seizure, almost non-frictional (coefficient of friction of the order of 0.001 to 0.03 which is less than that for the best hydrodynamic sliding bushings and ball bearings) [98].

In the classic tribological friction wear node, whose work is based on the principle of minimization of **destruction** of the rubbing surfaces, there occurs a forced, direct contact between the two materials, resulting in their mutual tendency to match, fit, wear-in and even fuse. A roughness, different from that formed by technology, is developed, and a real contact surface is created in which one material may be aggressive with respect to the other. On the other hand, non-wear friction bears a **constructive** character and is based on the principle of exchange of energy and matter between the friction pair and the environment, as well on the interaction of complexes and ions of copper in the formation of a copper layer which protects the rubbing surfaces from wear. This layer has been called (by A.A. Polyakov) **servovital layer** (from the Latin word *servovitae* - to save life) or **low friction layer** [99].The following analogies can be made between the human joint (e.g., forearm) and the friction node (e.g., bushing - journal): one rigid and hard element (forearm bone - journal) transmits the load to the other element (shoulder bone - bushing) by means of soft, constantly revitalized structures (cartilage - copper layer), separated by a liquid phase (organic substance - oil) [98].

In friction nodes matter is formed in the process of friction due to the action of a flux of energy; friction does not destroy the bushing but creates it. The microfriction layer is not subjected to fatigue wear during deformation. Covering surface asperities it absorbs the whole load, with the result that the rubbing surfaces (e.g., steel and bronze) do not participate in the friction process. Since one soft material (copper) interacts with another soft material (in this case, the same - copper), the load is distributed evenly on the surface, resulting in small unit pressures [99].

The formation of a layer of pure copper is made possible by the reducing (with respect to copper oxides) properties of glycerine and the occurrence of the Rebinder effect (conducive to the plasticization of the superficial layer), the carrying over of metal due to fusion, selective electrochemical dissolution of alloying elements from the alloys (so-called Kirkendal effect, in this case resulting in the formation of the copper layer), and heterogeneous catalysis at surfaces activated by friction [20, 100].

The mechanism of selective carrying over has not found a full explanation. Different concepts and different models of the process exist. One of the most interesting is that put forward by A.A. Polyakov [98], in which the friction process is localized within a thin metallic layer, capable of dissipating[1] energy and matter. In this model dissipation is presented as a process of mutual adsorption, oriented with diffusion fluxes in two opposite directions: one moving into the layer (flux of vacancies) and the other out of it (flux of dislocations and atoms). Vacancies are formed as the result of the action of the lubricant which causes the release of metal atoms from the surface of the low friction layer, while dislocations are formed as the result of layer deformation [101].

Ensuring long life without wear of the friction pair boils down to the creation of such conditions in which a low friction metallic layer (simplest is copper) is formed between the rubbing surfaces, separating one from the other [101].

The friction pair need not necessarily be steel and bronze. The phenomenon of selective carrying over occurs also during friction of steel on: steel, cast iron, alloys of precious metals (silver, gold and platinum) and colored metals (brasses, bronzes), sintered materials, metal-polymer composites, glass; of cast iron on cast iron, etc. [98]. It occurs with the application, as lubricants, of mineral and synthetic oils, plastics, sea and fresh water, hydraulic fluid, petroleum and its derivatives, mixtures of oil and freon, as well as different acids and alkalis [20].

Low friction metallic layer is formed from:

– metals which are components of one or both rubbing surfaces (independent elements or inserts to elements of the rubbing pair, e.g., in the form of cores or rings),

– electrolytic coatings or thin layers (2 to 4 μm) rubbed onto the surface of friction,

– powder (metal or metal oxide) or chemical compound introduced into the lubricant (including metal-organic compounds),

– powder (metal or metal oxide) introduced as filler (approximately 10%) to the synthetic material (so-called metal-polymers).

It should be emphasized that in many cases it is possible to create such conditions of selective transportation of e.g., copper, in which this effect also inhibits the formation and the detrimental action of hydrogen [20].

The effect of selective carrying over is utilized practically, although not fully investigated. It has been estimated that as the result of an increase of maximum feasible contact capacity of the rubbing pair by 1.5 to 2 times, it would be possible to reduce the mass of machines by 15 to 20% and even to achieve a severalfold increase in the service life of friction nodes, as well as to reduce the energy necessary to drive the machines [20].

---

[1] Dissipation, dispersion - the process of irreversible creation of e.g., thermal energy at the cost of other types of energy, accompanied by the production of entropy (heat absorbed by the system at thermodynamic temperature).

5.8.2.11 Limiting tribological wear

Tribological wear is usually defined by the intensity of wear (proportion of wear to time) which - to varying degrees for the various types of wear - depends chiefly on allowable velocities of relative motion of the rubbing surfaces (see Fig. 5.64) and allowable (i.e. not causing accelerated wear or seizure) unit loads.

Generally it can be stated that such manufacturing techniques should be used which enable work in conditions of non-wear friction and in cases of friction with wear - reduce its intensity, allow the sustaining of high unit loads, broaden the range of wear by oxidation and prohibit wear by fusion of the Ist and IInd kind [21], as well as by such shaping of the surface which makes them suitable to cooperate with the lubricant. This means that surfaces should be characterized by a high contact capacity with recesses fulfilling the role of lubricant reservoirs (e.g., smoothed and oscillation superfinished, burnished by oscillation or eccentric motion and shot peened) [31, 102, 103].

General directions in which efforts are made to minimize wear by friction are the following [21, 38]:

– selection of appropriate properties of superficial layers to conditions of tribological wear (load, type of work, environment);

– such constitution of superficial layers that the effective thickness of the hardened layer be greater than the depth at which maximum contact stresses occur (location of the Belayev point);

– appropriate matching of the friction pair;

– optimization of design of friction nodes in order to ensure continuous fluid or boundary lubrication;

– perfection of the oils and lubricants used;

– ensuring that superficial layers feature good cohesion, high creep resistance, heat resistance, good corrosion resistance, high compressive stresses with low gradient and other specific properties, appropriate for particular types of friction, as below.

abrasive:

– increasing hardness of the superficial layer above that of the abrasive grains (powder, dust, oxides) by surface hardening, thermo-chemical treatment (carburizing, nitriding, boriding), pad welding, thermal spray and electroplating;

– lubrication which separates the rubbing surfaces and reduces the coefficient of friction;

adhesive:

– increasing hardness and reducing ductility of the superficial layer;

– effective lubrication and reduction of unit loads;

– application of fusion-resisting superficial layers (sulfurizing, sulfonitriding, oxidizing) and electroplated coatings (copper, tin, cobalt), and, very seldom, enameled coatings.

As regards thermal wear, this is achieved by the application of lubricants which are resistant to elevated temperatures, e.g., molybdenum disulfide, deposition of metallic coatings and reduction of service parameters (lowering of friction velocity and unit loads and the application of cooling);

by oxidation:
– change of plastic properties of the superficial layer, affecting the intensity of oxide formation;
– change of chemical composition of oxides due to adsorption and diffusion;
by fatigue:
– raising of yield strength of the superficial layer by heat and thermochemical treatment;
– application of appropriate lubrication (not too much and not too little) as well as appropriate types of lubricants (e.g., solid in place of liquid).

5.8.3 Anti-corrosion properties

It would be desirable for superficial layers to be resistant to the corrosive effect of the environment under different conditions. In the manufacturing technology, the main aim of which is shaping of the object, anti-corrosion properties of the superficial layer play a very important role, particularly when the surface of the product is not to be coated. On the other hand, in the manufacturing technology whose main aim is to give the superficial layer appropriate service properties, predominantly tribological and fatigue, anti-corrosion properties play a minor role. In this case corrosion affects these properties - usually it enhances tribological wear and causes a deterioration of fatigue properties (Fig. 5.67; 5.68).

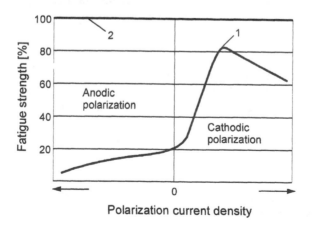

Fig. 5.67 Fatigue strength of 1045 steel in 3% solution of NaCl vs. type of polarizing current and its density (*1*) and a comparison of fatigue strength of the same steel grade in air (*2*).

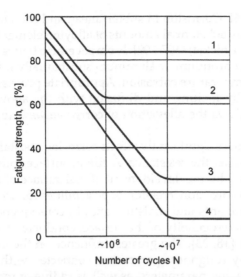

Fig. 5.68 Effect of type of environment on fatigue strength of 1045 steel: *1* - in air; *2* - in oil with adsorption-active additives; *3* - in hydrogen-liberating liquid; *4* - in strong corrosive medium (aqueous solution of sodium chloride).

Taking into account the large number of corrosion types, anti-corrosion properties may be different in different conditions. Types of corrosion will be discussed in greater detail in the chapter on coatings (Chapter 6), because the main task of coatings is appropriate protection against corrosion. In this place it will suffice to note that superficial layers should be resistant to chemical and electrochemical corrosion, including corrosion fatigue (caused by the simultaneous effect of a corrosive environment and variable stresses) and stress corrosion (caused by simultaneous action of a corrosive environment and static stresses), as well as intercrystalline corrosion (aided by the action of static and variable stresses).

The causes of corrosive wear are chemical and electrochemical reactions of the superficial layer with the surrounding environment, usually active from the point of view of corrosion (e.g., water, electrolytes, saline solutions), absorbed into the volume of the superficial layer (e.g., hydrogen or other elements forming solid solutions with metals) or with a non-active environment (e.g., surfactant substances, whose effect on the mechanical properties of metals or alloys is manifest mainly by adsorption and chemisorption [20]. The extent of corrosive wear depends on the properties of the corrosion-active environment, its concentration and temperature, as well as on the resistance of the material subjected to corrosion, which, in turn, depends mainly on its chemical composition and condition of the superficial layer [31].

The chemical composition of the superficial layer has a decisive influence on the extent of corrosive wear of steel; alloying additions in the form of sulfur and phosphorus enhance, while chromium, nickel, manganese, copper, molybdenum and aluminum retard the rate of atmospheric corrosion [104-106]. Corrosion in subterranean and sea water, in soil and in acids is retarded by substantial amounts of alloying elements (over 13% chromium and 5 to 12% nickel) [107-109]. From several to between ten and twenty percent of nickel, chromium or aluminum substantially raises the resistance of steel to high temperature corrosion. Appropriate processes of alloying the superficial layer by melting and diffusion with the above-mentioned and other elements allows the alleviation of corrosive wear (e.g., anti-corrosion nitriding) [110].

Surface roughness significantly affects corrosion resistance. The lower the roughness, the higher the resistance. A rise in surface roughness intensifies corrosion processes by the development of real surface of contact of the element subjected to corrosion, creation of possibilities for accumulation on the surface of contaminants and moisture and, in consequence, chemical and electrochemical heterogeneity of the surface, conducive to the formation of corrosion sources [18, 23]. The greatest influence on the extent of corrosive wear is exerted by roughness parameters connected with asperity height, some asperity spacing parameters, as well as radius of recess in the roughness profile [31].

The structure of the superficial layer also affects the extent of corrosive wear. Greatest resistance to the action of diluted acids is exhibited by martensite, lesser resistance by ferrite and pearlite [106]. Implanted layers are usually characterized by good anti-corrosion properties. Generally, treatment by an ion, photon or electron beam is conducive to a rise of corrosion resistance, especially so when it causes the formation of amorphous structures [31, 109, 111]; in other cases it may even exhibit the opposite effect [31].

Deformation, being the result of cold work, disturbs the crystalline structure of alloys and raises the value of surface energy of the superficial layer. The said disturbance is by way of a rise in the number of defects in the lattice, and enhancement of carbon and nitrogen atom mobility near lattice defects. Both single factors and their joint interaction favor the susceptibility of alloys to corrosion [69, 111]. Corrosion resistance depends on the degree of cold work. With the so-called critical cold work (for steels this is: 5 to 10%), corrosion resistance deteriorates very significantly, with lesser amounts of cold work - it deteriorates too but to a lesser degree [21].

No effect of hardness (or of differences in hardening) on corrosion resistance has been determined [21].

Stresses in the superficial layer, including residual stresses, affect corrosion resistance. Compressive residual stresses do not exhibit a detrimental effect and even slightly improve corrosion resistance of metal alloys [21]. Tensile residual stresses cause a deterioration of corrosion resistance, as do tensile stresses from external loads [31]. Generally it is thought that superficial layers with residual stresses are more susceptible to electrochemical corrosion in the presence of a corrosive environment than layers which are free of stresses. It is assumed that a difference in the state of stress causes the creation of differences of potential in the metal [21]. Stress corrosion causes the metal to crack [31, 106].

5.8.4 Decorative properties

The most significant decorative properties are color, luster and resistance to aging.

In the case of superficial layers, decorative properties play a secondary or even a less important role. Sometimes, color and luster allow an initial organoleptic evaluation of the condition of the superficial layer, a determination of the type of treatment to which the superficial layer was subjected and comparative tests of layers.

Decorative properties are of significantly greater importance to coatings. These will be discussed in greater detail in Chapter 6.

Decorative properties of superficial layers change with time of service under the influence of the environment. To a great extent they also depend on the type of lighting used to observe the superficial layer.

5.9 The significance of the superficial layer

The superficial layer plays a predominantly technical role. It is only very rarely that decorative values are desired of it (e.g. surgical and dental instruments). It has, however, high requirements placed on it regarding the enhancement of service life of machine components and tools designated for work in conditions of friction and fatigue loads, often with corrosive action of the environment.

Depending on working conditions, the technologically developed superficial layer allows:

– improvement of working conditions, reliability and service life, both from the point of view of tribology, as well as fatigue, less often improvement of corrosion resistance and decorative value of machine components and tools;

– lowering of mass, down time; frequency of replacement and of energy required for manufacturing and utilization of machine components and tools.

Appropriately used technological superficial layers allow, due to the formation of surfaces with optimum service properties, significant extension of service life of tools, machine components and appliances, especially with regard to tribological resistance.

References

1. **Beilby, G.**: *Aggregation and flow of solids*. The MacMillan Co., London 1921.
2. **Nowicki, B., Stefko, A., and Szulc, S.**: *Surface treatment. Giving machine components their service properties* (in Polish). PWN, Warsaw 1970.
3. **Kolman, R.**: *Mechanical hardening of machine component surfaces* (in Polish). WNT, Warsaw 1965.

4. **Szulc, L.:** *Structure and physico-mechanical properties of treated metal surfaces* (in Polish). Special edition of Warsaw Technical University, Warsaw, September 1965.
5. PN-73/M-04250 Polish Standard Specification. The superficial layer. Terms and definitions.
6. **Leszek, W.:** *Once more and somewhat differently about tribology* (in Polish). Published by Scientific Center for Utilization of Capital Equipment (in Polish), Radom, Poland 1994.
7. **Kruszyński, B.:** *Basic of the surface layer.* Science Periodicals of Łódź Technical University, Mechanics Series, Vol. 79, Łódź, 1990.
8. **Hebda, M., and Wachal, A.:** *Tribology* (in Polish). WNT, Warsaw 1980.
9. **Burakowski, T., Roliński, E., and Wierzchoń, T.:** *Metal surface engineering* (in Polish). Warsaw University of Technology Publications, Warsaw 1992.
10. **Okoniewski, S.:** *Fundamentals of mechanical technology* (Edition III) (in Polish). WNT, Warsaw 1971.
11. Superficial layer terminology (in Polish). Institute of Metal Machining. *Instruction Manuals* No. 78, 1968.
12. **Janecki, J., and Hebda, M.:** *Friction, lubrication and wear of machine components* (in Polish). WNT, Warsaw, 1969.
13. **Szulc, S., and Stefko, A.:** *Surface treatment of machine components.* Physical fundamentals and effect on service properties (in Polish). WNT, Warsaw 1976.
14. **Rowiński, E., and Lagiewka, E.:** Utilization of Auger electron spectroscopy to investigate surfaces of NiTi alloys (in Polish). Proc. *II All-Poland Conference on Surface Treatment,* 13-15 October, 1993, Kule, pp. 323-327.
15. **Nowicki, B.:** *Geometrical structure. Roughness and waviness of surface* (in Polish). WNT, Warsaw 1991.
16. **Solski, P., and Ziemba, S.:** *Problems of dry friction* (in Polish). PWN, Warsaw 1965.
17. **Cotrell, A.:** *Dislocation and plastic flow in crystals.* Oxford University Press, London 1956.
18. **Oiding, J.A.:** *Theory of dislocations in metals and its application* (translation from English), PWN, Warsaw, 1961.
19. **Leszek, W.:** Studies of problems of non-homogeneity of chemical composition of materials (in Polish). Poznań Technical University Periodicals, Poznań, 1973.
20. Joint report edited by **W. Zwierzycki:** *Selected problems of wear of materials in sliding nodes of machine components* (in Polish). PWN, Warsaw - Poznań 1990.
21. Joint report edited by **J. Łunarski:** *Surface treatment* (in Polish). Rzeszow Technical University, Rzeszów 1989.
22. **Burakowski, T.:** *Emissivity of resistance heating alloys* (in Polish). IMP, Warsaw 1976.
23. **Sala, A.:** *Emissivity of metals and alloys, as a function of their surface condition* (in Polish). IMP, Warsaw 1973.
24. **Burakowski, T.; Giziński, J., and Sala, A.:** *Infrared radiators* (in Polish). WNT, Warsaw 1970; (in Russian). Publ. Energia, Leningrad 1976.
25. **Sala, A.:** *Exchange of heat through radiation* (in Polish). WNT, Warsaw 1982.
26. **Wesołowski, K.:** *Metallurgy* (in Polish). PWT, Warsaw 1959.
27. **Dobrzański, L.A.:** *Metallurgy and heat treatment of metal alloys* (in Polish). Silesian Technical University Publications, Gliwice 1993.
28. **Błażewski, S., and Mikoszewski, J.:** *Metal hardness testing* (in Polish). WNT, Warsaw 1981
29. **Kortmann, W.:** Vergleichende Betrachtungen der gebräuchlisten Oberflächenbehandlungsvefahren. *Fachberichte Hüttenpraxis Metallverarbeitung.* Vol.24, No. 9, 1986, pp. 734-748.
30. **Frey, K., and Kienel, G.:** *Dünnschicht Technologie.* VDI Verlag, Düsseldorf 1987.
31. **Zielecki, W.:** Modification of technological and service properties of steel by the laser and electron beam. Ph.D. thesis. (in Polish). Rzeszów Technical University, Rzeszów 1993.
32. **Zandecki, R.:** Analysis of the effect of ion nitriding and nitrogen ion implantation on the tribological properties of steel grades 33H3MF and 36H3M. Ph.D. thesis (in Polish). Military Technical Academy, Warsaw 1993.
33. **Solski, P.:** *Metal wear by friction* (in Polish). WNT, Warsaw 1968.
34. **Górecka, R., and Polański, Z.:** *Metrology of the superficial layer* (in Polish). WNT, Warsaw 1983.
35. **Birger, I.A.:** *Residual stresses* (in Russian). Publ. Masgiz, Moscow 1963.
36. **Tietz, H.D.:** Entstehung und Einleitung von Eigenspannungen in Wertkstoffen. *Neue Hütte,* Vol. 25, No. 10, 1980, pp.375-377.

37. **Tubielewicz, K.:** *Analysis of stresses formed in the superficial layer during the burnishing process* (in Polish). Częstochowa Technical University Periodicals, Częstochowa 1993.
38. **Kula, J.:** *Sorption of hydrogen in the nitrided layer and its effect on friction and wear* (in Polish). Dissertation. Łódź Technical University Publications, Łódź 1994.
39. **Rose, A.:** Eigenspannungen als Ergebnis von Wärmebehandlung und Umwandlungsverhalten. *Härterei-Technische Mitteilungen*, Vol.21, No. 1, 1966, pp.1-6.
40. Joint report: *Engineering Manual: Heat treatment of ferrous alloys* (in Polish). WNT, Warsaw 1977.
41. **Janowski, S.:** Changes in residual stress state in elements with hardened superficial layers, as a result of subsequent heat treatment (in Polish). *Metaloznawstwo, Obróbka Cieplna, Inżynieria Powierzchni (Metallurgy, Heat Treatment, Surface Engineering)*, No. 99-100, 1989, pp. 54-58.
42. **Janowski, S.:** Effect of residual stresses on mechanical properties of structural steels (in Polish). *Metaloznawstwo, Obróbka Cieplna, Inżynieria Powierzchni (Metallurgy, Heat Treatment, Surface Engineering)*, No. 79, 1986, pp. 6-12.
43. **Svecev, V.D.:** Measurement of residual stresses in investigations carried out on the SMC-2 friction machine (in Russian). *Zavodskaya Laboratoria*, No. 4,1977, pp. 500-502.
44. **Roliński, E.:** Phenomena occurring at the interface between solid and gas phase in the ion nitriding process. Ph.D. thesis (in Polish). Warsaw Technical University, Warsaw 1978.
45. **Ościk, J.:** *Adsorption*. PWN, Warsaw 1983; E. Horwood Lim., Chichester 1982.
46. **Tompkins, F.C.:** *Chemisorption of gases on metals* (Translation from original English) PWN, Warsaw 1985.
47. **Brodski, A.:** *Physical chemistry* (in Polish). WNT, Warsaw 1974.
48. Joint report: *Physical chemistry* (in Polish). II Edition, PWN, Warsaw 1965.
49. **Przybyłowicz, K.:** Diffusion in surface treatments (in Polish). Proceedings: *Summer School on Surface Engineering*, Kielce, 6-9 September 1993, pp. 31-40.
50. **Mrowec, S.:** Selected topics from the chemistry of defects and theory of diffusion in the solid state (in Polish). *Geological Publications*, Warsaw 1974.
51. **Jarzębski, Z.:** *Diffusion in metals* (in Polish). Publ. Śląsk, Katowice 1975.
52. **Przybyłowicz, K.:** *Theoretical metallurgy* (in Polish). 5th Edition. Mining Academy, Krakow 1985.
53. Collection of lectures: *Summer School on Diffusion in Solids*, Kraków-Mogilany 1985, published by the Mining Academy Publications, no 113, 1985.
54. Joint Report: *Wärmebehandlung der Bau- und Werkzeugstahle*, BAZ Buchverlag Basel, 1978
55. Joint Report: *Metallurgy* (in Polish). Publ. Śląsk, Katowice 1979.
56. **Rudnik, S.:** *Metallurgy* (in Polish). PWN, Warsaw 1980.
57. **Wesołowski, K.:** *Metallurgy and heat treatment* (in Polish). WNT, Warsaw 1972.
58. **Adamson, A.W.:** *Physical chemistry of surface* (Polish translation from English). WNT, Warsaw, 1972.
59. **Senkara, J.:** Control of adhesion energy between molybdenum and tungsten and liquid metals in brazing processes. *Prace Naukowe Politechniki Warszawskiej (Scientific Reports of Warsaw Technical University)* (in Polish). Vol. 156, Mechanics, Warsaw, 1993.
60. **Germain. J.E.:** *Catalysis in heterogeneous systems* (in Polish). PWN, Warsaw 1962.
61. **Karapetjanc, M.Ch.:** *Introduction to theory of chemical processes* (Polish translation from Russian), PWN, Warsaw 1983.
62. **Bond, G.C.:** *Heterogeneous catalysis* (Polish translation from English) PWN, Warsaw, 1979.
63. **Prowans, S., and Jasiński, W.:** Mechanism of growth of TiC coating on a steel substrate (in Polish). Proc.: International Conference: *Carbides, Nitrides, Borides*, Poznań-Kołobrzeg 1981, pp. 320-326.
64. **Welles, A., and Yates, S.C.:** Chemical vapours deposition of titanium nitride on plasma nitrided steel. *Journal of Materials Science*, No. 23, 1988, pp. 1481-1484.
65. **Wierzchoń, T., Michalski, J., and Karpiński, T.:** Formation of TiN and composite layers under glow discharge conditions. Proc.: *First International Conference on Plasma Technology*, Garmisch Partenkirchen, Sept. 1988, pp. 177-181.
66. **Wierzchoń, T.:** *Formation of iron boride layer on steel in glow discharge conditions* (in Polish). Mechanika, p. 101, Warsaw Technical University Publications, Warsaw 1956.
67. **Morel, S., and Morel, S.:** Possibilities of carrying out processes of non-selective reduction of nitrogen oxides in the presence of catalytic coatings (in Polish). *Refractory Materials*, No. 4, 1992, pp. 114-118.

68. **Morel, S., and Cubała, J.**: Catalysing burn effect of rest CO and reduction emission NO_x by ceramic coatings. Proc.: *Research Problems in Heating Energy*, Warsaw, 8-10 December, 1993, p. 235-242.

69. **Sulima, A.M., Sulov, V.A., and Yagodnitski, Yu.D.**: *Superficial layer and service properties of machine components* (in Russian). Publ. Masinostroenye, Moscow 1988.

70. **Kocańda, S., and Szala, J.**: Fundamentals of fatigue calculations (in Polish). PWN, Warsaw, 1985.

71. **Katarzyński, S., Kocańda, S., and Zakrzewski, M.**: *Research of mechanical properties of metals* (in Polish). Edition III, WNT, Warsaw 1967.

72. **Dyląg, Z., and Orłoś, Z.**: *Fatigue strength of materials* (in Polish). WNT, Warsaw, 1961.

73. Joint Report: *Metal fatigue* (translation from English). WNT, Warsaw 1962.

74. ISO R 373-1964 General principles for fatigue testing.

75. **Przybylski, W.**: *Burnishing technology*. WNT, Warsaw 1987.

76. **Nakonieczny, A., and Szyrle, W.**: Fatigue strength, residual stresses and microstructure of carburized and shot-peened layer (in Polish). Proc.: II Polish Conference *Surface Treatment*, 13-15 October, 1993, Kule/Czestochowa, pp. 61-66.

77. **Gurnej, T.R.**: *Fatigue of welded structures* (translation from English). WNT, Warsaw 1973.

78. **Olszański, W., Sułkowski, I., Tacikowski, J., and Zyśk, J.**: Thermo-chemical treatment (*Heat treatment of metals*) (in Polish). Vol. 5, ODOK SIMP - IMP, Warsaw 1979.

79. DIN 50323 Tribologie. Begriffe. Deutsches Institut für Normung 1990.

80. **Senatorski, J.**: *Evaluation of materials for sliding friction nodes* (in Polish). Dissertation. Institute of Precision Mechanics, Warsaw 1994.

81. **Jastrzębski, Z.D.**: *The nature and properties of engineering materials*. John Wiley and Sons, New York 1976.

82. **Krzemiński-Freda, H.**: *Ball-bearings* (in Polish). PWN, Warsaw 1989.

83. **Łuszczak, A., Machel, M., and Wachal, A.**: *Tribology. Friction and lubrication of machine components* (in Polish). Vol. I and II. Military Technical Academy, Warsaw 1979.

84. **Iżycki, B., Maliszewski, J., Piwowar, S., and Wierzchoń, T.**: *Diffusion brazing* (in Polish). WNT, Warsaw 1974.

85. **Garkunov, D.N.**: *Selective transportation in heavily loaded friction nodes* (in Russian). Publ. Masinostroenye, Moscow 1982.

86. **Simons, E.N.**: *Metal wear: a brief outline*. Frederick Muller Limited, London 1972.

87. **Bowden, F.P., and Tabor, D.**: *Introduction to tribology* (translation from English). WNT, Warsaw 1980.

88. **Wranglen, G.**: *Fundamentals of corrosion and metal protection* (in Polish). WNT, Warsaw 1985.

89. **Burakowski, T., and Senatorski, J.**: Comparison of resistance to tribological wear of carburized and nitrided layers (in Polish), *Trybologia*, No. 3, 1984, pp. 4-8

90. **Burakowski, T., Senatorski, J., and Tacikowski, J.**: Effect of microstructure of diffusion layers on their tribological properties (in Polish). *Trybologia*, No. 6, 1986, pp. 4-8

91. **Senatorski, J., and Tacikowski, J.**: Tribological properties of diffusion layers on structural and tool steels. *Trybologia*, No. 2, 1988, pp. 11-13.

92. **Senatorski, J.**: *Problems related to increasing of tribological properties of parts by heat treatment* (in Russian). Series: Scientific-technological progress in machine-building, Edition 28. Publications of International Center for Scientific and Technical Information - A.A. Blagonravov Institute for Machine Building Research of the Academy of Sience of USSR, Moscow 1991.

93. **Rogalski, Z., and Senatorski, J.**: Über den Einfluss der thermochemischen Oberflächenbehandlung auf die Fressbeständigkeit von Konstruktionsstählen. *IfL-Mitteilungen*, 1967, pp. 11-16.

94. **Kostetski, B.I., Barmosenko, A.I., and Slaviskaya, L.V.**: The role of crystalline structure and orientation of monocrystals in the formation of the internal wear process (in Russian). *Metallofizika*, No. 40, 1972, pp. 24-30.

95. **Piaskowski, Z.**: The effect of surface deformation on the wear resistance of the superficial layer (in Polish). *Trybologia*, No. 6, 1987, pp. 13-15.

96. **Janowski, S., Senatorski, J., and Szyrle, W.**: Initial research of the effect of residual stresses on wear resistance (in Polish). Science Periodicals of Rzeszów Technical University, No. 82, Mechanika, vol. 28, 1991, pp. 127-132.

97. **Marczak, R.**: Progress in investigations of the Garkunov effect (in Polish). Proc.: *Problems of wear-free friction in machines*, Radom, 12-13 May, 1993, pp. 126-137.

98. **Krol, M.**: The application of metal polymers for improvement of efficiency and reliability of combustion engines (in Polish). Proc.: *Problems of wear-free friction in machines*, Radom, 12-13 May, 1993, pp. 196-204.
99. **Garkunov, D.N.**: Introductory presentation. Proc.: *Problems of wear-free friction in machines*, Radom, 12-13 May, 1993, pp. 6-11.
100. **Firkowski, A.**: The mechanism of selective transportation effect and its usable aspect (in Polish). Proc.: *XVI Fall School of Tribology*, Piła-Tuczno 1988, pp. 96-102.
101. **Marczak, R.**: The effect of wear-free friction (in Polish). Paper presented at the meeing of the Committee for Machine Building of the Polish Academy of Sciences, in Borków (Poland), 3-4 October, 1991.
102. **Korzyński, M.**: The application of burnishing to improve tribological properties of machine components (in Polish). Science Publications of Rzeszów Technical University, No. 36, Mechanika, vol. 15, 1987, pp. 131-133.
103. **Willis, E.**: Surface finish in relation to cylinder liners. *Wear*, No.109, 1986, pp. 351-366.
104. **Prowans, S.**: *Physical metallurgy* (in Polish). PWN, Warsaw 1988.
105. **Przybyłowicz, K.**: *Physical metallurgy* (in Polish). Edition II. WNT, Warsaw 1992.
106. **Uhlig, H.**: *Corrosion and protection against it* (translation from English). WNT, Warsaw 1976.
107. **Tomashov, D.N.**: *Theory of corrosion and metal protection* (Polish translation from Russian). PWN, Warsaw 1962.
108. **Akimov, G.W.**: *Fundamentals of corrosion science and metal protection* (Polish translation from Russian). PWT, Katowice 1952.
109. **Klinov, I.J.**: *Corrosion and structural materials* (Polish translation from Russian). WNT, Warsaw 1964.
110. **Tacikowski, J., and Zyśk, J.**: Modern methods of gas nitriding. Proc.: Monotheme Conference on *Nitriding and Related Processes*, Rzeszów, 26 June 1980, pp. 24-42.
111. *Surface treatment for improved performance and properties*. Edited by J.J. Burke, V.Weiss. Plenum Press, New York - London 1982.

chapter six

Coatings

6.1 The concept of the coatings

The concept of coatings most probably originated from the realm of anatomy where it means the cover of animals' bodies, isolating the organism from the environment and, at the same time, providing contact between the organism and that environment. In vertebrates the coating is constituted by the skin. This concept has been imported into the realm of technology.

The concept of coatings was given recognition in official documents and specifications in the 1950s. It must, however, be admitted that to this date there is no general terminological specification defining the coating and its types. There are, on the other hand, specifications related to corrosion and corrosion protection [1, 2], which though assume that the concept of "coating" is self-evident and define the various types of coatings.

It seems appropriate to put forward a general definition of the concept in the following form:

Coating - *a layer of material, formed naturally or synthetically or deposited artificially on the surface of an object made of another material, with the aim of obtaining required technical or decorative properties.*

The substrate, in other words, the coated object, or, in stricter terms, its superficial layer, constitutes one phase of the system. The coating constitutes the second phase. Between the coating and the substrate there exists an interface in the form of a layer of certain volume, with intermediate properties, usually facilitating adherence of the coating to the substrate. In the case of some coatings this layer bears the name of **intermediate**.

In some cases it is difficult to distinguish between the coating and the superficial layer, particularly in incremental diffusion layers.

6.2 Structure of the coating

For obvious reasons, the coating has a laminar structure. Because of the great variety of coatings, both from the point of view of material and technology, which stems from different designated uses, it is difficult to develop one universal model of coating structure. Specific models, pertaining to the particular types of coatings, are given and discussed in literature on the subject, usually with varying degrees of simplification.

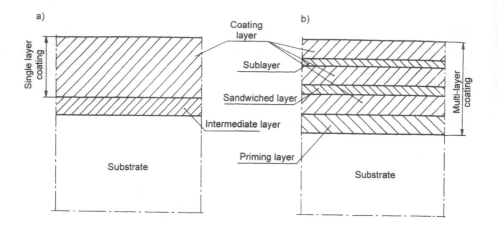

Fig. 6.1 Schematic representation of coating structure: a) single layer; b) multi-layer.

The general simplified model of coating structure is shown in Fig. 6.1, using the single layer and multi-layer coatings as examples.

The **single layer coating** (single coating, monolaminar) is a coating deposited on an appropriately prepared substrate in one process or technological operation, comprising one layer of material. Single layer coatings are divided into:

– **single constituent coatings** - consisting of one material (element, compound), e.g. chromium, titanium nitride TiN,

– **multi-component coatings** - consisting of several material components, e.g. alloying additives, titanium carbonitride Ti(C,N) or titanium aluminonitride Ti(Al,N).

The **multi-layer coating** is one which consists of two or more materials. These may be layers of the same material, separated by a sublayer, or they may be of different materials in which case a sublayer may but need not necessarily be applied. The aim of applying multi-layer coatings is the intensification of the protective or decorative or other function. The simplest kind of multi-layer coating is the double-layer nickel-chrome electroplated coating on automotive and motorcycle components, as well as some items of everyday use. The external chromium layer gives the coating a shiny appearance and does not tarnish. The internal nickel layer binds it to the substrate. Products made from steel and zinc alloys are sometimes coated with three layer coatings of copper-nickel-chromium or four layer coatings of nickel-chromium-nickel-chromium with a particularly high corrosion resistance [1, 3-8]. Heat-resistant spray coatings on steel or cast iron components may comprise double or triple layers. Thin and hard anti-wear and anti-reflection coatings, composed of several to more than ten layers, are also known. A 60-layer coating, obtained by PVD method, has also found practical application, and 150-layer coatings are also known. Usually, coatings are composed of several layers, very rarely more than ten and only in exceptional cases several tens.

The following modifications of multi-layer coatings are known:

– **multiple coating** - comprising two or more layers of the same material, deposited in technological environments that differ only slightly as to physico-chemical properties, e.g., in two or more electroplating baths. The simplest modification of a multiple coating is the **double** (two-layered) **coating**, e.g., a nickel coating, deposited from two different baths.

– **sandwich coating** - comprising several layers of different materials, with at least one of them occurring twice and not directly on top of same material, e.g., Ni-Cr-Ni-Cr [3].

– **self-stratifying paint** - deposited in the form of a liquid (or powder) mixture, stratifying during drying (or melting) into a bottom sublayer with strong adhesion to substrate which is usually metallic, and a top sublayer (surface), with high resistance to environmental hazards.

Between internal surfaces of layers in a multi-layer coating, usually **intermediate layers** are formed.

Some single and multi-layer coatings may be deposited on top of an earlier deposited primer.

The external surface of the coating (phase A), in contact with the surrounding medium (phase B), forms with it an inter-phase boundary (interface) which is simply the surface layer of the coating. To put it in stricter terms, it constitutes zones I-III of the developed, 8-zone model of the surface layer (see Chapter 5).

Coating thicknesses range from hundredths of a micrometer to several millimeters.

6.3 Types of coatings

Coatings may be divided in different ways, depending on the criteria used. The most significant division appears to be by material, by designation and by method of manufacture of the coating.

6.3.1 Division of coatings by material

From the point of view of material, coatings may be divided in two groups: metallic and non-metallic. Very often the name of the coating is derived from the coating material used.

6.3.1.1 Metallic coatings

Such coatings are made from different metals, metal alloys and metal composites, and deposited on substrates, most often themselves metallic, by different methods. In principle, coatings may be manufactured from all metals and composites. However, not all metals, alloys or composites are applicable in practice on account of their properties and technological difficulties in their manufacture. Most often used are:

– **Coating metals**: zinc, nickel, chromium, aluminum, tin, cadmium, copper, lead, silver, gold, iron, cobalt, indium, ruthenium, rhodium, palladium, and platinum. Often, refractory metals are also used, such as titanium, zirconium, hafnium, vanadium, niobium, tantalum, molybdenum and tungsten.

– **Alloys and coating composites**: steels (particularly alloyed, corrosion and heat resistant), brasses as well as alloys of metals: Pb-Sn-Cu, Sn-Ni, W-Co, W-Ni, Ni-Fe, Co-Mo, Zn-Al, Zn-Fe, Zn-Ni, Zn-Mn, Zn-Ce, Zn-Sn, Al-Si, Ni-Cr, Co-Cr, Ni-Al, Pb-Zn, Ni-B-Si, Ni-Cr-B-Si, Ni-Cr-B-Si-C, Co-Mo-Cr-Si,   Ni-Cr-Al-Y.

Metallic coatings are usually deposited on expensive, precision components, rather small in size, which may also be exposed to mechanical hazards in service conditions.

6.3.1.2 Non-metallic coatings

These coatings, numerous and varied, are made from organic materials (paint, rubber, plastic) and inorganic (enamels, ceramics), of natural and synthetic origin and bearing many different trade and chemical names. Most often used coating materials (or their basic constituents) are:

Paints - coating materials for covers, comprising suspensions of pigments (organic and inorganic coloring substances) in a binder (filmogenic substance) [9-15], manufactured in the form of liquids and powders. The dispersed phase is the pigment, while the binder is the dispersing agent. Typical inorganic pigments are zinc white, titanium white, chrome yellow, minium, graphite, etc., while organic pigments are divided into natural and synthetic, e.g., indigo, phtalocyanine blue [14, 15]. Depending on the binder used, we distinguish the following types of paints: oil, cellulose [10], emulsion, varnish [10], latex, limewash, silicate, thixotropic, ceramic (including mirror and muffle), minium (rust-inhibitive), oil-resin (e.g., antifouling) [9], fireproof, etc. [9-15]. They are most often deposited in the liquid form on a painted substrate, on which, after drying (in the case of powder paints, after remelting) they form the cover coating.

Varnishes - liquid (or powder) products, which are colloidal solutions of filmogenic substances: prepared drying oils, natural and synthetic resins, bitumins or esters of cellulose in organic solvents and dilutants, often with pigments, plasticizers (softeners used for high polymer substances to ensure their workability) and siccatives (quick dryers - chemical compounds accelerating the drying of varnish, e.g., naphthanates, resinates, oleates of manganese, cobalt and calcium). After being deposited in thin layers on surfaces, as the result of physical and chemical transformations they form permanent coatings, closely adhering to the surfaces of coated objects. From the point of view of chemical composition, they are divided into oil, resin, cellulose and bituminous. Best known and broadly used are varnishes such as asphalt, oil-asphalt, oil-resin, nitrocellulose, zapon, dope, fossil, polyester, silicate, spirits, teraphthalene, chlorinated rubber, alkyd, phenol, phenol-formaldehyde, melamine, urea, epoxy, epoxy-phenol and polyurethane [9-15].

Resins - amorphous organic substances of different chemical composition with consistence ranging from semi-liquid to vitreous, not soluble in water, soluble in organic solvents or totally insoluble and non-melting [16, 17]:

– *natural* - collected mainly from living plants (contemporary), less often effluents from living plants which found their way into the soil (young fossils) or from extinct plants (old fossils),

– *synthetic* - surrogates of natural resins for varnishing (malein resins) or vitreous and dense products of polycreation (acrylic, polystyrene, phenol, polyester and epoxy resins).

Varnishing enamels - pigment cover products with varnish binders, used to manufacture coatings for different applications. Similarly to varnishes, enamels are also subdivided, depending on the type of filmogenic substance in the binder. Thus, there can be enamels with an oil and cellulose etc. base [9, 10].

Oils - oily liquids or low-melting solids, not water-soluble. These can be divided into

– *vegetable*, which are mixtures of glycerine esters and fatty acids; the most common are walnut, tung (Chinese wood), soy, sunflower, sesame, rapeseed, castor, palm, olive, oitica, cotton seed, peanut, mustard, cocoa, coconut, poppy, linseed, hempseed;

– *mineral*, which are mixtures of hydrocarbons obtained from petroleum;

– *animal*, which are liquid fats of animal origin, with high content of unsaturated fatty acids. Among the most important in this category are neatsfoot (hoof) oil, bone oil and sperm oil.

In surface engineering, the majority of oils is used as components in the manufacture of paints, varnishes, enamels and oil varnishes; grease oils (used for lubricants in machine parts and mechanisms) and protecting oils (used for temporary protection of metal products against corrosion) [9,18].

Cements - binders, obtained synthetically from various materials. The following types of cements have found application in surface engineering: acid and alkali-resistant organic cements (with a natural resin base), acid and alkali-resistant bituminous and tar cements (in which the binder is constituted by high boiling point fractions of distillation of gas pitch and petroleum) as well as acid-resistant litharge-glycerin (PbO litharge + glycerin) [19, 20].

Waxes - a group of complex organic substances of natural origin (animal, plant and mineral) or synthetic (obtained by chemical transformation of natural wax and totally synthetic waxes) which in physical (less often chemical) properties resemble bee's wax. Waxes most often contain esters of higher fatty acids and alcohols, mainly monohydroxide, as well as acids, alcohols, ketones, ethers, hydrocarbons, etc. [20, 21].

Asphalts - also called bitumins - mixtures of high-molecular hydrocarbons (aliphatic, naphthene and aromatic) of natural origin or obtained from treatment of petroleum (pyrogenic), solid or semi-liquid, softening and melting upon warming, soluble in carbon disulfide. Of the great variety of asphalts, surface engineering uses only few, mainly for the manufacture of asphalt (bituminous) varnishes and for the insulation of pipelines [22, 23].

Lubricants - a general name given to all greasy substances which have the properties of diminishing friction between surfaces, as well as those which protect the surface. From the point of view of consistence, we distinguish solid, semi-liquid and oil lubricants; from the point of view of application - anti-friction lubricants and protective greases. All types of lubricants are used in surface engineering, lubricants being used in the overwhelming majority of cases, to alleviate friction of the superficial layer, while protective greases - to develop coatings, deposited on the superficial layer.

Solid lubricants, of solid or semi-solid (semi-liquid) consistence, comprise the lubricant oil and a densifier. The dispersing phase is oil, while the dispersed phase is the densifier, suspended in the form of grains, fiber or needles. Different synthetic and natural oils are used. Thickeners are mainly barium, lithium, aluminum, sodium, strontium and calcium soaps. Less often, inorganic thickeners are used: clays, silicates or organic compounds (e.g., acrylo-urea, phthalic acid nitrile). Lubricants with organic thickeners are most often used for corrosion protection. Sometimes used are upgrading additives, e.g., oxidation inhibitors, corrosion inhibitors, additives enhancing the stability of the film and resistance to high loads. There are also solid additives, so-called fillers, as asbestos, graphite, metal oxides, metal dust or flakes, metal sulfides (molybdenum disulfide, etc.) [23, 24].

Protective lubricants (including anti-corrosion) are usually solid, less often semi-liquid, and they serve to protect metallic surfaces from corrosion, including electro-chemical corrosion during storage and transport. For this reason they always contain a corrosion inhibitor as the basic constituent. Their protective action consists of partial insulation of the metal surface from the aggressive environment (they block the passage of vapors, moisture and gases), slowing down of the corrosion process and reduction of wettability of the metal by water. As opposed to varnishes and protective paints, they are easy to deposit and to remove from the surface. They do not require time for drying and adhere well to surfaces. They also exhibit anti-friction properties. For temporary protection, protective oils are used in the form of emulsions or oils with varying viscosity, while for permanent protection, lubricants are used. Some protective lubricants have the capacity to absorb substances which give rise to corrosion, while others react with the metal, creating protective layers. The most often used protective lubricants are natural and synthetic vaselines. They are also prepared from oils and from petrolatum (a by-product of deparaffination of paraffin oils), paraffin and ceresines. In many cases, additives enhancing certain properties are added to protective lubricants. Among the better known protective lubricants are anti-corrosion grease - for guns and small arms, for artillery shells, for steel wires, for aircraft and for tropic-proofing [24-27].

Vitreous enamels (enamel mixes) - opaque mixes, obtained by melting of natural rock materials, containing silica (sand, chalk, clay, quartz, feldspar, etc.) with fluxes which reduce melting point (borax, soda, potash, saltpeter, etc.) and with blurring materials and pigments which make the enamel opaque and give it color. After melting, they are ground and in powder form they are applied dry or wet to the dried and baked substrate. We distinguish artistic, technical, basic and covering enamels [28, 29].

Metal ceramics (cermets, sintered metal ceramics) - materials obtained by the combination, in different ways, of ceramic material with metals. Surface engineering uses materials in the form of thermally sprayed powder, or metal and non-metal compounds, deposited in vacuum. As binding metals, usually applied are iron, chromium, nickel, tantalum, titanium, aluminum, zirconium, tungsten, molybdenum, hafnium, vanadium and niobium, most often used in amounts of 10 to 30%. The ceramic components are usually oxides, carbides, borides, nitrides, silicides and sulfides. Most broadly used are heat resistant and refractory cermets and those resistant to sudden changes in temperature. These are carbides of tungsten, chromium, titanium, molybdenum, zirconium, hafnium, vanadium, niobium and tantalum; nitrides of titanium, vanadium, zirconium, niobium, hafnium, tantalum, tungsten, chromium and molybdenum; borides of titanium and zirconium; silicides of molybdenum and tungsten. Also used are mechanical mixtures of the above, as well as ternary and quadruple compounds, 'e.g., titanium carbonitride or titanium aluminonitride [30-32].

Ceramic materials - materials in the form of fine mineral grains, characterized by high hardness, brittleness, mechanical strength high melting point, sintering ability, low thermal conductivity and expansion, deposited on the surface of the substrate by:

– thermal spraying; among most frequently applied are oxides of aluminum, chromium, titanium and zirconium; boron and silicon carbides, silicon nitrides or their mixtures [31];

– vacuum deposition; mainly due to crystallization on the metallic substrate - deposited are titanium, chromium, aluminum and silicon nitrides; titanium, boron and silicon carbides, aluminum, zirconium, titanium and other oxides or combinations of the above [32];

– deposited by the Sol-Gel technique: aluminum, titanium, cerium, silicon, magnesium, boron, iron, iridium, tantanum, lead and other oxides or combinations of the above.

Compounds of substrate metals - materials obtained directly on the surface of the metal substrate (aluminum, zinc, cadmium, steel, copper and its alloys, magnesium alloys, silver, etc.) by subjecting it to artificially induced corrosion in different solutions in the form of oxides, phosphates, chromates, and oxalates of the substrate material; known accordingly as (oxide, etc.) conversion coatings [5, 33].

Synthetics (plastomers, plastic mixes) - among these are high polymer materials based on synthetic polymers (big molecule compounds, obtained from small molecule compounds - monomers, with physical and chemical properties differing from those of the original substance). Next, there are materials based on natural or modified polymers, with fillers, pigments and softeners which, at certain stages of manufacture exhibit a tendency to change, by way of simple physical or chemical processes, from the solid or liquid (melted, softened, dissolved) state to the solid (hardened) state or become plasticized (usually at elevated temperatures) irreversibly (duroplasts, thermosetting and chemosetting materials) or reversibly (thermoplastics). They are characterized by good wettability of the surface in the liquid state, low reactivity and, in consequence, high resistance to the action of chemicals (including electrolytes, solvents and the atmosphere), usually lowering with a rise of temperature. They are good insulators. They may be flammable and non-flammable, resistant to the action of light or photosensitive. They may be easily formed into a variety of objects, they are very well suited for coatings which tightly adhere to the substrate. The most often used form is powders and plastisols, i.e. powder suspensions of low volatility in a softener, or aqueous and other suspensions of coating plastics (filmogenic). These may be thermoplastic (softened for an unlimited time at elevated temperatures but setting upon cooling), e.g., polythene, polystyrene, PVC, polyamide, methyl methacrylate. Some may be chemo- and thermosetting (mainly various resins) [34-39].

Indiarubbers (latexes) - high molecule substances (unsaturated hydrocarbons) of natural origin (natural caoutchouk, obtained from latex, i.e. caoutchouk milk of indiarubber trees, mineral caoutchouk) or of synthetic origin (synthetic indiarubber, obtained by polymerization of organic compounds) with exceptional elastoplastic properties. Indiarubbers, particularly synthetic, are used for lining and in the manufacture of anti-corrosion coatings, e.g., butadiene-styrene, butyl and chloride indiarubber; and for anti-corrosion paints, e.g., chlorinated rubber rustproof varnishes, obtained by chlorination of natural indiarubber. It can also be used in the manufacture of some paints and varnishes [43-45].

Rubbers are products of vulcanization (curing - the process of hot transformation of indiarubber with an appropriate amount of sulfur and accelerants, or cold transformation in the presence of sulfur chloride). Rubbers are characterized by high elasticity and properties which depend to a high degree on the type and amount of components which are mixed in with the indiarubber. There are rich mixtures (soft rubbers) containing, besides indiarubber, only a small amount of accelerants, activators and sulfur, as well as lean mixtures (hard rubbers), containing a big amount of hardening fillers. Similarly to indiarubber, rubbers are characterized by good anti-corrosion properties, sunlight resistance, as well as resistance to temperature and atmospheric effects [43-45]. In surface engineering, rubber is used mainly as acid-resistant lining of reservoirs, pipelines, etc.

Non-metal coatings, particularly paints, varnishes and enamels, are used to coat rather big objects which are exposed to only small mechanical hazard. Exceptions to this rule are ceramic, enamel and metal ceramics which can sustain high mechanical loads.

6.3.2 Classification of coatings by application

From the point of view of application, coatings can be divided into four groups [1, 4]: protective, decorative, decorative-protective and technical.

6.3.2.1 Protective coatings

Protective coatings are those coatings whose task is exclusively protection of the object from the harmful effect of the environment, mainly the atmosphere and chemicals, as well as from mechanical hazards. These coatings may also feature properties other than anti-corrosion, e.g., higher hardness, resistance to tribological wear, or attractive appearance. These, however, are properties of secondary importance. Since they protect the object from different types of corrosion, they are sometimes referred to as **anti-corrosion coatings.**

From the point of view of **service life,** protective coatings may be classified as temporary and permanent protection coatings.

Coatings for **temporary protection** serve to secure (mostly metals) against corrosion during transport or storage, as well as during interoperational periods before the product goes into service [1]. Most often these are coatings easily removed by stripping, grinding down or washing off. Their composition is based on e.g., oils, greases, asphalts, waxes, some varnishes and some plastics [26, 27, 44, 45]. Temporary protection is also assured by tight packaging with the use of solid, liquid or gas corrosion inhibitors [26, 27].

Coatings for **permanent protection** serve to secure the surfaces of products during service. These are the majority of metallic coatings and many paints, varnishes and enamels, metal ceramic, ceramic, rubber and latex.

The protective function of coatings may consist either of only insulation of the coated object from the environment or by electrochemical interaction of the coating with the substrate.

In both cases the coating fulfills its function only when it adheres tightly to the substrate, has no cracks, pores and delaminations, scratches and other flaws, besides being mechanically strong. The substrate may be manufactured from any type of material. Likewise, the coating may be of any coating material.

In the second case, both the substrate and the coating must be metallic. Coatings reacting **electrochemically** with the substrate metal can be further subdivided from the point of view of protection mechanism as anodic and cathodic. Usually, these are coatings deposited by electroplating.

Anodic coatings (Fig. 6.2a) are made of metal which in a given environment exhibits a potential lower than that of the substrate. In other words, they are made of a less noble metal than the protected substrate

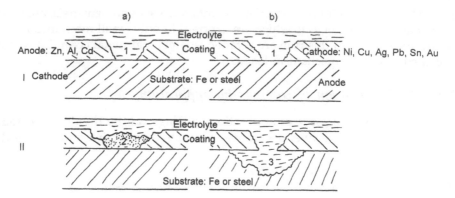

Fig. 6.2 Schematic representation of progress of coating corrosion: a) anodic coating; b) cathodic coating; I - beginning of corrosion; II - advanced corrosion; 1 - scratch or pore; 2 - product of corrosion; 3 - pit-type loss.

metal. This means that upon mechanical or other damage of the coating or in the case of the presence of porosity, in the presence of an electrolyte (when an elecrochemical cell is created) it is the substrate metal and not the coating that is subjected to corrosion. Anodic coatings protect the substrate electrochemically. Examples of anodic coatings are zinc, aluminum and cadmium plating on iron and steel [4, 43].

Cathodic coatings (Fig. 6.2b) are made of a metal whose electrode potential in given environmental conditions is more electropositive than the potential; of the substrate metal, i.e. made of a nobler metal than that of the substrate. Consequently, they protect the substrate metal only when they are totally tight (without porosity cracks, flaking, spalling and scratches). In the opposite event, as soon as the substrate metal is exposed a galvanic cell is formed in the presence of an electrolyte. The coating metal forms the cathode, while that of the substrate - the anode. In consequence, the electrochemical reaction causes an intensified corrosion process; the substrate metal dissolves under the coating and the coated object is destroyed. The rate of destruction may sometimes be so rapid that it exceeds the rate of damage of the uncoated substrate. For that reason, cathodic coatings constitute only mechanical protection of the substrate and therefore should not be too thin. Examples of cathodic coatings are nickel, copper, silver, lead, tin, and gold plating on iron and steel [4, 43].

The most often used coatings (anodic with respect to steel) are zinc coatings, particularly for products which are utilized in atmospheric conditions. For these coatings, the most aggressive is industrial atmosphere, less aggressive urban and seacoast, and least aggressive - rural. Passivation of zinc coatings positively affects their corrosion resistance which is approximately proportional to coating thickness [4].

Anodic cadmium coatings are less often used but exhibit an anti-corrosion resistance similar to that of zinc coatings. Their passivation also raises this resistance.

Tin, tin alloy, aluminum, aluminum alloy and lead coatings are also used on a fairly broad scale.

Paint coatings effectively protect the metallic substrate only when they are multi-layered. Typical are 4-layered coatings: primer, surfacer, enamel, varnish. The anti-corrosion function is fulfilled mainly by the primers, pigmented by elements with rustproof action, e.g., lead minium, zinc yellow and zinc dust. Pigments with rustproof action passivate steel and cast iron and are used exclusively as components of primers. Because primers are not sufficiently resistant to the effects of the external environment, they have to be coated by paints and varnishes which may also contain pigments, but ones not exhibiting rustproof action. The surfacer smoothes the irregularities of the coated surface [46]. Good protective properties are also exhibited by non-metal ceramic, plastic, enamel and cement coatings, the last especially in strongly corrosive environments.

Research has been continuing for the past 10 to 15 years into a new class of anti-corrosive paint materials - both solvent and powder - which constitute mixtures of resin vapors and certain solvents which cause self-stratification of the resins. This takes place mainly due to different surface energies. The solvents are a decisive factor in the direction of stratification and adhesion of both layers between themselves and between the bottom sublayer and the substrate. As a result, deposition of a resin mixture allows the obtaining of a double layer coating. The ongoing research has already yielded some application successes, i.e., painting of submarine hulls and other structures.

Good and even very good anti-corrosive properties are exhibited by vacuum deposited ceramic coatings (by PVD and CVD techniques) and those obtained by the Sol-Gel technique.

6.3.2.2 Decorative coatings

Decorative coatings, once called ornamental, serve predominantly to give the metal or non-metal object an aesthetic external appearance. This depends first of all on color, luster and resistance to tarnishing and also, perhaps, surface finish of the coating (hammer finish, webbling, crystal, crocodile skin), as well as shine properties (fluorescence, phosphorescence, radioactivity). It is evident that decorative coatings, in many cases, make good protective coatings. Decorative coatings may be both metallic and non-metallic.

Among metallic coatings, electroplated coatings have found broadest use. Each metal or alloy has its own characteristic color. Luster is dependent on surface smoothness or it is obtained by the addition of so-called lustering substances to the electrolyte bath. An alternate method is the periodic reversal of current flowing through the electrolyte [4, 47].

Among metals used in the manufacture of decorative coatings, the only ones which do not tarnish are chromium, gold [47], rhodium, palladium and platinum. Silver and nickel coatings become tarnished in some environments, e.g., in an atmosphere contaminated by sulfur compounds; silver surfaces are covered with a gray-black film of silver sulfide. Tarnishing of some coatings is counteracted by their passivation (giving the metal a more electropositive potential than the equilibrium potential, in order to render electrochemical corrosion impossible) or by depositing on them varnish coatings, e.g., a thin and invisible layer of varnish is used to top bronze or brass coatings. In some cases, tarnishing is prevented by the deposition of additional coatings of rhodium or palladium [4].

Among non-metallic coatings, paints are the most broadly used. Almost all paint coatings have decorative functions. The remaining non-metallic coatings, with the exception of enamels, some coatings in the form of deposited nitrided layers and some plastic coatings, are, as a rule, not used for decorative purposes.

In the case of some electroplated and paint coatings, used both for protective and decorative purposes, the latter are much thinner than the former. Usually, natural environment penetrates easier to the substrate metal through a thin layer (through pores, leaks and other flaws), but protection is not the main function of the decorative coating. Besides, such coatings are designed to be used in gentler conditions than their protective counterparts. For example, thicknesses of electroplated decorative coatings are small and range from 0.25 to 3 µm, while protective coatings often reach 60 µm [4]. Sometimes, decorative coatings are artificially aged.

6.3.2.3 Protective-decorative coatings

Protective-decorative coatings, sometimes also called decorative-protective [4], serve both to protect the object against corrosion and light mechanical damage, and to give the surface an aesthetic appearance.

Regarding electrolytic coatings, the protective-decorative role is fulfilled by nickel, chrome, and copper-nickel-chrome coatings on condition that they are sufficiently thick. It is accepted that their thickness should not be less than 25 µm [4]. Corrosion protection is assured by sandwiched layers of nickel or copper and nickel. Sometimes, sandwiched copper layers are replaced by brass but with the same thickness these offer poorer corrosion protection than copper. Nickel is used for the protection of steel, copper and zinc substrates, as well as those made of copper alloys or zinc alloys.

The external chrome coating in multi-layer coatings is very thin - its thickness is in the range of 0.25 to 1.5 µm. It protects the underlying layers against tarnishing, enhances wear resistance and makes the coating tighter.

Single chrome layers deposited on steel are usually not tight, on account of pores and cracks which occur in them. Only when their thickness approaches 50 µm do they protect well against corrosion [4]. Black chrome coatings of 5 to 8 µm thickness suffice to protect copper or its alloys against atmospheric corrosion [48].

A gold coating of 25 to 30 µm thickness insulates copper alloys well from aggressive atmospheric effects; such a coating is used on the elements of the clock face of the 17th century clock on the tower of the royal castle in Warsaw [48].

The overwhelming majority of non-metallic coatings functions as both protective and decorative, particularly those of plastic, enamel and paint [49, 50]. Oxide coatings synthetically obtained on metals not only offer good protection, but also give the object an attractive appearance, e.g., on copper or copper alloys, from a bright to a dark-brown color. Often, these coatings are fixed by organic coatings [48].

6.3.2.4 *Technical coatings*

Technical coatings serve to give the object certain physical properties: mechanical, electrical and thermal.

Coatings which enhance tribological properties. In the majority of cases better tribological properties are exhibited by harder than by softer metal coatings. The hardness of electroplated coatings is varied and ranges from the hardness of lead to that of rhodium. Moreover, hardness may also differ for coatings of the same material, depending on both composition and conditions of deposition [4].

Most frequently, as resistance coatings wear, hard chrome-plated layers of 10 to 30 μm thickness are used. For high sliding velocities and high unit loads, electroplated silver or indium coatings of 500 to 1500 μm, as well as porous chrome coatings, are used [4].

Among non-metallic coatings, very high hardness and excellent tribological properties are exhibited by nitride, oxide, carbide and boride coatings, deposited in vacuum by PVD and CVD techniques.

Coatings which enhance electrical properties. These coatings serve first and foremost to enhance electrical conductivity of terminals and are used in electrical and electronics applications. Since very good electrical conductivity is exhibited by silver, very often silver coatings are deposited on copper, brass and bronze substrates. In conditions of normal use of terminals made of these metals, the thickness of silver coatings applied is 12 μm, while in the presence of moisture and condensing water vapor this thickness is doubled and in the case of sliding contacts it is even greater. Protection against tarnishing of silver coatings insulates them electrically; additional very thin (0.2 to 1 μm) electrodeposited layers of gold, rhodium or indium are used, with chemical or electrochemical passivation as an alternate method [4].

Technological coatings. These coatings serve to enhance different properties of semi-finished products during the execution of the technological process of manufacture or to protect the product against diffusion by undesired elements.

Coatings which enhance solderability of joined surfaces are deposited by electroplating. These are tin and copper coatings, as well as alloys: tin-zinc or zinc-lead, less often cadmium and twin-layered cadmium-tin and copper-tin, deposited on brass and steel components.

Copper coatings of 2.5 to 7 μm, often protected by a layer of varnish against tarnishing, are applied to components immediately before soldering.

The thickness of tin coatings, which insignificantly oxidize and for that reason require the use of fluxes, is 5 to 15 μm.

Alloy coatings composed of 70% tin and 30% zinc are well suited for soldering and for service in tropical climates.

The best proportion of tin to lead in tin-lead coatings is 40:60 [4].

Coatings protecting against diffusion, particularly of carbon, nitrogen, the two elements together and less often of other elements in thermo-chemical treatment operations, are applied as electroplated or in the form of pastes. They fulfill the role of a blockade, stopping the passage of a given element to a given (coated) fragment of the component which is subjected to thermo-chemical treatment. Electroplated coatings are usu-ally copper or tin, as well as alloys of copper and tin with a thickness of up to 25 μm [4]. Paste coatings with varied chemical composition may reach thicknesses of over 1 mm.

Corrective coatings. Such coatings are used to re-create the initial dimensions or shape of a component partially worn in the course of service. In the case of re-creation of dimensions, especially when close tolerances are called, the coatings applied are electroplated: iron, chrome and nickel for the regeneration of steel components and copper - for the regeneration of components made of copper and its alloys. Shape correc-tion (always) and dimensional correction (quite often) are carried out with the aid of various metal and alloy coatings, thermally sprayed or applied by welding and laser beam techniques. Their thickness may reach several mm.

Catalytic coatings. These coatings serve to change the rate of reaction in the gaseous environment with which it is in contact, as well as to raise or lower the temperature at which the reactions occur. Since a large area of contact between the coating and the surrounding gas is required, the real surface should be developed as much as possible. This condition is best met by thermally sprayed coatings. Their real surface may in practice be even 5 times that of the nominal value.

The coating material may be metallic or metal ceramic (mixtures of oxides of cerium, copper, manganese, aluminum, nickel, cobalt, lantha-num, neodymium, etc.). These coatings serve various purposes, among others, to reduce the content of carbon monoxide or nitrides (NO_x-es) in gas and carbonaceous exhausts emitted to the atmosphere. They are also usually resistant to erosion from exhaust gases [51]. Their thickness does not, as a rule, exceed 1 mm.

Coatings enhancing selected thermophysical properties. Most often, these coatings are applied in order to enhance resistance to the effect of elevated temperatures, as well as emissivity and thermal conductivity properties, them-selves dependent on temperature. In most cases, these are metal ceramic or non-metal coatings, thermally sprayed on elements of steel mill equipment, heat exchangers, radiators, recuperators, walls of industrial furnaces and hearths. They are also applied to ballistic missile heads and to spacecraft components.

Coatings which enhance emissivity may be monolayer, but those which retard or enhance thermal conductivity, as well as heat-resistant coatings, are all multi-layered, most often comprising three layers, in order to improve adhesion to the substrate and thus to prevent cracking of the external layer due to excessive residual stresses which rise with a change in temperature. Metal ceramic and ceramic coatings are characterized by high emissivity, within the range of 0.6 to 0.95.

Coatings which retard thermal conductivity constitute a barrier for the flow of heat and change the heat flux by several to more than ten percent.

Ablation coatings.[1] These constitute one type of coating with special thermophysical properties. They are most often produced by thermal spraying of ceramic refractory materials (the main constituents of which are Al_2O_3 and ZrO_2 as well as silicides $ZrSiO_4$ and $MoSiO_4$) onto metallic or non-metallic surfaces with good thermal insulation, in order to protect them against the effect of elevated temperatures at which the surface may melt. Under the influence of heat they undergo ablation, thereby protecting the substrate material. Such coatings are deposited on gas turbine blades, components of high temperature equipment and, primarily, on short and long range ballistic rocket heads, as well as on external surfaces of space vehicles. In the latter case, they enable the vehicle to overcome the heat barrier during re-entry into the dense layers of the earth's atmosphere. As an example, at a speed of 6000 km per hour, at an altitude of several kilometers, the temperature of the vehicle surface, due to friction from atmosphere particles, reaches approximately 1600 K.

Optical coatings. These coatings may have different tasks. Steel and brass elements are coated with electroplated silver, chrome and rhodium, nickel-chrome and copper-nickel-chrome layers to enhance surface luster [4]. Thin multi-layered anti-reflection coatings are applied by PVD techniques to surfaces of glass and plastics. These coatings may absorb or reflect selected bands of thermal radiation, especially in the visible range; they may transmit radiation in one direction and they may counteract the accumulation of dust, gases, vapours, etc.

6.3.3 Classification of coatings by manufacturing methods

This classification is not totally strict but takes into account traditional names and applied distinctions. Thus, in this chapter we will present only general information regarding significant processes of coating manufacture. A more detailed discussion regarding coating and coated materials is given in Part II, Chapter 1 and following chapters of this book.

---

[1] Ablation (from the Latin: *ablatio* - take away) - the taking away of heat by evaporation or sublimation of the heated material.

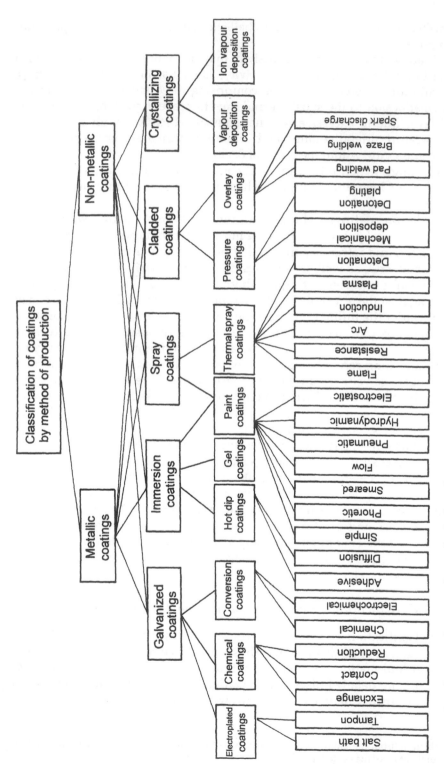

Fig. 6.3 Classification of coatings by method of production.

From the point of view of coating manufacture we can distinguish five groups (Fig. 6.3): electroplated, immersion, spray, cladded and crystallization.

6.3.3.1 Galvanizing

The manufacture of galvanized coatings is in the realm of galvanostegy, which is the most important branch of electroplating [55]. Coatings of this type are deposited directly from a galvanic bath or with the application of an external source of electric current [2]. The thickness of mono- or multi-layered coatings may even exceed 50 μm. The following groups of coatings belong to this category [2].

Electroplated (electrolitic) coatings. The electroplated coating is a metal coating applied in a process of electrolysis, with the application of an external source of electric current [1, 2]. It is usually direct current. In some of the most recent applications, pulsed alternating current may be used (*Electropulse plating*), which allows the obtaining of superior technological results, e.g., higher cathode current density, better penetration and good refinement of the crystallized coating grains. These coatings can be further subdivided into:

– **bath** - deposited on objects immersed in an electrolyte tank; plated objects connected with the negative pole of the current source constitute the cathode, while the anode is formed by plates of the throwing metal which replenishes the loss of that metal in the electrolyte, or by non-soluble materials (lead, graphite). In such cases, loss of metal in the electrolyte is replenished by the addition of appropriate salts [5]. The vast majority of coatings is manufactured by this method; most frequently zinc, nickel and chromium.

– **tampon** - applied locally on selected fragments of a fixed or moving object which forms a single electrode, with the aid of a tampon saturated by a warm or solid electrolyte or by brushes in contact with the other electrode by means of the electrolyte [54].

The quality of electroplated coatings (tightness, grain size, hardness), as well as rate of deposition depend on the type of electrolyte, its concentration, temperature and current density.

Chemical (electroless) coatings. These are metal coatings obtained by way of chemical reaction, without the participation of externally supplied electric current. Depending on the mechanism of reaction, they may be obtained by:

– **exchange**: the less noble metal (more electronegative) displaces the nobler metals (more electropositive) from the solution. For example, a steel object immersed in a solution of copper sulfate is covered with copper, the process continuing only until the moment of covering the entire object (i.e., moment when contact between bath and substrate ends). As a result, only very thin coatings (0.02 to 0.5 μm) are obtained. These are mainly used as decorative but sometimes as technical coatings [43];

– **contact**: the more noble metal (anode) is deposited on the surface of the less noble metal (cathode) in solution, as the result of contact between the coated object and the contact metal [2]. Contact coatings are significantly thicker than those obtained by exchange - their thickness is 1 to 2 μm - and they are well suited to be deposited on small objects [43];

– **chemical reduction**: a metallic element undergoes transition from the ion state (in solution) to the free state (metallic) as the result of attaching to the ion of an appropriate number of electrons from a substance which is capable of donating electrons, called a reducer. Usually, this is an organic compound. Simple salts are less frequently present in the solution. More often, the solution contains complex compounds of the metal used to coat the substrate surface. Reduction begins at the moment of addition of the reducer to the bath and occurs in the entire mass of the solution, of which only a small proportion is deposited on the substrate surface in the form of a coating. For this reason, the method is not very economical. It is used mostly in cases when other methods, notably electroplating, fails. More often it is applied by a spray gun, rather than by immersion. A modification of chemical reduction is catalytic coating, occurring when the bath (prepared from a complex compound and reducer) comes into contact with the catalyst which may be (but does not have to be) constituted by the metal substrate of the coated metal. The reducing reaction must be catalyzed by the metal being deposited. In this case autocatalysis occurs and deposition of the coating proceeds until it reaches the required thickness. The reducing reaction takes place only on the catalyst surface [43].

Conversion coatings. These are non-metallic coatings, obtained on metal surfaces in the form of compounds of the substrate metals, e.g., chromate coatings on zinc, cadmium and silver or oxide layers on steel and aluminum [1,2]. These may be divided into

– **chemical** - when the coating is formed by simple immersion of the metal in the solution (so-called oxidizing), e.g., oxide coatings on aluminum and its alloys, on steel (by the application of black oxide) and on copper and its alloys;

– **electrochemical** (anodic) - when the process is controlled by electric current flowing through the electroplating tank, while the coated metal is the anode (so-called anodic oxidizing or anodizing), e.g., oxide coatings on aluminum and its alloys, copper and its alloys, zinc, cadmium and steel.

Many conversion coatings may be obtained by chemical and electrochemical means.

Conversion coatings are predominantly protective, also used as underlayers, less often as technical coatings (mainly technological) and decorative by the obtaining of color effects. This is especially true of aluminum coloring or electrolytic patination of copper. Most frequently applied conversion coating thicknesses are several to several tens of micrometers [5, 33, 43].

6.3.3.2 *Immersion coatings*

Immersion coatings are obtained by immersion of the entire object or of its portion in a bath of the coating material. Depending on the type of coating material, its physical state may be solid or liquid. In the first case, before depositing the coating, the material must, obviously, be melted. After removing from the bath, the coating material dries on the object, or solidifies, forming the coating.

Immersion coatings are divided into three major groups: hot dip, gel and paint:

Hot dip coatings. Coating materials in the form of solids are melted in a pot or a tank furnace, using the energy of gas burned in burners, or alternately, by electric resistance heating. Since coating materials are predominantly metals and the melting point of used coating materials is usually several hundred degrees Celsius (not exceeding 1000şC), coatings made from them are traditionally called **hot dip**. Properties of hot dip coatings are decided by surface preparation, chemical composition of the metallic material (purity and composition of alloy) and its temperature, time of soaking of the objects in the bath and by the substrate material. Before dipping into the bath, the object is degreased, sand blasted and covered with fluxes by immersion or spraying. The fluxes (most often mixtures of zinc chloride and ammonium chloride with foaming agents - carbohydrates, glycerine, tallow) may also be added to the bath. Metallic materials are hot dip coated by tin, zinc, aluminum, lead and their alloys.

Gel coatings. These are manufactured in colloidal systems in the form of sol, transformed into gel (e.g., due to polycondensation) and next into a coating through e.g., firing. Gel coatings are most often deposited by immersion but they may also be deposited by other methods, i.e., by centriguge, spraying or simply by paintbrush.

Paint coatings. To manufacture paint coatings, the material is used is liquid (e.g., paints, varnishes) or pseudo-liquid (fluidized beds, suspensions of powdered materials, e.g., plastics in fluidizing gas [56]).

The painted coating has the form of a coherent, more or less hard, smooth, shiny substance, adhering to the substrate after drying of the layer of filmogenic material. Only some coatings may be applied as single layers; in most cases, these are multi-layer compositions.

Paint coatings are applied by painting, i.e. depositing the coating material onto the substrate (metallic, ceramic, wooden, etc.) by optional methods. The method selected affects the quality of the coating, as well as the effectiveness and economy of the operation. Some methods require special adjustments of the coating material. Two basic groups of methods of application are distinguished, i.e., immersion and spraying (see Section 6.3.3.3).

Paint coatings applied by immersion may be divided into common and phoretic.

Common immersion coatings are applied by very simple means and at low cost. High coating quality is achieved only with appropriate shapes of

objects, e.g., castings, tooling, components of agricultural machines and tractors, auto chassis and small mass-produced elements. The quality of the
coating depends on the rate of immersion, viscosity of the material and conditions of drip-drying [46].

Phoretic immersion coatings are applied with the utilization of
phoresis, i.e. movement of particles of a dispersed suspension or a colloidal system relative to the dispersing phase. Depending on the factor causing particle movement, we distinguish:

– *electrophoresis* (electrodeposition) in which the factor causing movement of electrically charged particles is an electric field, resulting in the
deposition of a water-dilutable coating material (dispersion of pigments
in a colloidal resin solution) on the painted surface. During electrophoretic
painting, the water-diluted coating material is broken up into cations and
anions. The anions are particles of the paint, while cations are amines or
ammonia, with the aid of which the varnish binder is neutralized to make
it water soluble. Under the influence of direct current, negatively charged
anions migrate to the positive pole (anode), where they lose their charge
and thereby undergo transition into the insoluble state. Positively charged
cations migrate to the negative pole (cathode) where they, too, lose their charges
and change to free amines [46]. Electrophoresis is used to apply mostly monolayer coatings of 15 to 35 μm thickness. If using electroconductive primers, it
is possible to apply double-layer electrophoretic coatings of 30 to 45 μm
thickness. Most frequently, anaphoretic paints deposited on the anode (coated
object) are used. More costly, cataphoretic paints, ensuring better durability,
are used especially in the automotive industry (to paint bodies and other
elements), agricultural machines and tractors, home appliances, etc.

A modification of electrophoresis is the powder process. First, the electrophoretic coating is deposited from the bath containing a suspension of powdered paints; next, a coating of normal electrophoretic paint is applied. An
advantage of electrophoretic coatings is their uniform thickness, unattainable with normal immersion painting, on condition, however, that the substrate is of one material; coatings on objects made from different metals have
varying thickness [46];

– *autophoresis*, in which deposition of the coating is accomplished
without electric current, but is based on a chemical reaction of the coating material with the substrate. Coating materials, in this case, are polymer solutions. In the acidic environment of an autophoretic bath, the
iron ions being released react with particles of latex of the polymers or
emulsifiers, causing destabilization of the solution and deposition of the
coating. Autophoretic coatings constitute a non-organic-organic protective layer of 20 to 25 μm thickness. They find application mainly in the
automotive industry, as primers for auto frames. Corrosion resistance of
autophoretic coatings is close to that of electrophoretic coatings.
Autophoretic coatings are not frequently applied and then only in few
available colors [46].

Depending on the nature of joining of coating with substrate, hot-dip
and paint coatings may be subdivided as follows:

– *adhesive* - to which all non-metal paint and metal hot-dip coatings belong, e.g., tin, lead and copper;
– *diffusion* - e.g., hot-dip zinc and aluminum coatings; the diffusion join is achieved by soaking the object in the hot bath or by its subsequent diffusion annealing.

The thickness of immersion coatings usually fluctuates from over ten to several tens of micrometers (paint) to several tenths of a millimeter (hot-dip). The majority of these are protective or protective-decorative coatings. They are applied continuously (by continuous passage of a wire or tape through the bath) or in batches (by single or double immersion in the bath of usually finished products).

6.3.3.3 *Spray coatings*

Spray coatings (sprayed coatings) are obtained by spraying the surfaces of different objects (metallic or non-metallic), depositing a layer of substance which is generated by dispersion (usually with the aid of a spray gun) into tiny particles of powdered material or already powdered material, applied with high kinetic energy. Owing to this energy, the particles, upon making contact with the sprayed surface, exert pressure which assures good adhesion of the coating to the substrate. Dispersion of particles is usually accomplished pneumatically, less frequently by hydraulic means.

Material particles may fall on the substrate as
– cold (e.g., particles of paints, varnishes and plastics) and adhere to the substrate either by forces of viscosity or electrostatic forces (powdered coating materials), or both (liquids) [55];
– hot (semi-plastic or plastic state or even heated to a temperature above melting point).

The first method is used for spraying paint materials and is traditionally counted among painting methods [49], while the second method is termed thermal spraying [50, 57, 58].

Paint coatings. Three basic types of sprayed paint coatings are distinguished: those applied pneumatically, hydrostatically and electrostatically. To these may be added coatings obtained by flows. Although quite diverse from the above, some common elements are also shared by coatings which may be called smeared.

Smeared coatings are applied by a brush, roller or sponge. Application of coatings by a brush is the most traditional painting technique, later improved by the use of the roller and, more recently, by a sponge. It is most frequently applied in repair and field jobs, in places with difficult access, in small projects involving different places, in small plants, especially for priming where the paint must be strongly rubbed into the substrate. Oil paints are smeared by cruciform movements, with strong rubbing in [50]. Paints based on synthetic resins which require physical drying (e.g., polyvinyl, chloro-latex) should not be reversed by the brush too many times because this disturbs the freshly deposited, fast-drying layer [55]. The decorative quality of smeared coatings, applied by brush, roller, or sponge, is generally low [46].

Flow coatings are obtained by pouring paint, flowing from many nozzles (so-called multi-stream flow) under low pressure, covering the contours of the painted object. A modification of this method is gravitational flow of flat objects by paint flowing out of a slit nozzle, and roller painting, used for painting long and thin metal sheet [49]. Flow coatings are smoother than immersion coatings; their thickness is in the range of 12 to 32 μm. The multi-stream flow method is used for deposition of prime coats or single layer coatings, while the roller method can be used for multi-layer coatings [46, 55].

Pneumatically sprayed coatings are obtained with the use of compressed air, dispersing the coating material. The pressure of the compressed air depends on material viscosity and on nozzle diameter. The paint material is sprayed cold or hot (heated to several tens of degrees Celsius). The thickness of a single deposition is 10 to 30 μm. The pneumatic method may be used for spraying all types of coating materials, although some limitations exist, mainly with respect to spraying of heated materials. This concerns mainly temperature of ignition [46, 50, 55].

Hydrodynamically sprayed coatings (without air, under pressure) are obtained by dispersion of the paint material without contact with a stream of air, but as the result of sudden decompression of a stream of paint material, fed under high pressure (up to 25 MPa) and flowing at high velocity (even exceeding 100 m/s) through a small diameter nozzle. Hydrodynamic spraying is used for the same paint materials as in the pneumatic method, applied hot and cold, the condition being very thorough refinement of the pigment. The only materials not suitable for this method are those with fibrous fillers and some chemo-setting substances. The hydrodynamic method is used for coatings on big, flat surfaces [46, 50, 55].

Electrostatically sprayed coatings are obtained by cladding the object with liquid or powdered paint materials after their prior electrical charging. (Electrization is carried out before or after spraying). Spraying may be either pneumatic or centrifugal. Between the nozzle of the electrostatic spray gun and the coated object an appropriately strong electrostatic field is created. For this type of painting, materials with special dielectric properties are used. Coatings obtained by this method are characterized by high quality. Electrostatic painting is used for depositing thick single layer coatings (for powdered materials thicknesses range from 60 to 90 μm) which correspond to a 2 to 3 μm coating composed of traditional solvent materials. This method is used for painting small objects in mass production [46, 50, 55].

Thermal spray coatings. Depending on the method of generating heat to melt or plastify the coating material, to be simultaneously or subsequently sprayed with a pneumatic drive, the following types of coatings are distinguished:
– **flame** - when the particles are heated by passing through a gas-oxygen flame,
– **resistance** - when the coating material is melted in a resistance heated pot,

– **arc** – when the coating material is melted in a zone of arc discharges (in arc plasma) or when the material is passed through such a zone,

– **induction** – when the powdered coating material is passed through a zone of induction plasma,

– **plasma** – when the coating material is melted in an electric arc strongly extended by the flow of plasmogenic gas or when the coating material is passed through the arc zone.

The most often applied is arc spraying; less often are flame and plasma which allow the obtaining of highest plasma temperatures - more than 10,000 K. In order to eliminate the oxidation of particles of the coating material during spraying, the spraying operation may be carried out in low pressure chambers (*LPPS - Low Pressure Plasma Spraying*), or in order to achieve the highest kinetic energy (for least porosity), the powder particles may be sprayed at supersonic speeds (*HVOF - High Velocity Oxygen Fuel*).

To the group of thermal spray coatings also belong **detonation (explosive) coatings (high energy rate coatings)**. Their deposition requires giving particles of coating material very high kinetic energy (highest of all achieved in thermal spraying). During the process, particles of powdered coating material heat up to a temperature lower than that of particles heated by plasma.

Because of the possibility of attaining temperatures within the range from several hundred to approx. fifteen hundred degrees, thermally sprayed coatings are predominantly ceramic and metal ceramic, with high hardness, heat and erosion resistance, and with controlled thermophysical properties. They are applied as corrective coatings (repair of worn machine components, casting defects and other faults caused by machining), wear resistant protective coatings, as well as technical coatings (e.g., conductive coatings on non-conductive materials [58], catalytic coatings), and less often as decorative [50, 57].

6.3.3.4 Cladded coatings

Cladded coatings have thicknesses reaching several millimeters. They are made of a metal which is more noble or more decorative than the substrate on which they are deposited. These coatings, also called plating [59], were applied to steel and non-ferrous metals, by coating with a single or double layer of noble metal, e.g., tableware [60, 61].

Presently, the concept of cladded coatings has been significantly broadened. They comprise two groups: pressure and overlay coatings.

Pressure coatings are obtained by joining the cladding material with the cladded material by exerting pressure, usually at appropriately elevated temperatures, in order to weld them. Most often, plating is accomplished by:

– **mechanical means**, by one-sided or two-sided hot rolling of sheet or strip from both metals, after prior cleaning of surface, mainly of oxides. The cladding material may also be deposited by casting, usually centrifugal (the cladded object is coated, after which the semi-finished product is subjected to further treatment, usually cold or hot work, by pressing or rolling). Other methods include shrinkage deposition (in which the prior made cladding is hot slid over the substrate material, press forming and broached. Metal joining is obtained by their welding, i.e., heating to dough consistency and pressed. Metals may also be joined by diffusion welding;

– **detonation action** - by generating very high pressure (even up to millions of atmospheres) as the result of detonation of explosive material and generating extremely high velocities of particles at the moment of contact with the substrate (from several to several tens km/s). Metal joining has the characteristics of diffusion, but in this case there is a significantly higher degree of mixing of plating material and substrate than that observed in other methods, a kind of mechanical alloying under pressure (pressure transportation of mass), enhancing substrate strength.

In both methods the substrate material is usually plain carbon steel (in which case the plating material is martensitic and austenitic stainless steel, copper, nickel, silver, titanium, aluminum and Monel metal), aluminum (in which case the plating material is pure aluminum or its alloys), copper, nickel or alloys of these two metals. The thickness of plated coatings varies, depending on the type and application, and ranges from 1.5 to 15% of the substrate thickness [59]. Pressure plated coatings are used predominantly for anti-corrosion purposes (particularly in the chemical industry), less often for decorative purposes, and sometimes as technical coatings with special electrical or thermophysical properties. Technical coatings deposited on thin substrates are referred to as bimetals (a name applied to coating and substrate together) [60, 61].

Overlay coatings are obtained by

– **pad welding** - i.e., melting the plating material on the substrate surface, using various sources of heat: welding torches, plasma burners and lasers. The pad welded material with a composition close to that of the substrate metal is melted and joins with it, aided by the use of an oxy-acetylene torch or electric arc; it can be achieved cold (without preheating of substrate) or hot (with preheating of the substrate usually to a temperature of 200 to 400°C or higher, close to the melting point of the substrate material, in order to avoid cracks). The overlays may be composed of stellites (stelliting), steel rods in the form of small pipes, filled with tungsten carbide grains (applied in corrective coating of machine components working in soil or rock, e.g., excavators, mining drills); brass (when overlaying is done on cast iron objects, i.e., pump pistons, water turbine blades, gears) and lead (overlaying of walls of steel reservoirs for the chemical industry);

– **braze welding** - by melting, with the aid of flux (e.g., mixture, comprising 70% borax, 20% kitchen salt, 10% boric acid), of coating material, which is mainly common brass or nickel-brass, without melting the substrate material. The latter is usually a high melting point material, e.g., cast iron, bronze, steel coated with zinc or tin. The operation is accomplished with the aid of an oxy-acetylene torch or, less often, electric arc. The overlaying material does not melt together with substrate material (similarly to brazing);

– **electro-spark discharge** - by deposition of material of an eroding electrode on the substrate, due to electric discharges (spark discharge) between electrode and substrate.

Plated overlay coatings require machining; in some cases prior heating is essential. Applications are mostly anti-corrosion and anti-wear. They may be made of the same materials as pressure coatings or different ones. Usually, they are deposited on smaller objects or fragments. These coatings may also be used for corrective purposes, e.g., repair of streetcar and railroad tracks, worn journals, road wheels, excavator teeth, cutting tools, dies, punches, etc.

6.3.3.5 Crystallizing coatings

Crystallizing coatings are obtained in conditions of technical vacuum, as the result of crystallization of usually ionized metal vapours on a cold substrate or one preheated to 200 to 500°C. The metal vapours form compounds with ionized elements of gases which constitute the plasma (nitrides, oxides) or with elements from the substrate (carbides). Crystallizing coatings are mainly compounds of non-metals with metals (nitrides, carbides, oxides, borides, silicides, carbonitrides and aluminonitrides) or pure metals (e.g., aluminum). Coatings formed by compounds of metals with non-metals have thicknesses of the order of several micrometers, are very hard, wear resistant, with an attractive golden or gray color. They are bound to the substrate by adhesion, very rarely by adhesion and diffusion. Depending on the method of deposition of the metal vapour, these coatings are divided into

– **vapour deposition** (electroless) - these are usually metal coatings, obtained by thermal vaporization of the coating material and its electroless deposition in the non-ionized state, e.g., silver, aluminum. They are used as reflectors in optical and thermal devices;

– **ion vapour deposition** - deposition of metal vapours from a gas which is electrically ionized; these are mainly composed of carbides, nitrides or borides, applied as technical coatings with high wear resistance, but also sometimes as decorative and optical.

6.4 Potential properties of coatings

Similarly to potential properties of surface layers, the potential properties of coatings may be divided into the following groups: geometrical, geometric-physico-chemical and physico-chemical. They are described by the same parameters as the properties of surface layers but, in addition, supplemented by other parameters, especially for coatings.

6.4.1 Geometrical parameters of coatings

Geometrical parameters of coatings include thickness, three-dimensional structure of the surface, unevenness and coating defects.

6.4.1.1 Thickness

Coating thickness is the basic parameter on which protective properties, decorative and technical properties significantly depend. Porosity, tightness, corrosion resistance and mechanical strength all depend on the appropriate coating thickness. Besides, for different types of coatings, this thickness has a different effect on usable properties. For example, protective properties of coatings increase with the rise of thickness, similarly to wear resistance. But thickness has the opposite effect on flexibility, impact strength, and sometimes even on adhesion (e.g., a coat of hard putty of 0.5 mm thickness has poorer adhesion than a coat of 0.1 mm) [46]. Fig. 6.4 shows thickness ranges for some coatings, as well as their corresponding hardness ranges.

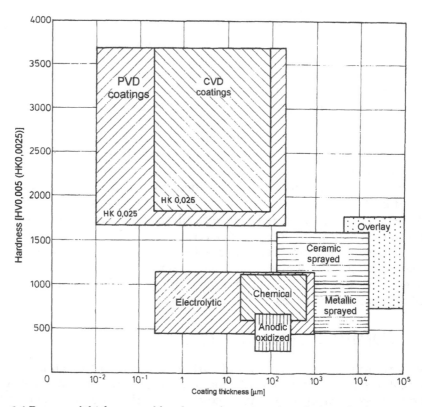

Fig. 6.4 Ranges of thickness and hardness of some coatings (orientation).

The following categories of coating thicknesses are distinguished:
– **point** - measured only once in one point location,

– **local** - measured usually as an average of several readings at a site of small dimensions (several to several tens of millimeters),

– **mean** - which is the arithmetical average of results of point or local readings (expressed in μm or mm) or the result of dividing the mass of the deposited coating by the surface area of the coated object, and expressed in mg/cm^2,

– **minimum** - measured at a site where the thinnest coating is expected, or the least value obtained in a series of readings.

Coating thickness varies over a very broad range - from hundredths of a micrometer to several millimeters.

6.4.1.2 Three-dimensional structure of the surface

The three-dimensional structure of the coating surface is the result of the coating process, dependent on the method or technique used, on defects formed during the deposition and on substrate roughness. It may constitute a more or less clear reflection of the coating gun movement. It may also be a deformed representation of the three-dimensional structure of the substrate. The most general classification is smooth and rough coatings, with sometimes an additional distinction of coarse coatings. Smooth coatings are the opposite of rough ones. The smoother they are, the lower their roughness. A measure of smoothness or roughness of coatings is given by various parameters describing the unevenness of the surface.

6.4.1.3 Surface unevenness

Surface unevenness of the coating is described by the same parameters as those used in the description of surface layers. Coatings may serve to deteriorate (e.g., ceramic sprayed and metallic immersion coatings) or to enhance surface unevenness (e.g., some electroplated coatings smooth the surface; paint coatings hide recesses). For the majority of coatings, particularly of the decorative category, it is important to obtain small surface unevenness (high smoothness).

6.4.1.4 Defects of the three-dimensional structure

Defects of the three-dimensional structure, which are simply coating defects, all have the same character as those of the surface layer (see Section 5.7.1.3). The most frequent defects, common to both surface layers and coatings, are blemishes, scratches, cracks and porosity.

Among the most important defects, common to all coatings, are non-uniform thickness (including object edges), delamination, exfoliation, incomplete coverage, pits, blisters, raising of the coating, brittleness, chipping, and sagging. Moreover, each type of coating exhibits certain own, characteristic defects. Among the most frequently encountered are:

– for metallic, electroplated coatings: burns, streaks, striation, haziness, roughness (here understood as major build-ups or surface contamination inclusions), spots, decoloration, excessive matting and runs [43];

– for organic paint coatings: bleaching, browning, webbling, blushing, spots, pinholing, uneven color, uneven luster, pock-marking, orange peel, crocodile skin, fading and runs [46];

– for sprayed inorganic coatings: irregularity of coating thickness (resulting from "spitting" by the spray gun), varying degree of melting of coating material particles.

6.4.2 Geometric and physico-chemical parameters of coatings

Among the most important geometric and physico-chemical parameters are those which describe coating properties, such as:

– relating to energy, mainly surface energy,

– relating to radiation, mainly reflection and emissivity, and significantly less often (only for selected types of coatings): radiation transmittance,

– catalytic (dependent on degree of surface development and coating components which accelerate or retard chemical reactions),

– thermophysical, mainly thermal conductivity and solderability.

6.4.3 Physico-chemical parameters of coatings

6.4.3.1 General characteristic

Physico-chemical properties of coatings differ from those of coating materials from which they are made. Properties of metallic materials in the bulk state, i.e. as obtained in metallurgical processes, are usually different from the properties of coatings manufactured from them. Usually, although not always, the parameters which describe these properties assume higher values for coatings than for bulk materials. Moreover, for the same materials, the differences in values of the same parameters depend on the method used in the manufacture of the coating.

Properties of metallic coatings classified as electroplated, chemical and crystallizing clearly differ from those of the bulk material. Properties of thermally sprayed coatings differ significantly less, similar to overlays (which may have properties poorer than those of the bulk material). Properties of immersion and plated coatings are either the same as those of the bulk material in the same condition (e.g., for a rolled plated coating and bulk material rolled to the same degree of deformation) or the differences are insignificant.

The reason for the occurrence of these differences in the values of parameters of coatings and bulk material is, first of all, different structure of coating from that of the bulk material obtained by crystallization from the liquid and possibly subjected later to cold working or heat treatment.

Table 6.1 lists selected properties of bulk metals and electroplated coatings obtained from these metals. Electroplated coatings feature higher hardness and strength but lower ductility and electrical conductivity than rolled bar from these same materials. In the case of electroplated coatings,

Table 6.1
Comparison of certain properties of metals deposited
by electroplating with properties of the same metals in the bulk state,
after 30% cold work (From *Kozłowski, A., et al.* [55]. With permission.)

State of material	Copper			Nickel			Silver		
	Tensile strength [MPa]	Elonga-tion [%]	Resisti-vity × 10^x [Ω·m]	Tensile strength [MPa]	Elonga-tion [%]	Resisti-vity × 10^x [Ω·m]	Tensile strength [MPa]	Elonga-tion [%]	Resisti-vity × 10^x [Ω·m]
bulk	350	45	1.67	650	30	6.48	250	50	1.59
electro-plated coating	180-660	3-35	1.7-4.6	350-1050	5-35	7.4-11.5	230-330	12-20	1.59-130

in particular copper, nickel and silver, an added characteristic feature is the scatter of parameter values, i.e. conditions of deposition, coating thickness, as well as presence of dopants and contaminants (including nonmetallic substances, especially inorganic) which are built into the coating during its growth [55].

6.4.3.2 Structure of metallic coatings

Metallic coatings may be generated by crystallization of coating materials from solutions or, in vacuum, from metal vapors, by crystallization of a metallic bath or of a drop of coating material on the substrate surface, by chemical deposition or by spraying earlier formed ductile particles on the substrate surface. In each case, the structure of the coating - even when depositing always the same coating material on the same substrate material - will differ, depending on method used and deposition parameters.

Generally, the structure of metallic coatings may be crystalline (similarly to most metals and alloys) or amorphous (similarly to a very small group of specially developed solid metals and alloys). However, as opposed to the crystalline structure of solid metals and alloys, the crystalline structure of coatings is characterized by both finer grain size and greater amount of defects. Moreover, metallic coatings feature frequent occurrence of non-equilibrium structures, i.e. disparity in type and proportion of phases between the coating and bulk material in a state of equilibrium (e.g., after casting and annealing).

During electroplating deposition in conditions conducive to the formation of continuous and cohesive coatings, **granularity**, expressed by the size of crystallites, i.e., crystals deformed by contact with other crystals, depends on the rate of formation and growth of crystallization nuclei. Greater value of deposition voltage than that of equilibrium, and greater current density (greater polarization) allow a reduction in the size of the critical nucleus and are conducive to obtaining fine-grained structures.

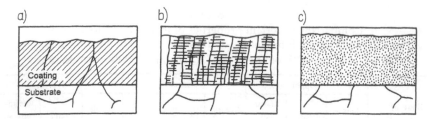

Fig. 6.5 Typical modifications of structures of electrolytic coatings: a) coating with structure recreating substrate structure; b) coating with columnar structure; c) coating with disordered fine-grain structure.

Some inhibitors introduced into the bath and adsorbed by the surface of the coated substrate also retard crystallite growth.

The structure of electroplated coatings, depending on conditions of deposition, may [55]:

– recreate the structure of the substrate (Fig. 6.5a). This case occurs with low current densities, when conditions of deposition deviate little from equilibrium, and with the absence of inhibition. In such conditions, few nuclei are formed, slowly growing parallel to the substrate surface. The rate of growth of remaining nuclei exceeds the rate of formation of new nuclei, and a preferred orientation is thus formed, corresponding to positioning parallel to the surface of crystallographic planes with closest packing. The resultant substrate structure is identical with that of the coating and lattice parameters of the substrate material and coating are similar (e.g. when depositing nickel on copper). The coating structure thus continues the structure of the substrate: crystallite dimensions are of the same order as in the substrate material and the coating structure is coarse-grained (coarse-crystalline) [43];

– form a columnar structure (Fig. 6.5b). This case occurs when the deviation of deposition conditions from equilibrium is greater (with greater polarization of the cathode, i.e., with greater current densities) and is with moderate inhibition. The nuclei formed are successively superimposed one on another, forming columns or fibers of crystallites. The preferred orientation accompanying this often corresponds to perpendicular orientation of crystallographic axes. Columnar structures are also formed in metals crystallizing out of the gas phase;

– form fine-grained structure (Fig. 6.5c). With strong polarization and strong inhibition, the rate of formation of new nuclei exceeds the rate of growth of nuclei already formed. Very soon, crystallites of random orientation are formed, leading to a fine-grained structure (e.g., copper, nickel and zinc coatings, deposited from a bath with brightening agents).

Electroplated alloy coatings and those deposited from baths containing organic compounds exhibit a layered structure.

The equivalent of strongly defective crystalline structure of the surface layer due to heavy plastic deformation is the generally high degree of **defectiveness of structure** of the coating, due to "freezing" of vacancies (unoccupied nodes of the crystal lattice) or dislocations, because the temperature of crystallization is too low to restore the state of equilibrium through diffusion [55].

The metallic layer of the coating may also contain inclusions of non-metallic products of electrode reactions (e.g., hydroxides) or organic substances (e.g. introduced into the bath in the form of wetting agents, brighteners or reducers of residual stresses in the coating) [55].

Non-equilibrium structures, characteristic of alloy coatings (multi-component), are formed as the result of new limits of stability of the particular phase which differ from those in conditions of equilibrium. In most cases there occurs the extension of ranges of existence of solid solutions, i.e. a homogenous structure of a solid solution is obtained in the coating even when the amount of the alloy component exceeds the limit of its solubility. For example, in alloys of copper and silver, the solubility of copper in silver does not exceed 8.8 wt.%, while in electrodeposited coatings it is possible to obtain homogenous solutions even with 20 wt.% of copper. However, some alloys, which upon crystallization form solutions with limited solubility, exhibit a non-homogenous, biphase structure when deposited electrolytically, particularly with significant cathode polarization. Examples of this behavior are Cu-Ni alloys, deposited from oxalate baths or Au-Cu alloys, deposited from cyanide baths. Finally, in some cases, electrolytic crystallization causes the formation of phases which do not normally exist in equilibrium (unstable and breaking down upon heating). As an example, in Sn-Ni alloys, with 31 to 43% Ni, homogenous monophase coatings are formed which, upon heating above 325°C, are broken down to intermetallic phases Ni_3Sn_3 and Ni_3Sn_2 [55].

When depositing by the chemical reduction or vacuum crystallization methods on substrates at low temperatures, there occurs a total atrophy of the crystalline structure. The coating then assumes an **amorphous** structure, characteristic of glass. Metal with amorphous structure is called **metal glass**.

The structure, itself dependent on chemical composition, affects many other properties of coatings, including those directly connected with service. Among these are

– mechanical (hardness, ductility, which is very important in the case of coatings, residual stresses, tribological wear resistance and fatigue strength),

– chemical (absorption, chemisorption, corrosion resistance, especially to chemical corrosion),

– electrochemical (resistance to electrochemical corrosion)

– thermophysical (thermal conductivity and expansion, physisorption, adhesion, solderability)

– radiation (emissivity, reflection, transmittance - only in the case of coatings optically transparent to radiation of varying wavelength),

– optical (color, luster),

– electrical (resistivity conductivity),

– magnetic (coercion, permeability).

Coatings, as opposed to surface layers, must also feature good adhesion to the substrate.

The majority of the above listed properties has been discussed in Section 5. For this reason, only those properties which are most characteristic of coatings will be discussed in detail below.

6.4.3.3 Residual stresses

Residual stresses are formed in coatings as the result of differences in thermal expansion coefficients of substrate and coating materials (stresses of the I kind), as well as significant defects in the structure of the coating material (stresses of the II and III kind). Residual stresses may play a positive or a negative role, depending on their character. Usually, compressive stresses are favorable, while tensile stresses are unfavorable. In order to reduce residual stresses of the I kind, multi-layered coatings are deposited, comprising a composition of layers with successively changing thermal expansion coefficients, relative to the substrate material. Residual stresses may be reduced by the selection of appropriate materials and process parameters. The sign (kind) of stresses depends on the coefficients of thermal expansion of substrate and coating layers and on the character of structural defects. If the layer contains more atoms in interstitial positions than there are vacancies in the lattice, tensile stresses are formed, while the reverse situation leads to the formation of compressive stresses [55].

In electroplated coatings residual stresses depend on the type of bath, concentration of hydrogen ions (pH), concentration of components, current density, temperature (its rise causes relaxation of stresses), type and concentration of brighteners and contaminants. In chrome-plated coatings, for examples, residual stresses may vary from −800 to 1000 MPa. Brighteners, in particular, significantly affect the rise of residual stresses, mainly in nickel coatings, in which an addition of saccharin or other organic compounds of sulfur reduces tensile stresses. This may be counteracted by the addition of anti-stress substances to the bath [3]. Without the use of brighteners, tensile stresses are generated in nickel, cobalt, iron, palladium, manganese and chrome-plated coatings, while compressive stresses are generated in zinc and cadmium coatings. Values of residual stresses vary, depending on layer thickness, e.g., in chrome and nickel-plated coatings; stresses diminish with a rise in layer thickness and achieve a constant value when the layer thickness reaches 30 μm. Very big residual stresses; typical of chrome, rhodium and some alloys, may relax spontaneously, which is manifest by cracks or a crack web, both tensile, as well as big compressive stresses being capable of such relaxation. Coatings with big tensile stresses, e.g., chrome and nickel-plated, exert an unfavorable effect on some mechanical properties of coated objects, especially on fatigue strength which may decrease by 20 to 70% as the result of coating deposition. Compressive stresses in coatings do not exhibit any significant effect on fatigue strength [55].

In coatings composed of metals with a low melting point (lead, cadmium, zinc) and with high plasticity at room temperature (aluminum, silver, gold, copper) small residual stresses, not exceeding 50 MPa, occur. Spontaneous relaxation (atrophy) of these stresses, due to the migration of structural defects, is also observed in these layers [55].

6.4.3.4 Adhesion

Adherence of coatings to the substrate or of layers of multi-layer coatings to each other, described by the force necessary to detach the coating from the substrate or layers from each other, reflects the character of the dominating bond:

- **metallurgical** - consisting of melt-mixing of the coating and subsurface zone materials (e.g., in cladded and alloying coatings);

- **epitaxial** - consisting in the formation of coating crystals, similarly oriented relative in each other, on the crystals of the substrate, on condition that the difference in lattice parameters will not be bigger than approx. 10%. Epitaxy plays a major role in the process of coating crystallization in need conditions (in electro-chemical processes and in PVD);

- **adhesive** - consisting of the utilization of adhesion of the coating to a well-cleaned substrate (e.g., in paint, electroplated and crystallization coatings); adhesion may be utilized in cold or hot bonding of coatings to substrates, combined with significant deformations; often adhesion follows the formation of a proadhesive compound (intermediate layer). On steel, for example, such a compound may be $FeTiO_3$;

- **diffusion** - consisting of mutual displacement, through diffusion of components of the coating and the substrate. This occurs very seldom. Examples are some immersion and crystallizing coatings;

- **mechanical** - consisting of creation of conditions in the substrate (e.g., by dovetail knurling) for mechanical anchoring of coating material. This version occurs in some thermal spray techniques.

Naturally, combinations of the above types of bonding are also possible, e.g., adhesion-diffusion or mechanical-adhesive.

In all cases, the condition essential for good adherence of coatings to their substrates is high purity of the substrate surface prior to deposition of coating. Surface purity is understood here not only as the absence of greases, dust and other contaminants, but also as the removal from the substrate of all adsorbed compounds and layers of a non-metallic character, i.e., oxides, sulfides and other products of corrosion. Only then will adherence of the coating to the substrate be optimum, since it is dependent on the interatomic character of binding forces.

Theoretically, best adherence can only be achieved when the distance between the coating and the substrate is comparable to lattice parameters, i.e. when the crystalline structure of the coating is an extension of that of the substrate. Such an "extension of structure" is possible when lattice parameters of the substrate and the coating differ by not more than −2.4 to 12.5%. An adherence approaching the optimum can be achieved with very thorough cleaning of the substrate before depositing the coating. If the substrate is very well cleaned mechanically and, in particular, chemically, the adherence of a coating may approach the value equal to that of tensile strength of the weaker of the two materials, e.g., for a copper coating on steel, this adherence would be 300 to 400 MPa, i.e., equal or higher than tensile strength of copper [43].

It is thought that the adherence of a coating to substrate is better when the mutual solid solubility of coating and substrate components allows the formation of thin diffusion layers between substrate and coating [55].

Adherence of coating to substrate is affected, besides surface purity, by: residual stresses, degree of surface development (higher roughness - better adherence), as well as by differences in the ductility of surface and substrate materials. This is especially significant when the coated object is subjected to strong mechanical deformations [4].

Good adherence of coatings prevents their scaling and detachment due to temperature, strong internal interaction of a mainly mechanical character and residual stresses.

6.4.3.5 Hardness

Hardness of coatings is one of the most often determined parameters. For different coatings, it obviously differs and depends on the coating material and its structure. It varies within a very broad range, from the hardness of soft rubber to that of diamond (Fig. 6.4). Similarly to surface lay-

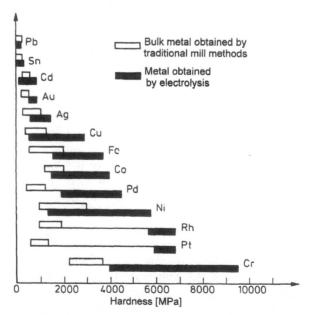

Fig. 6.6 Comparison of hardness of metals obtained by traditional metal mill methods (annealed or rolled) with that of electrolytically deposited. (From [43]. With permission.)

ers, we distinguish **macrohardness** and **microhardness** of coatings. Values obtained by both methods are not, as a rule, comparable. Usually, macrohardness constitutes a mean hardness value of a certain, quite big area; while microhardness refers to almost a point-size zone (the surface of a grain or a grain boundary).

Hardnesses of electroplated coatings are usually higher than those of same metals obtained by metallurgical means (Fig. 6.6). An increase in current density during deposition, as well as the presence of organic substances, including inhibitors which retard crystallization, refine grains and disturb microstructure, all cause a rise in coating hardness, while an increase of the temperature of deposition causes its drop [43]. The hardness of electroplated alloy coatings or of chemical amorphous coatings (e.g., Ni-P, Co-P) may be increased by heat treatment (aging). This is especially true of supersaturated solutions [55].

6.4.3.6 Ductility (elasticity)

The **ductility** of metal coatings, the equivalent of which in painted coatings is **elasticity**, is understood as the susceptibility of the coating material to plastic deformations without loss of cohesion (cracks, delaminations, etc.). The elasticity of a coating is of special importance in two cases: when the semi-finished coated product is subjected to deformation in subsequent operations or when the coated product is exposed to deformations during service. Coatings with low hardness and low brittleness are usually ductile.

6.4.3.7 Electrical properties

Electrical properties of coatings may vary and may pertain to conduction of electric current (electrical conductivity) or the ability to resist its flow (electrical resistivity), properties which insulate the object from the environment (electrical insulating power), resistance to electrical breakthrough and others. Metallic coatings are conductors or resistors, while non-metallic coatings are insulators. Most frequently, metallic coatings find application as the first two of the above.

Resistivity of metallic coatings depends on the absence of inclusions and on density. When they do not contain many contaminants, their resistivity is close to (differing by max. several percent) that of solid material, obtained metallurgically, as is the case with the resistivity of electroplated cadmium, gold, silver and copper coatings. On the other hand, resistivity of slightly contaminated chrome, lead, nickel and zinc-plated coatings is higher by 8 to 11%, cobalt coatings by 25% and rhodium coatings by as much as 90% than that of corresponding bulk material [55]. In practice, electroplated coatings, in particular, chemical, are deposited in the presence of various additives. For example, in nickel and chrome-plated coatings there are oxides, hydroxides and other non-organic coatings of these metals; in bright copper and silver coatings there occur absorbed organic compounds [55] which constitute the reason for the usually increased resistivity of the coatings.

Thin metal coatings, especially electroplated and chemical, are applied as conductive layers in microelectronic circuits or to coat electrical terminals (silver, palladium and gold, with a thickness of 0.5 to 2 µm) [55].

6.4.3.8 Magnetic properties

Magnetic properties are exhibited by some metallic coatings containing ferromagnetics, mainly iron, cobalt, nickel and their alloys with phosphorus. Binary alloys: Co-P, Ni-Fe, Ni-P and ternary: Ni-Co-P, Ni-Fe-P are characterized by high permeability and small coercion intensity. Alloy coatings like Ni-Fe containing 15 to 25% Fe are applied in computer memory components. Often, in the manufacture of magnetic coatings (by electroplating, chemical deposition or vacuum crystallization), the magnetic field is utilized in order to obtain a state of uniaxial anisotropy. This is the direction of easy magnetizing in which the state of saturation is obtained with lowest values of field intensity and is usually situated in the plane of the coating. Coatings with a preferred orientation of lowest energy of magnetizing are utilized in measuring instruments and in memory systems [55].

6.5 Service properties of coatings

The most important properties of coatings directly related to service are anti-corrosion, decorative and tribological. They are described by the same parameters as corresponding properties of the surface layer.

6.5.1 Anti-corrosion properties

6.5.1.1 Types of corrosion

Besides static and dynamic loads from extraneous forces, and the effect of friction, corrosion is one of the main hazards to which all structural materials are exposed during service. Unfortunately, it is the one type of hazard, as opposed to the two former, to which all materials are subjected, always and everywhere.

To this day, the most versatile structural materials most often used in technology are metals and their alloys. Among these, a privileged role is played by ferrous alloys, constituting more than 90% by weight of all produced metallic materials. The majority of these, particularly the alloys, do not occur naturally in the native state but in the form of compounds (metal ores), predominantly (although not limited to) with oxygen, and have a negative thermodynamic potential. This means that the majority of commonly used metals (iron, tin, nickel, cobalt, chromium, zinc, manganese, etc.) are thermodynamically unstable in conditions of practical usage. These metals and alloys exhibit a natural tendency to pass from a

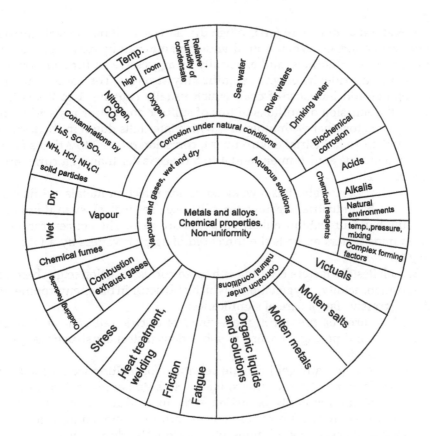

Fig. 6.7 Factors and environments causing spontaneous initiation of corrosion processes of metals and alloys. (From [62]. With permission.)

state of practical application, artificially created by man, to that of a thermodynamically stable one, i.e., such in which they occur in the earth's crust in the form of ores. The transition of the metal from the thermodynamically unstable to the stable state is a process of natural destruction of that which man created artificially and is known as **corrosion of metals**. Factors initiating or accelerating the process of corrosion may vary (Fig. 6.7) [62].

In the most general sense, **corrosion** is a process of gradual destruction of all materials, existing in all environments, as the result of chemical or electrochemical effect of the environment on these materials. All environments – air, gases, water, soil, particularly in the presence of oxygen, moisture and chemical compounds of sulfur, acids, alkalis, electric field and other factors - spontaneously act destructively (with very few exceptions) on all structural materials, i.e., metals and their alloys, concrete, rocks, ceramic, wood, plastics, etc.

Environments which most often affect structural materials, particularly metals, are

1. **Natural atmosphere (air)**: When dry and clean, it has a weak corroding effect on metals. When humid and contaminated by dusts (e.g., soot), gases, sulfates, chlorides, and especially by aggressive sulfur compounds, e.g., sulfur dioxide, its effect is significantly stronger. The combustion by households and industry of coal, which usually contains 1 to 2% sulfur, causes the emission of millions of tins of sulfuric and sulfurous acid to the atmosphere, together with moisture, which later falls down in the form of acid rain! In an industrial environment, corrosion rate is 2 to 3 times faster than in a marine environment and 4 to 5 times faster than in a rural environment [62].

2. **Water**: Pure water has a weak corrosion effect on metals, while contaminated by industrial wastewater, salts (e.g., sea water), even if not biologically harmful, exhibit enhanced aggressiveness. The concentration of SO_2 in mine waters exceeds the threshold of aggressiveness by a factor of ten [62].

3. **Soil**: Besides factors favoring enhanced corrosive action (e.g., chemical composition, acidity, oxygenation), which also occur in other environments, other aggressive factors are of significance:

– eddy currents: present in the vicinity of electric railroad and urban power lines and posing a hazard to subterranean steel constructions, pipelines, cables, etc.;

– salt: used in many cold climate countries for melting ice on roads, of particular danger to ferrous alloys, mainly to auto bodies and chassis.

4. **Special environments**: occurring in the form of gases, vapours or liquids, aggressive from the point of view of corrosion, and storaged in reservoirs (acids, alkalis, saline solutions, other chemicals), flowing through pipelines (gas, drinking and heating water, petroleum) or used for diverse technological purposes, also at elevated temperatures (furnaces, retorts, reactors, etc.).

With the development of civilization, the pollution of the natural environment increases through wastes, gases, dusts and smoke, enhancing corrosive aggressiveness with respect to material goods produced by man. The earth receives an annual average of approximately 12 tons/km^2 of precipitations of which much is acidic (mainly acid rain, but it may also be snow, hail and some fogs), because their acidity may sometimes reach 4 which is only slightly less than the acidity of a lemon. Acid precipitations began to occur with increasing frequency from the middle of this century in many European countries (Sweden, Germany, France, Czech Republic, Poland, Belgium and Norway), as a result of increased emission to the atmosphere of sulfur compounds (mainly SO_2), nitrogen compounds (NO_x) and volatile hydrocarbons, formed due to the ever-increasing combustion of coal, petroleum and its derivatives. The main "producer" of SO_2 are power stations and household fireplaces. In the case of NO_x the main culprit is automotive transport, as well as the already mentioned power stations. The emission of SO_2 has been reduced over recent years, due to a reduction in the consumption of coal and petroleum prodin terms of pure sulfur, annual precipitations reach several tons per km^2, while the emission of NO_x is on the increase due to ever-increasing traffic [63].

Most frequently, from a statistical point of view, metallic materials are subjected to atmospheric corrosion, less frequently, to water corrosion (including sea water) and to soil corrosion (including the effect of eddy currents). During service, metallic materials are also subjected to biological corrosion, caused by living organisms, intercrystalline corrosion (occurring along grain boundaries), stress corrosion (occurring as the result of simultaneous action of the environment and residual stresses), and fatigue corrosion (occurring as the result of simultaneous action of the environment and rapidly variable stresses induced by extraneous loads).

From a qualitative point of view, we distinguish the following types of corrosion [62]:

– **Chemical** - occurring as the result of direct action on metallic materials of dry gases, especially at elevated temperatures, or of liquid environments which do not conduct electricity;

– **Electrochemical** - caused by the action of short-circuited local corrosion sources, formed upon contact of metallic phases with an electrolyte.

Once initiated on the surface of a metal or alloy (surface layer or coating), chemical or electrochemical corrosion at first causes the creation of a thin layer of corrosion products (most frequently oxides or sulfides, less frequently nitrides, carbides, etc.) which with time increases in thickness and is often aided by other types of corrosion. This may lead to

– a total inhibition of further corrosion if the corroded layer covers the metal completely, does not dissolve in the surrounding environment, adheres tightly to the metal substrate and has a coefficient of expansion similar to that of the substrate. Such a mechanism occurs very seldom and then only in some metals (e.g., oxides on the surface of aluminum, protecting it from further oxidation);

– total destruction of the metal if the corroded layer does not meet the conditions quoted above. This is the case in the overwhelming majority of metals and alloys. Of the typically used metals, like lead, copper, nickel, zinc, iron and alloys like brass, bronze and steel, the least resistant to corrosion is the one used in most applications, on account of its strength and wear resistance, i.e., steel, primarily non-alloyed and low carbon. Its corrosion occurs in humid atmosphere.

6.5.1.2 Corrosion resistance

Ensuring protection against corrosion, i.e. imparting corrosion resistance, is the fundamental function of the majority of coatings. By corrosion resistance we understand it to be the ability of coatings to withstand the effects of different types of corrosion.

Taking into account that besides chemical corrosion there also occurs electrochemical corrosion, the corrosion resistance of coatings should be considered jointly with the substrate, with respect to which the coating may be either anodic or cathodic (see Section 6.3.2.1). Corrosion resistance of metals alone can only be considered when the bulk metals are thick and tight and have no surface defects which disturb their cohesiveness. When those conditions are met, in most cases the corrosion resistance of the coating is the same or better than that of the bulk metal [55].

The same object, coated with the same type of coating, exhibits different corrosion resistance in different environments. For that reason, a generalization of the problem of corrosion resistance is extremely difficult. It depends most significantly on: chemical composition, structure of the coating, three-dimensional structure of coating surface, on defects, residual stresses, type and condition of the substrate, type and intensity (temperature and concentration) of the corrosive medium and time of exposure.

In general, thick coatings offer better protection than thin ones. Moreover, the coatings should be tight and should ensure anodic or cathodic protection (depending on the material of substrate and coating and on the corrosive environment).

Corrosion resistance of coatings is determined experimentally by corrosion testing. All methods of corrosion resistance testing on specimens in laboratory (including accelerated testing methods) and natural conditions allow only an introductory evaluation of the behavior of coatings on real components in real service conditions, in a similar way as testing of the effect of the surface layer on the fatigue strength of specimens. An absolute indicator of corrosion resistance of a coating is the life of that coating on an object used in service in given conditions of external chemical, electrical, mechanical and other loads. The corrosion resistance of a coating which is not subjected to any loads may be good, but in given service conditions it may be subjected to constant, variable or impact-type loads, often in the presence of electrical or magnetic fields which significantly change the value of residual stresses and, in consequence, service life.

To a certain extent, the corrosion resistance of paint coatings depends on their tightness, permeability and resistance to swelling.

6.5.1.3 Porosity

Porosity is a characteristic of coatings, manifest by the existence in them of pores. It is usually determined by the ratio of joint volume of pores to the total volume of the coating.

Pores are understood as recesses in the coating in the form of narrow channels of diverse shapes and cross-sections, filled with substances which do not constitute coating material, like air and other gases, liquids, solids, etc. In a broader sense, cracks and scratches are also treated as pores, with the understanding that these are pores which are significantly extended, parallel to the surface.

From the point of view of size, pores may be *macroscopic* (visible with the unaided eye), *microscopic* (visible under a minimum 10 x magnification) and *submicroscopic* (invisible under an optical microscope) [4].

tion) and *submicroscopic* (invisible under an optical microscope) [4].

From the point of view of shape, the following types of pores are distinguished:

– *specific*: penetrating the coating from the substrate to the surface, where the pores may be perpendicular, inclined or curved relative to the coating surface;

– *masked* (blind): running in the coating from the substrate surface, narrowing down and closed or covered with the next coating layer; in particularly aggressive environments they can easily transform into specific pores;

– *superficial*: forming from the external surface of the coating and reaching inward but not as deep as the substrate.

In some coatings there may also occur *branched* pores [4] with irregular and complicated shapes.

Pores negatively affect tightness[1] of coatings, substantially reducing their corrosion resistance. This is true especially of coatings which are cathodic relative to the substrate metal and does not apply almost at all to anodic coatings. Porous coatings which are not tight do not assure total insulation from the surrounding corrosive environment, and do not totally inhibit the diffusion of aggressive agents through the coating which leads to the formation of local corrosion sources, sub-coating corrosion of the substrate, and blistering of the coating [4, 43].

The size and extent of porosity change during service. Only in exceptional situations do they close. In most cases they grow in a way which depends on the type of corrosive medium which enters the pores and cause accelerated bulging of paint coatings, as well their premature aging and loss of protective properties. Thicker coatings ensure better tightness because they contain proportionally more masked pores than thin coatings.

The most frequent causes of formation of pores in metallic coatings are defects of the metallic substrate, insufficiently clean substrate surface (contaminations in the form of oxides, sulfides, greases, oils, sand, dust, adsorbed gases, salts and polishing pastes), inappropriate technological processing during coating deposition, and chemical and mechanical effects (e.g., scratches) during deposition and service.

Some coatings, e.g., thermally sprayed, are porous, regardless of the method of spraying and their porosity stems from the very nature of spraying.

In paint coatings the number and size of pores depend on the amount of evaporated solvent and on the size of solvent particles. Coating materials with good fluidity exhibit a lesser tendency to formation of pores in the dried coating. Pores present at the moment of formation of the coating

---

[1] **Tightness** of coatings is the resistance to penetration of liquids and gases. A measure of tightness of coatings is the number of pores penetrating the coating to the substrate, per unit area. Coating tightness is a concept used mainly in electroplating [43].

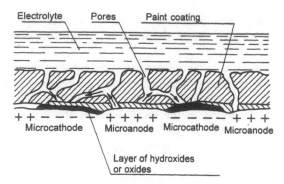

Fig. 6.8 Schematic representation of porous paint coating and a system of possible microcells caused by porosity at the surface of the metallic substrate. (From [46]. With permission.)

grow deeper and wider as the result of constant chemical decomposition of organic components of the coatings, due to the action of atmospheric oxygen and solar radiation. The porosity of paint coatings is the main cause of coating deterioration. All ionic reactions which cause corrosion of the metallic substrate occur as the result of the existence of pores which form a passage for the corrosive medium (including water) from the outside to the substrate (Fig. 6.8). Water coming in contact with an anodic metal substrate, e.g., iron, causes the transition of the iron to the solution which initiates corrosion. In the presence of moisture, the rust formed becomes a cathode relative to the iron and sub-coating corrosion progresses continuously, making the substrate non-homogenous. This favors the detachment of the coating. If the water comes in contact with the cathodic space, it reacts as an alkali and exerts a chemical effect on the paint coating [46].

6.5.1.4 *Bulging*

Bulging is the rise of volume of the paint coating due to absorption of liquids, most frequently of water. It will depend on the surface tension and the dielectric constant of the bulging liquid if dissolution, bulging, or solvation[1] will take place or not. Paint coatings constitute systems of macro-particles, connected into micelles (compounds of macro-particles) which under the influence of water and atmospheric moisture may solvatize, i.e., surround themselves with water particles or be subjected to the next stage of destruction, i.e., bulging. Water may pen-

---

[1] **Solvation** - the process of reaction of an ion or particle with particles of the solvent, resulting in the formation around the ion or particle of zones of loose groups of solvent particles with a smaller or greater degree of ordering. These zones are called solvates. When the solvent is water, this effect is called **hydration**. The amount of solvation depends on the charges and size of the ion or particle and on the type of solvent. The degree of solvation of the ion affects numerous properties of ion solutions, e.g., electrical conductivity, coefficient of diffusion.

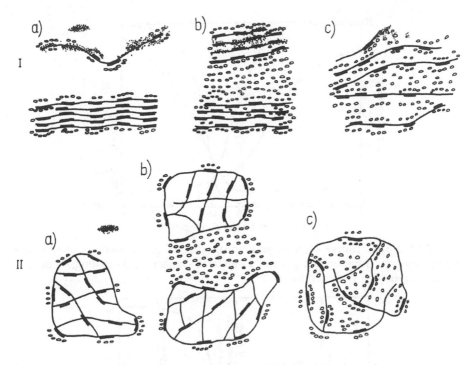

Fig. 6.9 Schematic representation of the effect of water on particles of chain (*I*) and on the microparticle or cluster (*II*): a) solvatation; b) externally micellar swelling; c) internally micellar swelling. (From [46]. With permission.)

etrate into the micelle and then the micelle swells (intramicellar bulging) or concentrates on its surface (extramicellar or intermicellar bulging). Extramicellar bulging is the first stage of absorprtion of the liquid (water) by the paint coating and is critical to the rate of diffusion of moisture in the coating. It may transform to intramicellar bulging (Fig. 6.9) [46]. The absorbed water may remain in the bulged coating causing its further degradation and creating an intermediate stage between solubility and non-solubility of the coating. If the absorbed water evaporates and the coating dries, its shrinkage occurs, but it will be smaller than the former volume increment caused by bulging. This absorption and expulsion of water (desorption) cause structural changes in the coating and its aging [46].

6.5.1.5 Permeability

Permeability is the ability to allow fluids to pass through a paint coating, due to porosity. It is closely connected with the ability of an organic coating to bulge. In the case of paint coatings, the focus is mainly on permeability of water vapour. Permeability of a coating by water vapor depends on air humidity and temperature and rises with their increase. The rate of diffusion rises by approximately 10%, with a rise of air temperature by 1 K, which causes that the humid tropical climate particularly favors permeation of water vapour into the coating. A greater permeability of the paint

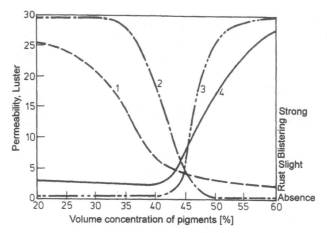

Fig. 6.10 Typical course of changes of properties of paint coatings, depending on volume concentration of pigments: *1* - luster; *2* - blistering; *3* - rusting; *4* - permeability. (From [46]. With permission.)

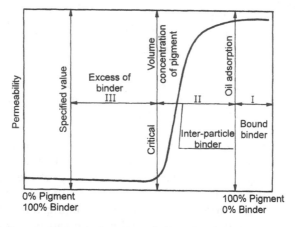

Fig. 6.11 Typical variation curve for permeability of water vapour within typical ranges of: I - adsorption of binder on pigment grains; II - spatial packing; III - free excess of binder. (From [46]. With permission.)

coating (which is opposite in concept to tightness of the electroplated coating) favors intensification of sub-coating corrosion.

The permeability of coatings depends on the type of coating substance. Polyvinyl and chlorolatex coatings are almost impermeable, on condition that all volatile components have been allowed to evaporate. On the other hand, oil and oil-resin coatings allow permeation of water vapour, depending on type of oil or resin.

Permeability of paint coatings depends strongly on the pigment content in the coating (Fig. 6.10) and on the appropriate quantitative and qualitative selection of the binder. Pigments act very favorably, limiting bulging of coatings exposed to moisture by making access of water vapour into the organic substance difficult (aluminum bronze, lead minium, lead litharge). Moreover, they passivate the metal surface (lead minium and chromate pigments), ensure electrochemical protection (zinc dust), neutralize acids permeating into the coating from the exterior or those formed within the coating, due to aging (alkaline pigments), and give the coatings their color. And for that reason, the percent proportion of pigments in the coating should be relatively high. A coating composed of almost only pigments would have high permeability, while a coating with almost only binder - very low permeability (Fig. 6.11). The binder, however, does not exhibit protective properties or ones that color the coating. Pigments are bonded in a stable manner to particles of the binder by Van der Waals forces. The remaining part of the binder fills free space between the particular particles of the pigments or their agglomerates in the case of close packing (so-called interparticle binder). The optimum volume proportion of pigments to binder is below the critical volume concentration of pigments, i.e., below the bend of the curve in Fig. 6.10.

6.5.2. Decorative properties

6.5.2.1. External appearance

All coatings, to a greater or lesser extent, feature decorative values, but special decorative properties are required of decorative coatings, as well as protective-decorative paint, electroplated and vacuum deposited coatings.

The basic criterion of a coating's decorative value is its external appearance. Of all the properties of coatings, external appearance is the easiest to evaluate because it can be done immediately, by visual means. Visual methods may also be used to evaluate not only the external appearance but also the quality of the coating and - before their deposition - the quality of the substrate or the particular layers (primers, intermediate layers, etc.)

Eyesight, while presenting the observer with aesthetic sensations, similarly to any organoleptic method of quality assessment, is a subjective factor. A subjective evaluation depends on visual acuity, absence of sight defects (in particular color-blindness), the effect of external factors (color and lighting intensity, presence of dust or smoke in the air).

The external appearance of almost all coatings deteriorates with time of service, causing coating aging. Exceptions to this rule are coatings made to look like the patina. The older they get, the more they resemble real old coatings covered by natural patina.

The most important factors taken into consideration when evaluating the external appearance of coatings are color, luster, smoothness (opposite of roughness) and the ability to cover the substrate. These properties can be not only evaluated visually but also measured in an objective way, similarly to the resistance of coatings to intense ultraviolet and infrared radiation, as well as to tarnishing.

6.5.2.2 Color

The concept of color has two meanings [65, 66]:

– that of a physical property of light from a coating illuminated by electromagnetic radiation in the visible range (of 0.36 to 0.76 μm),

– that of a psychological property of a visual sensation which allows
the observer to distinguish differences in light stimuli caused by differences
in the spectral distribution of the stimulus; visible radiation reflected by the
coating surface (or its external layer) enters the eye and stimulates photosensitive elements of the macula, giving a sensation of color [67].

The color may be treated as a subjective experience of the eye, caused
by light radiation reflected by the coating. The spectrum of visible radiation is composed of 6 basic colors (violet, blue, green, yellow, orange and
red) which may form the so-called color wheel by adding intermediate
colors. This color wheel is composed of 12 or 23 chromatic colors. Besides
these colors there are also achromatic colors like white, black and gray
with various shades [46].

The following color characteristics are distinguished:

– the **color** itself - dependent on the wavelength of light radiation, reflected by the coating. This is a qualitative characteristic, described by a name,
e.g., green, red, etc.

– **saturation** - dependent on the degree to which the color is closer to
white or black;

– **purity** - dependent on the width of spectral band, i.e., on additions of
other colors. The purity of a color is highest when the coating reflects radiation monochromatically (as one color);

– **brightness** - dependent on the intensity of radiation reflected by the
coating.

Coating colors stem from

– the nature of the components forming the metal or ceramic coating,
be it electroplated or deposited chemically, by immersion, spraying or
overlaying. In all these case the influence on color is small. Only some
metallic coatings may be colored. Also, different types of coatings may
have the same color, e.g., both gold and titanium nitrided coatings are
yellow;

– pigmentation of paint materials or ceramic enamels, i.e., introduction of
pigments into the coating composition. Pigments may be organic or inorganic coloring substances, practically insoluble in water and exhibiting the
ability to color paints and varnishes, as well as ceramic enamels in the undissolved condition. The ability to color paint materials increases with pigment refinement.

Besides offering aesthetic sensations, colors have their own way of affecting the human psyche, as well as the physiological and physical changes
which take place in the human organism. The force of color action is called
color dynamics. For example, the application of cold colors in hot industrial
production rooms and warm colors in cold rooms affects the sensing of temperature by the organism. Red color surrounding man from every side pro-

duces excitation and nervousness, yellow - brings on a happy mood, green may act depressively on neurotics, some shades of brown may cause a feeling of sadness; white retards the functions of the brain while black has an unfavorable effect on people who easily succumb to psychological depression [46].

For those reasons, as well as visibility, bodies and fixed components of machines are painted with such colors which attract least visual attention (light gray, light green, light green-gray). Moving parts are painted with colors which easily attract attention even in bad lighting conditions (yellow, canary and light orange). Stamps, levers and valves are usually painted with colors that strongly stand out and attract the eye (bright yellow, vermilion, orange or turquoise) [46].

6.5.2.3. Luster

Luster is a property of the surface of a smooth coating (or surface layer) consisting of oriented reflection of radiation falling on it in such a way that clear images of bright objects are formed in the field of vision of the observer. The smoother the surface, the more ordered is this reflection and the more equal is the angle of incidence to the angle of reflection. The more luster the coating has, the more mirrorlike it is [68].

The degree of luster is described by the ratio of the coefficient of oriented reflection of the observed surface to the coefficient of total reflection. The numerical value of this degree of luster varies from zero (ideally dispersive surface, practically non-existent with approximate properties exhibited by coarse, rough and matte surfaces, e.g., those obtained by thermal spraying) and unity (ideally reflecting surface, practically non-existent, with approximate properties exhibited by very smooth, polished surfaces, i.e., mirrorlike). An example of the latter is the surface of an electroplated coating with addition of brighteners, deposited on an ideally smooth surface [69].

The degree of luster of coatings decreases with time of service, as the result of aging and absorption of particles from the environment. Moreover, in subjective observation it depends on lighting conditions, angle of viewing, acuity of contrast of a visible object, seen as a reflection by the surface. Highest luster is exhibited by metallic electroplated coatings, mechanically and chemically polished, some vacuum deposited coatings, as well as by paint coatings. In the case of paint coatings it depends on the type and amount of pigment in the coating and on the degree and uniformity of its dispersion in the coating material.

The type and intensity of luster affect the psychological sensations caused by colors. They may be enhanced or weakened (Table 6.2). Luster deepens the vitality of colors of painted surfaces (particularly of the golden color), vitalizes gray tones, regarded as devoid of expression, attenuates the somber and depressing appearance of the black coating, gives green the feeling of coolness and peace [43].

Table 6.2
Effect of luster on man's psychological sensations (From [43]. With permission.)

Type of luster	Gives the colored surface
Strong	fiery appearance, vitality of color, majestic look, ability to create sensation of being flooded with light
Silky, shiny	vitality, lightness, transparency, brightness
Matte	serenity, gentleness, seriousness without causing depression
Typical of translucent coatings	glassiness, cold, serenity of stationary surface of water

6.5.2.4. Coverability

Coverability describes the degree of coverage of the substrate by the coating material and has two meanings:

– the ability of the organic (paint) coating or inorganic (enamel) coating to cover the substrate or the primer or an intermediate layer by the deposited coat of top paint. This is connected with the ability to selectively reflect light radiation falling on the coating, caused by differences in refraction by the structural components of the coating, mainly by pigments and the binder, relative to air. The greater these differences, the less light penetrates into the coating, due to full reflection on transition from an optically denser medium (pigment) to a less dense medium (binder). Coverability increases with a rise in refinement of pigments but only up to a point. When the refinement is excessive, coverage decreases. Pigments with a crystalline structure give better coverage, due to edge refraction;

– the ability of the electrolyte to deposit the coating metal in recesses of the substrate. It is described by the value of area not covered or by the value of current density needed to precipitate a coating on a cathode bent at an angle. Sometimes, coverability is understood as the ability to cover recesses by an electroplated coating with given decorative properties, e.g., glance.

Coverability of paint coatings undergoes insignificant changes with time of service (its color changes more and at a higher rate).

6.5.2.5. Specific decorative properties

Among specific decorative properties of coatings, pertaining primarily to paint coatings, are the following:

– ability of the surface coating to fold - characteristic of rippled and crystalline coatings, containing among its components tung oil, dehydrated castor oil, as well as alkyd resins, modified by these oils or by their fatty acids;

– ability to form dimples on the top surface of the coating - characteristic of a hammer finish (from fast-drying nitrocellulose or synthetic varnishes) or mosaic finish surface;

– ability to form the "crocodile skin" effect - characteristic of coatings with at least two layers: "rich" primer" (e.g., oil paint) and "lean" enamel with a high pigment content;

– ability to reflect in preferred orientations - characteristic of reflective coatings, containing glass pellets with diameters up to several tens of micrometers;

– fluorescence, phosphorescence, radioactivity - characteristic of coatings which feature fluorescent, phosphorescent or radioactive shine in which, in order to initiate the effect not only light is utilized but also radioactive substances, introduced into the coating composition;

– ability to dull (lose luster) - characteristic of matte coatings.

It should be noted that in principle, decorative effects, especially the above-mentioned specific ones, weaken protective properties of the coating.

6.6 Significance and directions of development of coatings

Coatings primarily play a protective role - by protecting the substrate material against various types of corrosion. They may also fulfill a decorative role as a sideline. They are only seldom applied for solely decorative purposes (if so, mainly in building construction). More often, they are used for technical purposes, mainly for enhancement of tribological properties and for repairs.

The significance of coatings in technology is derived mainly from their anti-corrosion role. Corrosion, by destroying materials, causes certain economic effects, classified as [62]:

1. Losses due to corrosion, including

– *direct losses* - stemming from a lack of protection, inappropriate or insufficient protection against corrosion, or those occurring despite good protection which, however, does not act infinitely. These losses comprise cost of components or objects physically destroyed, costs of repairing failures, overhauls and costs stemming from shortened life of components, devices and objects;

– *indirect losses* - stemming primarily from the need to remove the effects of corrosions, e.g., down-time of production installations, public utility plants (e.g., water works), contaminations of products and of the environment, fines to pay, etc.

2. Investment costs for anti-corrosion protection, comprising costs of materials, labor and machinery, cost of material stock to accommodate corrosion, cost of maintenance of already applied corrosion protection, as well as costs of research and development of corrosion protection.

The distribution of costs varies from country to country. On an average, in the productive sector, effects of corrosion amount to approximately 70% in losses and approximately 30% in expenses on anti-corrosion protection.

Strict computation of the economic effects of corrosion is extremely difficult, due to the very high cost of carrying out such research, the universal nature of corrosion, its many and varied effects, irrationality in the evaluation of corrosion damage, subjective assessments and practical impossibility of accurate quantification of all negative effects of corrosion. In some parts of the world, primarily in industrialized countries, such analyses have been made, even repeatedly. More or less approaching reality, such analyses allow the following conclusions:

– corrosion losses grow incessantly, especially dynamically in less industrialized countries; a rational approach and economical possibilities of highly industrialized countries allow a limitation or retardation of the growth rate of these losses, at the cost of a rise in expenditure on anti-corrosion protection,

– economic losses due to corrosion may even reach 5% of national income

– 15 to 35% of corrosion losses may be avoided by the application of appropriate anti-corrosion protection, especially of steel products.

Methods of counteracting corrosion are many and varied and, in general, comprise two areas of activity [63]:

– **Indirect**: involving **creation of conditions for maximum reduction corrosion hazard** for components, products and constructions used in production. These conditions may be reduced to the following groups of problems.

1. Reduction of pollution of the natural environment by precipitations, wastes, smoke, dusts of industrial, communal or household origin, and their utilization, often combined with recycling of components in short supply. For example, on an industrial scale in:

– **steelmaking** - it means reduction of pollution of the atmosphere by the application of filters which absorb solid particles from smoke, desulfurize exhaust gases and remove from them other components, and the application of catalytic coatings in heating installations, for the purpose of reducing the emission of harmful NO_x-es;

– **electromachine industry**, in which production processes used are usually burdensome to the environment, e.g., forging, heat treatment and machining, pickling, degreasing, washing, surface cleaning, abrasive treatment and polishing, electrolytic and electroless deposition of metals, production of conversion coatings, zinc plating, hot dip aluminizing, metal spraying, explosive cladding; painting which produced toxic wastes, both liquid (effluents and used technological solutions), solid (post-neutralization deposits, metals, etc.) and gaseous (different gaseous compounds) - it means purification and neutralization of liquid effluents, especially those containing cyanide and toxic heavy metals from electroplating and pickling, of used oils and emulsions from washing, degreasing, machining and heat treatment; paint wastes and hydrated deposits originating from neutralization of wastes from electro-plating and pickling shops [62, 70].

2. Replacement of energy-consuming technologies used in production by technologies which are energy-efficient, including high energy electron beam, glow, plasma and induction technologies, which, allowing a reduction of consumption of primary fuels (coal, petroleum, natural gas), also alleviates environmental pollution, especially by sulfur compounds, as well as reduces risks caused by acid rain. A similar role is played by the growing utilization of natural sources of energy (solar, wind, geothermal and hydroenergy).

3. Maximum degree of elimination of usage of these structural materials which are especially susceptible to corrosion (thus, naturally of steel), and their replacement by materials which are more resistant (e.g., aluminum, synthetic materials, various composites).

4. Use of structural materials amenable to low energy recycling after service which implies preference of aluminum over steel.

5. Creation of artificial anti-corrosive atmospheres by tight packaging of finished products, coupled with the introduction of various corrosion inhibitors between the protected object and the packaging, including objects destined for the tropical climate.

– **Direct**: involving design of appropriately resistant materials or **protecting** of components, products or structures by the deposition of protective surface layers, more resistant than the substrate material in the working environment. These can be summed up as

1. Application of structural materials which are resistant to the environment in which they work, e.g. use of austenitic stainless steel and special materials in the chemical industry and in nuclear energy. These materials should be so designed that the appropriate material exhibit maximum resistance to corrosion by sulfates, acids, pitting (chemical), welds, cavitation, fatigue, stresses, friction, contact or high energy, all this with retention of appropriate strength. This problem is not of a universal character and pertains only to special applications. Nevertheless, it is of great significance from a practical aspect. These are so-called tailor-made materials, custom designed for the user's needs [62, 63].

2. Development on high strength materials (usually steel) which are also not highly corrosion resistant, of surface layers of high corrosion resistance or coating them with corrosion-resistant coatings.

Organic coatings, primarily paint, play by far the most important role in corrosion protection. Of the approximately 95% of surfaces of steel structures protected against corrosion by protective coatings, as much as 90% are protected by paint. The life of these coatings ranges from several months to between ten and twenty years. In this area we can distinguish the following trends of development [70]:

– reduction of the percent share of paint materials based on traditional binders, i.e., organic solvents, particularly of costly and harmful aromatic hydrocarbons (xylene and toluene) and an increase in the percent share of paint materials based on synthetic resins (including water-soluble dispersive paints) [2];

– a rise in the application of paints with a high content of zinc dust to cover cold rolled auto body sheet, as a standard structural material in the automotive business;

– elimination of liquid substances from paints and varnishes and transition to the powder form; elimination of traditional techniques of application of paint with a paintbrush or the pneumatic varnish gun and their replacement with electrostatic pistols for depositing liquid and powder paints. This allows better material economy than with the use of pneumatic pistols, among other reasons, because of the possibility of its recycling. Other replacements should include coating rollers, allowing an almost 100% utilization of paint material, extruders, often coupled with drying or hardening by a radiator, electron beam or ultrasonic energy. With pneumatic painting the material efficiency (relative to dry mass) is approximately 50%, while the remaining 50% of the solvent evaporates and pollutes the atmosphere, and the utilization of the coating material is only 25% [2];

– rise in production and application of aluminum sheet, anodized and varnished in a continuous operation;

– increase and greater diversity in applications of coatings offering temporary protection against corrosion, especially in the case of different means of transportation, predominantly automotive, with underside chassis protection as well as of interiors of closed profiles.

Electroplated coatings take second place in widespread usage, after paint coatings, and are characterized by an exceptionally high diversity. Among these, of greatest significance from the aspect of corrosion protection are zinc coatings, decidedly replacing the until recently most popular three-layer copper-nickel-chrome coatings. The metal which is now most often used in electroplating in place of nickel is zinc [2].

Zinc galvanized sheet captures the interest of the automotive industry because, in numerous cases, hot dip zinc coatings are excessively thick and not always uniform. Electroplated zinc sheet has been manufactured by the industry throughout the world for many years. The aim is to broaden this production to include auto body sheet with alloy Zn-Fe, Zn-Ni and also recently Zn-Mn coatings. A good future is also predicted for zinc coatings with an addition of 0.5 to 1% Co and Sn to coat metallic components of firearms, as substitute of the commonly used cadmium coatings. Cadmium is suspected of having a carcinogenic effect!

Finally, there should also be a rise in the application of lead-base coatings, including lead alloys, primarily lead-tin, as well as the replacement of white tin-coated sheet by other coatings [70].

The application of double and triple component coatings, as well as amorphous coatings, allows control of surface properties, especially when modifying them by concentrated streams of photons (laser technique) or by ions (ion implantation).

Hot dip coatings come third, after organic and electroplated, in terms of widespread use. These include, primarily, hot dip galvanizing (zinc) and hot dip aluminizing, as well as lead and tin. In 1976, of all hot dip coatings deposited worldwide 89.6% were zinc, 5.7% were aluminum, 3.6% were lead and 1.3% tin. Presently, application of hot dip aluminum coatings (especially aluminum alloy) is on the rise and a further development in this area is expected. Among hot dip coating techniques, of special interest are the following four which are expected to find increasing application [63]:

– replacement of traditional hot dip galvanizing of sheet - by alloys of zinc with titanium, which clearly improves corrosion resistance of this type of coating to acid rain; sheet coated with zinc with small additions of titanium (approximately 1%) is commonly used in northern France and Belgium for roof making and other steel sheet jobs;

– continuous coating of sheet and strip with zinc alloys (4.7 to 6.2% Al; 0.03 to 0.10% Cr + Zn);

– zinc-aluminum coatings (55% Al; 43.4% Zn; 1.6% Si). These techniques are currently replacing straight zinc coatings, reaching approximately 10 million tons of mainly steel sheet annually;

– continuous and no-continuous aluminizing, allowing the formation of coatings with high corrosion resistance in water, industrial atmospheres and gases containing sulfur compounds, as well as very good heat resistance. Aluminized carbon steel exhibit lasting heat resistance at temperatures up to 850°C and temporary heat resistance up to 1000°C. Components which have been hot dip aluminized are particularly well suited for heat transmission ducts and for heating appliances (portable cookers, furnaces, recuperators).

Thermally sprayed coatings: deposited by flame, induction, arc, plasma or by explosive techniques, may be generated on the substrate surface or used for the repair of worn substrates or coatings, the life of objects regenerated in this way usually being longer than that of original ones. Thermal spraying methods may be used to deposit anti-corrosion coatings, especially zinc, and aluminum, and composed of their alloys with other materials. It is not, however, this function alone that creates expectations of further development for these techniques. It is the possibility of achieving results which are unattainable by other methods. These are the possibilities of making coatings which are resistant to high temperatures and also resistant to erosive and abrasive wear. This is made possible by the fact that after deposition they may be additionally sealed or alloyed by an electron, photon or ion beam. Among such applications are coatings resistant to gas corrosion, deposited on turbine blades in conventional and nuclear power plants and in aerospace (nickel, cobalt, aluminum and yttrium alloys), insulating coatings (usually oxide), coatings resistant to high temperature (zirconium oxide, nickel-chrome oxide, powder metals), and coatings resistant to abrasive wear (tungsten carbide). Particularly attractive properties may be obtained by plasma spray and explosive spray techniques and these methods are expected to find application in very special areas, although their use will not be as widespread as that of gas and arc spraying.

Enamel coatings are not expected to undergo intensive development and in some cases may even be replaced by synthetic and organic coatings [63].

Cladded (plated) coatings, used mainly for enhancing resistance to atmospheric and gas corrosion at elevated temperatures and in chemically aggressive environments, will not find an increasing number of applications [63].

Overlay coatings: pad welded (deposited with the application of welding techniques) and **melt** (deposited with the application of laser, electron or electrode discharge heating), as well as **alloy overlays** (by laser or electron beam) are finding increasingly broad application, predominantly in areas where extreme conditions are expected in service, e.g., high loading forces, high temperature, corrosion hazard. They are facing a favorable future and expected to undergo intensive technical development. There is a great diversity of materials applicable for these coatings, ranging from austenitic stainless steels to refractory metals and alloys to metal ceramic. An intensive development is expected in methods of spot melting, melting, and alloy overlays. This is particularly significant due to the possibility of further modifying properties of such layers and coatings obtained by additional ion implantation of practically any chosen element [63].

Thin and very hard coatings, vacuum deposited by CVD and PVD techniques, mainly to enhance service life of tools, particularly cutting tools (2 to 3 times on the average) and machine components, as well as to achieve significantly favorable decorative properties (handwatches, scissors, surgical instruments), also feature high resistance to corrosion in atmosphere, body fluid and some technologically used liquids. Although used since the early 1980s, these coatings have undergone significant development only in the last several years and are expected to develop rapidly, especially CVD. From the standpoint of tonnage or area of coated surface, they play a small role in corrosion protection but their technical significance is big. In future, a development in their application is expected both as single layers (carbides, nitrides, borides or oxides of iron, chromium, titanium, tantalum, aluminum and boron) and as complex layers (e.g., titanium nitride + titanium carbides or TiAlN), especially when combined with diffusion processes (e.g., nitrided layer + TiN or TiAlN layer). From a technical and technological point of view, we should note the current development in the direction of lowering CVD process temperatures due to the application of auxiliary heating, mainly by glow discharge, and raising of PVD process temperatures in order to achieve better binding of coating to substrate [63].

Table 6.3 lists the percentage share of costs of different methods of obtaining surface layers, with special emphasis on coatings, in the United Kingdom in 1991 [71].

Table 6.3
Percentage share of costs of obtaining different types of surface layers in the United Kingdom in 1991 (From *Gawne D.T., Christie I.R.* [71]. With permission.)

Technique	Share[%]
Industrial paint	50
Electroplating	20
Surface heat treatment	9
Galvanizing	6
Polymer powder coating	4
Phosphating, chromating	3
Thermal spraying	3
Anodizing	2
Blasting and peening	2
New technologies (PVD, CVD, laser, plasma, etc.)	1
Other (enameling, friction, mechanical, etc.)	1

References

1. ISO/R 2080-1971. Electroplating and related processes. Glossary of terms.
2. ISO/R 919-1969. Guide for the preparation of classified vocabularies.
3. **Tkaczyk, S.** (ed): *Protective coatings* (in Polish). Student script. Silesian Technical University, Gliwice, Poland 1994.
4. **Biestek, T., and Sękowski, S.**: *Methoden zur Prüfung metalischer Überzüge.* Vieweg Verlag, Braunschweig 1971.
5. **Biestek, T., Socha, J., and Weber, J.**: *Modern methods of producing conversion and metallic protective coatings* (in Polish). PWT, Warsaw 1960.
6. **Uhlig, H.H.**: *An introduction to corrosion science and engineering.* John Wiley & Sons Inc., New York 1971.
7. **Shreir, L.L.** (ed): *Corrosion,* Vol. 1 - *Corrosion of metals and alloys,* Vol. 2 - *Corrosion protection.* Georg Newnes Tld., London 1963.
8. **Tomashov, N.D.**: *Theory of corrosion and protection of metals* (in Russian). Publ. Masgiz, Moscow 1960.
9. **Porejko, S., Fejgin, J., and Zakrzewski, J.**: *Chemistry of macromolecular compounds* (in Polish). WNT, Warsaw 1968.
10. **Kacprzak, J.**: *Nitrocellulose varnishes and enamels* (in Polish). Warsaw 1954.
11. Joint report: *The varnisher's handbook* (in Polish), Warsaw 1964.

12. Joint report: *New developments in the field of paints and varnishes* (in Polish). Series: *New Technology*, Vol. 31, WNT, Warsaw 1964.
13. **Zassowski, J.**: *Manufacture of oil-based varnish products* (in Polish). PWT, Warsaw 1954.
14. **Dobrowolski, A.**: *Chemistry and technology of waxes and pigments* (in Polish). PWT, Warsaw 1963.
15. **Holtrop, W.**: *Iron oxide-based pigments* (in Polish). PWT, Warsaw 1952.
16. **Brojejr, Z., Herz, Z., and Penczek, S.**: *Epoxy resins* (in Polish). PWT, Warsaw 1960.
17. **Lazaryev, A.I., and Sorokin, M.F.**: *Synthetic resins* (Polish translation from Russian). PWT, Warsaw 1957.
18. Joint report: *The fat industry worker's manual* (in Polish). WPLiS, Warsaw 1958.
19. **Gajdek, S.**: *Acid and alkali-resistant cements and concretes* (in Polish). PWT, Warsaw 1955.
20. *Encyclopedia of Technology: Metallurgy* (in Polish). WNT, Warsaw 1969.
21. **Warth, A.H.**: *The chemistry and technology of waxes.* New York 1960.
22. **Losikow, B.W., and Łukaszewicz, J.P.**: *Petroleum products science* (in Polish). PWT, Warsaw 1953.
23. **Abraham, H.**: *Asphalts and allied substances.* D. von Nostrand Co., Canada 1962.
24. **Petryn, T., and Śmiechowski, R.**: *Low friction lubricants* (in Polish). WNT, Warsaw 1962.
25. **Solik, J., and Troszok, A.**: *The technology of lubrication.* Handbook (in Polish). PWT, Warsaw 1960.
26. **Zawadzki, J.**: *Protection and packaging of metal products* (in Polish). WNT, Warsaw 1962.
27. **Zawadzki, J.**: *Temporary protection of metals* (in Polish). WNT, Warsaw 1962.
28. **Lokshin, V.I.**: *The technology of enameling of metal components* (in Russian). Publ. Rozgizmestprom, Moscow 1963.
29. **Tomsia, S., and Zapytowski, B.**: *The technology of enamelling industry* (in Polish). WPLiS, Warsaw 1960.
30. **Gibas, T.**: *Sintered ceramics and cermetals* (in Polish). WNT, Warsaw 1961.
31. **Milewski, W.**: Treds in the development of thermal spraying (in Polish). Proc: Conference on *Techniques of producing surface layers on metals*, Rzeszow, Poland, June 1988, pp. 135-143.
32. **Burakowski, T., Miernik, K., and Walkowicz, J.**: Manufacturing techniques of thin tribological coatings with utilization of plasma (in Polish). *Metaloznawstwo, Obróbka Cieplna, Inżynieria Powierzchni (Metallurgy, Heat Treatment, Surface Engineering)*, No. 124-126, 1993, pp. 16-25.
33. **Biestek, T., and Weber, J.**: *Electrolytic and chemical conversion coatings.* PPL Redhall - WNT, Warsaw 1976.
34. **Chudzyński, S., and Krajewski, B.**: *Application of polymers in industry and in everyday life* (in Polish). PWT, Warsaw 1958.
35. **Brag, E.**: *The technology of polymers* (in Polish). PWN, Warsaw 1957.
36. **Surowiak, W., and Chudzyński, S.**: *Polymers in the machine building industry* (in Polish). PWT, Warsaw 1960.
37. **Bojarski, J., and Lindeman, J.**: *Polyethylene* (in Polish). WNT, Warsaw 1963.
38. **Mark, H., and Tobolsky, A.**: *Physical chemistry of polymers* (in Polish). PWN, Warsaw 1957.
39. **Kowalski, Z.**: *Polymer coatings* (in Polish). WNT, Warsaw 1961.
40. **Koszelew, F..**: *The technology of rubber* (in Polish). PWT, Warsaw 1956.
41. **Porayski, T.**: *The rubber industry calendar* (in Polish). PWT, Warsaw 1957.

42. **Boström, S. et al.**: *Kautschuk-Handbuch*. Berliner Union, Stuttgart 1959/60.
43. Joint report: *The electroplater's handbook* (in Polish). WNT, Warsaw 1957.
44. **Czarnecki, T., Dębicki, M., and Marczak, R.**: *Non-lubricant protection of metal products* (in Polish). Publ. MON, Warsaw 1961.
45. **Juchniewicz, R.**: *Problems of metal corosion* (in Polish). PWN, Warsaw 1965.
46. Joint report: *Paint and varnish coatings*. Handbook (in Polish). WNT, Warsaw 1983.
47. **Socha, J.**: *Gold electroplating* (in Polish). Institute of Precision Mechanics, Warsaw 1979.
48. **Socha, J., and Safarzyński, S.**: *Protective-decorative layers on copper and its alloys* (in Polish). Intitute of Precision Mechanics, Warsaw 1988.
49. Joint report: *The varnisher's handbook* (in Polish). WNT, Warsaw 1964.
50. Joint report: *The mechanical shop handbook*, 8th edition (in Polish). Chapter XXIV: *Corrosion of metals and protective coatings*. WNT, Warsaw 1981.
51. **Morel, S., and Morel, S.**: Catalysis with the aid of ceramic coating, of carbon monoxide combustion in gas exhausts (in Polish). Proc.: II International Conference on *Effect of technology on the condition of the superficial layer - WW '93* Gorzów, Poland, 20-22 Oct. 1993, pp. 291-300.
52. **Morel, S.**: Thermal spray deposition of coatings for extension of machine and appliance service life (in Polish). *Przegląd Mechaniczny* (*Mechanical Review*), No. 14, 1989, pp. 29-31.
53. **Lajner, W.I., and Kudryavtsev, N.T.**: *The fundamentals of galvanostegy*, Vol. 1-2 (in Polish). WNT, Warsaw 1955-60.
54. **Grześ, J.**: *Usable properties amd microstructure of selected Ni-W-Co coatings, obtained in the process of tampon deposition* (in Polish). Ph.D. thesis. Warsaw Technical University, Warsaw 1991.
55. **Kozłowski, A., Tymowski, J., and Żak, T.**: *Manufacturing techniques. Protective coatings* (in Polish). PWN, Warsaw 1978.
56. **Kowalski, Z.**: *New methods of producing coatings from polymers* (in Polish). PWT, Warsaw 1960.
57. Joint report: *Metal spraying handbook* (in Polish). PWT, Warsaw 1959.
58. **Kowalski, Z., and Bagdach, S.**: *Metallization of polymers and other non-conductors* (in Polish). WNT, Warsaw 1965.
59. **Wesołowski, K.**: *Metallurgy* (in Polish). Vol I - 1959, Vol. III - 1966, PWT, Warsaw 1966.
60. **Tsaruchina, R.E., et al.**: *Bimetallic joints* (in Russian). Publ. Metallurgia, Moscow 1970.
61. **Golovanenko, S.A., and Meandrov, L.V.**: *Conductivity of bimetallics* (in Russian). Publ. Metallurgia, Moscow 1966.
62. Joint report: *Corrosion protection*. Handbook (in Polish). Telecommunications Publishing, Warsaw 1986.
63. **Burakowski, T.**: Current state and development trends in corrosion protection (in Polish). *Metaloznawstwo, Obróbka Cieplna, Inżynieria Powierzchni* (*Metallurgy, Heat Treatment and Surface Engineering*). No. 115-117, 1992, pp. 43-50; Current state and development trends in corrosion protection. *Przegląd Mechaniczny* (*Mechanical Review*), No. 8, 1993, pp. 13-16 and 21-22; *Mechanik*, No. 8-9, 1993, pp. 309-314.
64. **Kortum, G.**: *Elektrochemia* (Polish translation from German). PWN, Warsaw 1971.
65. **Zausznica, A.**: *The science of color* (in Polish). PWN, Warsaw 1971.
66. **Felhorski, W., and Stanioch, W.**: *Colorimetry* (in Polish). WNT, Warsaw 1972.
67. **Starkiewicz, W.**: *The psychology of sight* (in Polish), PZWL, Warsaw 1972.

68. Joint report.: *Light* (in Polish). PWN, Warsaw 1972.
69. Joint report.: *Light technology*. Handbook (in Polish). PWT, Warsaw 1960.
70. **Kozłowski, A.**: *Scientific and technological problems critical to the development of protective coatings technology at the end of the XX century* (in Polish). Institute of Precision Mechanics, Warsaw 1987.
71. **Gawne, D.T., and Christie, I.R.**: The UK surface engineering industry. *Metals and Materials*, December 1992, pp.646-649.

The newest techniques of producing surface layers

chapter one

Formation of technological surface layers

1.1 Techniques of formation of technological surface layers

In the overwhelming majority of cases, surface layers are formed before the beginning of their service, by subjecting the object to a technological treatment process - these are *technological surface layers*. Only in exceptional cases are surface layers produced on objects during service, e.g., layers formed during low wear friction - these are *service-generated surface layers*. The following discussion relates to only the first kind of layers.

Depending on the type of effects utilized to form surface layers [1-8], all techniques[1] of formation may be generally divided into six groups (Fig. 1.1): mechanical, thermo-mechanical, thermal, thermo-chemical, electrochemical and chemical, and physical.

Each group of techniques allows the obtaining of a specific type of surface layer, of given thickness and application, and may be subdivided into many types. The same types may be accomplished in different ways (Fig. 1.2).

Surface layers may be formed either by one technique (the most frequent case) or by a combination of techniques (less frequent but fast growing) [3].

1.1.1 Mechanical techniques

In mechanical techniques, the utilized effect is the pressure of a tool or the kinetic energy of a tool or particles (burnishing) in order to strain harden the superficial layer of a metal or alloy at room temperature, or to obtain a coating on a cold metal substrate. This is accomplished by static burnishing, dynamic burnishing, explosive spraying (deposition) and by machining [4].

---

[1] Technique (in this case: of formation of surface layers) - a combination of actions and means, based on the utilization of same or similar effects, aimed at the accomplishment of a given task, e.g., formation of a surface layer, strengthening of the superficial layer, deposition of coatings.

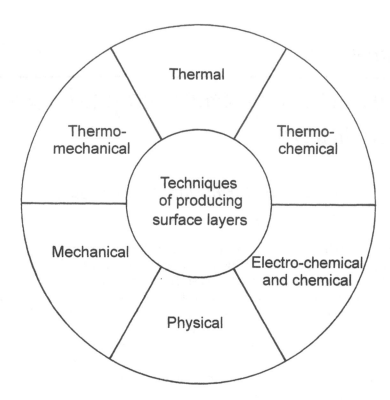

Fig. 1.1 Techniques of manufacture of surface layers.

Static burnishing - exertion of constant or variable pressure on the surface of the treated metal object with a tool made of a harder metal (less often of another material, e.g., diamond) with a smooth surface, i.e. free of cutting edges, in the form of a disk (surface rolling), shot or roller in order to cold harden the superficial layer. Static burnishing is carried out on objects made from alloys of iron with carbon and possibly other elements, as well as other metals and their alloys, primarily with the aim of enhancing fatigue strength. The thickness of the hardened layer may reach several millimeters.

Dynamic (impact) burnishing - the utilization of kinetic energy of steel, ceramic or glass particles in the form of shot or pellets, hurled by centrifugal force, stream of compressed air or energy of detonation of explosive gases, or the kinetic energy of a smooth tool (hammering) hitting the surface of the treated metal object in order to cold harden the superficial layer. The layer thickness is usually less than in static burnishing, while the applications are similar. In some cases, dynamic burnishing is also applied as a means of introduction of compressive stresses in order to deform thin-walled objects.

Detonation (explosive) spraying (deposition) - a process related to thermal spraying, consisting of the utilization of the kinetic energy of particles in the form of metal powders or ceramic, imparted to them in order to obtain a

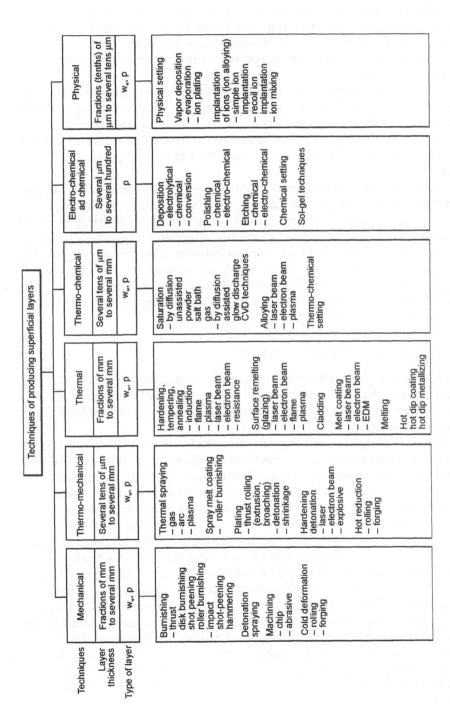

Fig. 1.2 Classification of surface layer manufacturing methods by method of layer generation.

coating with properties different from those of the metal substrate, coupled with an insignificant heating of the surface. Powders of coating materials - metals and their alloys, chemical compounds of metals (oxides, carbides, borides), metal-ceramic composites or ceramic - are usually finer than those used in thermal spraying. The thickness of deposited coatings is usually 0.3 to 0.4 mm, although in some cases may even reach 1 mm. The application is similar as in the case of thermally sprayed coatings, but explosive sprayed coatings exhibit better properties.

Machining - a process which has, as its aim, the shaping of the object, its dimensions and surface finish. Machining is usually accompanied by hardening of the superficial layer, although it is not this effect, but the obtaining of a desired surface smoothness that is the focus of surface engineering. Depending on the geometry of the cutting tool, we distinguish:

– *chip machining*, accomplished by a tool with a defined number and geometry of cutting edges, e.g. turning, milling, planing, slotting, etc.,

– *abrasive treatment*, accomplished by grains of an abrasive material with an undetermined number of cutting edges and a random geometry,

– *grinding* (reduction of surface unevenness with the aid of grinding wheels or by electric techniques: EDM, electrocontact, electrolytic), polishing (reduction of surface unevenness, usually following grinding, by soft abrasives, blasting by abrasives in a stream of liquid or by loose abrasives in the form of pellets), etc.

1.1.2 Thermo-mechanical techniques

Thermo-mechanical techniques utilize the combined effects of heat and pressure in order to obtain coatings or, less frequently, of superficial layers. The techniques used are [4] spraying, plating, explosive hardening and plastic deformation.

Thermal spraying - coating of different objects (usually metallic) with a layer of coating material, by pneumatic dispersion of tiny particles in a flame (usually generated by gas, electric arc or plasma), giving them velocity (even supersonic) in air, vacuum or protective atmosphere, and kinetic energy. This energy assures the exertion of pressure on the coated surface, allowing good adhesion of the sprayed coating to the substrate, with simultaneous heating up of the substrate to low temperatures (not exceeding 150°C). Coating materials usually applied by this technique are alloy steels, zinc, aluminum and its alloys, copper, tin, lead, nickel, brass, cadmium, bismuth, cobalt, chromium, tungsten, titanium, molybdenum, composites of Ni-Cr, Co-Cr, Ni-Al, Pb-Zn, tungsten carbides, Al_2O_3 and TiO_2 oxides, and synthetics. Layer thickness ranges from 50 to 1000 μm. Thermally sprayed coatings are applied mainly for the protection of machinery and steel structures against atmospheric and gas corrosion.

When metals are used as the coating material, the process is called **spray metallization**. Sprayed coatings may be subjected to heat treatment.

Spray padding - spray deposition of a metallic layer on a metallic substrate by means of welding, i.e. by surface melting of the substrate and the binder of a composition similar to that of the substrate. The bond between the molten layer and the substrate is of a metallurgical nature. Spray padding is carried out in order to restore or to improve tribological or anti-corrosion properties.

Plating - the coating of the substrate metal by another metal or alloy, sometimes with the use of intermediate layers - by means of exerting pressure on the coating material at an appropriately elevated temperature. We distinguish static plating (e.g., rolling, pressing, burnish broaching), detonation and shrinkage plating (e.g., utilizing casting shrinkage). Plating materials most often used are aluminum and its alloys, bismuth, steels containing chromium, nickel or both, tool steels, copper and its alloys, precious metals, Monel metal, Hastelloy, Invar, molybdenum and its alloys, brasses, niobium, nickel, tin, tantalum and titanium. Coating thickness varies within a broad range from several micrometers to several millimeters. Plating is used mainly for the purpose of enhancing resistance to atmospheric and gas corrosion at elevated temperatures and in chemically aggressive environments. In less frequent cases, they may be used for the enhancement of tribological, electrical and thermal properties or for decorative effects.

Hardening by detonation - hardening of a metal or alloy by a shock wave created by sudden evaporation of substrate material when acted upon by a strongly concentrated stream of electrons (electron beam hardening) or photons (laser beam hardening) with a rise of substrate temperature or detonation of an explosive mixture (explosive hardening). This type of hardening is, basically, still in the stage of laboratory research.

Plastic deformation - a process aimed at shaping or division of the treated material, effecting changes of its physico-chemical properties, structure and surface smoothness, and the creation of residual stresses. Depending on temperature, we distinguish:

– *hot deformation* - accomplished at temperatures at which recrystallization of the treated material takes place,

– *cold deformation* - accomplished at temperatures at which recrystallization does not occur, but only reduction and strain hardening (this process belongs to the mechanical techniques group).

Depending on the type of deformations taking place, we distinguish:

– *rolling* - when the material is plastically shaped between turning rollers,

– *forging* - when the material, in the form of a block, is plastically shaped by reduction, blows by a hammer or swaging machine, or by static pressure exerted by a press. Modifications of forging are extrusion, drawing and pressing.

1.1.3 Thermal techniques

Thermal techniques employ effects connected with the influence of heat on materials and are aimed at [1-8]:

– changing the microstructure of metallic materials in the solid state (hardening, tempering, annealing),

– change of state of aggregation,

– transition from the solid to the liquid and again to the solid state of a metal material of the substrate (partial melting or melt surfacing) or coating (pad welding, building up),

– obtaining a solid metal out of powdered coating material by melting,

– transition from the liquid to the solid state of the coating material (hot dip cladding and coating).

Hardening, tempering, annealing - involves changes of microstructure of the metallic material (in most cases of steel) in the solid state, in order to obtain desired changes in mechanical, chemical and physical properties of the superficial layer, without changes of the chemical composition. These processes are accomplished by induction, flame, plasma, laser, electron beam and resistance heating, followed by cooling or quenching at a required rate [9].

Melt surfacing - the smoothing of a surface of metal or sealing of a metal or non-metal coating, alternately obtaining of an amorphous structure (metal glass) of the melted layer with physical chemical properties different from those of the core, but without changes to chemical composition. Obtaining of an amorphous structure (vitrification) is possible only with extremely rapid heating and equally rapid cooling (*called splat-cooling*). This process is accomplished with laser beam, electron beam, plasma and flame heating [9].

Pad welding - a modification of surfacing, accomplished with the use of welding torches, for overlaying of the metal substrate with a layer of alloy material in order to obtain a coating with properties either:

– similar to those of the substrate to replenish worn material (repair),

– different to those of the substrate to enhance service life.

Pad welding causes some insignificant melting of the substrate material, allowing a metallurgical bond between the substrate and the coating. Pad welding is carried out with the utilization of welding techniques, mainly arc and flame (oxy-acetylene) heating. Materials used for pad welding and for the generation of coatings with special properties are carbon and low alloy steels, austenitic high manganese and chromium-nickel steels, chromium and chromium-tungsten steels, high speed steels, high chromium cast irons, alloys such as Co-Cr-W, Ni-Cr-B, Ni-Mo and sintered carbides. The thickness of pad welded layers usually reaches several millimeters. In the past, pad welding was considered a modification of plating.

Melt coating - utilization of laser beam, electron beam, spark discharge (also with ultrasound participation) heating to deposit a coating composed of metal (e.g., Al, Ni, Si), metal alloys (e.g., Cr-Ni, Cr-B-Ni), intermetallic compounds (borides, nitrides, carbides), ceramic (stellite) or metal ceramic, on the surface of a metal or alloy, the properties of substrate and coating being different from each other. These are usually coatings featuring heat resistance, acid and corrosion resistance, as well as resistance to high temperature erosion. As an example, cobalt alloys may be coated by nickel alloys to obtain corrosion-resistant superalloys while aluminum-silicon alloys are coated by silicon. The thickness of coatings may reach several millimeters and their quality is better than that of thermally sprayed coatings. Melt coating is sometimes considered to be a modification of plating.

Melting (firing, remelting) - transition of powdered enamel to a compact, glassy state, strongly adhering to the substrate. The metal substrate in this case is usually steel sheet, cast iron or cast steel. The powdered enamel is deposited "dry" or "wet", mainly in the form of metal oxides, e.g., fluorides, borates and silicates. This type of coating protects against corrosion and enhances the aesthetic value of the product. Melting is accomplished by firing the enamel mass at 850 to 950°C. Thickness of enamel coatings ranges from fractions of a millimeter to several millimeters. These coatings are deposited on kitchenware, sanitary equipment, reservoirs and apparatus for the production of chemicals, grocery and medication, in order to ensure resistance to the action of water, corrosive liquids (both acidic and alkaline), aggressive gases and high temperatures [1-11].

Hot dip coating (dip metallization) - involves the solidification of melted coating material into which the coated object is immersed. The melting point of the coated material must be higher than that of the coating. Solidification takes place on the object surface after removal from the bath in order to obtain an adhesive coating (tin coatings, and double-layer tin-lead, copper and cadmium coatings) [1-8].

1.1.4 Thermo-chemical techniques

Thermo-chemical techniques utilize the combined effects of [1-9]:

– heat and a medium chemically active with respect to the treated metal, in order to **saturate** it with a given element or elements, bringing about desired changes in chemical composition and microstructure of the superficial layer,

– heat and chemical factors (reaction of reticulation) acting on the coating material in order to **set** (harden) it.

Depending on the state of the chemically active medium, we distinguish the following techniques:

– powder pack (powdered solid),

– paste (powdered solid with binder),

– bath (a bath containing saturating components, e.g., salt bath for carburizing or nitriding or bath composed of molten saturating metals),

– gas (mixtures of hydrocarbons).

In the case of baths which are composed of molten saturating metals, when coating of the substrate metal takes place at a temperature higher than the melting point of the coating material (e.g., Al, Zn, Al + Zn), the techniques are termed **hot dip** or **immersion** (e.g., hot dip galvanizing [zinc], hot dip aluminizing). The surface layer typically comprises a layer of coating metal and an intermediate diffusion layer, usually multi-phase.

Saturation by diffusion (diffusion alloying) - a process of introduction to the superficial layer, by diffusion, of atoms or ions of metals or gases which increase its tribological properties, fatigue strength and corrosion resistance. This process is mainly dependent on temperature, time and concentration of the active (diffusion) medium [12].

Two types of diffusion saturation are distinguished [12]:

1) **unassisted** - occurring without the participation of additional factors which affect the process. Usually takes a long time (up to several tens of hours), while the active medium may be solids (powder packs and pastes), liquids (baths, usually salt) or gases. Typically, the saturating elements are carbon, nitrogen, chromium, titanium, silicon, sulfur, niobium, vanadium, aluminum and zinc. Saturated alloys are steels, cast iron and steel, less frequently single metals (nickel, cobalt, titanium, molybdenum, tungsten, tantalum). The thickness of diffusion layers (diffusion alloyed) is dependent on process temperature and time and usually ranges from 0.1 to approximately 3 mm (in most cases 0.3 to 1.5 mm). Unaided saturation by diffusion includes all traditional thermochemical treatments and usually calls for high temperatures. It is used primarily to increase the hardness of components and tooling to enhance wear resistance and for corrosion resistance;

2) **assisted** - occurring with the participation of a factor activating the process (by activating the surface and increasing adsorption of the material constituting the layer). Its duration is short, up to several hours, and it may only be accomplished in a gas phase. This term is given to some thermo-chemical treatments, modified by surface activation, and primarily to the so-called CVD techniques, which are carried out at temperatures lower than those of unaided processes, due to

– selection of appropriate gas atmospheres and the utilization of compounds characterized by lower temperatures at which chemical reactions occur, e.g., metallo-organic compounds;

– lowering of pressure to values approximately 500 to 1000 Pa; this is the so-called low pressure CVD technique, in use for quite a long time;

– electrical activation of the gaseous medium by means of glow discharge or high frequency currents. This is the so-called activated CVD technique, mastered on a semi-technical scale with respect to machine components and tooling (glow discharge) and applied practically in the electronics industry (obtaining of Si_3N_4, Al_2O_3 layers, activated by high frequency currents). The layers produced in this technique, with thicknesses usually in the range of 0.01 to 0.02 mm, may be single (carbides, nitrides, borides or oxides of iron, chromium, titanium, as well as titanium carbides) or double and even triple, composed of different single layers. They are used to coat tools (mainly inserts made of sintered carbides) or machine components for enhancement of wear resistance.

The process of saturation, involving mixing of the alloy element or elements with the thin melted surface layer of the substrate metal, coupled with partial diffusion, is called **melt alloying** (by laser, electron beam or plasma).

Thermo-chemical setting - irreversible transition of thermo-setting resins, deposited by any given method onto the surface of the coated substrate in the form of an adhering compact layer, from the liquid (or doughy) state to that of a solid, under the influence of heat (usually temperatures in the range of 20 to 200°C) and chemical reactions (polymerization, polycondensation, polyaddition), to obtain paint coatings.

1.1.5 Electrochemical and chemical techniques

In electrochemical (electroplating) and chemical techniques, several effects may be utilized. To deposit a metallic coating or to either deposit or set a non-metallic coating on the surface of a metal, alternately, to polish or clean (pickle) a metal surface, this effect is electrochemical reduction (for electrochemical and conversion coatings, electrolytic polishing and etching) or chemical reduction (for chemical and conversion coatings, and chemical polishing and etching). In the case of paint coatings, other chemical reactions are utilized. Coatings obtained in this way exhibit properties that are superior to those of the coated metal. These properties include: corrosion resistance, wear resistance and some physico-chemical ones, like color, luster and reflectivity [5-11]. Coatings are most commonly produced by the bath technique, i.e. immersion in an electrolyte, a chemical bath, paint or sol [5-11], by spraying and, less frequently, by tampons, centrifugal rinsing or spreading [13-16].

Electrolytic deposition (electroplating) enables the creation of metal or alloy coatings, as the result of reduction, by electric current at the cathode, of ions of the coating metal from electrolyte solutions. The obtained coatings may be single layered, with a thickness of 0.3 to 300 μm, or multilayered. The most frequently used coating metals (in descending order) are chrome, nickel, zinc, tin, cadmium, copper, lead, silver, gold, rhodium, palladium, platinum, ruthenium, iron, cobalt, indium, as well as alloys: Sn-Pb, Sn-Ni, Sn-Cd, Zn-Ni, Cu-Zn, Ni-Fe, Ni-Co, Ni-P, Co-P, Co-W, brasses, used single or in combination with other coatings, mainly for corrosion protection and for decorative purposes. In some branches of technology, e.g., repair of machine components, intermediate layers in electronics, and special military applications, electroplating is complementary to tampon deposition (instead of an electrolytic bath, a fabric tampon dipped in it or a saturated solid is used). The tampon technique allows selective deposition of microcrystalline metal or alloy coatings on fragments of even very big objects. With some modifications of the electrolyte (aqueous solutions of alkalis, acids and salts) as compared to immersion techniques, this technique also enables the obtaining of coatings with special properties on both metallic and non-metallic substrates, e.g. ceramic, glass, plastics. Besides a 10 to 20 times shorter deposition time, lower cost and material as well as energy economy, the tampon technique yields coatings which are harder than those obtained in the immersion technique and have a disordered or amorphous structure with a lower hydrogen content. Finally, the tampon technique also enables the deposition of composite ceramic-metal coatings [13].

Chemical deposition (electroless) enables the obtaining of metal or alloy coatings on metals or alloys, as the result of exchange, contact or reduction with or without the participation of a catalyst. Exchange and contact are used for the deposition of tin coatings; exchange, contact and reduction with the utilization of a catalyst, are the technique used for copper and nickel coatings. Contact and reduction, but without the catalyst, are used for the deposition of silver coatings, while gold is deposited by means of exchange and reduction with a catalyst. In most cases deposition is by chemical reduction without a catalyst (e.g., by sodium hypophosphite) in baths or by spraying. It is applied for the deposition of coatings, mainly nickel, on substrates which are otherwise difficult to electroplate (complex shapes, slender long holes, etc.) and on other coating metals: Ag on Cu, brasses or non-conductors, such as glass and plastics, Au on Cu or brasses, Co and Cu on plastics; Pd, Pt, Sn on Cu or on Al; Bi on steels; Ag on glass. Layer thickness is 5 to 20 µm. The main purpose of these applications is enhancement of corrosion resistance or creation of contact layers on cast iron and steel (Ni) prior to coating with enamel.

Conversion deposition is an artificially induced and controlled process of metal or alloy corrosion by chemical or electrochemical treatment. Its result is the formation on the surface of a coating which is practically insoluble in water or in the triggering environment, tightly bound to the substrate material and exhibiting dielectric properties. It is composed of compounds of substrate material with the reagent solution, in which objects are immersed or which is sprayed on the objects. It may be a chromate, phosphate, oxide, oxalate or a other coating [49]. Depending on the type of bath used and on the substrate material (e.g., aluminum, zinc, cadmium, steel, copper and its alloys, magnesium alloys, silver, etc.) the coatings may have different compositions, color and properties. The thickness of coatings ranges from several to several hundred micrometers. Conversion coatings are applied for corrosion resistance and

– to improve adherence of paint coatings to steel, zinc and aluminum,
– to improve the properties of other coatings,
– to activate the diffusion of nitrogen to steel, facilitate cold deformation of steel (broaching, extrusion, pressing), electrical insulation of substrate,
– to reduce friction (lubricating coatings),
– to enhance aesthetic value (decorative coatings).

Broadest application is found by phosphate coatings (coatings of steel sheet prior to painting) and oxide (oxidizing of machine components, tools and firearms).

Polishing - finishing treatment, carried out with the purpose of obtaining smoothness and luster of the object surface, and accomplished in an electrolyte or a chemical bath. It consists of selective dissolution of peaks of microasperities, while leaving microrecesses practically unchanged [9]:

– *chemical polishing* (electroless polishing, chemical brightening) - brightening and partial polishing, carried out by treatment of metals and alloys, most frequently aluminum and its alloys, in baths containing oxidizing agents (primarily acids, like orthophosphoric, nitric, sulfuric, acetic), in order to achieve an attractive appearance;

– *electrolytic* (electropolishing, electrochemical polishing) - smoothing (dissolution of asperities of heights greater than 1 µm) and brightening (reduction of asperities from above 1 µm to below 0.01 µm), carried out by treatment of metals and alloys (mainly aluminum), with appropriately selected electrolytes and current conditions. The polished object is the anode. The treatment does not alter the state of residual stresses in the surface layer. It is used to obtain high luster or as preparation of the substrate for protective-decorative processes (e.g., electroplating).

Etching - removal of layers of scale, rust, oxides or alkaline salts from the surface of metals and alloys, carried out before final pickling and deposition of electroplated coatings [9]. It can be carried out by

– *chemical means* (electroless) - by immersion in acidic solutions, reacting with metal oxides,

– *electrolysis* - in an electrolytic process, where the metal may be pickled by the anode or the cathode.

Chemical setting enables the production of paint coatings out of material deposited by any chosen technique, as the result of

– oxidation at ambient or elevated temperature, upon contact with oxygen from the air - by spontaneous oxidation or oxide polymerization of the filmogenic substance (drying oil or the product of its initial transformation);

– reticulation, without the participation of oxygen, of chemosetting resins at ambient temperature, due to polymerization, polycondensation or polyaddition, under the influence of reagents (catalysts, resinous co-reagents or other macromolecular substances).

Gelling or formation of coatings by the *sol-gel* technique - a low temperature synthesis of coating material by way of the sol[1] colloidal solution and multi-phase gel.[2] In stricter terms, this is a process of formation of sol, its subsequent transformation into gel and final treatment of gel [14]. The sol is a homogenous solution of an easily soluble precursor (e.g., aloxyl derivatives) in an organic solvent, mixed with a reagent, e.g., with water [15, 16]. After acid treatment (e.g., by water with HCl), the sol is transformed into gel by polycondensation [16]. The material thus obtained, prior deposited on the

---

[1] **Sol** - colloidal system in which the solvent phase is liquid (in excess) and the dispersed phase is the solid, of 1 to several hundred µm grain size.

[2] **Gel** - colloidal system which lost its liquidity due to an increase in the mutual interaction between sol particles.

substrate, is next dried and heated. The ceramic is fired at approx. 500şC [16] and forms a tight crystalline coating. It is possible to obtain coatings of different thickness from the same solution by multiple immersion of the coated material. It is also possible to form multi-layer coatings by multiple immersion (centrifuge or spreading) of the object in different solutions. The advantage of this technique is rather uncomplicated equipment, possibility of precise control of the microstructure of the deposited coating as well as of formation of different coatings, e.g., corrosion resistant (metal and non-metal oxides), anti-reflective, catalytic, dielectric and in the form of ceramic glass, etc. [16].

1.1.6 Physical techniques

In physical techniques, the production of organic coatings (setting) or metallic or ceramic coatings (deposition) on the surfaces of metals or non-metals, with adhesive or diffusion bonding, or the creation of a surface layer makes use of various physical effects. These may occur under atmospheric pressure (evaporation of solvent) or lowered pressure, in the majority of cases, with the participation of ions or elements of metals or non-metals [1-10].

Physical setting (drying) consists of a transition of coating substance, deposited by any chosen technique, from the liquid or doughy state to the solid state, as the result of evaporation of the solvent, carried out in order to produce a paint coating.

Physical vapour deposition (PVD techniques) of metals or ions in a vacuum consists of [1-9, 17]

– bringing the deposited metal (with a high melting point) to the vapour state, with the utilization of resistance, arc, electron and laser beam heating,

– introduction of gas,

– ionization of metal and gas vapours,

– deposition on the surface of a cold or insignificantly heated substrate, of a single metal, or compounds (e.g., nitrides, carbides, borides, silicides, oxides) of that metal with the gas or with the substrate metal. This is accomplished with the utilization of electrical effects (in a physical sense, PVD techniques constitute a crystallization of vapours of plasma), among others, of glow discharge. An example of this is the PAPVD technique, i.e., the PVD process, aided by glow discharge [17, 18].

When metal vapours crystallize on a cold substrate, the process is called simply **vapour deposition**, and if the crystallization of metal vapours is combined with the formation of its compounds with the gas or the substrate, the process is called **sputtering** or **ion plating**.

Ion implantation of metals and non-metals consists of ionization of metal or gas vapours and acceleration of positive ions by electric fields to such velocities where the kinetic energy of the ion is sufficient to penetrate the metal or non-metal to a depth of several or even more atomic layers. This is the implantation of primary electrons. Implantation of secondary ions takes place when secondary ions are sputtered out of the layer deposited on the implanted material. Ion implantation may take place in the presence of other physical phenomena, e.g., vaporization, vapour deposition and magnetotronic sputtering. In this case, the process is referred to as **ion mixing**.

Implanted ions change the structure and the chemical composition of the surface layer of the implanted material. The depth of implantation is 0.01 to 1 µm (for steels, in most cases: 0.2 to 0.3 µm) and may increase during the work of an implanted tool or machine component, due to migration of the implanted ions. Theoretically, any metal material may be implanted by any type of ions. In practice, the most common application is implantation by nitrogen, less frequently by boron, carbon, tin, cesium, silicon, chromium and palladium. Implantation is aimed at extending service life, by increasing hardness and wear resistance, of cutting and forming tools. In less frequent cases, machine components may be ion implanted. Ion implantation is sometimes referred to as **ion alloying** [9].

1.2 Classification of techniques of producing technological surface layers

Techniques of producing surface layers, presented in Section 1.1, embrace the whole group of problems related to surface engineering but do not draw the distinction between techniques of producing surface layers [9] and techniques of deposition of coatings [10, 11]. They do not combine the same groups of techniques (e.g., laser and electron beam techniques are used in many manufacturing applications) nor do they take into consideration their degree of their modernity [1-9].

Generation of surface layers may reduce, leave without changes or increase dimensions of the treated object (Fig. 1.3). Techniques of producing surface layers may, therefore, be divided into

– **decremental** - accomplished by decreasing the dimensions of the object, e.g. by machining or burnishing; decremental techniques are used to form surface layers,

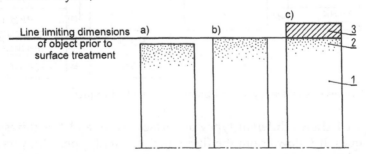

Fig. 1.3 Diagrams showing surface layers manufactured by various techniques: a) decremental (top layers); b) non-decremental (top layers); c) incremental (coatings on top of substrate with superficial layer); *1* - core or substrate; *2* - superficial layer; *3* - coating.

– **non-decremental** - accomplished without decreasing the dimensions of the object, e.g. by ion implantation. These techniques are also used in the production of surface layers,

– **incremental** - accomplished by increasing the dimensions of the object, e.g. by electroplating or by some thermo-chemical treatments. Incremental techniques are typically used in the deposition of coatings.

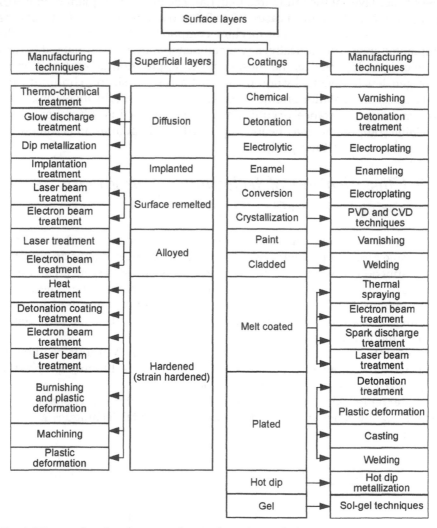

Fig. 1.4 Types of surface layers and manufacturing techniques.

Fig. 1.4 shows different types of surface layers and the corresponding techniques. It follows from the figure that different types of layers may be obtained by same techniques. These may be either commonly known and used for many years (traditional methods) or new methods, currently being implemented in industrial practice. It should also be emphasized that the techniques shown in the chart serve either exclusively surface engineering tasks (e.g., protective coatings, electroplating) or those and also other tasks (e.g., heat treatment, forging, casting).

For this reason, the domain of producing surface layers has been treated here in the form of groups of related techniques, basing on such factors as their modernity, technique of accomplishment, traditional classification and terminology, while singling out those techniques which are devoted exclusively to surface engineering tasks (Fig. 1.5).

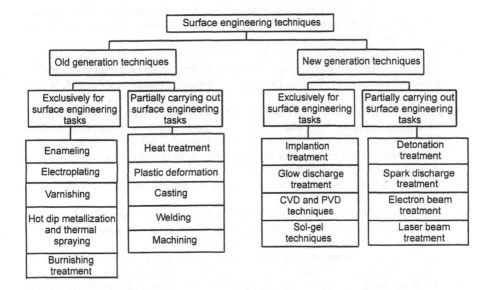

Fig. 1.5 Techniques fulfilling surface engineering tasks.

On account of the specific and broad nature of the problem, we have decided not to discuss techniques which are commonly known, used for many years and described in specialized technical literature, particularly those techniques which only partially fulfill surface engineering tasks. Thus, the techniques omitted from the following discussion are machining, forging, heat treatment, casting, enamel and varnish depositing, electroplating, welding, thermal spraying, dip metallization, spark discharge, sol-gel, etc.

In further considerations, not all techniques are discussed. The focus is on selected newest techniques of producing surface layers, all called collectively **new generation technology** and which in many cases have not yet been implemented practically in some countries. However, on account of their possibilities, these techniques appear very promising and are expected to develop to the point of full utilization. The less these techniques are known and used worldwide and the less they are presented in technical literature, the more space has been allotted to them in this book.

References

1. **Burakowski, T.**: Methods of producing surface layers - metal surface engineering (in Polish). Proc.: Conference on *Methods of Producing Surface Layers*, Rzeszów, Poland, 9-10 June 1988, pp. 5-27.
2. **Kortmann, W.**: Vergleichende Betrachtungen der gebrauchslisten Oberflächenbehandlungsverfahren. *Fachbereite Hüttenpraxis Metallverarbeitung*, Vol. 24, No. 9, 1986, pp. 734-748.
3. **Burakowski, T.**: Status quo and directions of development of surface engineering. Part I - Applied methods of producing surface layers and their classification by technique used (in Polish). *Przegląd Mechaniczny (Mechanical Review)*, No. 13, 1989, pp. 5-12.
4. **Burakowski, T.**: *Metal surface engineering - status and perspectives of development* (in Russian). Series: Scientific-technical progress in machine-building. Edition 20. Publications of International Center for Scientific and Technical Information - A.A. Blagonravov Institute for Machine Science Building Research of the Academy of Science of USSR, Moscow, 1990.
5. **Burakowski, T.**: Metal surface engineering (in Polish) *Standardization*, No. 12, 1990, pp. 17-25.
6. **Burakowski, T.**: Producing surface layers metal surface engineering (in Polish). *Metaloznawstwo, Obróbka Cieplna, Inżynieria Powierzchni (Metallurgy, Heat Treatment, Surface Engineering)*, No. 106-108, 1990, pp. 2-32.
7. **Burakowski, T., Roliński, E., and Wierzchoń, T.**: *Metal surface engineering* (in Polish). Warsaw University of Technology Publications, Warsaw 1992.
8. **Burakowski, T.**: Metal surface engineering - subject era and classification of superficial layer manufacturing method. Proc.: 4th International Seminar of IFHT on *Environmentally and Energy Efficient Heat Treatment Technologies*, Beijing, 15-17 September 1993.
9. **Burakowski, T.**: Techniques of shaping of the superficial layer (in Polish). Proc.: II International Conference on: *Effect of technology on the state of the superficial layer*, Gorzów-Lubniewice, Poland 20-22 October, 1993, Studies and Materials, papers, Vol. XII, No.1, pp.5-24.
10. **Burakowski, T.**: Present state and directions of development of corrosion protection (in Polish). *Metaloznawstwo, Obróbka Cieplna, Inżynieria Powierzchni (Metallurgy, Heat Treatment, Surface Engineering)*, No. 115-117, 1992, pp.43-50.
11. **Burakowski, T.**: Present state and directions of development of corrosion protection (in Polish). *Przegląd Mechaniczny (Mechanical Review)*, No. 8, 1993, pp. 13-16 and 21-22.
12. **Burakowski, T.**: Trends in the development of modern heat treatment. *Metaloznawstwo, Obróbka Cieplna, Inżynieria Powierzchni (Metallurgy, Heat Treatment, Surface Engineering)*, No. 85, 1987, pp. 3-11.
13. **Grześ, J.**: *Usable properties and microstructure of selected Ni-W-Co coatings obtained in a tampon process* (in Polish). Ph.D. dissertation, Warsaw Technical University 1992: *Prace ITME*, Vol. 38, publ. WEMA, Warsaw 1992.
14. **Brinker, G.J., and Scherer, G.W.**: *Sol-gel science. The physics and chemistry of sol-gel processing*. Academic Press, San Diego 1990.

15. **Strafford, K.N., Datta, P.K., and Gray J.S.** *(eds).: Surface engineering practice - process, fundamentals and application in corrosion and wear.* Ellis Horwood Ltd., New York-London-Toronto-Tokyo-Singapore 1990.
16. **Głuszek, J., and Zabrzeski, J.:** Ceramic protective, obtained by the sol-gel technique (in Polish). *Inżynieria Powierzchni (Surface Engineering),* No.3, 1996, pp. 16-21.
17. **Stafford, K.N., Smart, R.S.C., Sare, I., and Subramanian, Ch.** *(eds): Surface engineering - process and applications.* Technomic Publishing Co., Lancaster-Basel 1995.
18. **Tyrkiel, E. (General Editor), and Dearnley, P. (Consulting Editor):** *A guide to surface engineering technology.* The Institute of Materials in Association with the IFHI, Bourne Press, Bournemouth 1995.

chapter two

Electron beam technology

2.1 Advent and development of electron beam technology

In 1867, an Irish physicist G.J. Stoney, professor at Queen's College in Galway, first suggested the existence of an elementary electrical charge, the value of which he attempted to determine in 1874 and which in 1891 he called **electron** (from the Greek *elektron*, meaning amber - a fossil resin with strong electrostatic properties). The discovery of the first elementary particle in the modern meaning of the word, i.e., the electron, may be attributed to J.J. Thomson who in 1896 carried out experiments on the deflection of cathode rays in magnetic and electric fields and determined the ratio of the charge to the mass of the electron. Proof of the constant value of this ratio came later.

The first direct and fairly accurate measurement of the electron charge was made in 1911 by R.A. Millikan. Today we know that the electron has a rest mass of $9.109 \cdot 10^{-31}$ kg, i.e., approximately 1836 times less than the mass of the neutron or proton. It is equivalent to an energy of 511 keV and its electrical charge is $1.602 \cdot 10^{-19}$ C.

The emission of elementary charges, suggested by Stoney, from the glowing filament of an electric bulb was later accidentally discovered in 1883 by the self-taught genius and inventor T.A. Edison, during work on the improvement of his invention made four years earlier. Later research showed that electrons emitted by the glowing filament may be accelerated by an electrical field between the filament (cathode) and another metal element (anode, which could be, e.g., a furnace load) and the kinetic energy gained in this way may be passed over to that element in the form of heat.

In the beginnings of the 20th century, the majority of physicists were aware of the fact that electrons play a significant role in the structure of matter. Later, it was realized that electrons emitted from a glowing cathode and subsequently concentrated and accelerated may be used to research matter and to shape materials by means of radiation and thermal effects, occurring when the accelerated electrons come into contact with matter.

Practical utilization of a stream of electrons - which, after being shaped into a beam and appropriately deflected, has been termed **electron beam** - enabled an intensive development of electron optics[1] during the 1920s.

First practical cases of utilization of the electron beam took place in the 1930s:

1. Low current and low energy density beams, serving mainly research purposes, where effects other than thermal on material were utilized (so-called **non-thermal beams**):

– In 1931, in the Berlin Technical University, M. Knoll and E. Rusk built the first transmission electron microscope (the stream of electrons passes through the inspected material);

– In 1935 M. Knoll built the first model of the scanning electron microscope and in 1965 the first commercial, serial produced microscope was placed on the market by Cambridge Science Instrument Co.. In this case, the electron beam does not pass through the inspected material but generates different signals in the zone where the beam is dispersed.

2. High current and high power density beams, generating heat in the zone of electron dispersion, serving mainly technological purposes (so-called **thermal action** beams):

– In 1938 the electron beam was first used to make holes and to evaporate metals [1].

The beginning of a dynamic development of electron beam technology is associated with the early experiments of K.H. Steigerwald, who in 1948 to 1950 proved the technical possibility of utilizing the electron beam as a tool for repeatable production of small holes and for micromachining [1].

The first industrial-scale applications of electron beams concerned perforation and drilling of small holes. Later, in the 1950s, applications included welding and melting, and in the 1960s, enhancement of metal surfaces.

The 1960s see the beginning of a broad scale utilization of electron beam technology for melting, welding, and hole drilling in the US, ex-USSR, Japan, United Kingdom, France and Germany. For enhancement of surfaces, it began to be used on an industrial scale in some of these countries in the 1970s; the rest of the countries followed suit in the 1980s. In countries with leading technology, electron beam equipment with digital control has been used since the second half of the 1970s.

The first international conference dedicated to the application of electron beam technology was held in 1962.

---

[1] Electron optics - a branch of electronics, dealing with the effect of magnetic and electric fields on the path of electrons and the construction of equipment utilizing this effect.

2.2 Physical principles underlying the functioning of electron beam equipment

2.2.1 Electron emission

Electrons, together with protons and neutrons, are the basic constituents of atoms; they occur as so-called bound electrons.

Electrons may be freed from their atoms to let them act as free particles. This may be accomplished in two ways:

1) Supplying them with higher energy, e.g., by heating or illuminating the body,

2) Significantly lowering the potential barrier, e.g., by subjecting the body to the action of a strong electric field.

Depending on the method of freeing electrons from a solid, i.e., of causing electron emission, we distinguish the following:

– thermoelectron emission, occurring under the influence of heat,

– photoelectron emission, occurring under the influence of light radiation absorbed by atoms,

– field (autoelectron) emission, occurring under the influence of a strong electric field, generated at the surface of the body,

– secondary emission, occurring as the result of bombardment of the body by electrons or ions.

Electrons - in the form of β particles, i.e., of nuclear origin - are also emitted by radioactive bodies. In the case of gases, electron emission is obtained by ionization of the gas atoms: thermoionization, photoionization and collision ionization (occurring as the result of collisions of atoms with particles, in most cases with electrons).

Electron beam equipment used for surface enhancement purposes utilizes primarily thermoelectron and secondary electron emission, and in some, less frequent cases, gas ionization.

2.2.2 Thermoelectron emission

Thermoelectron emission (thermoemission) is a process consisting of freeing electrons from a solid by supplying it with thermal energy. As the result of absorbing thermal energy by the body, the electron, through collisions, gains kinetic energy. When this energy is sufficient, the electron leaves the body.

The dependence of thermoemission current density J_e on the absolute temperature T of the emitting body was discovered in 1914 by the English physicist, O.W. Richardson:

$$J_e = aT^2 e^{-\frac{A}{kT}} = aT^2 e^{-\frac{e\varphi}{kT}} \left[A/cm^2 \right] \qquad (2.1)$$

where: a - emission constant, dependent on material properties and on condition of surface emitting electrons, differing in practice from the theoretical value. $a_{pract.} = 30$ to 120 A/(cm$^2 \cdot$ K^2); $a_{theor.} = 4pemk^2/h^3 = 1.2 \cdot 10^6$ A/(cm$^2 \cdot$K^2); m - electron mass (m = 9.109$\cdot10^{-31}$ kg); h - Planck constant (h = 6.2 $\cdot$ 10^{-34} J$\cdot$s); e - elementary charge $(e$ = 1.602 $\cdot$ 10^{-19} C), e - base of natural logarithms (e = 2.71828); $A = e\varphi$ - work done by electrons exiting the emitting body, eV; j - exit potential, V; k - L. Boltzmann constant (k = 8$\cdot10^5$ eV/K); kT - mean energy of electrons.

The formula (2.1) is known as the Richardson or Richardson-Dushman equation.

In order to attain optimum density of the thermoemission current, which is approximately $(0.1$ to $1.5)\cdot10^{-8}$ A/cm^2, it is essential to generate a high temperature of the emitting body, of the order of 2400 to 2700°C [7]. Materials used for emitting electrons can only be those with a high melting point, i.e., pure metals, borides, oxides.

Thermal emission materials utilized in electron beam equipment designated for surface enhancement are most frequently tungsten, tungsten coated with thorium or with lanthanum hexaboride; in the second case, the essential current density can be attained at temperatures lowered to 1600 to 2000 K [1, 4-6].

In practice, the intensity of the thermoemission current depends on material and design of the element, as well as on treatment conditions. Its value is usually within the range of 10^{-6} to 5 A [1,6].

2.2.3 Utilization of plasma as a source of electrons

Plasma[1] is an electrically conductive, thinned and ionized gas with a sufficiently high concentration of charged particles, containing basically the same number of electrons and positive ions. It is a quasineutral mixture. Each substance may transcend to the state of plasma as the result of **thermal ionization**, occurring at an appropriately high temperature. For this reason, to this day plasma is sometimes called the fourth state of matter. With a rise in temperature there come about transitions: solid→liquid→gas→plasma. Plasma is also formed during **electrical discharges** in thinned gases subjected to high potentials. Electrical properties of plasma are similar to those of metals.

Electron beam methods of surface enhancement utilize plasma generated by electrical discharge in gases (so-called low temperature plasma), obtained under soft vacuum (at pressures of 10^{-3} to 10 A and lower, and at temperatures up to 10^4 K). In most cases, the utilized effect is **glow discharge** in gases - in neon, argon and nitrogen, occurring at low and medium pressure of gas (from several Pa to several kPa). It is characterized by a high potential drop

---

[1] The term: **plasma** (from Latin and Greek: *plasma* - something molded, something formed, something construed) was first used in 1929 by L. Tonks and I. Lagmuir from the Research Lab of General Electric, to denote a set of charged particles.

in the vicinity of the cathode, a strongly developed collision ionization, secondary emission of electrons from the cathode and a special distribution of gas glow in the space between electrodes, dependent on both type of gas and pressure. Detailed information about glow discharge plasma is given in Chapter 5.

By appropriate shaping of the zone of discharge and electric fields, it is possible to either extract electrons from discharge zone to form an electron beam, or to utilize them together with other particles to bombard the metal cathode and to sputter electrons out of it, i.e., to create secondary emission.

2.2.4 Acceleration of electrons

In order to impart velocity to an electron thermally emitted from a cathode or extracted from a glow discharge zone, it is necessary to supply it with a given amount of energy. The easiest way is to utilize an electric field [6–8]. This field acts on the electron with a force:

$$F_E = - eE \text{ [N]} \tag{2.2}$$

where: F_E - force of electric field, acting on an electron, N; E - intensity of electric field, V/m.

If the force acting on the electron has a constant value and sign, the electron will carry out a constantly accelerated movement in a direction perpendicular to equipotential lines, on condition that it will be in vacuum and that the intensity of the field is constant. In accordance with the principle of conservation of energy, the work of the electric field, expressed by the product eU, is transformed into kinetic energy of the electron:

$$eU = \frac{mv^2}{2} \tag{2.3}$$

where: U - potential difference along the electron's path or accelerating voltage, V; v - final velocity of electron, m/s.

A beam composed of n electrons has an energy equal to

$$E_W = neU \tag{2.4}$$

From equation (2.3) it is possible to determine the correlation between the velocity of the electron in the beam and the accelerating voltage:

$$v = \sqrt{2\frac{e}{m}U} = 5.932 \cdot 10^5 \sqrt{U} \approx 6 \cdot 10^5 \sqrt{U} \text{ [m/s]} \tag{2.5}$$

The energy of the electron, dependent on its velocity, depends, in turn, on the accelerating voltage. The velocity and energy are acquired by the electron in the electron gun. Accelerating voltage in such guns may reach a value of 200 kV, which allows electrons to reach velocities close to 0.7 of the speed of light. Equation (2.5) holds true when the velocity of electrons differs from that of light quite substantially, which is the case in the overwhelming majority of electron beam thermal equipment in which accelerating voltage usually oscillates within the range of 80 to 150 kV [7, 8].

Because an electron exhibits, besides corpuscular properties, also wave characteristics, the L.V. de Broglie hypothesis, formulated in 1924 (concerning the wave nature of particles and according to which all displacement of a particle may be described by wave motion equations), states that the de Broglie wavelength for an electron equals:

$$\lambda = \frac{h}{mv} \approx 10^{-10} \sqrt{\frac{150}{U}} \tag{2.6}$$

where: λ - wavelength, m; U - accelerating voltage, V.

De Broglie waves are very short, e.g., for an electron accelerated by voltage $U = 1500$ V, the wavelength is 10 pm (10^{-11} m).

2.2.5 Electron beam control

By changing the value of the accelerating voltage U, i.e., the potential difference between the cathode and the anode, it is possible to exert an effect on the value of the force acting on the electron, in accordance with equation (2.2). This force, however, is not the only one that can, and actually does act on the electron. Besides it, there can also be additional forces, generated by additional electric fields and with a different spatial orientation than the field with intensity E [8]. Additional electric fields in electron beam heating equipment form special electron-optical systems, among these, systems of electrodes called electron lenses [1]. As a result of their action, the electron will move along a path which is resultant of the joint action of all forces originating from electric fields in all points of this path. The electron beam may be focused or unfocused, bent, accelerated or retarded and even interrupted (pulsed beam). Electrical deflecting systems, however, exhibit low sensitivity, i.e., ratio of beam displacement to deflecting voltage [7].

Besides the electric field, the magnetic field also acts on the electron. The force of this action is called the H.A. Lorentz force and is expressed by the formula [7, 8]:

$$\vec{F}_M = -e(\vec{v} \times \vec{B}); \quad F_M = evB\sin\varphi \ \ [N] \tag{2.7}$$

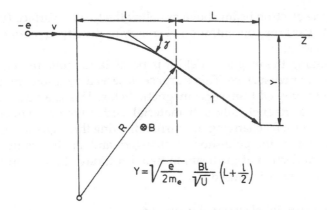

$$Y = \sqrt{\frac{e}{2m_e}} \; \frac{Bl}{\sqrt{U}} \left(L + \frac{l}{2}\right)$$

Fig. 2.1 Deflection of the electron beam in a magnetic field: 1) deflected electron beam; *l* - effective length of deflecting coil; *L* - distance from object to end of deflection zone; *γ* - angle of deflection of electron path; *R* - radius of curvature of electron path; *Y* - displacement of spot relative to object. (From *Oczoś, K.* [7]. With permission.)

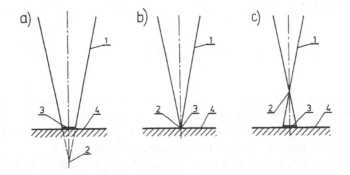

Fig. 2.2 Methods of focusing of the electron beam: a) focusing below surface of material; b) focusing on surface of material; c) focusing above surface of material; *1* - electron beam; *2* - beam focus point; *3* - electron spot; *4* - surface of treated material. (From *Oczoś, K.* [7]. With permission.)

where: F_M - force of magnetic field acting on electron, N; υ - electron velocity, m/s; B - magnetic induction, T; φ - angle between vector of electron velocity υ and vector of magnetic induction B.

Depending on the angle φ, at which the electron enters the magnetic field, it may move along a circular path (when $\varphi = 90\text{ş}$) or along a spiral (when $\varphi = 0\text{ş}$) (Fig. 2.1). Magnetic systems in the form of electron magnetic lenses are widely used in practice to focus and control the displacement of the beam, on account of its high sensitivity and low degree of dependence on electron energy [7].

The magnetic field with a rotating symmetry, generated in the axis of the beam, allows the focusing of electrons theoretically in one focus point. In practice, this is a small area called the **electron spot**. By changing the location of the spot along a line perpendicular to the treated surface, i.e., by focusing the electron beam above or below the treated surface (Fig. 2.2), it is possible to vary the concentration of energy or density of power supplied to the treated load.

By utilizing the magnetic field it is possible to move the electron beam across the treated surface. This purpose is served by a four-pole deflecting system with crossed transverse magnetic fields. This is a ring-shaped yoke, made from a magnetically soft material with four pole shoes on which coils are wound. By changing the current flowing through the coils, the magnetic potential of the pole-shoes is changed, and, as the result of magnetic induction, deflection of the beam in two directions, X and Y in the plane of the treated surface, is caused [7].

2.2.6 Vacuum in electron equipment

Electrons, upon collisions with gas particles, impart their energy to them and the electron beam is dispersed. According to the gas-kinetics theory, the mean length of free path of the electron in a gas is described by the formula [7, 9]:

$$\lambda = \frac{4\sqrt{2}}{\pi \, n\sigma^2} \; [m] \tag{2.8}$$

where: λ - free path of electron in gas, m; n - molar concentration of gas; σ - active cross-section of atom (particle) of gas to be ionized by the moving electron; σ depends on the energy of the electron and assumes maximum values within an electron energy range of 5 to 200 eV.

The molar concentration of gas depends on pressure. The lower it is, the lower the molar concentration of gas, and, in consequence, the longer the free path of the electron. In air at room temperature and at pressure $p = 133$ Pa, $\lambda = 0.266$ mm and at $p = 10^{-2}$ Pa, $\lambda = 2.66$ m [7].

It follows from the above that if one wants to accelerate an electron obtained as the result of thermal emission or glow discharge, it is necessary to create conditions for that electron, ensuring maximum free path. The longer the free path, the higher the energy the electron can assume [9].

Vacuum, therefore, is an essential functional feature of electron equipment. Minimum vacuum (maximum pressure) in electron equipment is several Pa. Soft vacuum (down to 133 Pa), easier to obtain and to maintain, has an undesirable effect on equipment life and may lead to breakdown of the interelectrode space, owing to the ionization of residual gases. For this reason, high pressure is most often used. Its typical range is 10^{-3} to 10^{-6} Pa [7].

Only an exceptionally strongly concentrated, high powered beam may be led from the vacuum chamber out into the atmosphere, in order to perform technological tasks. The path of the electron beam in air usually does not exceed 20 mm.

2.3 Electron beam heaters

2.3.1 Electron guns

The generation of an electron beam requires two sources of electrical energy. The first source serves the emission of electrons from the emitter (cathode), while the second source accelerates them. Both functions are accomplished by systems called **electron guns**. Electron guns constitute the fundamental functional element of electron heaters.

Depending on the type of emitter, two basic types of electron guns are distinguished [10–14]:
- thermal emission with a metal or non-metal thermo-cathode,
- plasma emission with a plasma cathode or cold metal cathode.

2.3.1.1 Thermal emission guns

A thermal emission gun is the oldest and the most frequently used type of gun. It comprises a thermal emission cathode (thermo-cathode), a control electrode and an anode (Fig. 2.3a). The source of electrons (emitter) is the thermo-cathode, placed in vacuum and built from a material with high electron emissivity and high melting point (1600 to 2900 K). In most cases, the metal thermo-cathode is made of tungsten or tantalum. The non-metal cathode is usually made from boron hexaboride LaB_6, sometimes doped with barium hexaboride BaB_6. The service temperature of the metal cathode is usually within 2300 to 2900 K, while in the case of the non-metal cathode it is 1700 to 2000 K [1, 2, 4–6, 10]. Thermo-cathodes may be glowed directly (by utilizing resistance heating) or indirectly, by bombardment by electrons emitted from an addition thermo-cathode and accelerated

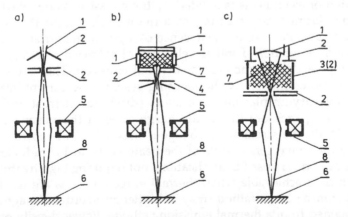

Fig. 2.3 Schematics of electron guns with different types of emitters: a) thermoemissive gun; b) plasma-emissive gun with plasma cathode; c) plasma-emissive gun with cold cathode; *1* - cathode; *2* - anode; *3* - controlling electrode; *4* - extracting electrode; *5* - magnetic lens; *6* - treated object; *7* - plasma; *8* - electron beam. (From *Denbnoweckij, S., et al.* [10]. With permission.)

by the electric field. Direct glowed are metal cathodes, indirect glowed are either metal or non-metal cathodes. By means of voltage applied between the cathode and the anode (30 to 150 kV), electrons emitted from the thermo-cathode are initially formed into a beam and accelerated to a velocity reaching 2/3 of the speed of light. Next, they are passed through an opening in the anode and, in the electro-optical system, containing one or (less frequently) two magnetic lenses, and they are finally formed into a beam with an angle of divergence of 10^{-3} to 10^{-1} rad. With the aid of additional systems, the beam may be deflected in any direction or in two mutually perpendicular directions, with a determined frequency. Thermo-cathodes may be of different design and shape but they all work in conditions of high vacuum (10^{-6} to 10^{-2} Pa). The average power of thermo-cathodes is from approx. 10 to above 100 kW. Extreme values may even be several times greater. Current density of the emission from thermo-cathodes is equal to several A/cm². Power densities obtained on the load may exceed 10^9 kW/m². The time of heating is very short (of the order of microseconds). Life of thermo-cathodes reaches 500 h [1, 2, 4–6, 10–16].

2.3.1.2 Plasma emission guns

The advent of the plasma emission gun came in the 1960s. It works at temperatures lower than those used in thermal emission guns and in conditions of softer vacuum. It is more resistant to the effect of atmosphere of the technological process and is characterized by long life (up to 5500 h), reliability and repeatability of beam parameters [10]. The emitter of electrons is, directly or indirectly, plasma, generated by glow discharge.

In **plasma cathode guns**, the direct emitter is plasma generated by glow discharge in nitrogen, argon, helium, hydrogen or methane. Electrons exit the plasma zone as the result of thermal movements (Fig. 2.3b). Extraction of electrons is facilitated by the emission diode. Next, the electrons are formed into a beam, in a manner similar to that in thermal emission guns. Because of the absence of a potential barrier at the plasma boundary, the scatter of initial velocities of extracted electrons is significantly higher than in the case of thermal emission. This is conducive to errors in representation by the focusing system. The current beam is controlled by varying plasma parameters (discharge current and voltage). The working pressure in the discharge zone of the plasma cathode is 10^{-3} to 10^{-1} Pa, the accelerating voltage is max. 60 kV, the power reaches 10 kW and the convergence angle of the beam is 10^{-2} to 10^{-1} rad. Guns with a plasma cathode are used in applications not requiring high treatment precision which is achievable when thermal emission guns are used. Current density from a plasma cathode may be higher by an order of magnitude than that obtained from a thermal emission cathode. Power density on the load may reach 10^7 kW/cm² [10].

In **cold cathode guns**, plasma at 0.1 to 10 Pa is the indirect emitter of electrons and the source of positive ions, while secondary emission of electrons from the cold metal cathode takes place as the result of its bom-

bardment by ions and high velocity neutral particles, created due to collisions of ions with gas particles (Fig. 2.3c). In most cases the cold cathode is
made of aluminum because of the high coefficient of secondary emission of
that material. The working gas is usually air. Current density of cold cathode emission is lowest and does not exceed tenths of an ampere per cm²,
which causes the necessity of using cathodes with a well-developed surface in order to obtain the appropriate power density on the load (up to 10^6
kW/cm²). Usually, the lateral dimension (or diameter) of the beam is big
and ranges from 10 to 20 mm. The angle of convergence of the beam is
relatively big (0.1 to 1 rad). The power of cold cathodes is approxi-

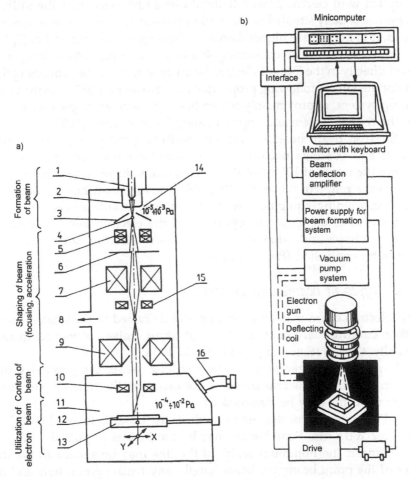

Fig. 2.4 Electron beam heater: a) design schematic; b) schematic of heater controlled by
minicomputer; *1* - thermocathode; *2* - controlling electrode; *3* - anode; *4* - electron-
optical lens; *5* - adjustment and centering system; *6* - aperture; *7* - adjustment lens; *8* -
vacuum pump; *9* - condensing lens; *10* - deflecting system; *11* - work chamber; *12* -
treated object; *13* - x-y stage; *14* - electron gun; *15* - stigmatizer; *16* - viewing port. (Fig.
a - from *Oczoś, K.* [7], Fig. b - from *Sayegh, G., and Burkett, J.* [15]. With permission.)

mately similar to that of plasma cathodes and the accelerating voltage is 10 to 20% lower. Cold cathode guns are especially suited for flat cathodes and are applied in processes not requiring high precision [1, 10-14].

2.3.2 Design of electron beam heaters

The electron beam heater comprises four basic functional systems: beam generation (with the electron gun); beam formation (focusing, acceleration), beam control (beam deflection), and beam utilization (rotating table or x-y stage with the load) (Fig.2.4a). These systems are appropriately supplied with electric power. It should be emphasized that the utilization system is usually situated in the working chamber where the pressure is 1 to 2 orders of magnitude higher than in the beam generating chamber [7, 11-16].

In an electron beam consuming 40 kW input power, 50% of that power is used directly in the form of electron beam power, while the remaining 50% is distributed in the following proportions: approximately 38% - on the vacuum pump system, approximately 5% on beam generation, approximately 3.5% on the control system and approximately 5% on cooling [108].

Electron beam heaters used for modifying the properties of surface layers and coatings usually work in a pulse mode and their power is within the range of 10^3 to 10^5 W. Accelerating voltage varies and is usually within the range of 0.1 to 23 MV [110]. Electron beam heaters are usually electron welding equipment with appropriate modifications.

Because of the great susceptibility of the electron beam to control of power, shape and other parameters, many electron beam heaters are computer controlled (Fig. 2.4b) [15].

2.3.3 Types of beams and patterns

The electron beam may be generated and delivered to the treated material either **continuously** or in the form of **short pulses** of varied duration. Usually, the duration of a pulse is 10^{-9} to 10^{-4} s.

Depending on geometry or, in a stricter sense, the geometry of the pattern left by the lateral shape of the beam on the heated load surface, electron beams may be classified as:

– **Point**: minimum diameter on which the beam is focused (**focal spot**) may reach 0.5 nm. Point beams may be continuous or pulsed.

– **Linear**: the minimum width of the line may be similar to the diameter of the point beam; the beam length may reach several tens and more millimeters. Usually, linear beams are continuous. A pseudo-linear beam may be obtained by a very rapidly deflected point beam (deflection frequency: 10^3 to 10^6 Hz).

– **Ring-shaped**: the diameter and thickness of the ring depend on the technological process. Usually, ring-shaped beams are pulsed.

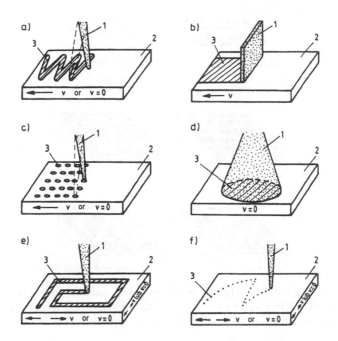

Fig. 2.5 Scanning patterns: a) sequential; b) strip; c) point; d) island (island-surface); e, f) free pattern; *1* - electron beam; *2* - treated object (*v* - feed velocity); *3* - beam track (electron path). (Fig. a, b, c, d - from *Szymański, H., et al.* [1]. With permission.)

– **Surface**: in the form of a circle or rectangle with dimensions up to several tens of millimeters and sometimes more. These are usually pulsed beams with pulse duration of the order of nanoseconds.

For practical purposes, the combined action of beam movement and feed or rotation of the treated object is utilized. This combined movement forms the so-called **pattern** which is a representation of the trace of the electron beam across the treated surface or a map of heated zones. Five basic types of patterns (Fig. 2.5) are distinguished [16]. These depend on the method of heating [1] or scanning [17]:

1. *Scanned Electron Beam (SEB):* a point electron beam, either continuous or pulsed, scans the treated surface with a given frequency (in most cases above 1 kHz) in a direction transverse to feed or rotation. The scanning amplitude is constant (Fig. 2.5a) or variable.

2. *Swept Line Electron Beam (SLEB):* a linear, continuous electron beam with constant or controlled thickness, directed at an object which is moved in a direction transverse to the beam, heats a strip of the surface (Figs. 2.5b and 2.6a). The pseudo-strip pattern may be obtained by deflecting the point beam as in the SEB method, with a very high density (close packing) of scanning traces [1, 12-14, 19].

3. *Pulsed Electron Beam (PEB):* a point, pulsed electron beam, with a constant or variable diameter of the focal spot, heats successive points of the load, changing its position by leaps (e.g., with leap time of 10 µs) and

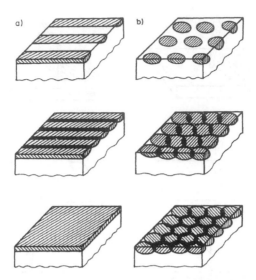

Fig. 2.6 Scanning patterns: a) strip; b) island - with different overlapping of heated zones; darkened zones - due to double or triple heating, exhibit changed different structure (i.e. tempered) than the lightly shaded zones (e.g., hardened).

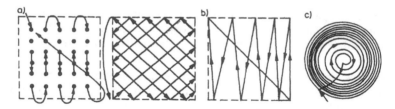

Fig. 2.7 Free scanning patterns: a) square or rectangular on flat surfaces; b) variable surface power density on flat or cylindrical surfaces; c) spiral on flat surface (From [17]. With permission.)

remaining fixed at each successive spot for a duration of e.g., 20 μs (Fig. 2.5c). The distance between the successive spots may be controlled within a range of tenths of a millimeter [1]. When treatment is called for of a surface exceeding several cm², additional movement of the object is applied [19].

4. *Island (island-surface)*: a surface or ring-shaped, pulsed electron beam, with dimensions of the focal spot either fixed or variable from several to several tens of millimeters, heats either successive zones of the load (less frequent case), as shown in Fig. 2.6b, or simultaneously the entire fixed load. This last case is more frequent, e.g., post-implantation heating of semi-conductors (Fig. 2.5d) [1, 11-14, 20].

5. *Custom shaped*: a point electron beam, either continuous or pulsed, scans the surface of the treated object according to program, dependent on the beam size and parameters, as well as the requirements of the technological process, as shown in Figs. 2.5e, f, 2.7 [16].

2.4 Physical fundamentals of interaction of electron beam with treated material

2.4.1 Mechanism of interaction of electron beam with treated material

Independent of type of beam or method of scanning, the character of interaction of the electron beam with the treated material is the same. The beam electrons are subjected to reflection (dispersion), absorption or transmission. They may evoke secondary electron emission from the bombarded material or cause excitement and ionization of atoms. They may also evoke the emission of X-ray and gamma radiation. Proportions between the scale of phenomena occurring depend on electron energy and on the nature of the bombarded material.

In all cases, the kinetic energy of electrons is transformed into other forms of energy. If it is transformed predominantly into thermal energy, we refer to a **thermal process**. This is the type of energy transformation that is primarily utilized in surface engineering. In other technological processes discussed further (see Section 2.5) only the thermal effect of the beam on treated material is utilized.

Accelerated electrons, on reaching the surface of the treated material, penetrate it. Upon penetration, they are rapidly decelerated. A single electron interacts both on the crystalline lattice of the material, as well as on single atoms, particles and electrons within that lattice. As a result of this interaction, electric fields of these particles are disturbed, causing migration of atoms and particles and a rise in the amplitude of their vibration. This is manifest by a significant rise in temperature. The treated material is heated in the zone where the electron beam interacts.

The so-called primary electrons penetrating the material meet electrons belonging to the material along their path. The electrons thus met may be either free electrons or electrons bound within a crystalline lattice. Penetrating electrons with high energy may collide with electrons belonging to the material, as a result of which some electrons, so-called secondary, may be expelled from that material (Fig. 2.8). This effect is called **secondary emission**. The primary electrons are those emitted by the cathode. Other electrons, due to collisions, may be displaced in the atoms. They may pass over to orbits further away from the nucleus; on passing back to orbits closer to the nucleus, the electrons emit electromagnetic radiation, among others, within the X-ray range [8].

Heating of the material occurs as the result of absorption of energy of the electron beam, due to non-elastic and elastic collisions of electrons with the material's crystalline lattice. The zone of energy exchange during this absorption is situated at the surface and immediately under the surface of the bombarded material. The size of that zone depends on condi-

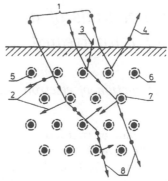

Fig. 2.8 Diagram showing interaction of the electron beam with material: *1* - electron beam (simple implantation); *2* - secondary electrons (recoil implantation); *3* - emission of secondary electrons; *4* - recoiled electron; *5* - atom nucleus; *6* - electron shell; *7* - thermal excitation; *8* - penetration (diffusion) of electrons. (From *Oczoś, K.* [7]. With permission.)

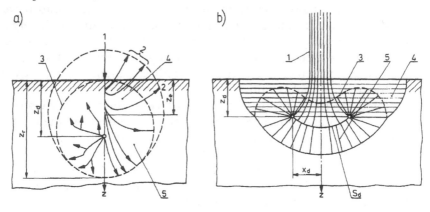

Fig. 2.9 Models of zones of interaction of the electron beam with the material: a) model with very small electron spot; b) model with big electron spot; *1* - electron beam; *2* -recoiled electrons; *3* - zone of dispersion of beam electrons to small angles upon collision; *4* - zone of dispersion of electrons to big angles upon collisions; *5* - zone of diffusion dispersion of beam electrons; z_d - situation of maximum dispersion of beam electrons; z_r - greatest depth of penetration of beam electrons; z_e - depth corresponding to greatest loss of energy by beam electrons; S_d - surface of maximum dispersion of electrons and its coordinates, z_d and x_d. (Fig. a - from *Oczoś, K.* [7], Fig. b - from *Boriskina, L.V., et al.* [21]. With permission.)

tions of dispersion of electrons in the material. The initially high energy of the electrons is reduced as the result of collisions. In the initial portion of the penetration path, dispersion of the beam is small. This dispersion increases with the drop in electron energy, reaching maximum at a depth z_d. The greatest portion of electron energy is absorbed at a depth z_e under the surface of the material (Fig. 2.9). Due to multiple collisions of electrons belonging to the beam with the crystalline lattice of the material, the mean energy of the electrons decreases and reaches the value equal to the mean energy of electrons belonging to the treated material. In the case of a beam with a focal spot approaching a point, the beam electrons totally lose their surplus energy, relative to material electrons within a zone of diameter z_r (Fig. 2.9a). Theoretically, this surface limits the zone of interaction of the electron beam with the treated material [7]. This surface takes on a slightly different form in the case of an electron beam with finite dimensions and a relatively big electron spot (Fig. 2.9b).

Assuming that the electron beam is characterized by a given wavelength, the electron beam can be treated as an electromagnetic wave of a given length [8]. Taking this into account, it is possible to calculate its depth of penetration into the material, similarly to the case of induction or microwave heating. Because the length of the electron wave is very small, the depth of penetration of the beam into the material is also very small. Practically, the total energy of the beam is transformed to heat in the subsurface layer of the material. The thickness of this layer has been determined empirically by B. Schönland [7, 22, 37, 38]:

$$z_r = k \cdot 10^{-12} \frac{U^2}{\rho} \tag{2.9}$$

where: z_r - depth of penetration of electrons, cm; k - empirical coefficient ($k = 2.1$ according to [7], $k = 2.35$ according to [17, 37, 38]); ρ - density of heated material, kg/m³; U - accelerating voltage, V.

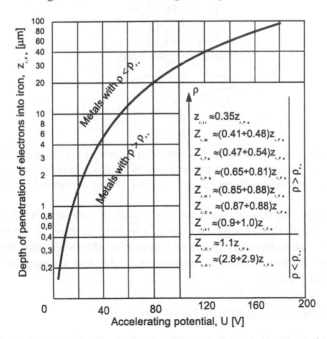

Fig. 2.10 Dependence of depth of penetration of electrons into iron on the accelerating potential.

The depth of penetration of electrons into the material increases with a rise of accelerating voltage and decreases with a rise of material density (Fig. 2.10). If U = 10, 20, 50 and 100 kV, the corresponding values for steel are: $z_{r(st)}$ = 0.3, 1.05, 6.1 and 27 µm, and $z_{r(Al)}$ = 0.8, 3.1, 19.4 and 80 µm for aluminum.

When the most frequent range of accelerating voltages is used, i.e., U = 30 to 150 kV, the depth of penetration of electrons into ferrous alloys is from several to over 40 µm [24]. With higher voltages, there is a possibility of substantially deeper penetration of electrons or of their passing through a heated or melted metal to a depth of several or even more millimeters. This is an experimentally determined phenomenon of deep penetration, as yet without a satisfactory explanation [16].

The distribution of current density I_w in the cross-section of the electron beam corresponds, in the majority of cases, to a normal Gaussian curve. Distribution of power density in the beam is proportional to the distribution of current density.

With a certain degree of approximation, the distribution of density of the energy released at the depth of penetration of electrons into material z may be assumed as being close to a normal Gaussian curve, with the maximum at depth z_e [7].

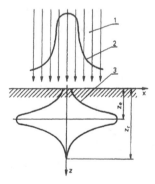

Fig. 2.11 Schematic representation of heat source formed in material due to the action of the electron beam: *1* - electron beam; *2* - distribution of current (and power) density in beam; *3* - distribution of energy (and power) density dissipated in material.

The source of heat created as the result of interaction of the electron beam on the material may be, therefore, said to have a normal surface distribution with the maximum in the axis of the beam and normal volumetric distribution with the maximum under the surface of the heated material (Fig. 2.11).

2.4.2 Efficiency of electron beam heating

Power carried by the electron beam is determined from formulas developed for

– a continuous beam

$$P_{cb} = IU, \text{ and} \tag{2.10}$$

– a pulsed beam

$$P_{pb} = IU\tau_p f_p \tag{2.11}$$

where: P_b - power of continuous (P_{cb}) or pulsed (P_{pb}) beam, W; I - intensity of beam current, A; U - accelerating voltage, V; τ_p - pulse duration, s; f_p - pulse frequency, 1/s.

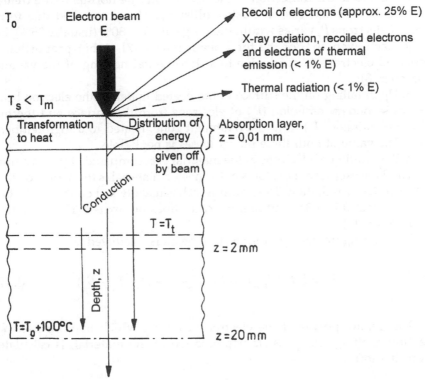

Fig. 2.12 Interaction of the electron beam with the surface of the treated material and the mean distribution of energy carried by the beam: T_s - surface temperature; T_m - melting point; T_t - transformation temperature; T_o - ambient temperature. (From *Zenker, R., and Müller, M.* [18]. With permission.)

As stated earlier, the electron beam is reflected, absorbed or transmitted upon striking the material surface. Since the aim of using the electron beam in surface engineering is the heating of treated material, the beam is used to heat such materials which have a thickness at least severalfold greater than the depth of penetration of electrons, so that the electrons do not pass through heated material (Fig. 2.12).

Thus, the power released on the load is the difference between the power carried by the electron beam (UI) and power losses, i.e power lost on unnecessary effects (not connected with heating).

The non-heating effects caused by the electron beam may be the following:

1. Power lost along the path of the beam from the gun to the heated material, due to collisions of electrons with particles of gases, vapors and carriers of electrical charges. The value of this power depends on the pressure in the vacuum chamber and on the length of the electron path, and varies within the limits: P_p = 1 to 10% [8].

2. Not all electrons falling on the surface are absorbed by the treated material. A portion, proportional to the atomic number of the absorbing material and to the deflection of the beam from the normal to the treated surface, is reflected due to elastic collisions. The power lost due to reflected electrons P_r varies within the range of 10 to 30% (usually 25%), but may reach even 40% of the total beam power [7]. A big proportion of reflected electrons is conducive to a detrimental heating of the vacuum chamber [8].

3. Upon falling on the surface of the heated material, the electron beam causes secondary emission (P_e) of electrons from the surface and thermo-electron emission (P_t) from the treatment zone, heated to a high temperature. The value of both these power losses is negligible.

4. Retardation of electrons in the material is accompanied by X-ray emission and the excitement of atoms of the heated material is the source of characteristic X-ray radiation. The power which is usually lost on these effects P_x varies within 0.1 to 3%, but in most cases does not exceed 1% of the total beam power [7].

The **heating power** P_h of the electron beam is calculated as

$$P_h = IU - P_{loss} = IU - (P_p + P_r + P_e + P_t + P_x) \qquad (2.12)$$

The **usable power** of the electron beam P_u, which may be used for heating, melting and possible vaporization of the material, is calculated from the formula:

$$P_u = P_h - (P_c + P_r + P_{i+ex}) \qquad (2.13)$$

where: P_c - power lost due to thermal conduction of the heated material from the heated zone to the core, W (on account of the big volume of the cold material, relative to the heated material, this value is usually very small); P_r - power lost due to radiation of the heated zone to the surroundings, W (usually, this power loss does not exceed 1% of total beam power); P_{i+ex} - power lost due to ionization and excitement of vaporized atoms which happened to be in the path of the electrons, W.

All three categories of power loss are negligible, compared to the heating power of the electron beam. When heating by a pulsed beam, the power during the pulse may be very high, while the mean power may be relatively low. The pulsed beam allows the achievement of very high local temperatures, with small losses on thermal conduction [7, 34–36].

Heating efficiency of the electron beam, i.e., efficiency of heat release in the heated material, is

$$\eta_h = \frac{P_b - P_{loss}}{P_b} = \frac{IU - (P_p + P_r + P_e + P_t + P_x)}{IU} \qquad (2.14)$$

Usable efficiency of electron beam heating is

$$\eta_u = \frac{P_h - P_{loss}}{P_h} = \frac{P_h - (P_c + P_r + P_{i+ex})}{P_h} \qquad (2.15)$$

This efficiency may vary within the range of 0.4 to 0.9, but in most cases values are between 0.7 and 0.8.

2.4.3 Rate of heating and cooling

Because the electron beam is a source of very high energy (usually several tens of kilowatts), its concentration on a small surface ensures achievement of heating rates as high as 10^3 to 10^5 K/s and allows not only practically immediate heating but also melting of the surface layer and its immediate cooling [25, 34–39]. The heated surface area usually has a diameter of several millimeters but minimum diameters of the electron spot may be as small as 0.5 mm [12].

Cooling of the load is without the use of additional cooling agents but the load mass is utilized. Owing to good thermal conductivity of the material, heat energy is conducted away very quickly from the heated zone to zones situated deeper. This so-called **self-cooling** [5] or **cooling by load mass** enables the achievement of cooling rates comparable with those of heating on condition that the volume of the cold material is 5 to 8 times greater than that of the heated zone [17]. This renders practically possible the heating of very thin elements [11–14] with thicknesses at least four times greater than the depth of the heated zone [3].

The process of very fast heating and cooling of metals is accompanied by many physical phenomena (see Table 2.1), which are characteristic not only of electron beam heating but also laser heating, allowing a modification of the properties of the surface layer, difficult or impossible to achieve by any other means [3].

The type of effect obtained depends mainly on power density of the beam and on the time of acting on the material.

Power density may be calculated from the formula:

Table 2.1
Physical phenomena accompanying the process of rapid heating
and cooling of materials (From *Frey, H., and Kienel, G.* [110]. With permission.)

Phenomenon	Application
Epitaxial growth of crystal	Removal of implantation effects (post-implantation heating)
Grain growth in thin polycrystalline layers	Increasing electrical conductivity of thin polycrystalline layers (e.g., silicon)
Grain refinement	Enhancement of ductility and hardness
Dissolution of alloying additives and microsegregation	Homogenization
Structural and phase transformations	Surface hardening
Melting-in of additional materials; formation of supersaturated solutions	Metal property enhancement through surface remelting
Introduction of insoluble or weakly dissolving additives	Formation of composite materials by surface remelting

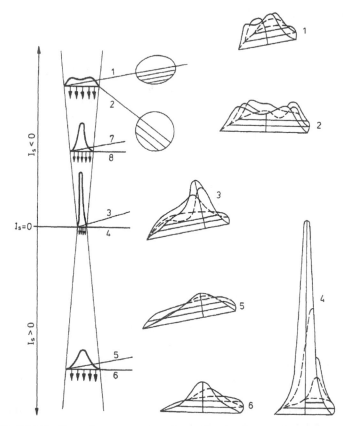

Fig. 2.13 Distribution of power energy in an electron beam, depending on method of focusing: *2, 4, 6, 8* - beam situated perpendicular to treated surface; *1, 3, 5, 7* - deflected beam; I_s - focusing current; $I_s>0$ - focusing above surface; $I_s<0$ - focusing below treated surface. (From *Zielecki, W.* [26]. With permission.)

$$q = \frac{P_b}{d_{es}}\left[W/cm^2\right] \quad \text{or [38]:} \quad q = \frac{4IU}{\pi\, d_{es}^2} = \frac{1}{\pi}\left(\frac{2}{S}\right)^2 \cdot I^{\frac{1}{4}} \cdot U^{\frac{7}{4}} \qquad (2.16)$$

where: q - power density, W/cm^2; d_{es} - diameter of electron spot, cm^2; S - electron-optics system constant.

Because the distribution of power density in the beam cross-section is not constant, it usually exhibits at least one maximum and is usually similar to a normal Gaussian curve; the formula (2.16) expresses the **mean power density** of the beam.

With constant power, the power density of the incident beam may be controlled by varying the distance of the focal spot from the material surface (Fig. 2.13).

Time of heating (exposure) t may vary considerably and range from a second down to microseconds. The time of heating is controlled by:
– varying the frequency and amplitude of beam deflection,
– varying the rate of feed or rotation of the heated object.

2.5 Electron beam techniques

Depending on the combination of parameters - power density q and time of heating t - it is possible to heat, melt, bring to boiling point and vaporize the material. The combination of these parameters is critical to the technological effect of electron beam heating (Fig. 2.14).

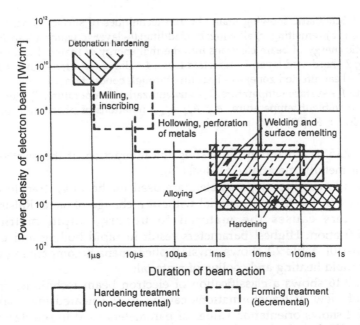

Fig. 2.14 Power density vs. time of action of electron beam of 100 keV energy on metal layer of 0.1 mm thickness, for different technological processes.

Surface of heated material

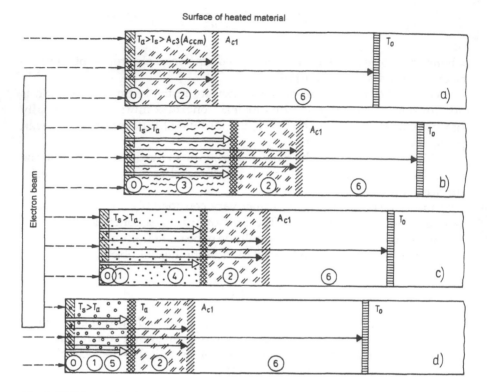

Fig. 2.15 Schematic showing main electron techniques of surface enhancement: a) hardening; b) remelting; c) alloying; d) cladding; *0* - layer in which a transformation of kinetic energy of beam electrons into thermal energy of material takes place; *1* - coating; *2* - hardened layer; *3* - remelted layer; *4* - alloyed layer; *5* - surface remelted layer; *6* - heat affected zone; → - heat flux through conduction; ⇒ - flux of alloying material; T_s - surface temperature; T_a - austenitization temperature; T_0 - temperature of core = ambient temperature; A_{c1}; A_{c3}; A_{ccm} - notations of phase transformations. (From *Zenker, R.* [93]. With permission.)

Fig. 2.15 is a diagram illustrating the interaction of the electron beam with the metallic material, e.g., steel [93].

When lower energy parameters are used for heating, transformations take place in the treated material without a change of state. Raising these parameters causes the material to undergo rapid melting and resolidification. Highest parameters result in rapid boiling and evaporation [28-110]. Table 2.1 shows physical phenomena accompanying the process of rapid heating and cooling of materials.

Fig. 2.16 shows a classification of electron beam techniques from the point of view of transformations caused in the heated material, while Table 2.2 shows orientation values of parameters used in selected electron beam techniques of surface modification.

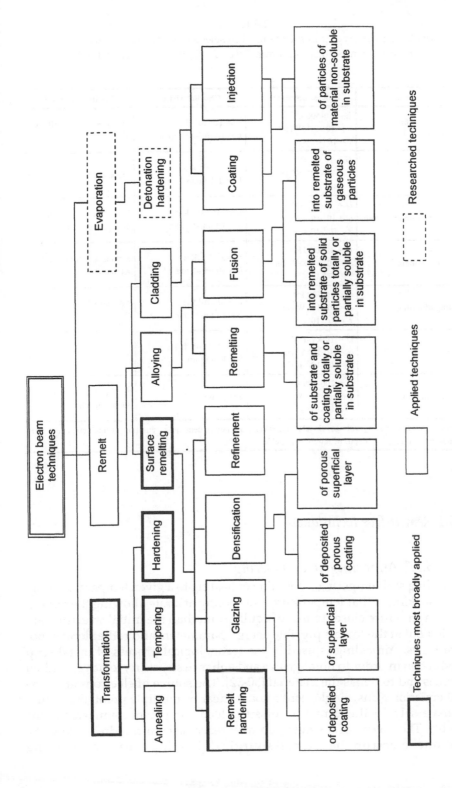

Fig. 2.16 Classification of electron beam techniques by transformations in heated material.

Table 2.2
Orientation values of parameters used in selected electron beam techniques
of surface modification (From *Pobol, I.L.* [95]. With permission.)

Method	Process parameters	Characteristics	Treated materials
Hardening without remelting	U = 10-500 kV q = 10^3-10^9 W/cm^2 τ = 10^{-8}-10 s	v_s = 10^3-10^7 K/s h = 10 μm-2.5 mm	Steels, cast irons, alloys of Ti
Refinement	U = 10-60 kV q = 10^2-10^4 W/cm^2 τ = 1-10^2s	v_s = 1-10^2 K/s h = 1-10 mm	Metals, alloys
Hardening with remelting	U = 10-500 kV q = 10^3-10^7 W/cm^2 τ = 10^{-5}-10 s	v_s = 10^2-10^7 K/s h = 10 μm-10 mm	Steels, cast irons, alloys of Ti, Al and Cu
Alloying, cladding	U = 10 kV-1MV q = 10^3-5·10^9 W/cm^2 τ = 10^{-8}-10 s	v_s = 1-10^7 K/s h = 10 μm-1 mm	Metals, alloys of Fe, Ti, Al and Cu
Sealing of coatings	U = 10-500 kV q = 10^3-10^5 W/cm^2 τ = 10^{-3}-10 s	v_s = 1÷10^7 K/s h = 10 μm-1 mm	Alloys of Fe, Ti, Al and Cu
Glazing	U = 10 kV-1 MV q = 10^5-10^9 W/cm^2 τ = 10^{-8}-10^{-3} s	v_s = 10^5-10^{12} K/s h = 1 -400 μm	Steels, alloys
Shock hardening	U = 200-300 kV q = 10^8-10^{12} W/cm^2 τ = 10^{-8}-10^{-6} s	h = 1-100 μm	Fe, Ti, Mo, Cu, steels, Al and Al alloys

Notation: U - accelerating voltage; E - electron energy; q - power density; τ - duration of interaction; v_s - rate of material cooling; h - depth of modification effect

2.5.1 Remelt-free techniques

2.5.1.1 Annealing and tempering

Annealing and tempering, consisting of heating to a given temperature (Fig. 2.17), soaking at that temperature and cooling at a rate allowing the obtaining of a structure closer to that of equilibrium than the initial structure, are carried out at the lowest possible energy parameters and rather long exposure times. Annealing is used in the technological process of metal strip production in order to homogenize and enhance structure, remove residual stresses and to degas the material [30-32]. The source of electrons are thermal emission guns, single, continuous, linear or multiple point, oscillating transversely to the direction of strip feed, or plasma emission guns [11-14]. The power of a single gun may reach several hundred kilowatts. On one electron beam metallurgical heating unit for heating

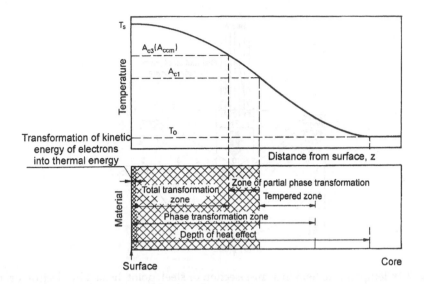

Fig. 2.17 Profile of temperature distribution along Z - axis, perpendicular to surface of material annealed by the electron beam. (From *Zenker, R.* [94]. With permission.)

strips of 1 mm thickness, fed at linear speed of 75 m/s, to 1000°C, 12 emission guns are used, each of 500 kW power [6]. Attempts are also made to apply plasma emission guns to anneal strips made of stainless steel powdered metal [1, 33]. Plasma emission guns are coming into industrial scale use for diffusion processes (electron beam heating in thinned diffusion atmosphere), e.g. for simple electron beam carburizing or for electron beam carburizing after vacuum carburizing, in order to achieve selectively differentiated depths of diffusion [27]. Among the advantages of the electron beam technique of strip annealing are, primarily, good degassing of strip material and absence of surface oxidation, due to vacuum of 10^{-4} hPa, while the process features high efficiency [6].

Electron beam tempering is used most often after electron beam hardening, as well as after electron beam welding of joints.

The electron beam is also used to preheat the joint zone, prior to welding [109].

Finally, the electron beam technique is used to heat electronic junctions [109].

2.5.1.2 Remelt-free hardening

Transformation hardening is, chronologically, the first method of electron beam hardening of the surface layer and consists of its short-duration heating, lasting approx. from 1 ms to 1 s, at a rate of 10^3 to 3×10^3 K/s, to a temperature exceeding that of martensitic transformation (Figs. 2.18, 2.19 and 2.20) but lower than the melting point. The usual power density applied is approximately several kW/m². Due to rapid cooling at a rate of 10^4 to

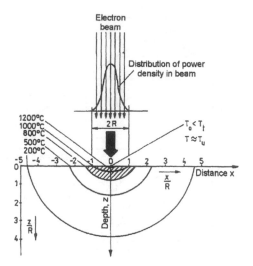

Fig. 2.18 Temperature field in a cross-section of steel, point heated by electron beam and with gaussian distribution of power density in the beam, vs. beam diameter 2R: isotherms in the shape of semi-ellipses; heating without surface remelting; shaded area - hardened layer. (From *Zenker, R., and Müller, M.* [88]. With permission.)

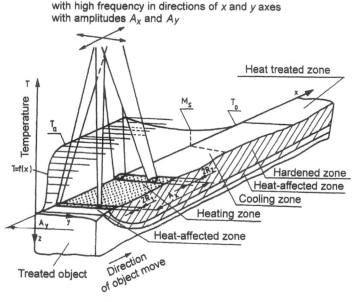

Fig. 2.19 Schematic representation of heating and cooling (quenching) of steel with two-directional scanning by the electron beam and variable focusing: greater (smaller diameter of focal spot $2R_1$ - greater supplied power density) at the beginning of heating; smaller (greater diameter of focal spot $2R_2$ - lower supplied power density) at the end of heating. T_a - austenitization temperature; M_s - temperature corresponding to beginning of martensitic transformation; T_o - initial temperature (of environment, core). (From *Zenker, R., and Müller, M.* [88]. With permission.)

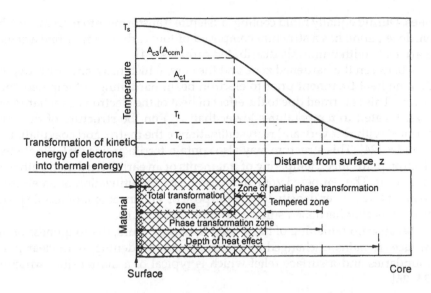

Fig. 2.20 Profile of temperature distribution in axis perpendicular to electron beam heated surface of material in the case of hardening and tempering. (From *Zenker, R.* [94]. With permission.)

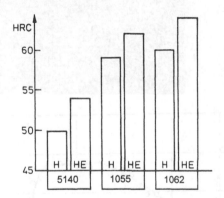

Fig. 2.21 Comparison of hardness of steels (5140, 1055 and 1062), traditionally hardened (H) with those hardened by the electron beam (HE). (From *Zielecki, W.* [26]. With permission.)

more than 10^5 K/s [24, 95], a very fine structure is obtained, higher by several HRC points (Fig. 2.21) than that obtained by conventional methods, i.e., induction, flame or plasma heating which is substantially slower, with rates of heating not exceeding 1000 K/s [11–14].

Steel hardened by the electron beam without remelting has a two-zone structure:

– from the surface end there is a hardened zone, formed as the result of heating to temperatures higher than A_{c3}, which enables the occurrence of austenitic transformation without diffusion, due to very rapid heating rates. During equally rapid cooling, austenite transforms into martensite. The hardened zone has a structure composed of low carbon lath or fine acicular martensite with uniformly distributed carbides [24],

– between the hardened zone and the core, which has a structure dependent on heat treatment prior to electron beam hardening, a tempered zone occurs. This is formed due to the effect of heat of the electron beam hardened layer (heated to temperatures lower than A_{c3}) on the structure of the core, coupled with recovery and recrystallization of the matrix and coagulation of carbides [24]. This structure may also contain ferrite grains [95]. Where two electron paths meet, tempering of fragments of an earlier hardened path may also occur. The tempered zone may also be by the utilization of an electron beam of low power density for the purpose of enhancement of usable properties of earlier hardened surfaces (see Section 2.5.1.1.).

Because no remelting of the treated material occurs, the roughness of the surface remains unchanged by electron beam hardening, with clear grain boundaries and a surface relief which is typical of a martensitic structure [24, 26].

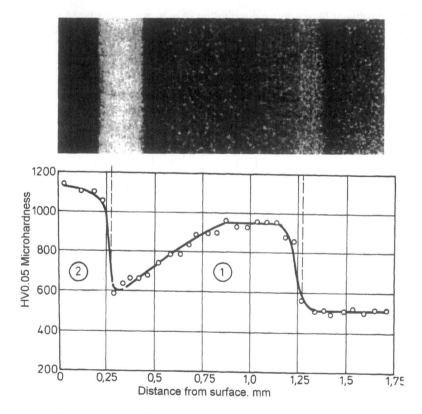

Fig. 2.22 Hardness profile for double-layer (sandwich) hardening of 6150H steel (prior hardened and tempered): *1* - greater energy density and longer time of beam action (2000 W·s/cm², 0.82 s, 1 cm/s); 2 - lower energy density and longer time of beam action (450 W·s/cm², 0.04 s, 5 cm/s). (From *Zenker, R., and Müller, M.* [88]. With permission.)

Most frequently, **single layer** hardening is carried out, i.e. a process in which only a single electron beam scan is made of a given zone. The result is a hardness profile with the maximum at or near the surface of the hardened object.

Less frequently, **multi-layer,** or *sandwich* hardening, is carried out by multiple electron beam scanning of the same zone. When it is, it is usually a double layer process. First, the inner portions hardened to a lower hardness, and next the outer portions to a higher hardness. Subsequent scans of the beam differ from the first one by heating parameters (lower power density, shorter heating time). The hardness profile obtained exhibits a second clearly marked maximum at a certain depth under the surface (Fig. 2.22). The sandwich structure of the surface layer is advantageous by diminishing the gradient of residual stresses and by yielding enhanced tribological properties [88].

The hardness distribution varies for different zones of the electron beam path and depends mainly on energy parameters of heating (Fig. 2.23) and on the direction of part feed relative to the direction of beam scan. Biggest stresses occur along the axis of the electron beam path [87].

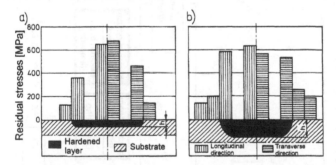

Fig. 2.23 Distribution of surface residual stresses after electron beam hardening of prior normalized 6150H steel, with different energy densities: a - 760 W·s/cm^2, $h = 0.45$ mm, b - 2400 W·s/cm^2, $h = 1.45$ mm. (From *Zenker, R. et al.* [87]. With permission.)

By causing a rise in hardness (up to 3.7 times for annealed steels and up to 1.7 times in comparison with conventionally hardened steels) [110] electron beam hardening brings about a significant improvement of tribological properties of structural and tool steels. The coefficient of dry friction may rise by 20 to 50%, while wear resistance by 70 to 100% [26].

The effect of electron beam hardening on fatigue strength has not been clearly explained. It usually causes its rise but effects in the opposite direction are sometimes known to occur.

Electron beam hardening may be applied as **simple** (single) or **complex,** when combined with other techniques, used in surface engineering. Fig. 2.24 shows possible techniques of remelt-free electron beam hardening, in combination with other methods used in conventional heat , thermochemical and thermo-mechanical treatment. It is obvious that the results of electron beam hardening obtained depend on prior heat treatment (most

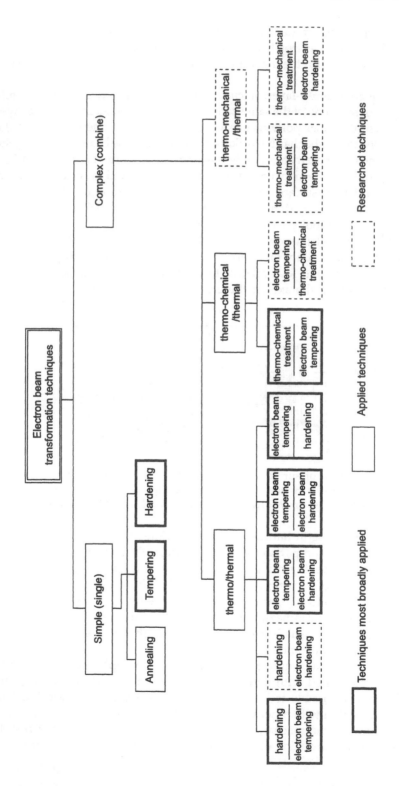

Fig. 2.24 Electron beam techniques - simple and complex. (From *Zenker, R., and Müller, M.* [88]. With permission.)

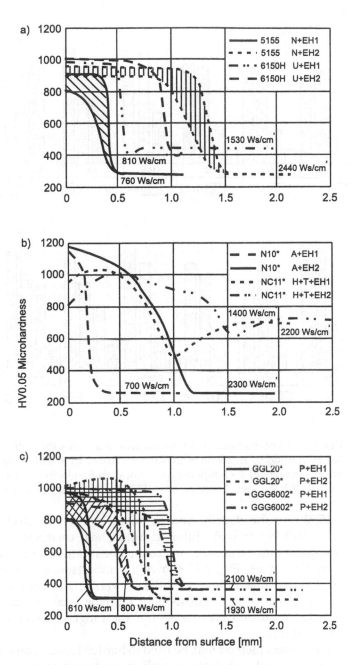

* Grade steel from cited original

Fig. 2.25 Hardness profiles of ferrous alloys with different prior heat treatment (*N* - normalized, *U* - hardened and tempered, *A* - annealed, *H* - hardened only, *T* - tempered, *P* - pearlitic annealed) and subsequently electron beam hardened (*EH*) with different energy densities; a) structural steels; b) tool steels; c) gray cast iron. (From *Zenker, R.* [94]. With permission.)

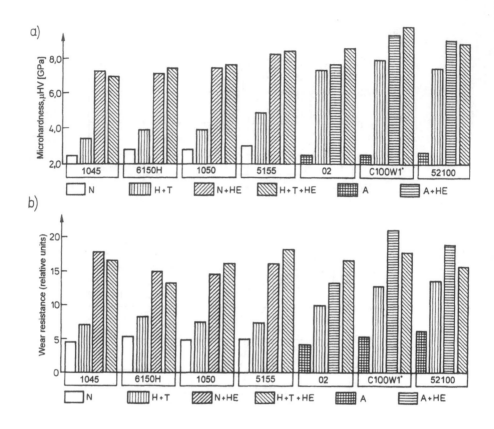

Fig. 2.26 Comparison of microhardness (a) and wear resistance (b) of different steels, heat treated by traditional methods with those enhanced by the electron beam; N - normalized, H - hardened; T - tempered, A - annealed; E - electron beam hardened. (From *Zenker, R.* [83, 86], *and Zenker, R., et al.* [89, 91, 92]. With permission.)

Fig. 2.25 shows the distribution of microhardness of electron beam hardened surface layers with different initial heat treatment, while Fig. 2.26 shows maximum achievable microhardness values and wear resistance for similar layers. Fig. 2.27 shows a comparison of microhardness values for different materials after different versions of thermo-chemical treatment, with and without subsequent electron beam hardening. The thickness of the hardened layer is within the range of several μm to as much as several mm.

The electron beam method may be used to harden low carbon and alloyed structural, bearing and tool steels, as well as white and gray cast irons [42-95]. All techniques are used, utilizing all types of beams.

Besides a higher hardness than that achieved by conventional hardening, the electron beam allows the precise heating of selected spots, even of very small dimensions, while maintaining very close tolerances of hardened layer thickness and lower quench stresses, occurring only in small zones or even microzones of the treated material. This method enables

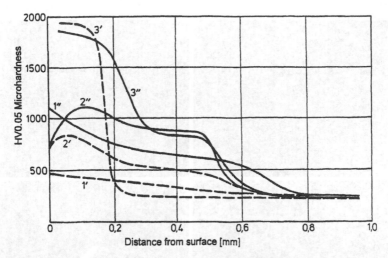

Fig. 2.27 Hardness profiles for: *1* - 5115 tool steel (*1'* - carburized, *1"* - carburized and electron beam hardened); *2* - 1045 heat treatable steel (*2'* - carbonitrided, *2"* - carbonitrided and electron beam hardened); *3* - 1045 steel (*3'* - borided, *3"* - borided and electron beam hardened). (From *Spies, H.J., et al.* [108]. With permission.)

hardening of finished components without dimensional changes. The advantages of electron beam hardening cause it to be the most widely used process of electron beam surface enhancement [16]. The cost of hardening of 1 cm² of surface is within the range of $0.001 to $0.01.

2.5.2 Remelt techniques

2.5.2.1 Surface remelting
Remelting is a development of surface hardening. It is carried with the application of power densities higher than those used for hardening, and heating rates reaching 10^4 K/s. It consists of rapid melting of a very thin surface layer of substrate material or a coating deposited on it, and an equally rapid subsequent crystallization.

Remelting is divided into four groups (see Fig. 2.16): remelt hardening, glazing, densifying, as well as refinement and defect removal.

Remelt hardening may be accomplished using different energy parameters of the electron beam. Therefore, we distinguish (Fig. 2.28) [26]:
– **Surface melting,** resulting in the retention of the three-dimensional structure of the earlier formed surface layer, with clearly defined grain boundaries and a relief typical of martensitic structure. The melted surface features numerous small craters which are the nodes of grain boundaries of dendrites, crystallizing from the melted subsurface layer of metal.
– **Remelting,** resulting in a changed structure of the surface, with clearly defined boundaries of dendrites formed during crystallization of the subsurface layer of the metal, remelted deeper than in the case of remelting.

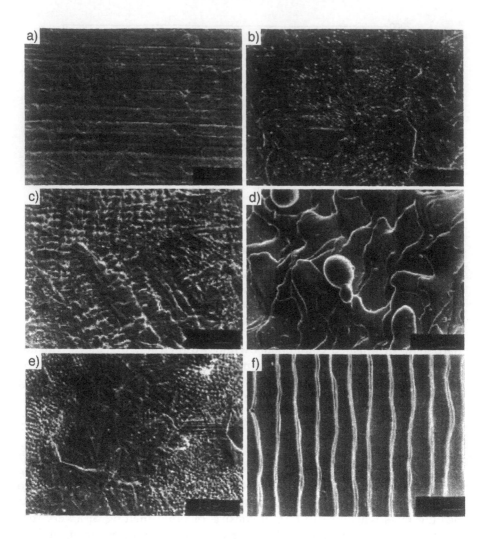

Fig. 2.28 Micrographs of the surface of N135M steel, after electron beam treatment: a) transformation hardened; b) surface remelted; c) remelted; d) intensive remelted; e) with non-homogenous structure; f) with traces of electron beam path. (From *Zielecki, W.* [26]. With permission.)

– **Intensive remelting**, resulting in a clearly deteriorated three-dimensional structure of the surface, primarily due to the formation of runs of remelted material.

– **Very intensive remelting**, resulting in a very clear deterioration of surface structure (increase of waviness and unevenness) with clearly visible electron paths.

It is also possible to obtain surface states which are intermediate between remelting and intensive remelting.

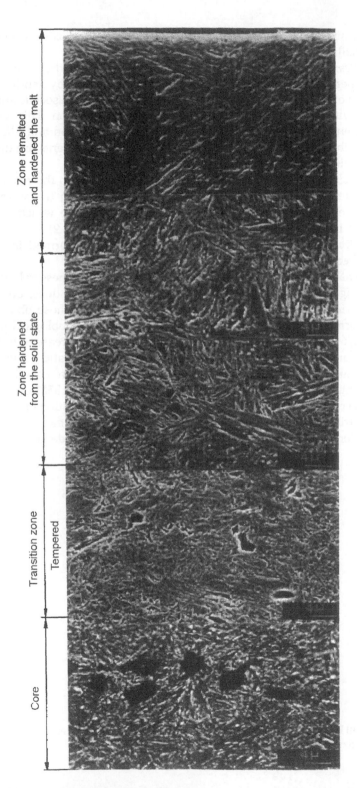

Fig. 2.29 Three-zone structure of remelt hardened superficial layer of N135M steel. Magn. 2500×. (From *Zielecki, R.* [26]. With permission.)

The surface layer obtained by remelting has a three-zone structure (Fig. 2.29) [26]:

1) *Remelted and hardened from the melt zone*, formed as the result of heating to temperatures higher than the melting point, followed by dendritic crystallization of the remelted steel. In consequence, carbides dissociate and carbon, along with other alloying elements, passes into solution. This zone has a homogenous martensitic structure, containing all carbon and alloying elements; the carbides are more refined and alloying elements are distributed more uniformly;

2) *Subsurface zone, hardened from the solid*, formed as the result of heating to temperatures above A_{c3}, allowing non-diffusion transformation of austenite to martensite. This zone exhibits a structure which is like that formed in transformation hardening (see Section 2.5.1.2).

3) *Transition tempered zone, close to core* which is formed in a manner described in Section 2.5.1.2.

Remelting causes a deterioration of surface roughness relative to initial roughness, especially after intensive remelting. It does, on the other hand, yield service properties which are better than those obtained after transformation hardening. This is true especially of tribological properties [11–14, 61, 62, 72–99]. The main reason for this is an increase of hardness or microhardness by between ten and several tens percent and a favorable distribution of residual stresses. The structures obtained are usually corrosion resistant. For example, the fatigue strength of Nitralloy 135M may be higher by several tens percent (depending on treatment conditions may be up to 40%), tribological wear may be down by 70% and so may be loss due to corrosion (a 65 to 80% decrease of passivation current density is obtained) [26].

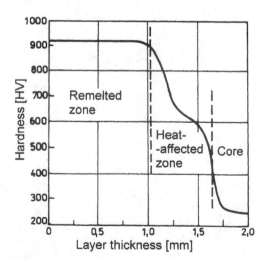

Fig. 2.30 Hardness distribution in remelted and self-cooled layer of nodular cast iron. (From *Szymański, H., et al.* [1]. With permission.)

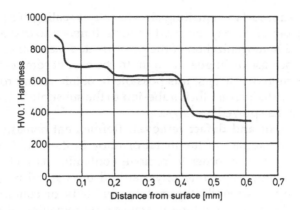

Fig. 2.31 Hardness distribution in remelt hardened and nitrided GGL25* cast iron. (From *Spies, H.J., et al.* [108]. With permission.)

Hardness and wear resistance of tool steels, which may even be three times higher [11–14], cause a 2.5 - to 3 - fold increase in service life of cold forming dies and an 80–90% increase in the life of turning tools [26, 95]. Microhardness of eutectic and hypereutectic aluminum alloys may increase by 30 to 500% and with alloying, even by 600% [95]. This is one way of hardening piston rings [95]. The hardness is increased and a martensitic or ledeburitic structure is obtained on nodular cast iron after remelting (Figs. 2.30 and 2.31). The hardness of sintered carbides (based on TiC) is increased by 12–30% [3] or even by 25–30% [1]. Coarse grained tungsten obtains a fine crystalline structure [1].

The assortment of components hardened by remelting is the same as that on which transformation hardening is carried out.

Glazing. Glazing (vitrification, amorphisation) is a modification of surface remelting.[1] In the case of remelting of very thin layers of some alloys or of very thin coatings and their equally rapid cooling (usually in excess of 10^7 K/s), it is possible to obtain amorphous structures - metallic glazes. These have the same chemical composition as that of the initial surface layer or coating, but a new set of properties, including electrical, magnetic (lower magnetic loss), mechanical (high hardness and tensile strength with the retention of ductility, high wear resistance), or chemical (corrosion resistance) [95]. It is for this reason that glazing is sometimes wrongly identified with remelting [11–14]. Electron beam glazing is applied to nickel and iron base alloys. Obtainable layer thicknesses are 10–40 μm [109]; in rare cases they may exceed 100 μm [95].

Densifying (healing). Densifying, alternately, sealing of porous material of either the surface layer or of a coating, consists of remelting the surface layer to a certain depth or of partial or total remelting of the coating, in order to make it very well sealed and to increase its density.

---

[1] Metal glazes may also be obtained through heating of some alloys by an electron beam at low temperatures [95].

Electron beam surface remelting always causes sealing of a porous substrate or coating but often may cause side effects. It may also enable the removal of defects and homogenization of the prior treated material's structure, resulting in an increase of fatigue strength. In the case of some coatings (e.g., titanium or sintered powders) it is possible to obtain an improvement of their structure and to increase their adhesion to the substrate. It is effectively used to seal plasma sprayed coatings.

Refinement and defect removal. Refinement consists of brief maintaining of the surface of the metal or alloy in the liquid state in order to degas in vacuum, in order to remove contaminants and non-metallic inclusions, thereby improving physical and mechanical properties, such as density, impact strength, thermal conductivity or contact strength. It is also possible at the same time to remove by remelting of mechanical and other defects, like casting flaws, scratches, cracks and blisters [95]. Although the process is physically very similar to vacuum refinement of metals or alloys in electron beam metallurgical furnaces, in the case of superficial layers or coatings it is still in the research phase [18].

Productivity of remelting processes is estimated at approximately 250 cm²/min. [3].

2.5.2.2 Alloying

Alloying, consisting of saturation of surface layers by alloying constituents which are totally or partially soluble in the substrate material, is carried out with power densities greater than those employed in hardening and with longer heating times.

Alloying causes a deterioration of surface roughness relative to the initial condition; after alloying the surface roughness, R_z depends to a great meàsure on the thickness of the alloyed layer, z_{alloy}. Usually, $R_z \approx (0.05 \text{ to } 0.1)z_{alloy}$ [3]. By the application of appropriate alloying constituents it is possible to obtain significant enhancement of corrosion resistance [102] and tribological properties.

Two types of alloying are distinguished, i.e., **remelting** and **fusion** (Fig. 2.32).

Remelting. The first type of alloying consists of remelting of coating as well as of the surface layer to a certain depth (Fig. 2.32a). The thickness of the alloying coating, z_{coat}, is approximately equal to the thickness of the remelted layer, i.e., the mixing coefficient is $k_m \approx 0.5$.[1] The coating may be deposited by any means (e.g., by electrolysis or thermal spray) on the substrate, either sealed (e.g., as foil, strip or electroplating) or porous (e.g., in the form of paste or powder). With the remelting of both layers, their mixing occurs and the alloying material partially or totally dissolves in the substrate material. After resolidification of the mixture, a different

---

[1] Coefficient of mixing k_m - ratio of cross-section of molten substrate material to total area of cross-section of molten material; approximate formula: $k_m \approx z_{m. \, subst}/z_{alloy}$.

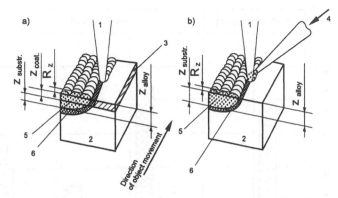

Fig. 2.32 Schematic representation of pulsed electron beam alloying by: a) remelting; b) fusion; 1 - electron beam; 2 - substrate; 3 - coating of alloying material; 4 - solid particles or alloying gas; 5 - alloyed zone; 6 - heat-affected zone.

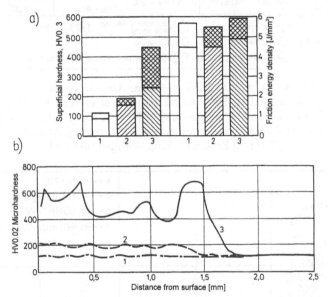

Fig. 2.33 Superficial hardness and density of wear energy (a) and hardness profile (b) for AlCu4Mg1* aluminum alloy: 1 - solution annealed and aged; 2 - alloyed by nickel; 3 - alloyed by iron. (From *Zenker, R.* [107]. With permission.)

structure and chemical composition to that of the substrate is obtained. This is now an alloy of both materials. This method is used for alloying of aluminum alloys, to a depth of several millimeters, with nickel, iron, copper, titanium, and silicon or of carbon and allow alloy steels with alloying elements [3], e.g., molybdenum, chromium, vanadium, tungsten and titanium [101]. In the case of aluminum alloyed with iron, the hardness obtained is higher by a factor of four, relative to the initial material, while alloying with nickel causes a two-fold rise in hardness. This, however, does not result in an improvement of tribological properties (Fig. 2.33)

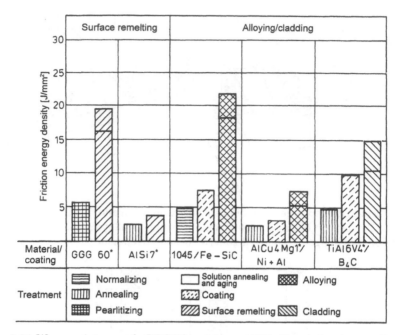

Fig. 2.34 Wear resistance of: GGG60˚ cast iron; AlSi7 silicon-aluminum alloy, 1045 structural steel (uncoated, SiC coating only, steel alloyed by Fe-SiC mixture), AlCu4Mg1˚ alloy˚, alloyed by arc spraying of Ni + Al (alloy only, coating only, alloy with coating), TiAl6V4 alloy (without coating, B₄C coating only, arc sprayed, and alloy cladded by B₄C. (From *Zenker, R.* [106]. With permission.)

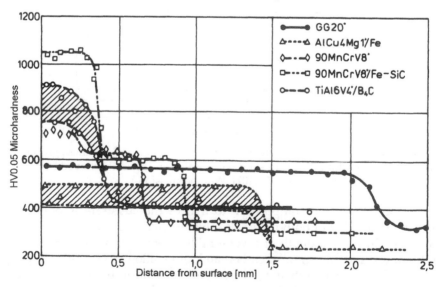

Fig. 2.35 Hardness profiles for different materials after electron beam treatment: surface remelted GG20˚ cast iron, AlCu4Mg˚1 alloyed by iron, 90MnCrV8 cold work tool steel, alloyed by Fe-SiC and TiAl6V4˚ alloy with B₄C cladding. (From *Zenker, R.* [106]. With permission.)

[104]. Enhancement of anti-corrosion and, especially, tribological properties is brought about by alloying of steel with nickel and chromium (electrodeposited or thermally sprayed), as well as by boron carbides (B_4C) and silicon carbides (SiC), plasma or arc sprayed (Figs. 2.34 and 2.35). Besides the above mentioned, other substances may be used as alloying materials, e.g. stainless steels, copper alloys, metal oxides, nitrides, borides and intermetallic compounds [18, 95].

Fusion. The second type of alloying consists of injecting of solid particles or blowing in of gas particles of the alloying material into the melted pool of the substrate material. Similarly to remelting, total or partial dissolution of the alloying material in the substrate takes place, along with mixing of the two materials ($k_m \approx 1$). The alloying solid particles can be e.g., carbides and other compounds, while alloying gas can be e.g., nitrogen (nitrogen alloying), carbon monoxide or acetylene (alloying with carbon).

2.5.2.3 Cladding

Cladding (*hardfacing, embedding, plating*) consists of remelting of a coating, deposited on a substrate, or of a mechanically fed wire, or by injection into the electron beam spot of particles of the coating material which are insoluble in the substrate, e.g., particles of ceramic. The substrate may be subject to only small amount of remelting ($k_m \approx 0.1$) or the coating may adhere to the substrate. In concept, hardfacing is a process similar to overlaying or spray melting, with the difference that instead of a welding torch or a metallizing gun, the source of heat is an electron beam and that the cladding material does not dissolve in the substrate. This method is used to produce heat, corrosion and wear-resistant coatings (e.g., in hydraulic components) and to repair worn machine components, like the working surface of turbine blades.

2.5.3 Evaporation techniques

Electron beam heating coupled with evaporation (vaporization) of the treated material may be utilized in the process of producing hard layers by PVD, as well as in detonation hardening.

Electron beam material evaporation consists of bringing the material to the volatile state in the form of vapors and of deposition of these vapors by PVD methods on a substrate (see Chapter 6, Part II).

Detonation (explosive, impact) hardening consists of very rapid heating of the treated material by an electron beam of highest power density, causing the material to vaporize rapidly. A shock wave is formed and its action on the treated material causes it to harden by impact [12]. Complex structures are obtained, with different densities and different distribution of deformations, with microhardness which can be 3 to 5 times higher than in the initial material but can also be lower. These microstructures may contain traces of hardening, recrystallization and other effects [109].

This type of treatment has not, up to now, been implemented on an industrial scale [11–14].

2.5.4 Applications of electron beam heating in surface engineering

For the past approx. 15 years, electron beam heating has been used successfully in highly industrialized countries, primarily to improve tribological properties, less often to enhance corrosion resistance or strength [18]. As an example, the German company Sächsische Elektronenstrahl GmbH in Chemnitz uses this method for surface enhancement of 120 different types of components, with a productivity of approximately 1 million parts annually [108].

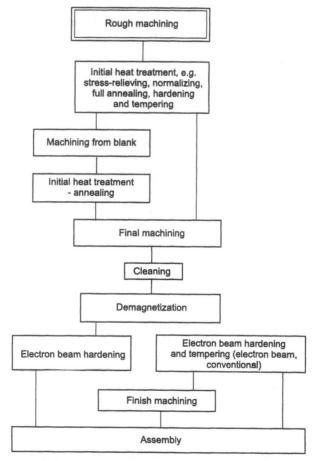

Fig. 2.36 Placement of electron beam hardening within the manufacturing sequence. (From *Zenker, R., et al.* [90]. With permission.)

Electron beam heating is used within a given technological cycle. Fig. 2.36 shows the location of electron beam hardening, relative to the entire component production cycle. Special attention should be paid to the need

for demagnetization of parts prior to electron beam treatment. Non-re-
melting techniques, as a rule, do not require final finishing treatment.
Techniques in which remelting occurs, on the other hand, do usually require mechanical finishing treatment in order to give the treated surfaces
appropriate smoothness.

Electron beam treatment, both pulsed and continuous, may be applied
to parts of different surface roughness and shape and to different fragments of components. The roughness of electron beam treated surfaces
should not exceed 40 μm. The shape should be such that the treated surface may be held perpendicular to the electron beam. Best cases are those
of long and flat surfaces or ones with rotational symmetry (Fig. 2.37).

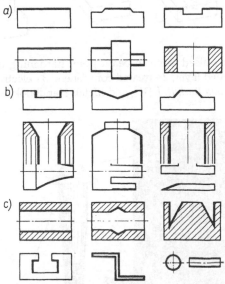

Fig. 2.37 Desired (a), partially desired (b) and undesired (c) shapes of parts for electron
beam heating. (From *Zenker, R., et al.* [90]. With permission.)

Electron beam heating of surface situated not perpendicular to the
beam is also possible, on condition that deviation does not exceed several
degrees [90, 106]. Examples of reaching different surfaces with the electron beam are shown in Fig. 2.38. Fig. 2.39 shows an example of local
hardening of a pin with a pulsed beam. In order to facilitate the hardening process, manufacturers of electron beam heaters develop diagrams for
various materials, correlating the desired hardening depth with the appropriate power density and heating time (Fig. 2.40).

Typical examples of electron beam hardened components are fragments of
automotive and agricultural machine parts, machine tool components (Fig.
2.41) or tools, ball bearing races, including big size, piston rings, articulated
joints, gears, crankshafts, camshafts, cams, flanges, rocker arms, rings, turbine blades, saw cutting edges, cutting edges of stamping dies, milling cutters turning tools, drills, etc. [18].

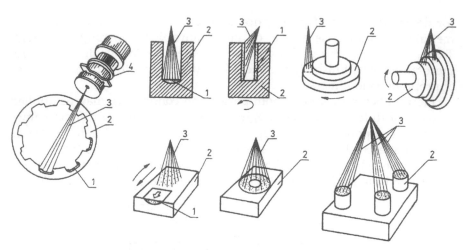

Fig. 2.38 Examples of pulsed electron beam point hardening of different machine components: *1* - hardened layer; *2* - treated component; *3* - electron beam; *4* - electron gun. (From *Sayegh, G., and Burkett, J.* [15], and Metals Handbook [65]. With permission.)

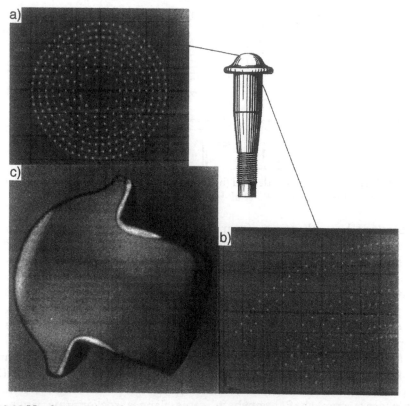

Fig. 2.39 Hardening of pin by pulsed electron point beam; a) point scan for hardening of pin head; b) point scan for hardening the cylindrical surface; c) microphotograph of cross-section of hardened pin. (From *Sayegh, G., and Burkett, J.* [15]. With permission.)

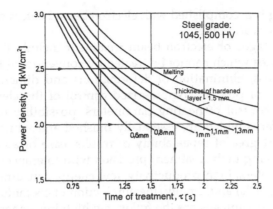

Fig. 2.40 Diagram showing correlation between depth of point hardening by pulsed electron beam of power density q and treatment time t; example of obtaining same hardening depth (1 mm) to 500 H IV hardness for 1045 grade steel, with different heating parameters (q = 2.5 W/cm², τ = 1.05 s, gun power 7.5 kW or q = 2 kW/cm², τ = 1.7 s, gun power 6 kW) and with a 300-point scan. (From *Sayegh, G., and Burkett, J.* [15]. With permission.)

Fig. 2.41 Examples of electron beam hardened automotive parts. (From *Sayegh, G., and Burkett, J.* [15]. With permission.)

Hardening is accomplished with electron beam heaters of several to several tens of kilowatt power.

The advantages of electron beam treatment include the possibility of treating surfaces which cannot be treated by conventional techniques [103, 105], cleanliness, elimination of deformations and dimensional changes, the possibility of precise, computerized control of the electron beam [17, 103], precise control of heating parameters, possibility of treating fragments of surfaces which are essentially finished and which have complex shapes, high degree of repeatability of results, ease of automation, possibility of achieving high treatment precision with tolerances of the order of several millimeters, high productivity, low energy consumption (efficiency reaching 80 to 90%) and, finally, the elimination of coolants.

Among disadvantages are the following: high investment cost of equipment, limitation of application to selected shapes and relatively small loads, usually not exceeding the length of several meters, the necessity of using vacuum and to protect against X-ray radiation when the accelerating voltage used is high - approximately 150 kV [11–14].

From the point of view of treatment quality, electron beam techniques are comparable with laser techniques.

References

1. **Szymański, H., Friedel, K., and Słówko, W.**: *Electron beam equipment* (in Polish). WNT, Warsaw 1990.
2. **Barwicz, W., Mulak, A., and Szymański, H.**: *Application of electron optics* (in Polish). WKiL, Warsaw 1969.
3. **Bielawski, M.**: Application of the electron beam to metal superficial layer modification techniques (in Polish). Proceedings: Conference on *The technology of formation of superficial layers on metals*, Rzeszów, Poland, 9-10 June 1988, pp. 126-134.
4. **Barwicz, W.**: More important applications of electron beams (in Polish), 1990. Proceedings: *First Polish Conference on Applications of the Electron Beam*, September 1972, Karpacz, Poland. *Transactions of The Institute for Electron Technology of Wroclaw Technical University*, No. 1, 1973, pp. 11-46.
5. **Barwicz, W.**: *Application of electron beams in industry and research* (in Polish). OBREP, Warsaw 1974.
6. **Barwicz, W.**: The electron beam in industry (in Polish). WNT, Warsaw 1990.
7. **Oczoś, K.**: *The shaping of materials by concentrated fluxes of energy* (in Polish). Publications of the Rzeszów Technical University, Rzeszów 1988.
8. **Gozdecki, T., Hering, M., and Łobodziński, W.**: *Electronic heating equipment* (in Polish). WSiP, Wrsaw 1979.
9. **Groszkowski, J.**: *High vacuum technology* (in Polish). PWN, Warsaw 1979.
10. **Denbnoweckij, S., Felba, J., Halas, A., Melnik, W., and Lubiniec, G.**: Technological electron beam guns with different types of emitters (in Polish). Proceed-

ings: *First Conference on Elecron Beam Techniques*, Wrocław-Karpacz, September 1982, pp. 520-531.

11. **Burakowski. T.**: The electron beam and possibilities of its utilization to enhance properties of metal surfaces (in Polish). *Przegląd Mechaniczny (Mechanical Review)*, No. 14, 1993, pp. 14-19.

12. **Burakowski, T.**: The electron beam and possibilities of its utilization to enhance properties of surfaces (in Polish). *Mechanik (Mechanicien)*, No. 8-9, 1992, pp. 281-284.

13. **Burakowski, T., Rolinski, E., and Wierzchoń, W.**: *Metal surface engineering* (in Polish). Warsaw University of Technology Publications, Warsaw 1992.

14. **Burakowski, T.**: Formation of superficial layers - metal surface engineering (in Polish). *Metaloznawstwo, Obróbka Cieplna, Inżynieria Powierzchni (Metallurgy, Heat Treatment, Surface Engineering)*, No. 106-108, 1990, pp. 2-32.

15. **Sayegh, G., and Burkett, J.**: Principe et application del'emploi des faisceaux d'electrons commande par mini-calculateurs dans le traitement thermique superficiel des metaux. *Traitement Thermique*, 1979, Vol. 36, No. 136, pp. 75-89.

16. **Sayegh, G.**: High energy density beams (electron beam and laser beam) for heat treatment of metals. 3rd International Congress on: *Heat Treatment of Materials*. Shanghai, November 1983, pp. 8.30-8.40.

17. Modern methods of enhancement of machine component surfaces (in Russian). Vol. 9 of series: Scientific and technical progress in machine building. Publ. International Center for Scientific and Industrial Information - A.A. Blagonravov Institute of the Soviet Academy of Sciences, Moscow 1989, pp. 121-133 and 157-174.

18. **Sipko, A.A., Pobol, I.J., and Urban, I.G.**: Strengthening of steels and alloys by the application of electron beam heating (in Russian). Publ. Nauka i Technika, Minsk, 1995.

19. **Bielawski, M., and Friedel, K.**: The electron beam as a source of heat in the process of surface hardening (in Polish). *Wiadomości Hutnicze (Metalmaking News)*, No. 3, 1985, pp. 67-71.

20. Cahiers techniques Sciaky: Emploi des faisceaux d'electrons commandes par mini-calculateurs dans le traitement thermique superficiel des metaux.

21. **Boriskina, L.V., Kabanov, A.N., and Judaev, V.N.**: About the electron beam diffusion by material during of electron beam tratment (in Russian). *Fizika i Khimia Obrabotki Materialov (Physics and Chemistry of Material Treatment)*, No. 5, 1974, pp. 78-86.

22. **Ryzkov, F.N., Baskakov, A.V., and Uglov, A.A.**: The amplitude of electron beam oscillation and its effect on the shape and size of the penetration zone (in Russian). *Physics and Chemistry of Material Treatment*, No. 5, 1874, pp. 93-99.

23. **Taniguchi, N.**: Research and development of energy beam processing of materials in Japan. *Bulletin of Japan Society of Precision Engineering*, No. 2, 1984, pp. 117-125.

24. **Łunarski, J., and Zielecki, W.**: Modification of the condition of the technological superficial layer and its properties by the electron beam (in Polish). *Postępy Technologii Maszyn i Urządzeń (Progress in Machine and Equipment Technology)*, Vol. 2, 1991, pp. 3-14.

25. **Skubich, J., and Stöckermann, T.**: Laser- und Elektronenstrahlbearbeitung in der Fertigung. *Werkstatt und Betrieb*, No. 7, 1975, pp. 425-440.

26. **Zielecki, W.**: *Modification of technological and service properties of steel by the laser and electron beams* (in Polish). Ph.D. Thesis. Rzeszów Technical University 1993.

27. **Warren, P.H., and Johnson, R.H.**: Selected areas of thermochemical treatment using glow discharge electron beams. Proc.: *Heat Treatment '84*, London, May 2-4, 1984, pp. 47.1-47.6.

28. **Artinger, I., Korach, M., and Pachomova, N.A.**: Changes in material's structure during surface treatment by high energy sources. Proc.: *International Congress on Heat Treatment of Materials*, October 1986, Budapest, pp. 1533-1542.

29. **Bakish, R.**: Electron beam melting, refining and surface treatment: II. *Industrial Heating*, September 1985, pp. 26-28.

30. **Burakowski, T.**: *Directions of development in heat treatment* (in Polish). Publ.: Technology Section of the Machine Building Committee of the Polish Academy of Sciences - Institute of Precision Mechanics, Warsaw 1989.

31. **Burakowski, T.**: Present state and directions of development of surface engineering, Part IV - Characteristics and range of applications of beam techniques, as well as possibilities of utilization of these processes in industry (in Polish).' *Przegląd Mechaniczny* (*Mechanical Review*), No. 16, pp. 26-35.

32. **Burakowski, T.**: Techniques for producing surface layers - metal surface engineering. Proc.: *Techniques of producing metal surface layers*, Rzeszów, 9-10 June 1988, pp. 5-27.

33. **Dyos, G.T., Warren, P.H., Winstanley, R., and Donnely, M.**: The development of a glow discharge electron beam heater for powder strip. Proc.: *10th Congress of Electroheat*, June 18-22, 1984, Stockholm, Sweden, paper No. 6.7.

34. **Friedel, K.**: Effect of the electron beam on the solid in deep penetration (in Polish). *Transactions of the Institute of Electron Technology*. Monograph series. Wrocław Technical University, Wrocław 1983.

35. **Friedel, K.**: Thermal interaction of the electron beam with the material (in Polish). *Transactions of the Institute of Electron Technology* of the Wrocław Technical University, No. 18, Conference series No. 3, 1979, pp. 18-25.

36. **Friedel, K.**: The electron beam as a source of heat - physical processes (in Polish). Proc.: *Implementation of electron beam welding in the machine industry*, Rzeszów, 1977.

37. **Rykalin, N.N., Zuev, I.V., and Uglov, A.A.**: *Basics of electron beam treatment of materials* (in Russian). Publ. Masinostroenye, Moscow 1978.

38. **Rykalin, N.N., Uglov, A.A., Zuev, I.V., and Kokora, A.N.**: *Laser and electron beam treatment of materials* (in Russian). Handbook. Publ. Masinostroenye, Moscow 1985.

39. **Giziński, J.**: Investigation of possibility of utilization of the electron beam to hardening heat treatment (in Polish). Report no. 109.00.0278 by Institute of Precision Mechanics, Warsaw 1989.

40. **Hansen, R.C.**: A comparison of high energy beam systems - electron beam/laser beam. Proc.: *The Lasers vs. the Electron Beam in Welding, Cutting and Surface Treatment. State of the Art - 1985*. Reno, part II. Edited by R. Bakish, Bakish Materials Corp., Englewood, N.Y., 1985, pp. 255-259.

41. **Hansen, R.C.**: Emerging technical developments in electron beam heat treatment. Proc.: *Electron Beam Melting and Refining - State of Art 1984*. Edited by: R. Bakish, Bakish Materials Corp, Englewood, N.Y. 1984, p.220.

42. **Barwicz, W.**: *Heat treatment of steel by the electron beam* (in Polish). Transactions of UNITRA OBREP, Publ. WNT, Vol. 3, No. 6, Warsaw 1975, pp. 3-11.

43. **Bielawski, M., Capanidis, D., Friedel, K., and Olszewska-Mateja, B.**: Application of the electron beam to the heat treatment of contact rings in electromagnetic clutches (in Polish). Proc.: *Termoobróbka 86* (*Thermal Treating 86*), Jaszowiec (Poland), May 1986.

44. **Carley, L.W.**: Electron beam heat treating. *Heat Treating*, 1977, Vol. 4, pp. 18-24.

45. **Ciurapiński, A., Waliś, L., and Kominek, J.**: An analysis of possibilities of utilization of isotope techniques to study migration of elements in materials treated

by PVD and low temperature CVD techniques. Part 1.2. Electron beam treatment (in Polish). Internal report No. 73/I/86. Institute of Nuclear Chemistry and Technology, Warsaw 1986.

46. **Dreger, D.R.:** Pinpoint hardening by electron beam. *Machine Design*, No. 10, pp. 89-93.

47. **Ebner, R., Pfleger, E., Jeglitsch, F., Leban, K., Goldschmied, G., and Schuler, A.:** Möglichkeiten der Oberflächenbehandlung metallischer Werkstoffe mit Elektronenstrahlen am Beispiel hochlegierter Stähle. *Practical Metallography*, 1988, No. 25, pp. 467-487.

48. *Fiorletta C.A.:* In-line electron beam system does through surface hardening job. *Heat Treating*, December 1980, Vol. 12, No. 12, pp. 28-32.

49. **Gridnev, V.N., Meskov, J.J., Oskaderov, S.P., and Trefilov, V.I.:** *Physical principles of electrothermal hardening of steel* (in Russian). Publ. Naukova Dumka, Kiev, 1973.

50. **Gruhl, W., Grzemba, G., Ibe, G., and Hiller, W.:** Durcissement superficiel par fussion d'aliages d'aluminium avec un faisceau d'electrons. *Metall*, 1878, 32, No. 6, pp. 549-554.

51. **Hałas, A.:** Application of electron and ion beams (in Polish). *Transactions of Research center for Vacuum Electronics*, Vol. 4, No. 9, 1976, pp. 24-26.

52. **Hick, A.J.:** Rapid surface treatment - a review of laser and ion beam hardening. *Heat Treatment of Metals*, 1983, Vol. 10, No. 1, pp.3-11.

53. **Hiller, W., König, D., and Ibe, G.:** Gezielte Beeinflussung der Gefügeeigenschaften an der Oberfläche metallischer Werkstücke durch Behandeln mit dem Elektronenstrahl. *DVS-Berichte Strahltechnik*, 1977, VIII, pp. 84-87.

54. **Hiller, W., König, D., and Stolz, H.:** Neue Möglichkeiten der thermischen Behandlung von Eisenwerkstoffen mittels Elektronenstrahles. *Härterei-Technische Mitteilungen*, Vol. 27, No. 2, pp. 85-91.

55. **Jenkins, J.E.:** Dynamic electron beam hardening cycles. *Metal Progress*, July 1981, Vol.120, No. 2, pp. 38-41.

56. **Jenkins, E.:** Electron beam surface hardening. *Tooling & Production*, 1978, No. 12, pp. 76-77.

57. **Keitel, S., Schultze, K.R., and Sobich, G.:** Lokale Oberflächenmodifikation mit dem Elektronenstrahl. *ZIS-Mitteilungen*, 1986, Vol. 28, No. 1, pp. 53-61.

58. **King, R.I.:** Heat treating with electron beam. Proc.: *4th ASM Heat Treating Conference/Workshop*, Chicago, October 1978.

59. **Krasnoscenkov, M.M., Spesinskij, E.E., and Makovski, E.A.:** Electron-thermal hardening of 1045 steel (in Russian). *Elektronnaya Obrabotka Metallov*, 1976, No. 5, pp. 25-28.

60. **Kulkinski, P.:** Investigation of the effect of multiple brief austenitization on the mechanical properties of 50CrV4 steel. *Neue Hütte*, 1977, Vol. 22, No. 12. pp. 669-672.

61. **Łunarski, J., Marszałek, J., and Zielecki, W.:** Superficial layer on 38HMJ steel after electron treatment surface hardening (in Polish). *Transactions of the Rzeszow Technical University - Mechanics*, Vol. 4, 1987, pp. 117-124.

62. **Łunarski, J., and Zielecki, W.:** Improvement of properties of the superficial layer by electron treatment (in Polish). *Transactions of the Rzeszow Technical University - Mechanics*, Vol. 15, 1987, pp. 19-24.

63. **Malajan, S.W., Venkataraman, G., and Mallik, A.K.:** Grain refinement of steel by cyclic rapid heating. *Metallography*, Great Britain, 1973, No. 6, pp. 337-345.

64. **Mawella, K.J.A., and Honeycombe, R.W.K.**: Electron beam rapid quenching of an ultrahigh strength alloy steel. *Journal of Materials Science*, 1984, No. 19, pp. 3760-3766.

65. *Metals Handbook, Desk Editon, Part III – Processing*, ch. 28- Heat Treating. ASM, International, Materials Park, OH 44073-0002 (formerly the American Society for Metals, Metals Park, OH 44073), Ohio, 1998, p. 28·51 (Fig. 12).

66. **Mulot A., and Badeau, J.P.**: Influence de la structure initiale et de la composition chimique sur les caractéristiques des couches durcies obtenues par trempe superficielles, bombardement électronique et laser. *Traitement Thermique*, 1979, No. 136, pp. 47-63.

67. **Müller, M., and Zenker, R.**: Randschichthärten mit Elektronenstrahlen. *Schweisstechnik*, 1986, Vol. 36, No. 11, pp. 484-486; Surface hardening by electron beam, *Welding International*, 1988, No. 2, pp. 180-183.

68. Rheinisch-Westfälisches Elektrizitätswerk (RWE): Die industriellen Elektrowärmeverfahren, p. 29 - Elektronenstrahlerwärmung.

69. Leybold Heraeus borochure: Electron beam technology in vacuum metallurgy; Electron beam special heat treatment.

70. **Samoila, C., Tonescu, M.S., and Druga, L.**: Technologii si utilaje moderne de incalzire in metalurgie (in Rumanian). Editure Technica, Bucharest 1986.

71. **Sayegh, G.**: Principles and applications of electron beam heat treatment. *Heat Treatment of Metals*, 1980, Vol. 7, pp. 5-10.

72. **Schiller, S., Hesig, U., and Panzer, S.**: *Elektronenstrahltechnologie.* Verlag Technik, Berlin 1976; *Elektronenstrahltechnologie*, Wissenschaftliche Verlagsgesellschaft 1977; *Electron beam technology*, John Wiley and Sons, New York, 1982.

73. **Schiller, S., and Panzer, S.**: Härten von Oberflächenbahnen mit Elektronenstrahlen. Teil I: Verfahrenstechnische Grundlagen. *Härterei-Technishe Mitteilungen*, 1987, Vol. 42, No. 5, pp. 293-300.

74. **Schiller, S., and Panzer, S.**: Härten von Oberflächen mit Elektronenstrahlen. Teil II: Experimentelle Ergebnisse und Anwendung. *Härterei-Technishe Mitteilungen*, 1988, Vol. 43, No. 2, pp. 103-111.

75. **Schiller, S., and Panzer, S.**: Surface modifications by electron beams. *Thin Solid Films*, 1984, Vol. 118, No. 1, pp. 85-92.

76. **Schiller, S., and Panzer, S.**: Oberflächenmodifikationen metallischer Bauteile mit Elektronenstrahlen. *Metall*, 1985, No. 39, pp. 227-232.

77. **Schiller, S., and Panzer, S.**: Thermal surface modifications by HF-deflected electron beams. Proc.: *The Lasers vs the Electron Beam in Welding, Cutting and Surface Treatment. State of Art - 1985*, Reno, NV, Part II. edited by R. Bakish, Englewood, N.Y., Bakish Materials Corp. 1985, pp. 16-32.

78. **Schiller, S., Panzer, S., and Müller, M.**: Advances in the use of thermal surface modification by electron beams. Proc.: *Electron Beam Melting and Refining - State of Art 1984*, San Diego, edited by R. Bakish, Englewood, N.Y., Bakish Materials Corp. 1984, pp. 252-261.

79. **Schirmer, W., and Zenker, R.**: Hochgeschwindigkeitshärten mittels Laser - und Elektronenstrahl. *Wissenschaftliche Zeitschrifts des Technische Universität Karl--Marx-Stadt*, 1986, Vol. 23, No. 6, pp. 786-792

80. **Schurath, H., Frost, H., and Panzer, S.**: Oberflächenhärten mit dem Elektronenstrahl. *Kraftwerkstechnik*, 1986, Vol. 26, No. 1, pp. 49-53.

81. **Tosto, S., and Nenci, F.**: Surface cladding and alloying of AISI 316 stainless steel on C40 plain carbon steel by electron beam. *Memoires et Etudes Scientifiques Revue de Metallurgie*, June 1987, pp. 311-320.

82. **Stähli, G.**: Die hochenergetische Kurzzeit-Oberflächenhärtung von Stahl mittels Elektronenstrahl, Hochfrequenz- und Reibimpulsen. *Härterei-Technishe Mitteilungen*, 1974, Vol. 29, No. 2 pp. 55-67.

83. **Zenker, R.**: Stand, Ergebnisse und Entwicklungsrichtungen auf dem Gebiet der Elektronenstrahl - Randschichtveredelung. Proc.: *3 Wärmebehandlungstagung - Grundlagen und Anwendung moderner Wärmebehandlungstechnologien für Eisenwerkstoffe*, 31 May- 22 June 1988, Karl-Marx-Stadt, Technische Universität Karl-Marx-Stadt, pp. 93-103.

84. **Stähli, G.**: Traitement thermique rapide. *Traitement Thermique*, 1986. No. 199, pp. 43-52; Kurzeit-Wärmebehandlung, *Härterei-Technishe Mitteilungen*, 1984, Vol. 39, No. 3, pp.81-89.

85. **Steigerwald, K.H., and Hiller, W.**: Thermal and structural effects of electron beam heating and welding processes on metals. Proc.: *18th International Conference on Heat Treatment of Materials*, Detroit, May 1980, pp. 363-373.

86. **Zenker, R.**: Elektronenstrahl - Randschichtwärmebehandlung - Technologische Möglichkeiten und Anwndungsprinzipien. Proc.: *12 Fachtagung Wärmebehandlungs und Werkstofftechnik*, 12-14 Dec, 1988, Gera, Kammer der Technik, pp. 68-81.

87. **Zenker, R., John, W., Kämpfe, B., Rathjen, D., and Fritsche, G.**: Gefüge- und Eigenschaftänderungen ausgewählter Stähle beim Elektronenstahlhärten. *Wissenschaftsliche Zeitschrifts des Technischen Universität Karl-Marx Stadt*, 1988, 30, No. 2, pp. 165-182.

88. **Zenker, R., and Müller, M.**: Electron beam hardening, part I - Principles, process technology and prospects. *Heat Treatment of Metals*, No. 4, 1988, pp. 79-88.

89. **Zenker, R., and Müller, M.**: Randschichthärten mit Elektronenstrahlen - verfahrenstechnische Möglichkeiten und werstofftechnische Effekte. *Neue Hütte*, 1987, Vol. 32, No. 4, pp. 127-134.

90. **Zenker, R., Müller, M., and Furchheim, B.**: Wärmebehandlungstechnologische und verfahrenstechnishe Aspekte und Anwendungsbeispiele des Elektronenstrahlhärtens. Proc.: Tagungsband: *11 Fachtagung Wärmebehandlungs- und Wertkstofftechnik*, Gera 1986. Kammer der Technik, Vortrag No. 13, pp. 98-114.

91. **Zenker, R., and Panzer, S.**: Stand, Ergebnisse und Perspektiven des Elektronenstrahl-Randschichthärtens. *Freiberger Forschungshefte*, Reihe B, Bergakademie Freiberg, 1987, pp. 13-26.

92. **Zenker, R., and Schirmer, W.**: Zum Einfluss des Hochgeschwindigkeitshärtens auf Struktur, Gefüge und Eigenschaften ausgewählter Stähle. Proc.: *5th International Congress on Heat Treatment of Materials*, October 1986, Budapest.

93. **Zenker, R.**: Electron beam surface modification - results and perspectives. Proc.: *7th International Congress on Heat Treatment of Materials*. Moscow, 11-14 Dec., 1990, pp. 281-289.

94. **Zenker, R.**: Gefüge- und Eigenschaftsgradienten beim Elektronenstrahlhärten. *Härterei-Technische Mitteilungen*, No. 5, 1990, pp.307-319.

95. **Pobol, I.L.**: Worldwide tendencies in applications of high energy electron beams to metal treatment (in Polish). *Elektronika (Electronics)*, 1993, Vol. 34, No. 8-9, pp. 41-47.

96. **Hiller, W.**: Traitement de surface par refusion des materiaux metalliques l'aide d'un faisceau d'electron. *Batelle Information*, 1986, No. 3, pp. 22-23.

97. **Kear, B.H., and Strutt, P.R.**: Rapid solidification of surface modification of materials. Proc.: *Electron Beam Melting and Refining - State of Art 1984*, San Diego. Edited by R. Bakish, Englewood, N.Y., Bakish Materials Corp., 1984, p. 234.

98. **Friedel, K.**: Electron beam technology in industrial applications (in Polish). *Elektronika*, No. 7-9, 1990, pp.25-27.

99. **Stutt, P.P.**: Formation of rapidly melted surface layers by electron beam scanning. *Materials Science Engineering*, 1981, No. 49, pp. 87-91.

100. **Nestler, M.C., Spies, H.J., Panzer, S., and Müller, H.**: Erzeugung von Verschleissshutzschichten durch Randschichtumschmeltzlegierungen mit energiereicher Strahlung. Proc.: Härtereitechnische Fachtagung *Härtereitechnik 1989*, Kammer der Technik, Suhl, 8-10 Nov. 1989, pp. 113-119.
101. **Łunarski, J., and Zielecki, W.**: Investigations of possibility of molybdenum alloying of steel by the electron beam (in Polish). *Postępy Technologii Maszyn i Urządzeń (Progress in Machine and Equipment Technology)*, Vol. 2, 1992, pp. 3-17.
102. **Zielecki, W., and Sęp, J.**: Electrochemical properties of 38HMJ steel, molybdenum alloyed by the electron beam (in Polish). *Transaction of Rzeszów Technical University - Mechanics*, Vol. 34, 1992, pp. 77-82.
103. **Gilbert, G.L.**: Computerized control of electron beam for precise surface hardening. *Industrial Heating*, 1978, Vol. 45, No. 1, pp. 16-18.
104. **Dietrich, W., Stephan, H., and Fischoff, J.**: Application of electron beam technology in the metallurgical industries. *Bulletin d'Information U.I.E.*, Dec. 1978, No. 27, pp.4-7.
105. **Demidov, B.A., Kriznik, G.S., and Tomaschik, J.F.**: Changes in metal structure after introduction of intensive fluxes of electrons of nanosecond frequency (in Russian). *Fizika i Chimia Obrabotki Materialov*, 1982, No. 4, pp. 114-117.
106. **Zenker, R.**: Electron beam surface modification - state of art. *Materials Science Forum*, 1992, No. 102-104, pp. 459-476.
107. **Zenker, R.**: Materials aspects of surface modification by electron beams and industrial applications today. *Surface Treatment - Solid State*, ECLAT-90, pp. 237-249.
108. **Spies, H.J., Zenker, R., and Nestler M.C.**: Electron beam treatment of surface layer. *Journal of Advanced Science*, 1993, Vol. 5, No. 2, pp. 50-60.
109. **Bielawski, M.**: Modification of surface of metals by the electron beam (in Polish). *Elektronika (Electronics)*, 1993, Vol. 34, No. 8-9, pp. 48-50.
110. **Frey, H., and Kienel, G.**: *Dünschicht Technologie*. VDI Verlag, Düsseldorf 1987.

chapter three

Laser technology

3.1 Development of laser technology

The history of laser technology is over 40 years old; lasers have been known for over 30 years and used in practical applications for more than 25 years. The scientific basis of laser technology lies in the realm of atomic physics, more strictly speaking, foundations were laid by the Danish physicist Niels Bohr (1913 - theory of the structure of the hydrogen atom) and the German Albert Einstein (1916 - introduction of the concept of stimulated emission) [1, 2].

In 1950, A. Kastler from France proposed optical pumping (creation of changes in the distribution of filling of different atomic energy levels as a result of excitation by light radiation) which earned him the Nobel Prize in physics in 1966 [2].

In the years 1953 to 1954, American scientists from Columbia University, Ch. H. Townes and J. Weber, and Soviet researchers N. G. Basov and A. M. Prokhorov, working independently at the Lebedev Institute of Physics, proposed the application of stimulated emission to amplify microwaves. For this achievement, Townes, Basov and Prokhorov received the Nobel Prize in physics in 1964 [1-10].

In 1954, Townes, together with co-workers J. Gorgon and H. Zeiger, applied the concept in practice, utilizing ammonia as the active medium and building the world's first wave amplifier in the microwave range (emitting radiation of wavelength 12.7 mm) which they called **maser**. This term is derived from the acronym of *Microwave Amplification by Stimulated Emission of Radiation* [1].

In 1958, Ch. H. Townes and A. L. Schavlov predicted the possibility of building a maser for light radiation but the first attempt at its construction in 1959 was unsuccessful [5]. In 1981, A. L. Schavlov received the Nobel Prize in physics for his overall contribution to the development of lasers [2].

It was only in May of 1960 that a young American physicist, T. H. Maiman, working in the laboratory of Hughes Research Aircraft Co., built the world's first maser, operating in the range of light radiation, initially called **optical maser**. The name was changed later to **laser** (*Light Amplification by Stimulated Emission of Radiation*). This was a pulse ruby laser, generating visible radiation of red color (of wavelength $l = 0.694$ µm) [1-10].

The construction of a laser based on the ruby crystal initiated the so-called solid crystal laser series. In 1961 F. Snitzer constructed the first laser on neodymium glass and three years later, a young physicist, I.E. Guesic, together with his co-workers at the Korad Department Laboratory in the US, implemented the first laser based on an Nd-YAG crystal, emitting short-wave infrared radiation ($l = 2.0641$ μm) [8].

The first gas laser operating continuously, in which a mixture of helium and neon replaced ruby as the active medium, was built in the Bell Telephone Laboratories in the United States in 1961 by A. Javan, W.R. Bennet Jr. and D.R. Herriote, according to a suggestion published two years earlier by A. Javan. This is today the most popular type of laser [5, 8].

In 1962, F.J. McClung and R.W. Hellwarth from Hughes Aircraft Laboratory (US) implemented the operation of the first laser with an active bandwidth modulation which later made possible the obtaining of high power and very short duration laser pulses, so-called *gigantic pulses* [5].

In 1964, the American physicist C.K.N. Patel, working at the Bell Telephone Laboratories built the world's first gas laser based on carbon dioxide, emitting continuous infrared radiation of wavelength $l = 10.59$ μm, which later found greatest application in industry [5].

The first excimer laser of the ultra-violet range (xenon, with a wavelength $l = 0.183$ μm) was made in 1972 (H.A. Köhler et al.); nine years earlier, in 1963, the first nitrogen-based gas laser emitting UV radiation was built by H.G. Hard [5].

From the moment of invention of the first laser, a tumultuous development of laser technology has taken place, recognized, not without reason, as one of the foremost achievements of our times in the field of science and technology. As a result, today there are several hundred different designs of lasers, i.e., quantum optical generators of almost coherent electromagnetic radiation for a spectrum range from UV to far IR [11].

Lasers have found application in many domains of everyday life and technology, where they have proven themselves to be of priceless service. They are successfully utilized in medicine, surveying and cartography, in rocket and space technology, in military and civilian applications. To this day, unfortunately, what triggers their further development are military requirements. In such applications as the so-called star-wars, lasers are to be the basic weapon destroying the enemy's weaponry (satellites, cosmic vehicles and rocket heads). Laser designs of very high pulse power or energy are known [11].

Somewhat overshadowed by these applications, although with equal intensity, we observe the development of design and application of lasers for industrial purposes, so-called **technological lasers**. These are mainly lasers operating with carbon dioxide as the active medium [11].

Technological lasers allow continuous operation or by repeated or single pulses of extremely short duration, i.e., within 10^{-3} to 10^{-12} s. They enable high precision delivery to selected sites of treated materials of great power densities (up to 10^{20} W/m^2), power of the order of terawatts, energy of hundreds of kilojoules and heating rates up to 10^{15} K/s [6].

It is estimated that in 1985, the industries of different countries of the world employed over 2000 technological lasers, of which approximately one third found application in the metal industry [11].

3.2 Physical fundamentals of lasers

3.2.1 Spontaneous and stimulated emission

All atom systems which go to make up the bodies surrounding us, as well as ourselves, exist in certain quantum states, characterized by given values of energy, in other words, by given energy levels. Each change of this state can only take place in the form of a non-continuous jump **transition** of an electron from the basic state to the excited state or reverse, which is accompanied by absorption or emission of a strictly defined portion of energy. The smallest such portion by which a system may change its energy is called **quantum** (from the Latin *quantum*, meaning: how much). Lasers utilize electron transitions between energy levels of particles - atoms, ions or particles which form solids, liquids and gases. Transitions of electrons are accompanied by changes of the energy level of the atom system.

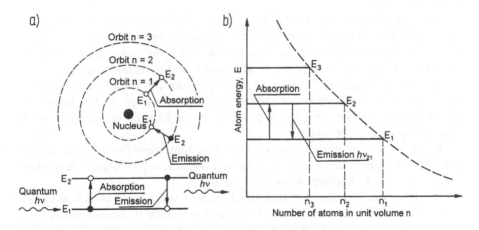

Fig. 3.1 Diagrams showing emission and absorption of energy: a) in an atom; b) in a set of atoms. (Fig. a – from *Oczoś, K.* [2]. With permission.)

The simplest quantum system is the two-level one, i.e., such a microsystem in which processes of emission and absorption of radiation take place between two discrete energy levels: basic (level 1 with energy E_1) and excited (level 2 with energy E_2) (Fig. 3.1). For simplification it can be assumed that energy levels are infinitely narrow, although in real systems they have a defined width.

The transition of such an isolated quantum system from one energy level to another may be of a **radiant** nature, in which case the energy absorbed or emitted by the quantum system takes the form of electromagnetic radiation.

The transition of such a quantum system but one that is part of a set of other quantum systems, from one level to another, may also be of a **non-radiant** nature, in which case the absorbed or emitted energy is passed over to a different atom system. Such non-radiant transitions of relaxation are those occurring with the exchange of energy between particles of gases, liquids or solids and they are accompanied by a change in temperature.

In accordance with the basic quantum correlation, established in 1913 by N. Bohr, radiant transitions obey the rule:

$$hv = E_2 - E_1 = h\frac{c}{\lambda} \tag{3.1}$$

where: hv - value of a quantum of radiation (infra-red, visible, ultraviolet, X-ray, gamma); $E_2 - E_1$ - difference in energy levels, between which quantum transition occurred; h - Planck's constant ($h = 6.62517 \cdot 10^{-34}$ Js); v - frequency of emitted or absorbed radiation, Hz; λ - radiation wavelength, µm; c - rate of propagation of light in vacuum (speed of light) $c = 2.998 \cdot 10^8$ m/s.

The transition of a system from a lower energy level E_1 to a higher one E_2 occurs after delivery, from an external source, to the system of a quantum of radiation (photon, from Greek *phos* - light) of hv value. The system absorbs the delivered energy and **absorption transitions** take place.

When the system undergoes a transition from a higher energy level E_2 to a lower one E_1, it gives off (emits) its surplus energy in the form of a quantum of radiation, the value of which is hv. In such conditions, **emission transitions** take place.

If the level of energy in the quantum system considered is the lowest possible, as shown in Fig. 3.1, it is termed **basic level** (or **state**). Any other level, e.g., E_2 is an **excitation level** (or **state**).

When an excited electron finds itself at an energy level which is higher than basic, there always occurs the natural tendency to spontaneous transition to the basic level which is the stable state of the system. Naturally, such spontaneous transition is accompanied exclusively by the emission of a quantum of radiation. This effect is termed **spontaneous emission**.

In the case of a set of different atomic systems, with different numbers of electrons orbiting atomic nuclei at different levels, atrophy of excitation of atoms or particles is of a random character. Photons are emitted by particles independently and, besides, different particles emit radiation of different frequency, corresponding to different wavelength. This is chaotic radiation, **non-coherent** with relation to either itself, time or space. For a given body it depends only on the degree of excitation, which itself depends mainly on body temperature. The spectrum of such radiation bears a continuous character and is described by the Stefan-Boltzmann and Planck laws. This is the manner in which radiation is emitted spontaneously by all bodies, including light sources.

In lasers, however, the emission which is utilized is not spontaneous but stimulated, although in all quantum effects spontaneous emission plays a significant role. This is manifest in the so-called background noise. It initiates the processes of amplification and excitation of vibrations and - together with non-radiant relaxation transitions - it participates in the formation and sustaining of a thermally unstable state of generation [6]. Stimulated emission always accompanies absorption and spontaneous emission because if it did not, it would be impossible to reach the state of thermodynamic equilibrium of many particles emitting and absorbing radiant energy [8].

In a set of atomic systems subjected to electromagnetic radiation of a frequency determined by eq. (3.1), two mechanisms of interaction of the photon (quantum of energy) with the particle may take place:

– if the particle is at a lower energy level - the particle passes to a higher level as the result of absorption of radiation [2];

– if the particle is already at a higher (excited) energy level - under the influence of an external stimulus (collision with a photon), the excited particle returns to its basic state: the electron drops to the basic energy level (to an orbit closer to the nucleus), emitting a photon of same energy $h\nu$ as the falling photon (Fig. 3.2); this is the so-called *resonance stimulation*.

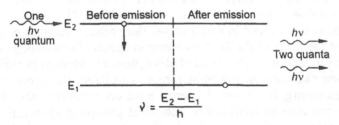

Fig. 3.2 Diagram showing forced emission of quanta of radiation - photons. (From *Oczoś, K.* [2]. With permission.)

This process is named **stimulated emission**. Instead of one photon entering an excited atomic system, two photons of equal energy (equal frequency of corresponding wavelength) exit the system. A process of **amplified radiation** thereby occurs. The probability of such a process taking place is proportional to the number of photons at the incoming end, i.e., to the power density of the stimulating radiation [13].

If in spontaneous emission both directions, as well as frequencies, phases and polarization planes of radiation are not the same, in stimulated emission these parameters for both forcing and forced radiation (i.e., external electromagnetic field and the field formed by stimulated transitions) are the same. Frequencies, phases, polarization planes and directions of propagation are mutually indistinguishable. The radiation of a set of only particles and atoms exhibits properties of radiation by a single quantum system: it propagates in the exact same direction, has the same frequency, it is in phase agreement and polarized the same. Such radiation is termed **coherent** and this is the type of radiation emitted by lasers.

3.2.2 Laser action

In conditions of thermodynamic equilibrium, the electrons of a set of quantum systems in atoms and particles occupy energy levels which are closer to the nucleus; occupation of higher energy levels is less. In order for laser action to take place, i.e., for the set of quantum systems to emit coherent radiation, it is necessary to fulfill two conditions:
- inversion of occupation of energy levels,
- creation of conditions favoring the occurrence of resonance stimulation.

3.2.2.1 Inversion of occupation of energy levels
Inversion of occupation of energy levels consists of inversion of the energy structure of the set of quantum systems, appropriate for thermodynamic equilibrium. The set should contain a predominance of excited particles because only in those conditions is it possible to achieve a surplus of emitted photons over absorbed ones, i.e., achieve amplification of radiation. It is therefore necessary to effect an **inversion of site occupations**, i.e., to energetically amplify the set of quantum systems which is called the **active medium** of the laser. Presently, over a million laser transitions are known which enable the achievement of site occupation inversion [7].

Inversion is achieved in many ways. Very often it consists of subjecting the active medium of the laser to electromagnetic (stimulating) radiation. Achieving inversion as the result of absorption of radiation is called **pumping**. When radiation in the light range is utilized, the process is called **optical pumping**. Inversion of energy level occupation of the laser active medium can also be achieved by electrical pumping: electrical discharge in gases (glow, spark or arc), bombardment by a stream of electrons, by utilization of the conducting current in semiconductor materials by chemical reactions, etc. [1-13]. The source of energy serving to attain the desired energy levels is named **pumping source**.

The effectiveness of optical pumping is relatively low because it is usually difficult to fit the spectrum range of work of pumping valves to the desired spectrum range of absorption of the active medium. This leads to high losses of light energy on heating the active medium. Optical pumping is most often used in solid and liquid lasers [7].

The effectiveness of electrical pumping taking place during electrical discharge in gases, attained as a result of collisions of active particles between themselves and between them and free electrons, is substantially higher. It depends on gas pressure and on the intensity of the electric field [7]. This type of pumping is used in gas lasers.

In high power gas lasers **gas-dynamic** pumping is also employed, utilizing the difference in times of relaxation of the lower and higher energy level of active medium particles, occurring during rapid decompression of a prior heated gas, characterized by thermodynamic equilibrium at the initial temperature. This type of excitation enables direct exchange of thermal energy to the radiant energy of a laser beam [7].

Fig. 3.3 shows schematics of three- and four-level optical pumping.

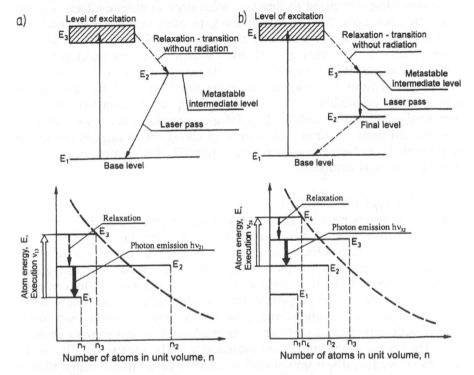

Fig. 3.3 Schematic representation of pumping systems: a) three-level; b) four-level. (From *Oczoś, K.* [2]. With permission.)

In the three-level system it consists of transporting particles from the basic level *1* to the level of excitation *3*, also called the pumping band. From this level they rapidly pass without radiation to a metastable intermediate level. Transition to the intermediate level is accompanied by a loss of a portion of the energy by the particles, this loss being used up by raising the temperature of the system, e.g., causing vibrations of the crystalline lattice of trivalent chromium ions in the ruby laser which must be cooled. In a three-level active medium, inversion may be achieved on condition that at least one half of the active centers is excited. Generation of radiation in such a medium requires intensive excitation by high power radiation [7].

The four-level system is free of these faults. Examples of this are neodymium ions in crystals or glass, as well as particles of CO_2 and CO. In such a system, the particles are transported from the basic level to the excitation level 4, while laser action takes place during transition from level 3 to level 2. When level 2 is far from the basic level 1, occupation of level 2 will be very small. In this case inversion of occupation relative to the final level 2 (Fig. 3.3b) requires less pumping energy than inversion of occupation in the three-level system, relative to level 1 (Fig. 3.3a).

In the intermediate stage, at the metastable level, particles may remain relatively long, compared to times of occurrence of atomic effects, e.g. for the ruby laser, up to 3 ms. In this time it is possible to bring many particles of the active medium to a state of excitation in which, by way of sponta-neous emission, they may give off their energy in a very short period of time. As the result of pumping, the particular particles of the active me-dium do not reach the intermediate state simultaneously but they attain the potential possibility of simultaneous giving off of surplus energy. This potential possibility is made real by designing the laser in such a way as to create conditions for almost simultaneous giving off of surplus energy, in a time of the order of several nanoseconds [7]. Such a possibility is ob-tained in optical resonators.

3.2.2.2 Optical resonator

The optical resonator, also known as the laser resonator or the resonance chamber, serves to contain the active medium (sometimes the active me-dium itself constitutes the resonator) and to amplify stimulated radiation by causing multiple transition of that radiation through the active me-dium. The basic element of the optical resonator is a set usually compris-ing two mirrors, placed perpendicularly to the axis of the resonator. Mul-tiple reflections of radiation from these mirrors may not only react along a long path with excited particles of the medium, but also increase its den-sity. The power of the stimulated radiation must be greater than its losses due to diffraction, dispersion or undesired reflection. The resonator allows, therefore, the accomplishment of positive optical feedback.

Laser action in the form of an avalanche of photon emission causes only radiation along the optical axis of the resonator or one insignifi-cantly deviated from it. Radiation propagating in other directions does not have the possibility of appropriate amplification with the help of stimu-lated emission and thus exits the active medium.

The properties of optical resonators depend mainly on the type of mir-rors, their geometry and distance between them. Depending on the method of exiting the resonator by laser radiation, stable and unstable resonators are distinguished.

In **stable** resonators, laser radiation is conducted out of the resonator through one of the mirrors which for this purpose is made as partially permeable. Usually, its permeability to infrared radiation of 10.6 µm wave-length is 30 to 35%. Most often resonators are designed as flat and parallel, with flat and strictly parallel circular mirrors (Fig. 3.4). Precision of set-ting of mirror parallelism should not be inferior to 5 to 10 µrad [15]. Such a

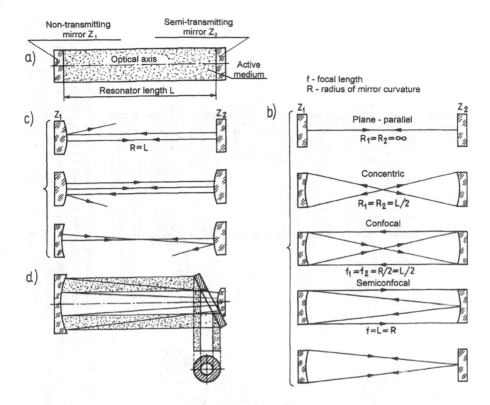

Fig. 3.4 Schematics of most often used optical resonators: a) resonator design; b) stable resonators; c) unstable resonators; d) lead-out of laser flux in unstable resonator. (Fig. b and c - from *Oczoś, K.* [2], Fig. d - from *Trzęsowski, Z.* [15]. With permission.)

resonator allows the obtaining of a beam of laser radiation with relatively small divergence but is sensitive to maladjustment. Often used are semiconfocal lasers; very seldom hemispherical with concave mirrors, condensing radiation in the vicinity of the optical axis of the resonator which counteracts undesirable escape of radiation. Their disadvantage is that the effectiveness of utilization of the active medium is not high and is not uniform within the whole volume of the resonator. It is highest around the optical axis and lowest around the side surface of the resonator cylinder. The utilization of a partially transmitting mirror heated by the laser radiation which passes through it limits the application of stable resonators to only those which generate radiation of relatively low power density [7].

In **unstable** resonators, the beginning of generation occurs in the region near the axis but successive reflections from the non-permeable mirrors cause the displacement of the laser beam in a direction perpendicular to the resonator axis. Next, with the aid of an additional mirror ring, an almost parallel beam exits the resonator in the region outside of the exit mirror (missing its edge), thus assuming the shape of a ring (Fig. 3.4d).

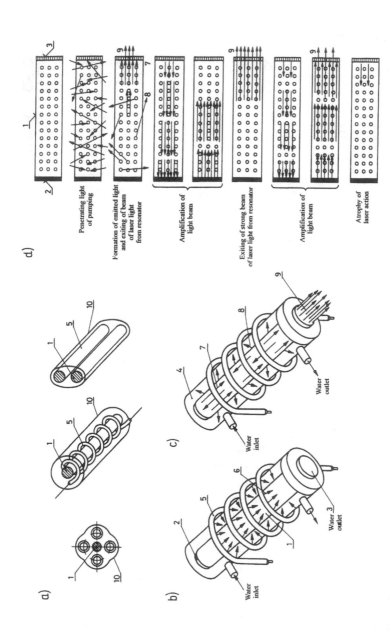

Fig. 3.5 Schematics showing the generation of a laser beam in an optical ruby laser resonator: a) optical pumping systems used in ruby lasers; b) pumping of active medium; c) laser action in a ruby rod; d) growth of axial laser beam; 1 - ruby rod, doped by C_2O_3 which is the active medium; 2 - non-permeating mirror (totally reflecting); 3 - semi-permeable mirror; 4 - cooling jacket; 5 - xenon pumping flash lamp; 6 - photons of flash lamp penetrating into the ruby rod; 7 - photons parallel to optical axis of rod; 8 - photons exiting rod with active medium; 9 - laser beam pulses; 10 - reflector surface of flash lamp. (Fig. a - from *Gozdecki, T., et al.* [7], Fig. b and c - from *Oczoś, K.* [2]. With permission.)

Unstable resonators are characterized by greater diffraction losses but allow the utilization of active media with a high degree of amplification and filling of the entire volume of the active medium with radiation [2, 7]. They find application in lasers generating radiation of high power density.

The course of laser action, as exemplified by the now classical ruby laser, is the following: the active medium in the ruby crystal (crystalline Al_2O_3 corundum) is a 0.05% coloring additive of Cr_2O_3. Chromium ions, which number 5000 times less than the remaining atoms, are excited. In a ruby rod of 10 mm diameter and 100 mm length, the number of chromium atoms is approximately 10^{19}. The excitation of chromium ions consists of irradiation of the ruby rod usually by blue light of a xenon photo flash lamp in the form of a pipe wrapped around the rod or in the form of pipes situated parallel to the ruby rod. The radiation from the lamp is aimed directly onto the rod and by way of reflectors (Fig. 3.5a). After initializing stimulated emission of the active medium (Fig. 3.5b, c), radiation in the form of wave beams propagates in the ruby rod in diverse directions (Fig. 3.5d). A portion of the radiation exits through the side surface. During the first phase of the process the portion of radiation which does not exit the rod is that which propagates parallel or almost parallel to the rod's axis because its end surfaces are silver-plated and form two mirrors, one of which is partially permeable. The rod with mirrors forms a flat-parallel resonator. The power of radiation exiting through the side surface decreases gradually. On the other hand, the power of radiation propagating parallel to the resonator axis increases because photons moving parallel to the rod's axis sputter other photons from excited chromium ions which leads to avalanche amplification of radiation. A radiant beam falling on the impermeable mirror is reflected and on its path to the partially permeable mirror its power increases as the result of liberating successive photons. Upon reaching the partially permeable mirror, a portion of the radiation exits the laser, while the remaining portion is reflected, travels along the rod, is again reflected by the impermeable mirror and again travels back. Thus the process is repeated while the power of the exiting beam increases. After some successive pass, this power begins to decrease because the number of excited atoms is steadily reduced. The entire process is then repeated: a new flash of the lamp excites the chromium atoms, etc. As a result, the laser emits pulses of coherent, monochromatic and non-divergent radiation of 0.694 mm wavelength, which is equal to the wavelength of the exciting radiation. The energy emitted in the form of one pulse reaches several hundred Joules, while the energy supplied by the flash lamp in the form of non-coherent radiation must be at least 100 times greater. The efficiency of a ruby laser is very low - 0.1 to 1%. Because the ruby rod becomes hot and must be cooled, the frequency of pulse repetition, despite the cooling, may not exceed 1 to 3 s [7].

3.2.3 Single-mode and multi-mode laser beams

In optical resonators there occur **standing waves**, as the result of interference of plane waves of light radiation of same amplitudes and periods, propagating along the resonator axis but in opposite directions, due to reflection from mirrors (Fig. 3.6). A condition for proper functioning of

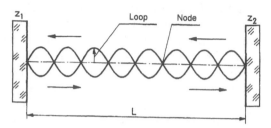

Fig. 3.6 Formation of standing wave in a plane-parallel optical resonator. (From *Oczoś, K.* [2]. With permission.)

the resonator is precise maintenance of such a distance L between mirrors which equals an integral number n of half wavelengths λ [2, 3, 6, 8].

$$L = n\frac{\lambda}{2} \tag{3.2}$$

Meeting this condition allows the formation of wave nodes on mirror surfaces of the resonator.

Usually the value of L is very big relative to λ. For this reason, in the optical resonator it is possible to obtain several types of **resonance vibrations** or **longitudinal modes**, fulfilling the condition:

$$L = q_k \frac{\lambda_k}{2} \tag{3.3}$$

where: $k = 1, ..., n$; q_k - number of half-waves.

The range of wavelengths or corresponding frequencies forms a spectrum (frequency spectrum) of resonance waves of the active medium, in other words the **laser radiation spectrum**. The spectrum composition of this radiation depends on longitudinal modes.

Diffraction occurs at mirror edges, giving rise to changes of amplitude and phase of the waves at mirror surfaces. The result of this is the occurrence of **transverse vibrations (modes)** or changes in the distribution of radiation intensity at the mirror surfaces and, consequently, in the cross-section of the laser beam after it exits the resonator, i.e., in the plane parallel to the mirrors.

The spatial distributions of laser radiation intensity depend on transverse modes which are denoted by symbols TEM$_{mn}$ (Transverse Electro-

Magnetic). The subscripts m and n are positive integers (0, 1, 2, ...), denoting the order of transverse vibrations. Fig. 3.7 shows examples of distribution of radiation intensity of rectangular (a) and axial (circular) symmetry (b). Digits denote the number of observed minima of radiation intensity in the beam's cross-section. For example, in the rectangular system, the TEM_{00} mode does not exhibit any minima (white area) either in the x or the y axis, while TEM_{20} exhibits two minima in the x-axis direction and TEM_{11} one minimum each in the x and y axis directions. On the other hand, in the axial symmetry system, the first digit denotes the number of minima along the radius while the second digit denotes half of the number of minima of radiation intensity in the azimuth direction φ. Modes with an asterisk constitute a superposition of two same modes but rotated relative to each other through 90° (about the optical axis of the beam). As an example, mode TEM_{01}^* is formed as a combination of mode TEM_{01} and TEM_{10} and bears the name of **toroidal** [14].

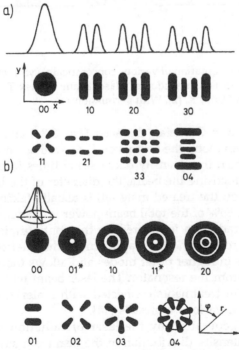

Fig. 3.7 Transverse modes with: a) rectangular; b) axial symmetry. (From *Rykalin, N.N. et al.* [14]. With permission.)

Laser radiation with different distribution of longitudinal and transverse modes is used for different technological purposes - theoretically best developed and possibly the most often used is TEM_{00} laser radiation of axial symmetry. This is **one mode** radiation, and the TEM_{00} mode is termed the **basic mode** because work in this mode makes possible optimum focusing of the laser beam. The distribution of radiation energy I in

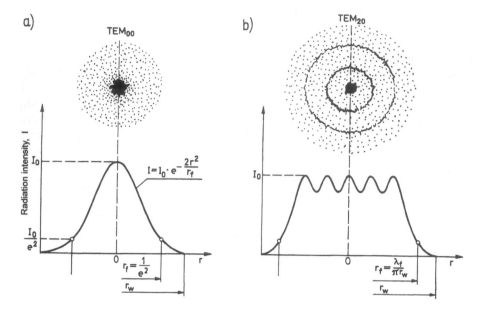

Fig. 3.8 Schematic representation of various concentrically symmetrical distributions of radiation intensity in a cross-section of a laser beam: a) basic TEM_{00} mode; b) multimode TEM_{20}. (From *Oczoś, K.* [2]. With permission.)

the TEM_{00} beam is of a Gaussian character (Fig. 3.8a) and depends on the intensity of radiation along the beam axis I_0, as well as on the radius r and radius r_f, along which the intensity decreases e^2 times in comparison with intensity I_0. When focusing the beam, the diameter of the laser spot (diameter of laser beam on the treated material) is usually taken to be the value $2r_f$. In such a spot, 85% of the total beam power is condensed. The generation of one-mode radiation is favored by the configuration of non-stable resonators. The introduction of a diaphragm to the interior of stable resonators forces losses in higher order modes and allows the exiting of a one-mode laser beam from the resonator. The laser beam with the basic mode is utilized mainly in treatments connected with material loss and in cutting and welding of various materials [14].

In the case of generation by the laser of radiation of two or more modes, the joint intensity distribution in the beam is a sum (superimposition) of fields of the particular modes. Such a beam is termed multimode. It is often very difficult to describe theoretically because it does not exhibit a stable character. Fig. 3.8b shows the distribution of radiation intensity in a beam of axial symmetry TEM_{20}. Multi-mode laser beams are utilized mainly in surface engineering applications.

In the case of pulse generation of laser radiation, the simplest type of generation is **free generation**, which yields radiation pulses with time of duration corresponding to the time of excitation of the active medium. Shorter pulses but of higher power, so-called **gigantic**, are obtained with the help of special elements modulating losses in the resonator, e.g. Pockles cells, non-linear dyes, etc. [7].

3.3 Lasers and laser heaters

3.3.1 General design of lasers

All lasers, regardless of design and function, are made of the following elements [1-14]:

– active medium, comprising a set of atoms, ions or particles which, upon excitation, is capable of stimulated radiation emission,

– a pumping system serving to excite the active medium, i.e., to create a state of inversion of occupation of energy levels,

– an optical resonator, serving to house the active medium, amplify the radiation and to initially form a beam,

– a system for cooling the active medium which sometimes, especially in high power lasers, is equipped with pumps forcing the flow of gaseous medium through the resonator and through a heat exchanger,

– an electrical system, serving to continuously supply energy to the pumping system and to other functional and control elements,

– supporting structure with housing.

Depending on the type of **active medium**, the following types of lasers are distinguished [3-5]:

1) **gas** (in which the active medium is gas, gas mixture or a mixture of gases and metal vapours):

– atom (e.g., helium-neon laser),

– ion (e.g., argon, cadmium, tin, zinc or selenium laser),

– metal vapor (e.g., copper),

– molecular (e.g., carbon dioxide laser, TEA [*Transversely Excited Atmospheric*] which is a CO_2 laser with transverse excitation by spark and pressure close to atmospheric, nitrogen laser, and lasers working in the submillimeter and millimeter range: H_2O, HCN, BrCN, ICN),

– excimer[1] (e.g., ArF, KrCl, KrF, XeCl, XeF lasers);

2) **solid** (in which the active medium is a dielectric crystal or glass, activated by e.g., ions of rare earth elements, actinide series or transition metals):

– crystalline (e.g., the ruby laser, the YAG - a single crystal yttrium-aluminum garnet $Y_3Al_5O_{12}$, CaF_2, SrF_2, BaF_2, $PbMoO_4$, $SrWO_4$, LaF_3),

– crystalline with color centers (e.g., lasers with centers of the F, F_A, F_2 and F_2^+ types),

– glass (e.g., the neodymium laser),

– semiconductor (e.g., InP, InS, GaAs, GaAlAs, GaSb, PbTe);

---

[1] Excimer - particle which does not exist in the basic state.

3) **liquid** (in which the active medium is formed by active centers suspended in a liquid):

– dye (e.g., lasers with rhodamine solutions, with fluorescein or with rhodulin blue),

– chemical (e.g., hydrogen chloride laser, laser utilizing the synthesis of excited HF or DF to excite the active medium or the gigawatt photochemical iodine laser);

4) **other types**:

– the FEL -*Free Electron Laser*- laser which generates radiation in the process of changing of velocities of relativistic electrons, passing through a specially shaped magnetic field),

– X-ray and gamma radiation lasers (lasers which utilize radiation from other lasers to stimulate emission of X-ray or gamma radiation);

Depending on the type and design of the laser, the emitted radiation may be

a) **continuous** - with power ranging from several tens of microwatts to several tens of kilowatts (the biggest may reach 1000 kW). Such lasers are called continuous;

b) **pulsed** (so-called pulse lasers) in the form of

– single pulses with duration ranging from milliseconds to femtoseconds (10^{-15} s) and power accordingly from watts to terawatts (even up 10^{15} W),

– a series of pulses with frequency of repetition ranging from several Hz to several tens of MHz, including pulses superimposed on the background of continuous radiation.

Of the abovementioned groups of lasers only a very few have found practical industrial application. For technological applications, the most often used lasers are those operating in the infrared range [2, 4, 5, 8, 11, 14–16]:

– continuous: molecular gas CO_2 and Nd-YAG,

– pulse: ruby, neodymium, glass and Nd-YAG, molecular CO_2 and excimer.

Surface engineering utilizes both continuous and pulse lasers [11]. The most often used are molecular CO_2 and solid Nd-YAG lasers. Broad perspectives for future applications are predicted for iodine-oxygen lasers (laboratory-scale models have 1 to 50 kW power and wavelength $\lambda = 1.315$ μm) as well as excimer lasers emitting UV radiation [15].

3.3.2 Molecular CO_2 lasers

3.3.2.1 General characteristics

In molecular lasers the active medium is a mixture of gases composed of 5 to 10% carbon dioxide, 15 to 35% nitrogen and 60 to 80% helium under pressure lower than atmospheric [$p \approx (3$ to $20) \times 10^3$ Pa].

For this reason they are sometimes called subatmospheric. Particles of CO_2 are excited as the result of collisions occurring between them and accelerated electrons (which originate from electrical discharges), as well as particles of N_2, the latter also excited due to collisions. Helium present in the mixture raises the thermal conductivity of the gas mixture and improves its internal diffusion cooling. Excited particles of carbon dioxide, upon their return to the basic state, emit infra-red radiation of 10.63 µm wavelength. A condition for obtaining a high quality laser beam is continuous removal of contaminations which are produced during operation (oxygen and carbon monoxide as reaction products, electrode burn debris, vapors of oil from the pump bearings, oxygen entering through leaks in the gas system, and nitrogen oxides NO_2, NO and N_2O). This is accomplished by replacing a portion of the gas mixture with a new or a regenerated one [19, 20].

The active medium of the laser is excited either by an electric field formed due to high direct current voltage on the electrodes (10 to 20 kV), or by a very high frequency (13.56 MHz) magnetic field. The latter type of excitation is more favorable because the electrical discharge is, in this case, more homogenous and stable in time, while the power achieved is higher than that achieved with direct current excitation. Moreover, it causes less contamination of the active medium and enables an almost unlimited modulation of the laser.

The efficiency of CO_2 molecular lasers is relatively high and ranges from 10 to 20% [2, 14, 15, 17, 19]. This means that 80 to 90% of the supplied energy is converted to heat and only 10 to 20% to usable radiation energy. This conversion to thermal energy takes place within the active medium. Raising the power of the laser causes an increase in the amount of heat dissipated, thus, a rise of the temperature of the gas. This rise is admissible but only up to the so-called critical temperature which, depending on the gas mixture composition ranges within the limits of 600 to 700 K. When this temperature is exceeded, the rate of relaxation of the upper laser level rapidly rises and thermal occupation of the lower laser level takes place, causing the amplification of laser radiation to drop [15].

It is precisely for this reason that the active medium requires intense cooling, so as not to allow the medium to reach critical temperature. The means of cooling constitutes a basis of division of molecular CO_2 lasers into: those diffusion cooled by thermal conductivity of the laser gas and those cooled by forced convection (so-called flow cooling) of the laser gas. The latter are themselves divided into two groups, i.e., with longitudinal and transverse flow.

3.3.2.2 Lasers with slow longitudinal flow (diffusion cooled)

This is the oldest type of molecular laser, now regarded as classical. Its resonator is built like the resonator of any gas laser: a simple glass or corundum or beryllium ceramic discharge pipe with sunk-in electrodes, filled with a gas mixture, closed from one end by a non-permeable mirror, from the opposite end by a partially permeable mirror. The gas mixture flows through the discharge pipe of internal diameter 5 to 25 mm, with a

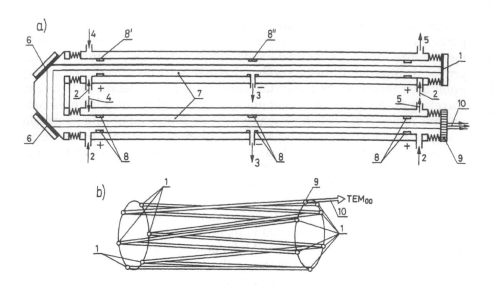

Fig. 3.9 Schematic of a CO_2 laser with slow longitudinal (axial) flow: a) schematic of a bellows-type resonator with 2 parallel discharge pipes; b) schematic of a segment-type, 16-pipe resonator; *1* - totally reflecting mirror; *2* - inflow of gas mixture; *3* - outlet of gas mixture; *4* - inlet for cooling water; *5* - outlet for cooling water; *6* - deflecting mirror; *7* - bellows-type resonator; *8* - electrodes (*8'* - anode; *8''* - cathode); *9* - partially transmitting mirror; *10* - laser beam. (Fig. a - from *Oczoś, K.* [2], Fig. b - from *Trzęsowski, Z.* [15]. With permission.)

velocity of approx. 1 m/s (Fig. 3.9). The active medium, heated in the discharge zone, gives off its heat to the resonator walls which are water cooled. For this reason, this type of laser is also called **diffusion cooled** or **laser with diffusion stabilization of discharge**. Because of the intensive heating of the active medium, the power of the laser may reach a maximum of approx. 100 W per 1 meter of the resonator length.

The ongoing effort to achieve higher power dictates the necessity of building long resonators. The length of simple resonators practically seldom exceeds 10 m with a diameter of up to 10 cm. The length of **segmented** resonators, composed of simple discharge pipes of up to several meters length and of mirrors changing the direction of radiation from several to several tens times is only slightly longer than segments of discharge pipes.

The power of molecular CO_2 lasers with slow (diffusion cooled) transverse flow usually does not exceed 1 kW. In most cases it is several hundred W, the common range being 400 to 600 W. The active medium in the resonator allows easy modulation and obtaining of a stable distribution of radiation intensity. A single pipe laser usually emits continuous radiation with basic mode.

Putting together several to several tens of long parallel resonators and concentrating the radiation from them into one beam allows the obtaining of a multi-mode beam of several kilowatt power, but such a system exhibits a tendency to slip out of adjustment settings [2, 14, 15, 17, 19].

High stability of power density distribution in the laser beam, its small divergence, great diversity of types of operation (continuous, pulsed, both with long and gigantic pulses), simplicity of design, high reliability and ease of operation are among the factors which make these lasers popular in those technological applications where the requirements are high precision, moderate power density and effectiveness [15].

3.3.2.3 Lasers with fast longitudinal flow

The design of resonators of these lasers is similar to that of conventional ones. What makes them differ from the latter is the mechanism of heat extraction from the active medium. The dominant mechanism here is not heat conduction to the walls of the discharge pipe as in lasers with slow longitudinal flow, but forced convection due to transportation of the hot active medium away from the discharge zone to the cooler. The velocity of axial flow of the medium in the resonator of such a laser is approximately 500 m/s, which enables the cooling of a gas mixture in double heat exchangers, built into the gas system (Fig. 3.10). In these, the central exchanger cools the hot gas mixture while the remaining two exchangers cool the active medium which is heated by compression in the blower. The achievable power can reach 1000 W per meter of resonator length. Radiation is

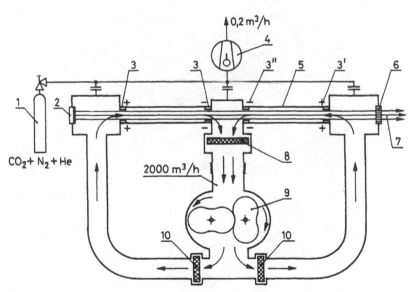

Fig. 3.10 Schematic of CO_2 laser with fast longitudinal (axial) flow: *1* - inflow of gas mixture; *2* - totally reflecting mirror; *3* - electrodes (*3'* - anode, *3"* - cathode); *4* - vacuum pump for suction of used gas; *5* - discharge pipe; *6* - partially transmitting mirror; *7* - laser beam; *8* - heat exchanger; *9* - Roots pump; *10* - heat exchanger. (From *Oczoś, K.* [2], and from *Trzęsowski, Z.* [15]. With permission.)

emitted in the basic mode (seldom in a lower order mode) as continuous or pulsed [2, 14, 15, 17, 19]. Usually, the power of such lasers does not exceed 5 kW. In most cases its value is within the range of 1 to 3 kW. Such lasers make up approximately 70% of all molecular CO_2 lasers in use. In the late 1980s, their cost was approximately $100 per 1 W for low power equipment from 1 kW upwards and approximately $40 to 60 per 1 W for equipment with over 2 kW power [15].

3.3.2.4 Lasers with transverse flow

In these lasers, the first of which was built in 1969, the flow of the active medium is perpendicular to the direction of generated laser radiation which, in turn, is perpendicular to the direction of the electric discharge field (Fig. 3.11). The hot medium is cooled in a heat exchanger and once cold, it is blown through the discharge zone, situated in the resonator. Circulation cooling allows the extraction of big amounts of heat. Many times repeated passage of the radiant beam through the unstable resonator allows the achievement of higher power than in stable resonators with fast longitudinal flow. Because of the instability of the system, the laser beam is not strictly coherent. These lasers usually emit continuous multi-mode

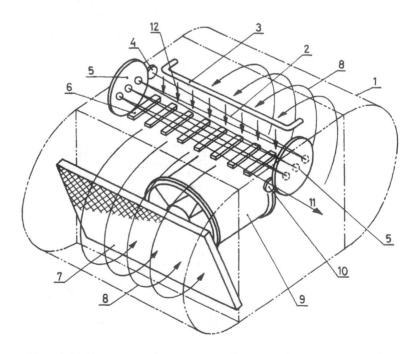

Fig. 3.11 Schematic of CO_2 laser with transverse flow: *1* - vacuum body; *2* - discharge zone; *3* - cathode; *4* - totally reflecting mirror; *5* - multi-reflecting, deflecting mirror; *6* - anode; *7* - heat exchanger; *8* - direction of flow of gas mixture flux; *9* - blower; *10* - exit mirror; *11* - laser beam; *12* - lines of force of electric field. (From *Oczoś, K.* [2]. With permission.)

beams of a relatively big diameter. The output power of such a beam reaches 25 kW in continuous operation and several hundred MW in pulsed operation. Energies up to several hundred J are achieved in industrial lasers when the frequency of pulse repetition is between several Hz and 2 kHz [8]. Industrial laboratory models feature powers reaching 50 kW, military models (like the gas-dynamic CO_2 laser) - 400 kW [6]. An additional advantage of this type of laser is its compact design [2, 14, 15, 17, 19]. The latest lasers belonging to this group feature, besides transverse spark excitation, pressure which is higher than atmospheric. These are the so-called TEA *Transversely Excited Atmospheric* lasers [6].

Operating costs of technological molecular CO_2 lasers range from \$1.5 to \$2.5 per kWh of laser radiant energy [15].

3.3.3 Solid Nd-YAG lasers

In these lasers the active medium is a rod made of yttrium-aluminum garnet ($Y_3Al_5O_{12}$), activated by trivalent neodymium ions Nd^{3+}, built into the crystalline lattice containing 0.8 to 1.5 wt.% Nd_2O_3, forming a four-level quantum system. These lasers usually operate at 1.0641 μm wavelength (infrared) or - in the case of using a non-linear crystal in the resonator and transforming the radiation to the second harmonic - at 0.53 μm (visual radiation range).

The design of the Nd-YAG laser is very similar to that of the ruby laser. In latest models, instead of xenon flash lamps (used for pulse operation) or krypton arc lamps (used for continuous operation) optical pumping is often accomplished with the aid of a semiconductor laser (e.g., CaAlAs of λ = 0.79 to 0.82 μm wavelength) which enhances pumping effectiveness.

When operating continuously, the laser usually emits a multi-mode beam with an input power of up to 2000 W or a TEM_{00} beam of 40 W power. When operating with pulsed multi-mode beam, the average power is usually 500 to 2000 W and may even reach 5000 W. For the TEM_{00} beam the average power is 40 W with pulse energy 0.1 to 60 J. The duration of pulses is controlled and ranges from 0.1 to 10 ms with frequency of repetition from a tenth of 1 Hz to more than 25 Hz.

In the Nd-YAG laser with continuous excitation it is possible to achieve stimulated pulsed work by commutation of the resonator gain bandwidth product. This consists of pumping the active medium at lowered or zeroed resonator gain bandwidth product, i.e., at limited or blocked generating power, this due to the introduction of an electro or acoustic-optical switch into the beam axis. After accumulating sufficiently high energy in the rod, there occurs a sudden rise in the resonator's gain bandwidth product, causing a sudden release of this energy in the form of a narrow laser pulse of megawatt power and several to several tens nanoseconds. The efficiency of Nd-YAG lasers is on the average 2%; for the TEM_{00} beam approximately 0.5%. Maximum efficiency reaches 5% [2].

3.3.4 Continuous and pulse laser operation

During **continuous** operation there usually exists the possibility to control the power output. From the moment of putting into operation, the laser power rises linearly to the value of nominal power P_1 (Fig. 3.12b). On the other hand, the average power P_{avg} is less than the continuous power P_1 and its value depends on pulse duration and on the gap between pulses. The average power is naturally the effective power of the laser. The value of average power may be controlled electronically and in the best of solutions may be conditioned during laser operation to the requirements of the technological process.

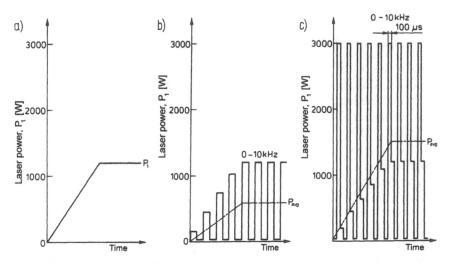

Fig. 3.12 Types of operation of CO_2 molecular lasers: a) continuous operation; b) pulsed operation; c) superpulsed operation. (From *Oczoś, K.* [2]. With permission.)

During **superpulse** operation the power of a single, so-called gigantic, pulse exceeds the value of continuous power P_1 by a factor ranging from 4 to 10 (Fig. 3.12c). In molecular CO_2 lasers this power increase is achieved by means of a rapid rise of the discharge current. For the same value of the average power (P_{avg}) the gap between pulses is greater than that of normal pulse operation [2].

Fig. 3.13 shows the shapes of power density distribution vs. time of emission, i.e. the shapes of laser pulses. Continuous radiation is obtained by continuous excitation of the active medium and thus continuous emission of radiation. Radiation in the form of normal pulses is obtained by continuous excitation (e.g., optical pumping) of the active medium, as, for example, the laser rod, while the radiant energy is emitted after reaching the threshold condition which is different for each laser. Gigantic pulses, so-called Q-s, constitute an envelope of a series of pulse peaks which are obtained by the optical cutting-off of the active medium from the mirrors with the aid of a rotating mirror, Kerr cell or an absorbing element [6].

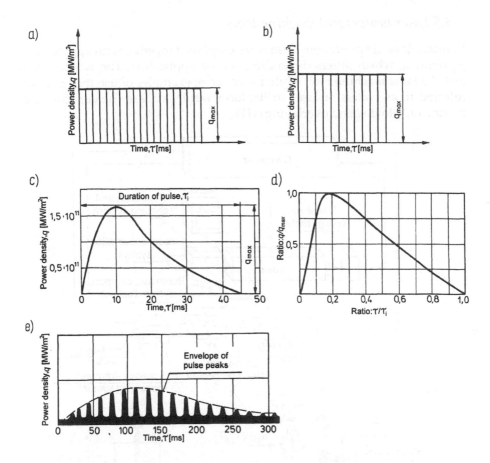

Fig. 3.13 Power density during: a) continuous operation of laser; b) during pulsed operation (rectangular pulse); c) gigantic pulse; d) normal pulse; e) structure of gigantic pulse.

Continuous operation allows gentler heating of the load than pulse operation and it is applied in processes requiring lower power (or energy) density, as well as longer exposure times (mainly for hardening but also for alloying and overlaying). Continuous operation yields power densities up to 10^{10} W/cm² [35].

In industrial applications, especially in surface engineering, pulse laser operation is used less often than continuous, mainly for processes less sensitive to sharp temperature gradients (contact pitting, coating deposition, much less often hardening), requiring a power density which is higher than that of continuous operation. Maximum power densities of industrial lasers reach 10^8 W/cm² and in extreme conditions even 10^{16} W/cm² [5, 11, 35]. The duration of normal pulses in CO_2 lasers is usually 30 to 100 µs at 5 Hz frequency [15].

3.3.5 *Laser heaters and machine tools*

In technological applications, lasers are employed together with appropriate equipment, which differs depending on the application. The whole set is called a **laser machine tool**, while for concrete examples of use they may be referred to as cutters or laser drills, laser welder or laser heater (for heat treatment, alloying and overlaying) [11].

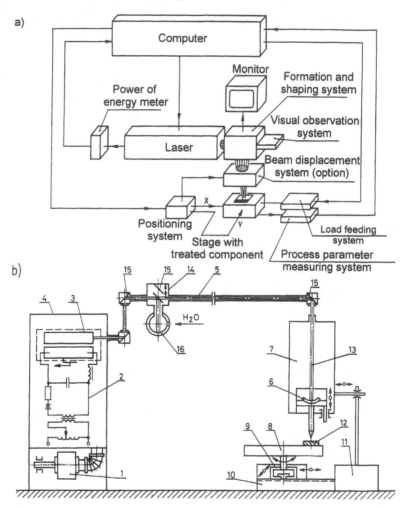

Fig. 3.14 Laser heater: a) block diagram; b) design schematic of CO_2 laser heater; *1* - pumping system; *2* - electrical supply system; *3* - work chamber with resonator; *4* - laser head; *5* - protective piping; *6* - focusing objective lens with gas nozzle; *7* - screening of optical system with automatic displacement of objective lens; *8* - rotating stage; *9* - slide rails for longitudinal stage movement; *10* - slide rails for transverse stage movement; *11* - bed; *12* - load; *13* - beam of laser radiation; *14* - system for cutting off of laser beam; *15* - mirror changing direction of laser beam; *16* - Ulbricht sphere (photometric globe - absorbing radiation). (Fig. a - from *Dubik, A.* [8], Fig. b - from *Burakowski, T., et al.* [11]. With permission.)

A **laser heater** comprises the following functional elements (Fig. 3.14) [11]:

– A laser, e.g. a CO_2 laser which itself comprises a work chamber with the active medium, elements for excitation, and systems of forced circulation, gas cooling, as well as of electrical supply.

– A system for transmission and displacement of the laser beam which itself comprises a system for beam shaping; sets of mirrors for changing the direction of the laser beam; a focusing system, usually in the form of short focal length objective lenses or mirrors allowing the obtaining in the focal spot of a laser beam diminished by several orders of magnitude [35]; an automatic system of displacement for the objective lens focusing the beam; a system for measuring the distance between the objective and the load and, possibly, gas blower protecting the objective lens against contaminations from the technological process (a so-called optical insulator). The beam shaping system is essential especially when it is necessary to obtain a homogenous spatial distribution of power density. It often takes the form of a multi-segment condensing mirror, built up of many plane mirrors situated inside a concave surface.

– A load manipulating system for positioning, comprising an X-Y or rotating stage, load fixtures and a system for measuring positioning parameters; in some cases an industrial robot may be employed. The precision of load positioning should not exceed 0.1 mm, while the accuracy of changes of load feed rate should not exceed several percent of the feed rate value.

– A system for cutting off the laser beam, comprising a moving mirror which changes the direction of the laser beam and a photometric globe with the inside radiation reflecting surface cooled by water. This system operates when the laser emits radiation while not heating the load.

– An automatic control system to control the operation of the laser, the focusing and load positioning systems, usually comprising a computer (with a monitor) which determines the rate of treatment (machining) and the duration of pauses between operations and analyzes the results of operation parameter measurements with necessary feedback.

– A system for visual observation.

The most modern laser heaters are equipped with changeable optics allowing the obtaining of different power densities and its different distribution. Moreover, they are equipped with microcomputer controllers, allowing the almost custom programming of operating conditions and their in-process change [14].

Laser machine tools are finding an ever-broadening technological application. In the 1980s, more than 100 companies manufactured laser machine tools while an annual growth of laser machine tools numbered 100 to 400 annually worldwide, the higher numbers attained in the later years of the decade. In 1984 approximately 1675 laser machine tools operated worldwide, including 804 in the U.S., of which 50 to 60% were of the molecular CO_2 type. It is estimated that in the group of laser machine tools with power above 0.5 kW, more than 90% were those with molecular lasers. In 1988 approximately 2650 laser machine tools were in operation. Almost 60% of these were those for metal treatment and slightly above 40% for the treatment of non-metals. Approximately 51% of the equipment is used for various cutting operations (e.g., straight cutting, round cutting, hollowing) of which as many as 38% for cutting of non-metals and 13% for cutting of metals. In laser treatment of metals, the foremost techniques are welding (16%), cutting (13%) and brazing (6%). Surface engineering techniques, mainly in the field of heat treatment, cover only 4% of all applications [5, 11].

According to the periodical "Industrial Laser Review" (January 1987), in the year 1986 approx. 2045 industrial lasers were sold in the western world, of which 1284 were molecular CO_2 lasers. These were broken down as follows: 7 with power above 5 kW, 39 with power ranging from 2 to 5 kW, 743 with power from 0.1 to 2 kW and 495 with power less than 0.1 kW. The remaining 761 lasers were of the solid type. Of the 1005 lasers designated for metal treatment only 21 were for heat treatment, while 648 were for cutting and 336 for welding.

The differences in the number of lasers operating worldwide stem from the fact that lasers with power of 0.1 kW and lower are not classified as machine tools and are used for metal treatment only in exceptional cases.

3.4 Physical fundamentals of laser heating

3.4.1 Properties of laser heating

Lasers serve to generate mainly light radiation (visible and non-visible - infrared and ultraviolet) but also X-ray and gamma radiation, coherent, practically monochromatic, of very minute beam divergence, small diameter of laser spot and very high power density.

Technological lasers, especially those applied in surface engineering, generate mainly infrared radiation of the near and medium range. The width of the spectral spot of laser radiation, i.e. the range of wavelengths of the waves emitted by the laser, is usually very small and may reach 10^{-6} µm [7].

The divergence of a beam emitted by molecular lasers operating with carbon dioxide is only 1 to 10 mrad, while that of a beam emitted by Nd-YAG lasers is 0.6 to 15 mrad [8].

Laser radiation leaving the optical resonator is condensed with the aid of lenses and mirrors made from special optical materials into a beam of down to 20 mm and even, in some case, to 3 to 10 µm. The dimensions of the beam spot from lasers operating with carbon dioxide range from 1.5 to 50 mm (continuous) and 4.5 to 20 mm or 20×30 mm (pulse). The power density of the beam which is defined as the ratio of beam power to the cross-section of the beam (size or diameter of the spot) ranges from 10^3 to 10^8 W/cm².

Concerning pulsed beams, the energy of a molecular CO_2 laser pulse reaches 5000 MJ (average power up to 5 kW) at pulse repetition frequency 12 to 2500 Hz and maximum pulse duration usually from 7.5×10^{-5} to 2×10^{-3} s [8].

3.4.2 The role of surface absorption in laser heating

The flux of laser radiation, falling on the surface of an opaque material, undergoes partial reflection and partial absorption, depending on the wavelength of the incident radiation and on physical properties of the material and its surface. The amount of heat absorbed by the material at the point of incidence (called the laser spot) depends on:

– surface absorption coefficient A ($A = 1 - R$ where R is coefficient of reflection),

   – wavelength of laser radiation,

   – power density of radiation q_o incident on the surface,

   – exposure time (time of interaction of beam with material) τ.

When the absorption coefficient A of the heated surface is constant, it is possible to obtain approximately similar effects (within certain bounds) by applying lower values of q_o and higher values of t or vice-versa [11].

The surface coefficient of absorption depends on the wavelength of the incident radiation (i.e., on the type of laser which emits radiation of a strictly defined wavelength), as well as on the absorption properties of the surface. These, in turn, depend on the type of material, its state of aggregation, the degree of surface oxidation and on roughness (Fig. 3.15). Metallic materials, especially those with a smooth and shiny surface (e.g., gold, silver, copper, aluminum and bronzes) absorb laser radiation very weakly. This is especially true of the range of near and medium infrared. Metallic materials with a dark and a rough surface (tungsten, molybdenum, chromium, tantalum, titanium, zirconium, iron, nickel, tin) exhibit much better absorption characteristics, although even here the value of absorption rarely exceeds 10% [22]. In the case of molecular CO_2 lasers, the reflection coefficient of a polished surface may reach 98% [22].

The rise of temperature of the surface of a heated material causes a rise of radiation absorption. For metals treated in air, the rise in absorption also occurs as the result of surface oxidation. Oxides usually absorb radiation better than metals. Further, very rapid growth of absorption is caused by the transition of the state of the heated material from solid to liquid (melting) [11].

In order to increase the efficiency of laser heating, it is necessary to increase absorption of the treated metal surfaces by:

– roughening (e.g., sand-blasting, shot peening, knurling or sanding with emery paper) which allows a rise in absorption rate by 30 to 40% [22–24];

– oxidation, causing a rise in absorption rate by 30 to 40% on the average, in rarer case up to 70%;

– raising surface temperature (preheating by any means of the surface to be laser treated) which raises the absorption rate by 10 to 30%;

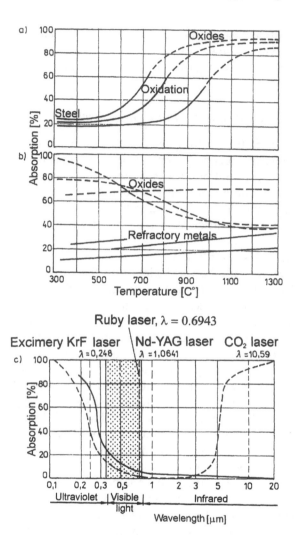

Fig. 3.15 Typical correlations of the absorption coefficient for metals (continuous curves) and oxides (dashed curves): a) with temperature for pure and oxidized steel; b) with temperature for the majority of refractory metals and oxides; c) with wavelength of laser radiation incident on the heated material. (From *Burakowski, T., et al.* [11]. With permission.)

– deposition of an absorbing coating which in most cases also fulfills the role of an anti-reflection coating. This can be a paint coating in the form of paints, varnishes and colloids, powder, paste, electrodeposit, chemical or any other coating which exhibits good absorption of radiation. Absorption coatings are most often produced from saline compositions (e.g., magnesium and zinc phosphates), metal oxides (e.g., zinc, titanium, silicon, chromium, iron, aluminum), non-metals, sulfides, carbides (e.g., silicon and molybdenum carbides), graphite, soot, blackening, etc. They increase the ab-to 50% in the higher temperature range (approx. 1500°C). As an example, paints of the gouache type increase absorption rates by 60 to 90% [25–27], magnesium zinc $Zn_3(PO_4)_2$ or magnesium phosphate coatings $Mn_3(PO_4)_2$ up to 55 to 90%, iron sulfide coatings (Fe_2S_3) up to 20 to 40%, graphite coatings up to 40 to 70% [11, 23–27]. An interesting development is that of exothermic oxide-metal coatings (e.g., FeO+Al, FeO+Si) which, besides raising the absorption rate of radiation, also cause the emission of heat. This is achieved by a exothermic reaction between the components of the coating, initiated by laser radiation. Such coatings may, within certain limits, intensify the effectiveness of laser heating, conducive to a decrease of the power density of the beam, thus to using lower power lasers. With appropriately selected composition, the coating may also prevent its constituents from migrating to the substrate [24].

It is also possible to employ combinations of the above methods.

When selecting the method of enhancing the absorption rate of the surface, one should take into consideration that increasing surface roughness causes a deterioration its three-dimensional quality. It does not, however, cause changes of the chemical composition of the metal's heated zone which usually, although to a negligible degree, takes place in the case of absorbing coatings. Unless remelting of the substrate material takes place, such coatings either do not cause deterioration or cause only insignificant deterioration of surface quality. Good absorbing coatings should, moreover, feature good adhesion to the substrate, stability of thermophysical properties, homogeneity of properties and of thickness, as well as ease of deposition and stripping [24]. The thickness of such coatings usually does not exceed 0.1 mm.

Since laser heating is usually carried out in air, a protective atmosphere is introduced, usually nitrogen or argon at $(0.5 \text{ to } 1.0) \times 10^5$ Pa pressure [5, 11]. This is done to protect against oxidation of the substrate which was not preoxidized, if it is detrimental to the technological process, but predominantly to protect the laser lenses against radiation.

3.4.3 Depth of penetration of photons into the metal

The initial phase of laser heating consists of absorption of photons which are not reflected from the surface by free or bound electrons of the heated material. Because of the high frequency of laser radiation, exceeding many times the frequencies of electromagnetic radiation used for induction heating, the depth of penetration of photons into the heated material is small. It can be determined from a formula analogous to that used for induction heating [7]:

$$\delta = 503 \sqrt{\frac{\rho}{\mu f}} \tag{3.4}$$

where: d - depth of penetration of laser radiation into the heated material, m; ρ - resistivity (reciprocal of conductivity) of heated material, $\Omega \cdot m$; μ -

relative magnetic permeability of heated material; f - frequency of changes of electromagnetic field, bound to laser radiation wavelength by the correlation: $f = c/\lambda$, where c - velocity of light (wavelength of 1 μm corresponds to a frequency of 3.33×10^{-15} Hz).

According to equation (3.4), the radiation of molecular CO_2 lasers penetrates the material deeper than that of Nd-YAG lasers.

To simplify the problem it can be said that photons from infrared radiation penetrate materials that are not transparent to this range of frequencies to a depth not exceeding 10^{-6} to 10^{-5} cm (1 to 10 μm), i.e., a depth which is comparable to the wavelength of the incident radiation [6, 7].

Photons are absorbed by free or bound electrons of the heated material and cause a rise of their energy. Electrons with higher energy interact with the crystalline lattice and with other electrons and cause passing of energy deeper into the heated material which is manifest by a rise of temperature. From the zone heated by the penetrating photons the heat is therefore passed on to the colder zones of the material by means of heat conduction at a rate which depends on the conductivity coefficient of the given material which is significantly higher for metals than for non-metals.

Usually materials with high thermal conductivity are characterized by high electrical conductivity and, at the same time, high reflectivity. These are properties which characterize metals. For non-metals there is a generally opposite rule. For this reason, the intensity of conducting heat into the material from the directly heated surface of the metal is greater than for an absorptive coating deposited on the metal substrate.

Assuming that the diameter of the laser spot is 20 mm and the depth of photon penetration is approximately 10 μm, the volume of material affected by photon penetration is approximately 3 mm³. In this volume there occurs absorption of 0.5 to 10 kW of radiant power and its transformation into heat energy. Such great powers evolved in such small volumes cause almost instantaneous heating of that volume of material. The time of energy transfer by electron collisions is approximately 10^{-12} to 10^{-14} s, while the time of energy transfer to the crystalline lattice is approximately 10^{-11} to 10^{-12} s [6]. For normal and gigantic laser pulse durations of 10^{-3} to 10^{-7} s it is safe to assume that during one pulse, electrons which absorbed photons collide many times with other electrons and with the lattice. For this reason it can be assumed that radiant energy is instantly transformed into heat energy at the site where radiation is absorbed. Thus it is possible to obtain uncommonly short heating times for laser heating, ranging from 10^{-1} to 10^{-8} s [4, 27, 29]. Transfer of heat from heated spots by penetration of photons (limited to a layer of a fraction of a micrometer to several micrometers at the most) to the remaining portion of the load is significantly slower and equal to the rate of heat conduction by the given material.

In laser heating, the highest temperature is attained by the thin subsurface layer in the zone of incidence of the laser beam (i.e., in the laser spot); in stricter terms, in the zone of the greatest power density. Surface temperature distribution approximately corresponds to the distribution of

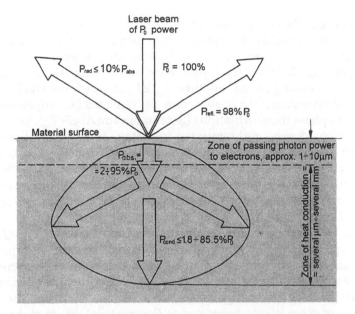

Fig. 3.16 Schematic representation of the interaction of laser radiation with material.

power density in the cross-section of the beam at the site of the laser spot [30].

Heat accumulated in the surface layer heated by laser radiation (over 90% of the total heat evolved) is passed on concentrically into the material by way of heat conduction. Thus, cooling rates are very high, compared to rates of heating. Only a small portion of the heat delivered to the material (max. 10%) is given off by radiation of the material's surface layer (Fig. 3.16).

The efficiency of laser heating is very low. This efficiency depends on the type and design of the laser heater and on the absorption coefficient of the heated surface. On the average, the efficiency of heating a solid by laser pulses ranges from 2 to 3%, while that of continuously operating gas lasers ranges from 5 to 7%.

3.4.4 Laser heating stages

For given values of exposure time t and absorbed power density $q = q_0 A$, the heated material, e.g., steel, may be subjected to successive heating stages to a higher temperature [11, 30] with a rise of q and t. Each such stage includes the previous heating stage (Fig. 3.18).

Stage 0. Heating to temperature $T_{max} < A_1$ (per iron-carbon phase diagram) at which no phase transformations take place in solid steel. The heated zone includes material at temperature $T < T_{max}$, practically $T < 150°C$.

Stage I. Heating to temperature $T_{max} < T_m$ (where T_m is melting point) at which phase transformations occur in the solid state. Due to the rise of surface temperature, emission of electrons and photons begins. The heated zone comprises the usable zone (at temperature $A_1 < T < T_m$), i.e., that in which the $\alpha \leftrightarrow \gamma$ transformation takes place; next - further away from the laser spot - tempering, and a unusable zone, i.e., that which is at temperature $T < A_1$ according to Stage 0. This is the typical zone for carrying out heat treatment.

Stage II. Heating to temperature $T_{max} \geq T_m$ at which the steel melts. The heated zone comprises: the usable zone, i.e., remelted (at temperature $T \geq T_m$), the zone of phase transformations (at temperature $A_1 < T < T_m$), i.e., hardened - after cooling down - and tempered (per Stage I), as well as the unusable zone (per Stage 0). During heating the surface which constitutes the boundary of melted material propagates inward, thereby increasing the zone of melted material to the moment of breaking off of heating or equalizing the amount of heat delivered with the amount of heat conducted to deeper, unmelted layers.

Stage III. Heating to the temperature of vaporization $T_{max} > T_v$ (where T_v is temperature of vaporization) at which steel undergoes rapid vaporization and plasma is formed which impedes or even stops (screens) the influx of radiation to the material by absorbing it. In such conditions it is possible - besides hollowing, cutting or welding - to harden the material by means of shock waves generated in that material due to sudden pulsed impact of propagating plasma. The heated zone comprises the vaporized zone (a crater) at temperature $T \geq T_v$ and the zone per Stage II.

Heating per Stage 0 is, from a technical standpoint, of no practical significance and it will not be included in further considerations.

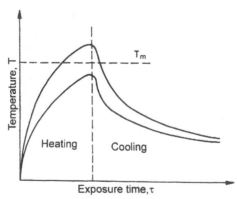

Fig. 3.17 Typical correlation between temperature and time of exposure in heating of metal by pulsed laser radiation. (From *Burakov, V.A., et al.* [30]. With permission.)

Because of very high densities of absorbed power in laser heating, steel temperature rises rapidly with time of exposure and after cutting off the influx of radiation drops equally rapidly (Fig. 3.17), mainly due to heat conduction into the material (so-called cooling by load mass) and partially due to radiation of heat to the environment [30, 31].

Maximum heating rate may even exceed 10^6 K/s and occurs in the surface layer of the heated material at the beginning of the heating pulse. It

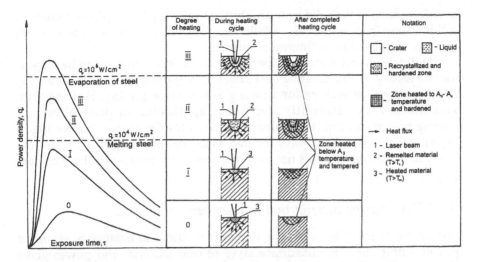

Fig. 3.18 Schematics showing laser heating with structural changes taking place during heating and after cooling. (From *Burakowski, T., et al.* [11], *Kusiński, J.* [33]. With permission.)

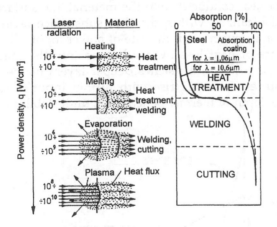

Fig. 3.19 Portrayal of thermal processes occurring at the surface of metals influenced by power density of continuous laser radiation.

later stabilizes at approximately $2 \cdot 10^5$ K/s [30]. This rate is greater by three orders of magnitude than that obtained in traditional methods of rapid heating. The rate of cooling by load mass is generally close to that of the heating rate, although in some stages of cooling it may even be greater by an order of magnitude and even more [29], ranging from 10^6 to 10^8 K/s [32].

As the result of joint processes of rapid heating and cooling there may occur processes and transformations in steel similar to those occurring in traditional treatments, although significantly differing in quality. Due to the very short duration of heating and cooling, precipitation and material homogenization processes are retarded or do not occur at all. Temperatures of transformation in the solid state are shifted upward while rapid and uniform solidification (big number of crystallization nuclei) allows non-epitaxial grain growth. The structure is thus more refined.

Heating without remelting (per Stage I) is applied in cases of heat treatment. Heating with remelting (per Stage II) is applied for welding, cutting, heat treatment as well as for alloying and surface plating. Heating with vaporization (per Stage III) has found application in perforation and cutting operations but could also be utilized for detonation hardening of crater walls by the shockwave formed as the result of rapid vaporization. Detonation hardening has not, however, been practically utilized to date (Fig. 3.19).

3.4.5 Temperature distribution in laser-heated material

The effectiveness of laser heating depends on the heat balance of absorbed power within the thin subsurface layer of cold material and power given off to the environment by way of radiation of the heated surface. For simplicity, the latter value maybe neglected.

If the rate of radiant power input (or power density q_1) is high relative to the power of heat conducted into the material, the surface temperature T_1 is high and localized within the thin subsurface layer. If the input power density q_2 is lower, i.e., if the rate of power input is less and comparable to the rate of heat loss by conduction into the material, surface temperature T_2 is lower and reaches deeper layers of the material (Fig. 3.20). temperatures T_1 and T_2 rise with further input of radiant power, similarly to the depth of thermal penetration, until stabilization is reached.

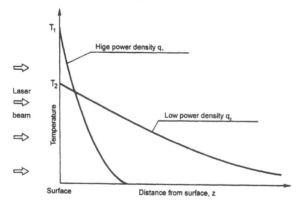

Fig. 3.20 Dependence of temperature profile on absorbed power density.

At the moment when radiant power input from the laser is interrupted (e.g., expiration of the laser pulse or displacement of laser beam to another zone) there occurs cooling of heated sites, including cooling by conduction to deeper-lying material bulk. In most cases, the cold bulk of the material may be treated as semi-infinite which rapidly consumes thermal power input. In such cases load mass cooling or autocooling occurs. When heating thin elements with low mass (as in e.g., knife edges, edges of circular saw teeth, thin sheet, etc.) which do not ensure rapid autocooling required for local hardening of an element, additional spray cooling is applied [33].

It appears from the above that the degree of heating and character of temperature distribution depend on physical properties, or, more strictly, on thermophysical properties of the heated material and on the shape and value of absorbed power distribution.

The distribution of temperature in laser-heated materials, especially in metals and predominantly in steels, has been the topic of theoretical and experimental determination by many researchers [33–46].

In the case of a fixed beam (one that is not displaced relative to material) the distribution of temperature in the heated material may be determined from heat conduction equations for point, linear or area heating. For a Gaussian distribution of power density the following correlation holds true [33, 34]:

$$T(r,z,\tau) = \frac{2P_{abs}}{\rho c(4\pi a)^{3/2}} \int_0^{\tau} \frac{1}{(\tau+\tau_f)\sqrt{\tau}} \exp\left(-\frac{z^2}{4a\tau} - \frac{r^2}{4a(\tau+\tau_f)}\right) d\tau \qquad (3.5)$$

where: $T(r,z,\tau)$ - temperature at radius r and depth z from the focal point of heating after time τ; P_{abs} - power of absorbed radiation; τ_f - time necessary for flow of heat through distance r_f ($\tau_f = r^2/4a$); r_f - distance from beam axis at which power density decreases to $1/e^2$ of value of maximum power in beam axis; z - perpendicular coordinate to surface of heated material in direction of heat propagation (into the material); r - radial coordinate, perpendicular to coordinate z; c - specific heat of material; r - material density; a - coefficient of thermal conductivity of material.

In laser heating practice the case of heating with a fixed beam occurs only very seldom. Usually the laser beam or the treated material is displaced relative to each other with a given velocity u. This case may be described by the equation of temperature distribution in the feed (scanning) plane of the beam along axis x, which holds true for the Gaussian distribution of power density in the beam [33]:

$$T(x,z,\tau) = T_0 + \frac{P_{abs}}{2\pi\lambda\upsilon[\tau(\tau+\tau_f)]^{1/2}} \exp\left\{-\left[\frac{(z-z_0)}{4a\tau} + \frac{x^2}{4a(\tau+\tau_f)}\right]\right\} \qquad (3.6)$$

where: υ - beam feed rate; λ - thermal conductivity; z_0 - depth to which heat penetrates into the material during the action of the laser beam $\tau = r_f/\upsilon$; x - coordinate perpendicular to coordinate z, in agreement with the direction of beam displacement; T_0 - temperature of environment.

For short times of beam action ($\tau << \tau_f$) which take place when the values of u are high and the diameter of the laser spot is small, the value of z_0 equals [33]:

$$z_0' = \sqrt{\frac{\pi a r_f}{2 e v}}$$

(3.7)

while the maximum temperature in the heating cycle is equal to:

$$T_{max}' = T_0 + \sqrt{\frac{2}{e}} \frac{P_{abs}}{\pi \rho c r_f (z + z_0')}$$

(3.8)

For longer times of laser beam action ($\tau >> \tau_f$) which take place when the values of u are low and the diameter of the laser spot is big, the following correlations apply [33]:

$$z_0'' = \left(\frac{r_f}{e}\right)^{\frac{1}{2}} \left(\frac{\pi a r_f}{v}\right)^{\frac{1}{4}}$$

(3.9)

$$T_{max}'' = T_0 + \frac{2 P_{abs}}{\pi e \rho c v (z + z_0'')^2}$$

(3.10)

The correlation (3.10) holds true only to the point at which the surface temperature T_{max} attains a value equal to that of melting temperature T_m.

Fig. 3.21 shows the temperature field for the case of the point source, across the surface of a semi-infinite body, while Fig. 3.22 shows the same for the case of a very strong and rapidly moving point source [29].

For many practical cases, the most interesting parameter to know is maximum temperature and its distribution along the depth axis z (i.e., along the axis of the laser beam).

For the case of **pulse heating with phase transformations without remelting,** by a beam with spot diameter d, incident perpendicular to the surface of the treated material, the temperature T_z at any point along z (to $0 \leq z \leq z_h$, where z_h is depth at which the allotropic transformation $Fe_\alpha \leftrightarrow Fe_\gamma$ takes place) is described by the following formula [14, 30, 31]:

$$T_z = \frac{2q}{\lambda} \sqrt{\frac{a \tau}{\pi}} \left(1 - \sqrt{\pi} \frac{z}{2\sqrt{a\tau}} + \frac{z^2}{4 a \tau} + \frac{z^4}{96(a\tau)^2}\right)$$

(3.11)

where: a - coefficient of temperature conductivity of heated steel, cm²/s; λ - coefficient of thermal conductivity of heated steel, J/(s×cm×K).

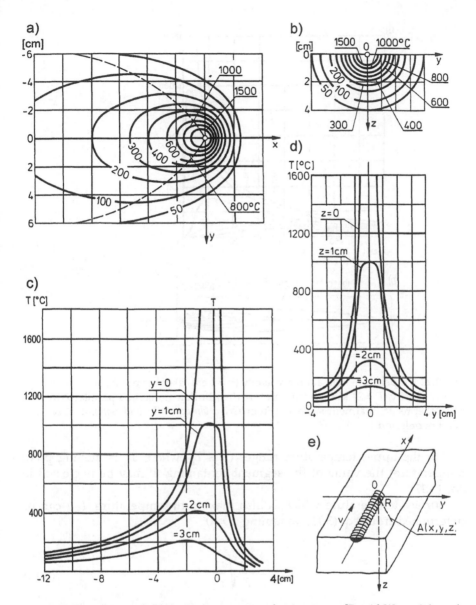

Fig. 3.21 Temperature field during movement of point source [P = 4 kW, v = 0.1 cm/s, a = 0.1 cm²/s, λ = 0.4 J/(cm·s·K)]: a) isotherms at treatment surface x0y; b) isotherms in y0z plane, perpendicular to treatment plane, passing through center of laser spot; c) temperature distribution along lines parallel to 0x axis, at the treatment surface, for different y values; d) temperature distribution along lines parallel to 0y axis, belonging to x0z plane, for different z values; e) diagram of coordinate system. (From *Grigoryantz, A.G., and Safonov, A.N.* [29]. With permission.)

Equation (3.11) gives a sufficiently good description of thermal conditions in the zone of heat input, provided the condition $d \gg d(at)\degree$ is met. It is therefore true for short exposure times t and for a defocused beam.

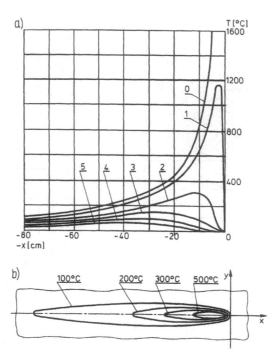

Fig. 3.22 Temperature field at high velocity ($v = 1$ cm/s) of the point source with high power density ($P = 21$ kW): a) temperature distribution along lines parallel to $0x$ axis; b) isotherms at treatment surface. (From *Grigoryantz, A.G., and Safonov, A.N.* [29]. With permission.)

For the upper temperature ranges ($T \leq T_m$, where T_m is melting point temperature) the value of the segment containing z^4 may be neglected in eq. (3.11).

Surface temperature which is also maximum temperature T_{max} can be determined from eq. (3.11), assuming $z = 0$:

$$T_{max} = \frac{2q}{\lambda} \sqrt{\frac{a\tau}{\pi}} \tag{3.12}$$

The depth of the layer z_h at which the allotropic transformation $Fe_\alpha \leftrightarrow Fe_\gamma$ takes place (i.e., the boundary of non-diffusion transition into the austenite range) is determined by the formula:

$$z_h = 2\sqrt{a\tau} \left(\sqrt{\frac{T_{max}}{T_h} - 0.215} - \frac{\pi}{2} \right) \tag{3.13}$$

where: T_h - value of temperature corresponding to the z_h boundary.

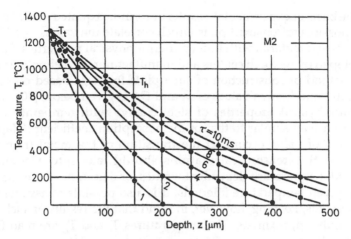

Fig. 3.23 Effect of duration of laser pulse t on temperature distribution T_z along axis perpendicular to heated surface (for M2 high speed steel). (From *Burakov, V.A., et al.* [30,31]. With permission.)

Fig. 3.23 shows the correlation $T_z = f(z)$ for different values of t, with the isotherm T_h for M2 grade high speed steel indicated ($T_m = 1310°C$) [30, 31].

The case of **pulse heating with remelting and phase transformations** is more complicated by the fact that calculations must take into consideration additional energy losses for latent heat of melting Q_m of the heated material.

Based on report [46], as well as on certain simplifying assumptions and indirect calculations, it is possible to determine the temperature of the surface and the depth of remelted zone z_m by the following equations:

$$T_{\max} = T_m \left(\frac{q^2 a \tau}{\lambda^2 \pi T_m^2} + 0.75 \right) \qquad (3.14)$$

$$z_m = 0.06 c T_m \frac{\sqrt{a \tau \pi}}{Q_m} \left(\frac{T_{\max}}{T_m} - \frac{T_m}{T_{\max}} \right) \qquad (3.15)$$

where: c - specific heat, J/(g·K).

For surface temperature $T_{max} = T_v$ (where T_v - temperature of vaporization, $T_v \approx 2900°C$) the depth of remelting is denoted by z_{mE}.

The total depth of the hardened layer z_{ht} is equal to the depth of remelting z_m and the depth of hardening z_h (determined from formula (3.15) for $T_{max} = T_m$ and for that value, denoted by z_{hL}).

$$z_{ht} = z_t + z_{hL} \qquad (3.16)$$

On the basis of experimental results or formulas quoted in literature [33–46], it is possible to construct plots which correlate the values of z, T, q and τ for any chosen material with known thermophysical properties. Such plots are developed by research centers and manufacturers of technological lasers.

The method of construction of such plots for steel, based on equations (3.11 to 3.16) and known values of λ and Q_m, is given below [30, 31].

Thermophysical properties of steel change with temperature during heating, in a way similar to that of the absorption coefficient A (an almost leap change of value occurs at the transition of the metal from solid to liquid). For this reason, calculations should take into account either tabularized, temperature-dependent values, or determine them experimentally for given temperature intervals. It is also possible to assume average constant values, keeping, however, in mind that the results of such calculations are only approximate. The temperatures T_h and T_m are read from the steel phase diagrams.

The plot is constructed in the following way [30, 31]: upon appropriating values of surface temperature T_{max} (within the range of $A_3 \leq T_{max} \leq T_m$) and time of exposure τ (e.g., within the range of 1 to 10 ms), the value of absorbed power density q necessary to achieve the assumed T_{max} temperatures is determined from the formula (3.12). Next, assuming a given value of the absorption coefficient A, the correlation $q = q_0 A$ is used to determine the input power density q_0, i.e., the power density of the beam as it exits the laser.

Next, for the same values of T_{max} and τ, the formula (3.13) is used to determine the value of the depth of hardened layer z_h, placing $T_h = 912°C$, per the iron-carbon phase diagram.

Using equations (3.14) and (3.15) for heating with remelting, and taking the value $T_m = 1310°C$ from the iron-carbon phase diagram and T_m within the range: $T_m < T_{max} < T_e$, an analogous method is pursued.

With mutual correlations between z, T, q and τ determined, the diagram is plotted. Fig. 3.24 shows such a plot for M2 grade high speed steel. The correctness of the assumed methodology has been tested in practice by a measurement of the depth of zones z_h and z_m for different conditions of laser treatment on metallographic mounts [30, 31].

The accuracy of the approximate formulas given above has been estimated for pulse heating of tool steels [42, 43] and for carbon steels [41] as ±5 to 10% in the case of heating to temperature T_m. Treatment with remelting of the surface lowers the accuracy of calculations. When heating a surface to above 2000°C, the experimental data differ from calculations by as much as ±20%. More accurate determination of the real situation in the high temperature zone of the diagram is possible after e.g., taking into account changes in absorption with the rise of surface temperature T_{max}. In this case the probability of an error may be lowered to ±10 to 15% both for carbon steels [43] and for tool steels [30, 31].

By using such plots, given in literature for continuous and pulse laser treatment, it is possible to control the conditions of heating of the surface

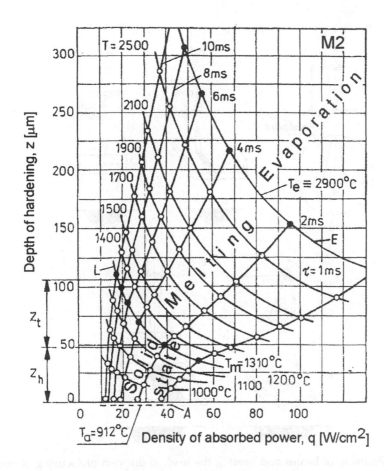

Fig. 3.24 Dependence of depth of hardening of high speed steel by pulsed laser on density of absorbed power q and on exposure time τ. T_a - temperature of austenitization; T_m - melting temperature; T_e - temperature of evaporation; z_h - example of hardening depth without remelting of superficial layer; z_{nn} - example of depth of remelting of superficial layer. (Curves representing beginning of: A - austenitization; L - melting; E - evaporation). (From *Burakov, V.A., et al.* [30,31]. With permission.)

and, based on that, to predict the depth of the heat affected zone and the degree of accomplishment of programmed processes. It is moreover possible to determine the temperature of the surface of the treated material, as well as one of the unknown parameters, t or q, after first having determined the depth of the hardened or remelted layer.

3.4.6 Laser beam control

Laser radiation may be condensed with the aid of special optical systems. This allows transmitting the radiation along great distances or focusing on very small surfaces which, in turn, allows the obtaining of high power densities at the surface, reaching 10^{14} W/cm^2 in the pulse mode [5, 11].

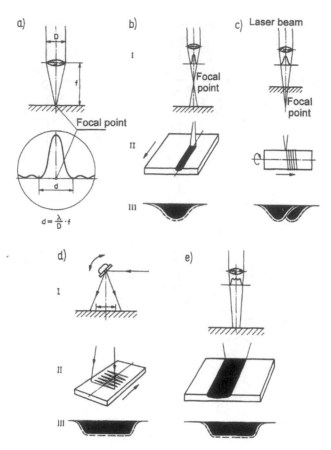

Fig. 3.25 Focusing of beams and heating the load: a) diagram of focusing of laser beam with Gaussian power density distribution (e.g. for laser beam of diameter $D \approx 11$ mm, emitted by molecular CO_2 laser of wavelength $\lambda = 10.6$ μm and focal length $f = 50$ mm, the diameter of the laser spot $d \approx 50$ mm; b) single mode defocused laser beam - load surface behind focal constriction, single path heating; c) single mode continuous defocused beam - load surface before focal constriction, multi-path heating, heated layers overlap, causing possibility of tempering of prior hardened layer; d) single mode beam, oscillating in one plane, multi-path heating; e) multi-mode continuous beam; I - schematic of focusing or oscillation; II - schematic of heating and displacement of load; III - cross-section of hardened layers, approximately corresponding to power density distribution in beam.

Laser heating of a surface of the treated material may be carried out in a manner similar to that with the electron beam. However, due to the specifics of the laser beam, the most often used type of heating is sequential or point-insular with beam deflection or beam transverse movement across precision *x-y* stages, rotation and movement of the load or a combination of the above. The input power density on which thermal processes taking place at the surface depend are controlled by focusing and defocusing of the beam (Fig. 3.25) and by a change of exposure to radiation time. Depth of heated or remelted layer for a given power density depends mainly on exposure time and fluctuates within the range from several micrometers to several millimeters. In most common cases it is of the order of several tens of micrometers [14, 19, 22, 30–36].

As opposed to stock-removing treatments (cutting, hollowing, perforating), forming (blanking, inscribing, bending) or joining (welding), in metal surface engineering the one-mode Gaussian power density distribution in the beam is not desired. What is desired is a multi-mode distribution, as far as possible close to uniform.

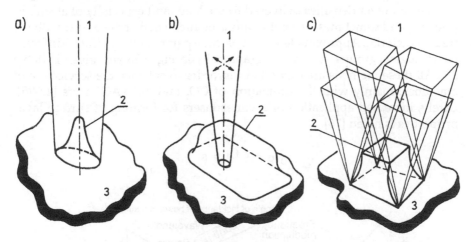

Fig. 3.26 Diagrams showing different power density distributions in load: a) single mode distribution, obtained by condensing lens and non-oscillating mirror; b) distribution close to rectangular, obtained by mirror oscillating in two perpendicular directions; c) rectangular distribution obtained by condensing several radiation beams; 1 - laser beam; 2 - shape of power density distribution; 3 - treated object. (From *Burakowski, T., et al.* [11]. With permission.)

Usually, the spatial distribution of power density in high power continuous and pulse laser beams is irregular and for very high power lasers, this irregularity is even greater. This distribution may be smoothed out by the application, outside of the laser, of a set of lenses or apertures, collimating of laser beams, or oscillation scanning by the beam (with the aid of a vibrating mirror). Focusing also smoothes out the distribution of power density in the beam. Usually, however, it is not possible to achieve a fixed power density distribution throughout the cross-section of the laser beam. In the outer zones of the beam, power density is usually lower than in the remaining zone. Two methods of obtaining a uniform distribution of power density are shown in Fig. 3.26 [11]. Most often, the beams are circular in cross-section, with a diameter of over 30 mm; in rarer cases rectangular, with dimensions of several tens of millimeters per side.

Scan patterns obtained with the aid of a laser beam are similar, as in the case of an electron beam. In general, laser beams are less flexible than electron beams.

3.5 Laser techniques

The utilization of laser radiation in surface engineering depends on a great many factors, of which the most important are different properties of materials subjected to laser treatment, as well as different properties of the laser beam (Fig. 3.27). Combinations of different properties of material and beam allow the utilization of *Laser Beam Machining* [36] in approx. 200 different areas [32]. Of these, surface engineering utilizes only some - up to 20%. They differ among one another - like in electron beam heating - mainly by combinations of power densities delivered to the load (and especially of absorbed power q) and sometimes by combinations of interaction of the beam t with the treated material. Approximate values of these parameters, typical of different technological groups, are shown graphically in Fig. 3.28 and given in Table 3.1. All these methods are carried out in order to enhance the service life of machines, mainly with the utilization of CO_2 and Nd-YAG lasers [47-75]. Lately, increased application of excimer lasers for forming of surface layer properties is noted [76].

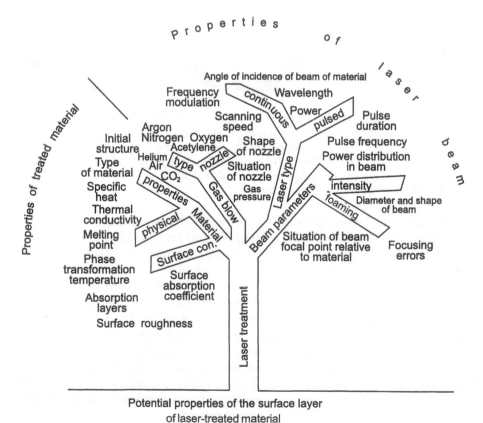

Fig. 3.27 Graphic portrayal of factors affecting the results of laser treatment. (From *Kusiński, J.* [33]. With permission.)

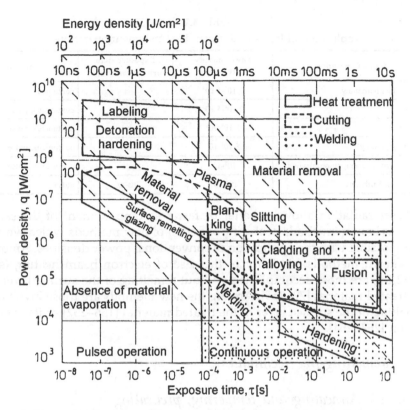

Fig. 3.28 Dependence of thermal processes taking place during heating of metals by CO_2 laser radiation on density of absorbed power and on time of exposure (duration of action of beam on material).

Thanks to the possibility of concentration of great power densities on selected fragments of treated objects in very short times, laser treatment of surfaces in most cases allows the obtaining of structures which differ from those obtained in equilibrium conditions. This concerns supersaturated solid solutions, metastable phases, fine-grain, dendritic and amorphic structures, etc. and, in consequence, allows an increase in hardness and enhancement of strength and tribological properties, and sometimes anti-corrosion properties, usually accompanied by a deterioration of surface quality. It should, however, always be kept in mind that changes of layer properties are limited only to sites affected by the laser beam, i.e., only beam paths. Thus the properties of whole big laser-treated surfaces depend on the pattern which is similar to that obtained by electron beam heating, i.e., these properties depend on the degree of coverage of the surface by these paths and their mutual locations. Unfortunately, in the area of fatigue strength and tribology research, literature lacks information whether the presented test results pertain only to areas with a changed state of physico-chemical properties or constitute an average of properties of laser-treated and untreated areas of the surface.

Table 3.1

Application of lasers to different heat treatment operations

Type of operation	Power density [W/cm²]	Exposure time (orientation range)	Laser work mode
Remelt-free hardening	$3 \cdot 10^2 - 5 \cdot 10^4$	s	continuous
Fusion	$2 \cdot 10^4 - 5 \cdot 10^5$	s	continuous, pulse
Alloying	$5 \cdot 10^4 - 10^6$	ms	continuous, pulse
Cladding (plating)	$5 \cdot 10^4 - 10^6$	ms	continuous, pulse
Glazing (vitrification, amorphisation)	$10^5 - 10^7$	μs	continuous, pulse
Detonation hardening	$10^8 - 10^{10}$	μs-ns	pulse

Laser radiation allows a better or faster accomplishment of the same goals as known methods or the introduction of new methods which would be impossible with the application of conventional power densities. Among the known methods, similarly to the case of electron beam heating (see Fig. 2.16), are those accomplished without remelting of the surface layer of the treated material [77–155]. Among the new methods are those involving remelting of the surface layer of the treated material [156–252] or its vaporization [253, 254].

3.5.1 Remelting-free techniques

3.5.1.1 Annealing and tempering, preheating

Because of the fact that laser beam spots cannot achieve big dimensions, and on account of technical difficulties connected with the utilization of several laser heads for simultaneous heating of the surface, laser annealing and tempering are not applied to the simultaneous treatment of big surfaces but to selected fragments, often following laser hardening. The range of practical applications of laser annealing, tempering and preheating is rather narrow.

Annealing. The main aim of laser annealing is homogenization and the obtaining of a structure closer to equilibrium in comparison with the initial structure. The new structure should be characterized by greater ductility and higher hardness. After ending laser heating, the rate of cooling should be significantly slower than critical. Laser annealing is applied in precision treatment of structures of semiconductor instruments [79]. The most often used is recrystallization annealing of thin layers vacuum deposited from the gas phase or from vapours. Such layers have a defective structure, sometimes amorphous. As the result of laser annealing, the defects of the crystal structure recede, density and conductivity increase and physico-chemical properties change. For example, a germanium layer, vacuum deposited from the vapour phase, comprises grains of several tens of nanometers dimension and is characterized by poor vacancy conductivity. After laser annealing, grain dimensions grow even to hundreds of nanometers while electron conductivity is strongly enhanced.

Tests of laser annealing carried out on layers of arsenic, silicon, chromium, aluminum, graphite, etc. on different substrates have shown that their properties very strongly depend on annealing parameters and on the initial condition. Laser annealing is applied for correcting the resistivity of metal and metal-ceramic resistors (e.g., $Cs-SiO_2$, $Au-Cr_2O_3$), for forming of ohm contacts in semiconductor instruments and for the formation of current paths [29].

A significant utilization of laser annealing, especially in the area of machine building, is local lowering of hardness and increase of ductility in order to facilitate subsequent deformation of those sites or to increase fatigue strength. Rapid heating of cold-formed steels causes shifting of beginning and end of primary crystallization to the high temperature zone and this, in turn, causes a rise in the number of crystallization nuclei and significant refinement of grains in comparison with slow heating. The greatest impact of heating rate on grain size is observed when the coefficient of deformation is low, close to the critical value [29].

In order to lower hardness and to increase ductility of metals, besides recrystallization annealing laser annealing with phase transcrystallization is used. This can also be utilized in the heat treatment of thin strip, sheet and wire [29].

Tempering. This process is used after hardening of steel and consists of heating it to a temperature lower than eutectoid and subsequent cooling in order to enhance ductility and to reduce brittleness at the cost of a drop in hardness. During rapid tempering by laser, steel undergoes the same phase and structural transformations as during slow conventional tempering (in a furnace). The formation of tempered martensite, decomposition of retained austenite and the formation of a mixture of ferrite and cementite all take place but the temperatures of all these transformations are shifted upward, especially the temperature of decomposition of retained austenite.

The short time of action of the laser beam yields very fine carbides and a significantly higher hardness than that obtained by conventional tempering. Moreover, it results in higher static and impact strength, higher cold brittleness and temper brittleness. Laser tempering of carbon steels allows the obtaining of mechanical properties comparable to those of conventionally tempered steel with chromium and vanadium alloy additives. The structure and properties of laser tempered steel may be controlled by a change of energy parameters (power density, exposure time) and by retempering, employing different parameters.

Laser tempering is used in those cases where there is a need to locally increase the ductility of or impact strength at e.g., join sites of different components after prior induction hardening. Laser tempering is also used on tools which were hardened and tempered at a low temperature [29].

Laser tempering allows a change in the distribution and lowering of the value of quenching stresses formed as the result of intensive local hardening, e.g., by laser, electron beam or plasma, especially in cases of remelting where a high stress concentration is formed [78].

No significant structural differences are noted when comparing laser tempering with conventional tempering of carbon and alloy steels [81, 82]. The recommended practice is to apply high temperature laser tempering for components working under dynamic loads at low temperatures when it is important for the material to avoid stress concentrations, formed e.g., during laser hardening [82]. The laser beam may also be utilized for annealing and tempering of welds and padwelds [83].

Preheating. The laser beam may also be used to increase the ductility of different materials by preheating the surface immediately before the next surface treatment or forming operation, e.g., before machining of materials which are difficult to machine (sintered carbides, creep-resistant steels and alloys, hardened alloys) [2]. Other applications include preheating prior to forging, also in order to achieve fine-grained structure [29], before welding [50], prior to laser treatment, e.g., plating or hardening, in order to avoid cracks [51] or prior to laser hardening in order to aid the heating process when using a high powered laser, by preheating up to 200°C [84]. Very advantageous is multiple laser heating in order to facilitate bending of even thick metals [14]. By laser heating a material to a temperature close to the melting point directly prior to subjecting it to a cutting tool or a welding torch, hardness and cutting resistance are reduced, bringing about a reduction in the deformation of components being treated or joined, an extension of the life of the cutting tool and shortening of treatment time [2, 8].

3.5.1.2 Remelt-free hardening

Remelt-free hardening consists of heating a metallic material to a temperature at which a known phase transformation takes place (e.g., for steel to several K above A_{c3}) and self-cooling, resulting in a structure which is less stable than the initial one and usually harder. In the case of steel, this structure is usually martensitic.

Hardening parameters are selected in such a way as to allow the occurrence of the phase transformation (in steels - the martensitic transformation) without the occurrence of remelting, even remelting of a thin surface layer. Remelting is usually accompanied (after self-cooling) by the obtaining of a less hard surface of the layer than that obtained without remelting. This loss in hardness may even reach 20 to 30%. If remelting occurs, retained austenite appears in the remelted and resolidified structure of high alloy tool steels. This retained austenite is difficult to remove even by cryogenic treatment [50]. Usually, for remelting-free hardening the applied power density is 10^2 to $2 \cdot 10^4$ kW/cm^2 and the time of exposure is $10^{-2} \div 1$ s. The typical heating rate is approximately 10^6 K/s and cooling rate is approximately 10^4 K/s [11].

In remelting-free laser hardening phase transformations do not, however, occur at the same temperatures as in conventional hardening, in accordance with phase diagrams for the particular elements forming the given alloy. Due to the very big heating and cooling rates, exceeding conventional rates by several orders of magnitude, phase transformations at temperatures which are appropriate for conventional hardening "lag behind." Such phase transformations require temperatures higher by several tens K [85].

In comparison with normally employed methods of conventional surface hardening (induction, flame or plasma), laser hardening yields harder surface layers with a more refined grain structure and thinner (thicknesses usually within 0.25 to 2.5 mm). Obtaining of thicker layers requires a longer exposure time, of the order of even whole seconds. The thickness of the hardened layer is usually 0.25 to 0.75 mm for steel, approximately 0.5 mm for gray cast iron and approximately 1.0 mm for nodular cast iron [11].

Similarly to the case of remelt-free hardening by laser beam (see Fig. 2.24) laser hardening can also be applied as a single layer and multi-layer version, as well as simple or complex.

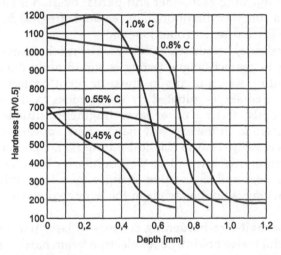

Fig. 3.29 Different hardness distribution in continuous laser-hardened (with different hardening parameters) superficial layer on carbon steel.

Simple remelt-free hardening consists of laser hardening of a material, the initial state of which is the result of a previous forming, surface or machining operation. On account of cooling rates which are much faster than in conventional hardening, it is possible to harden steels and alloys which are non-pardonable by conventional means, e.g., steel with a carbon content of 0.2% [50, 51]. Fig. 3.29 shows hardness profiles on low carbon steels. As a general rule, an increase in carbon content causes a rise in hardness [91, 92] and thickness of the hardened layer for the same laser treatment parameters. This is mainly the result of a rise in hardenability and of the lowering of austenitizing temperature [50, 51]. For steels which are normally hardenable by conventional means, hardnesses obtained by laser treatment are approximately 30% higher than those obtained by conventional hardening. This is connected, among other factors, with a more refined microstructure (carbides do not have the time required to dissolve during austenitization). Increments in hardness are accompanied by appropriate increments of yield strength, ultimate tensile strength and wear resistance [50, 51].

Good results are also attained in hardening of gray cast iron, with a mainly pearlitic matrix. This enhancement is most pronounced in the rise of tribological properties, with relatively small layer thicknesses [50, 51, 86, 89, 90].

The residual stresses formed (both thermal and structural) depend on the energy and geometry conditions of laser radiation. Along hardened beam paths, compressive stresses almost always prevail, while between paths, tensile stresses dominate [93, 123]. In martensitic stainless steels, compressive stresses prevail near the path surface while tensile stresses are present near the core [233]. A decrease of distance between two paths down to their touching each other and partial overlap is conducive to the formation of a complex distribution of residual stresses, due to local tempering of already hardened zones [93-96].

As a general rule, fatigue strength rises. The mechanism of fatigue damage occurring in contact fatigue testing is significantly different from that of classical pitting which occurs in linear or point contact of mating surfaces. Fatigue damage is initiated by microcracks in the hardened path and, subsequently, flaking off of layer fragments situated between the microcrack network. In the exposed matrix there appear secondary cracks which, by developing, lead to classical pitting [97]. Upon bending, fatigue cracks are temporarily arrested at the boundaries between the hardened paths and the core and their value changes by a leap. Within the paths, the rate of cracking is not big, while in the tempered zones it is even lower [98–101].

Complex remelt-free hardening consists of laser hardening as shown in Fig. 2.24, which also holds true for electron beam hardening. Thus it can be carried out after conventional heat treatment, e.g., hardening, tempering, annealing, or surface diffusion treatment, e.g., carburizing, nitriding, or after mechanical strengthening operations. Besides, in addition to what is shown in Fig. 2.24, this operation can be carried out after the deposition of coatings by different methods (e.g., electro-vibration, concealed arc [102], thermal spray or by PVD). It is also possible to apply some of these treatments after laser hardening or to apply them simultaneously. As an example, simultaneous arc and laser heating of normalized 1045 steel allows obtaining 2.5 to 3 times greater thickness of the hardened layer and a high hardness of 56 to 62 HRC in comparison with results obtained by laser heating alone [103].

In laser hardening of steel martensitic transformation occurs. Depending, however, on energy and geometric parameters of the laser beam, the properties of the steel and the type of prior (pre-laser) treatment, different martensitic structures, as well as different proportions of other residual phases (carbides, ferrite, etc.) are obtained.

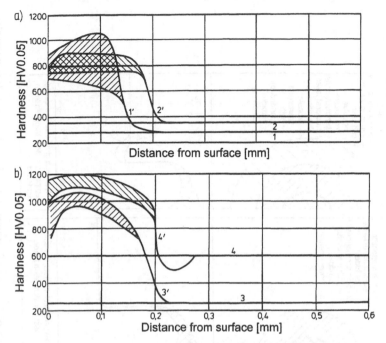

Fig. 3.30 Hardness distribution in direction perpendicular to center of laser path: a) 1045 steel, normalized (1) and hardened and tempered (2) without laser hardening and with transformation hardening (correspondingly: 1′ and 2′); b) 52100 steel, annealed (3) or hardened and tempered (4) without laser hardening and with transformation hardening (correspondingly 3′ and 4′). (From *Zenker, R., et al.* [104]. With permission.)

In hypoeutectoid steels, e.g., 6150H: 0.47%C; 0.26%Si; 0.64%Mn; 0.003%P; 0.018%S; 0.98%Cr; 0.09%V, normalization, carried out with high beam power density, causes lath martensite in big, regular clusters, to form in the subsurface zone, in a way similar to that occurring during conventional hardening. In deeper zones, increasingly finer, irregular martensite is found. It is so fine that it may be impossible to see with the resolution of an optical microscope. Often, traces of pearlite may be observed in the microstructure. The greater the pearlitic zones in the initial condition of the material, the more often such traces occur in the final microstructure. The presence of pearlite is an indication that the temperature at that particular site did not exceed A_{c3} [88, 104].

In hypoeutectoid steels, e.g., the 90Cr3* grade, after softening annealing with comparable energy and geometrical parameters of the laser beam, lath martensite and residual austenite occur in the subsurface zone. In the transition zone the predominant component phase is very fine acicular or lath martensite, as well as numerous undissolved carbides [104].

Adhering directly to the laser hardened zone in steel or to a conventionally hardened and tempered zone which was additionally laser hardened is a more or less broad zone of tempered material with a hardness

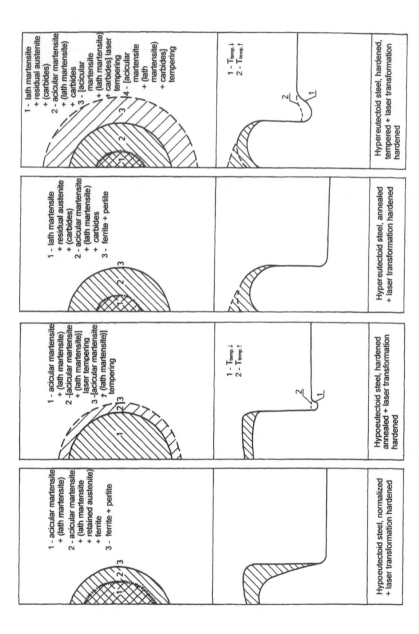

Fig. 3.31 Products of martensitic transformation and hardness distributions of hypo- and hypereutectoid steels, laser transformation remelt-free hardened, after different prior heat treatment: a) zone of action of laser beam on object surface; b) hardness profile. (From *Zenker, R., et al.* [104]. With permission.)

significantly lowered relative to that of hardened sites. The effect of tempering is manifest especially in steels with low temper resistance, e.g. in 52100 grade steel, and is characterized by a low increase of hardness (relative to initial) [104].

In steel which was laser hardened with prior hardening and tempering, hardness distribution is uniform, while in steel that was laser hardened after prior normalizing, hardness fluctuates more than in conditions of other initial heat treatment. It first increases from the surface in the direction of the core and then drops off to the level of core hardness (Fig. 3.30). A lowering of hardness at the surface of 52100 steel is probably due to the big proportion of retained austenite in this zone [104].

Fig. 3.31 shows the products of martensitic transformation and hardening profiles obtained during remelt-free laser hardening of hypo- and hypereutectoid steels, subjected to different prior heat treatments [104].

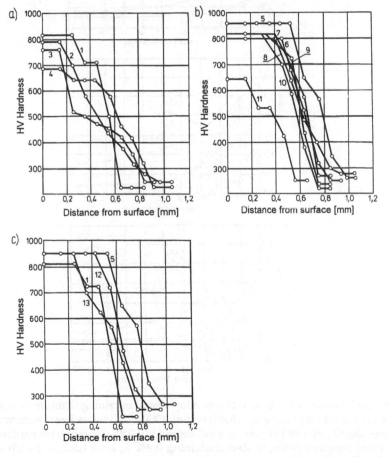

Fig. 3.32 Hardness profiles of superficial layers following annealing of laser hardened Russian steels: a) plain carbon; b) alloyed; c) eutectoid; *1* - U8ʹ; *2* - U10ʹ; *3* - U12ʹ; *4* - C1045; *5* - 60S2ʹ; *6* - HWSGʹ; *7* - 9HSʹ; *8* - HWGʹ; *9* - 12H1ʹ; *10* - Hʹ; *11* - H12Mʹ; *12* - C1062; *13* - 7H3ʹ. (From *Kremnev, L.S., et al.* [105]. With permission.)

If the condition of not remelting of the surface layer is kept, lowering of power density of the laser beam and shortening of the time of its action is not conducive to qualitatively different processes but only to a displacement of the said zones in the direction of the material surface. With the same radiation parameters, the depth of hardening rises with a rise of thermal conductivity of the treated material. The hardness of laser hardened steels drops with a rise in the content of retained austenite. Fig. 3.32 shows some

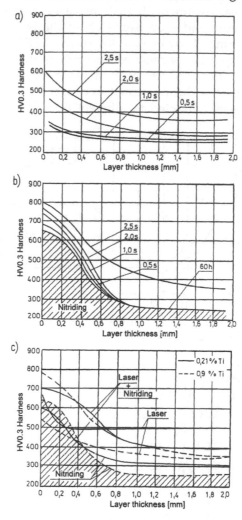

Fig. 3.33 Hardness distribution of low alloy steels containing titanium (initial condition: hardened and tempered), hardened by CO_2 continuous laser beam with fixed power density of 5.88 kW/cm²: a) steel containing 0.9% Ti, only laser hardened with different exposure times; b) steel containing 0.9% Ti, laser hardened with different exposure times and subsequently glow discharge nitrided 60 h at 520°C; c) steel containing 0.21% Ti and 0.9% Ti, only glow discharge nitrided (60 h at 520°C), laser hardened (exposure time: 2.5 s) and laser hardened (exposure time: 2.5 s) followed by glow discharge nitriding (60 h at 520°C).

hardness profiles for different steels: carbon (characterized by high thermal conductivity), alloy (with thermal conductivity, on an average half that of carbon steels) and eutectoid (with lowest amount of retained austenite). If maximum thickness of the hardened layer using same laser heating parameters is the desired goal, it is recommended to use eutectoid steel, alloyed by elements which raise thermal conductivity, and to subject it to prior heat treatment which yields a troostite structure [105].

A change in the hardness profile and a change of the content of the element alloying the surface layer by diffusion may be obtained by utilizing a combination of prior thermo-chemical treatment and laser hardening [132]. Fig. 3.33 shows the distribution of hardness on alloy steels containing titanium: laser hardened only and laser hardened, followed by glow discharge nitrided [11]. Such a combination enhances the steel's nitridability, resulting in thicker and harder nitrided layers [132]. Fig. 3.34 shows a diagram of correlations at different sites of the laser path between microstructure, nitrogen content and hardness at different distances from the surface [106–110].

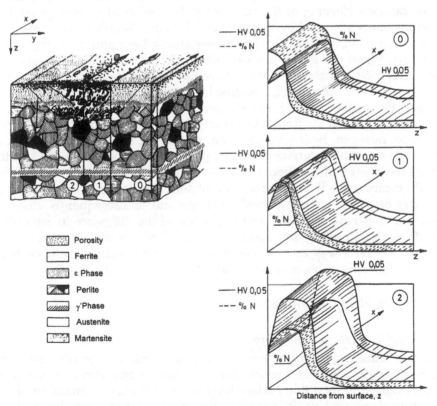

Fig. 3.34 Graphic representation of co-dependence between microstructure in the laser hardened zone, nitrogen content and hardness distribution in normalized and laser hardened 6150H steel, gas carbonitrided (4-8 h, 843 K). (From *Zenker, R., and Zenker, U.* [109]. With permission.)

Generally, remelt-free laser hardening increases the following: hardness (to a degree higher than that obtainable by conventional methods), static and fatigue strength, as the result of formation of favorable compressive stresses [113], impact strength and ductility, corrosion resistance and wear resistance [111, 118, 131]. As an example, wear of 1045 grade steel is 10% less after laser hardening than after conventional hardening [112]. Surface roughness after laser treatment does not change relative to pre-treatment condition.

Remelt-free laser hardening is applied to harden cold work tool steels [139, 142], high speed steels [127, 139, 148], structural steels, including those containing chromium [96] and manganese [134]; low carbon steels [87, 117, 128, 140] containing less than 0.2%C, medium carbon steels [147], alloy steels, steels used for armor plates, corrosion resistant steels [122], bearing steels, cast iron (especially gray) and for copper alloys [129].

3.5.1.3 Surface cleaning

Laser radiation, with shallow penetration into solids while heating them, also causes a cleaning of the surface by desorption of vapors, gases and contaminants adsorbed by that surface. The effect is enhanced by the employment of vacuum as the protective atmosphere and by using pulse heating. When the pulses are very short, the gaseous contaminants and atoms of foreign inclusions do not have the time needed to diffuse into the surface. On the other hand, combined action of temperature and vacuum cleans the surface quickly and effectively. Laser cleaning of materials in vacuum is used in electronics to remove inclusions and adsorbed atoms of oxygen, nitrogen, hydrogen, etc. from crystals of silicon, nickel, etc. It is possible to obtain a surface of atomic purity in times substantially shorter than those used in traditional cleaning methods [29].

In combination with the formation of slits, it is possible, with the aid of a laser beam, to remove varnish coatings and to subsequently etch the surface in etching solutions. The utilization of the difference in expansion coefficients of coating and substrate material allows the use of the laser beam in such a way as to cause so-called thermo-peeling of the coating, due to the generation of substantial shear stresses at the interface [29].

3.5.2 Remelting techniques

3.5.2.1 Surface remelting

Surface remelting is a name given to a group of methods carried out with the application of greater power densities and heating rates, consisting of rapid remelting of a thin surface layer of the substrate material or of the coating deposited on it, followed by equally rapid crystallization or amorphization (Fig. 3.35). It is sometimes called pure remelting. Remelting always causes a change of surface roughness. In the laser-affected zone a cloud of gaseous plasma is always formed.

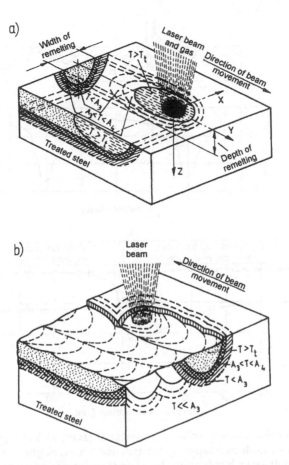

Fig. 3.35 Schematics of laser surface melting of steel: a) continuous heating; b) pulsed heating (T_m - melting point; A_3 and A_4 - temperatures per iron-carbon phase diagram; T - temperature of heated object). (From *Kusiński, J.* [33]. With permission.)

Similarly to electron beam surface melting (see Fig. 2.16) laser surface melting can also be divided into remelt hardening, glazing and densification, depending on the effects obtained. Moreover, smoothing can also be accomplished.

Remelt hardening. Depending on the energy parameters of the laser beam, in a way similar to that in electron beam remelt hardening, we can distinguish: surface remelting, remelting, intensive and very intensive remelting. With a rise of energy parameters, an increase of surface roughness takes place. The main aim of remelt hardening is the modification of the initial structure of the material, in particular, to obtain fine dispersion.

From Fig. 3.36 it follows that after remelt hardening the surface layer, similarly to the case of electron beam hardening, has a three-zone structure, extended by a compact external zone, formed as the result of interaction of the surrounding atmosphere (oxidizing or reducing) with the treated material. In the case of steel, the following zones are obtained [23]:

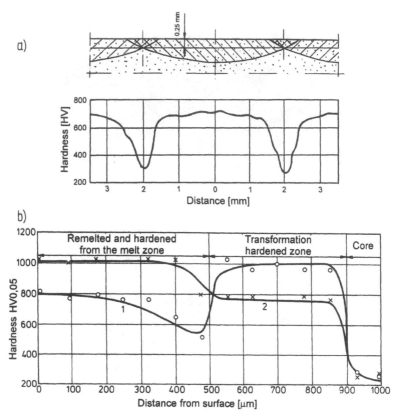

Fig. 3.36 Microhardness profile in superficial layer: a) 1045 grade steel, laser beam hardened with overlapping of hardened zones; b) gray pearlitic cast iron; 1 - after remelt laser hardening; 2 - after transformation laser hardening and tempering at 450°C for 1 h. Fig. b - from *Straus, J., and Burakowski, T.* [51]. With permission.)

– *surface layer of non-metallic phases* composed mainly of metal oxides and compounds formed as the result of chemical interaction of the atmosphere and thermal reaction of the laser beam with the steel, with gases dissolved in it and with components of the absorption coating; its layer usually does not exceed several micrometers;

– *remelted and hardened from the melt surface layer*, dendritic, with a martensitic structure; carbides present in the steel underwent total or partial melting; within the area adhering to the intermetallic phase layer, the melted zone has a diminished carbon content;

– *hardened from the solid phase subsurface layer* with a non-uniform structure: martensitic with retained austenite and carbides in the vicinity of the remelted zone and martensitic with elements of initial structure near the core. These elements are ferrite in hypoeutectoid steels and cementite in hypereutectoid steels. Throughout the layer, a dispersion of martensite occurs, 1.5 to 2.0 times greater than after conventional hardening;

– *tempered core zone*, also called intermediate, with a structure of tempered martensite or sorbite.

Remelt hardening by laser beam causes similar effects as experienced in electron beam hardening: deterioration of surface roughness (the surface after remelting has the appearance of a weld or overlay) and an enhancement of service properties such as: tribological, fatigue and anti-corrosion.

The microhardness profile along the laser beam path is approximately uniform, with clearly lowered hardness due to the tempering effect of beam overlap (Fig. 3.36a).

In the remelted zone of martensitic stainless steels and tool steels, residual tensile stresses prevail after resolidification [233].

By remelting the surface layer of the material it is possible to obtain a fine-grained structure and partial or total dissolution of precipitation phases and contaminations in the form of carbides, graphite or oxides which are usually present in the microstructure. Rapid crystallization (with cooling rates reaching 10^5 K/s) causes that after dissolution they do not precipitate again or precipitate in a different form. Strongly oversaturated solutions are obtained. For this reason, pure surface remelting is usually accompanied by a strong refinement of dispersive phases, e.g., ledeburite, as well as cleaning of grain boundaries. The latter effect is of particular significance to corrosion resistance [50].

Remelt hardening has found application primarily in the treatment of gray cast irons [50, 150–154], as well as stainless and tool steels [151–161].

Laser remelting of gray cast iron causes total dissolution of graphite and the occurrence of hard spots in the surface layer. A layer of fine-grained and non-etching quasi-ledeburite is formed on the surface. It is composed of very fine carbide precipitations, austenite and martensite, as opposed to ledeburite in gray cast irons which is composed of pearlite and carbide precipitations. Under the hardened layer there is an intermediate zone with only partially dissolved graphite flakes and a hardened layer [60, 153].

As the result of remelting of the surface of gray cast iron, a lower hardness is obtained in the zone hardened from the melt than in that hardened from the solid phase (see Fig. 3.36b, curve 1). After tempering at approximately 400°C, an increase of hardness is obtained in the zone hardened from the melt (Fig. 3.36b, curve 2) [50].

The depth of remelt hardening may reach several millimeters. Surface hardness of cast iron may even reach 1200 HV0.1. In the hardened condition the cast iron is resistant to wear (Fig. 3.37) [152] and to corrosion [151]. In the hardened layers compressive stresses usually prevail [154, 155].

Remelt hardening of gray cast iron has been broadly utilized in the automotive industry to harden slip rings and engine cylinders, components of turbines, cams and gears, obtaining a severalfold extension of service life [50, 150, 154].

Remelt hardening offers clear advantages in the case of chromium-bearing medium carbon steels [156], tool steels [157], including high speed

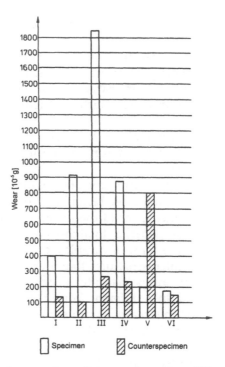

Fig. 3.37 Comparison of wear of pearlitic cast iron after different heat treatment: *I* - induction hardened and tempered 270°C to 42 to 46 HRC; *II* - gas sulfonitrided, 450°C/1.5 h; *III* - gas sulfonitrided, 450°C/2.5 h; *IV* - gas sulfonitrided, 450°C/4.5 h; *V* - glow discharge nitrided 520 C/12 h; *VI* - laser remelt hardened (continuous CO_2 laser of 1.5 kW power, q = 8.5×10^4 W/cm², beam movement speed 1 m/s, surface covered with parallel 50% laser paths). (From *Straus, J., and Burakowski, T.* [51]. With permission.)

steels [158–164], as well as structural stainless [60, 61] and bearing steels [158]. Laser remelting, combined with heat treatment, causes a significant rise in the hardness of the surface layer; e.g., some martensitic and ferritic stainless steels, by increasing hardness by a factor ranging from 1.5 to 4, feature a severalfold extension of service life. This extension may, in extreme cases, reach 10 [61]. High speed steels, subjected to remelt hardening, combined with preheat treatment, exhibit a 1.5 to 3 times increase in abrasive wear resistance following laser beam hardening [161], which underscores the significance of the issue, taking into account that more than half of all tool steels are of the high speed type. Tungsten-free high speed steels made of carbide powder (sintered P/M high speed steels) may also be remelt hardened [165], in a manner similar to titanium carbide-steel P/M sintered tools [166]. In all cases the hardened steel is more uniform chemically and more refined. Besides, the steel exhibits higher impact strength [167–169], a favorable distribution of residual stresses [170, 171] and fatigue strength. The life of some hot work tool steels is extended by as much as 250%, following remelt hardening [11].

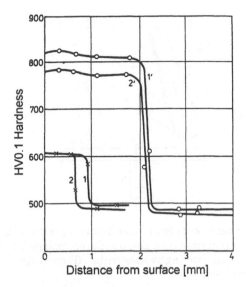

Fig. 3.38 Hardness distributions in superficial layer on low carbon structural steel: 1 and 2 - carburized; 1' and 2' - carburized and laser remelt hardened. (From *Straus, J., and Burakowski, T.* [51]. With permission.)

Interesting results have been obtained from remelting hardening of low carbon structural steel (0.14% C; 1% Cr; 4% Ni) after carburizing to a depth of 1 mm (Fig. 3.38): a 25% increase in hardness and a doubling of layer depth [60]. Laser remelting has also been applied to titanium after prior glow discharge nitriding [172].

Laser remelt hardening tests have also been conducted on carbon and low alloy steels, containing up to 0.2% C, carburized to a level of 0.7 to 0.9% C and coated with a TiN layer, thereby obtaining good adhesion and a 0.5 mm layer hardened to 880 to 900 HV1 [173].

After remelt hardening, carbon steel grades 1020 and 1045 exhibit a lowering of the fatigue limit, quite opposite to the effect after remelt-free hardening.

Remelting of pure nickel [174] and aluminum alloys containing silicon, titanium, manganese, nickel and iron [175] has also been researched, yielding an improvement of abrasive wear resistance.

Remelt hardening causes insignificant deterioration of surface roughness. Where roughness with $Ra<10$ to 15 μm is required, laser treatment should be carried out prior to grinding [5, 11, 29, 51].

Glazing. Laser glazing processes (vitrification, amorphization) are among the least researched. Glazing, i.e. obtaining of amorphous layers, requires cooling rates which are greater by an order of magnitude than those typically obtained by continuous CO_2 lasers. For that reason, Nd-YAG or excimer pulse lasers are often used, hence the obtained amorphous layers are very thin (typically: 20 to 40 μm) and the surface relief is not uniform. The application of special cooling methods makes utilization of continuous CO_2 lasers

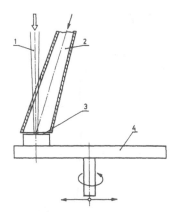

Fig. 3.39 Schematic representation of continuous operation laser glazing: *1* - laser beam; *2* - introduction of protective gas; *3* - glazed material; *4* - rotating and sliding stage (ensuring required covering by laser paths). (From *Grigoryantz, A.G., and Safonov A.N.* [29]. With permission.)

possible (Fig. 3.39) which, in turn, allows better surface and a greater depth of the vitrified metal layer. The power densities employed here are 10^6 W/cm^2 and treatment rates of 1 m/s and higher [29].

By ensuring very high cooling rates, the viscosity of the molten metal may be caused to rise sufficiently high to prevent the formation of crystallization nuclei. The alloy does not crystallize but solidifies in a disordered form, thus becoming amorphous with properties of a glassy mass.

Not all alloys exhibit a tendency to amorphization [176]. Those that do must have a certain chemical composition and exhibit an amorphization rate, related to that composition. For example, for the PdSiCu alloy this rate is approximately 100 K/s, while for pure germanium or nickel, it is approximately 10^{10} K/s. The amorphization rate is determined by a correlation between viscosity and temperature, the situation of the crystallization range, the rate of crystal nucleus growth and other factors [29].

The obtaining of the amorphous or fine-crystalline state is possible in the following cases [29]:

1) Alloys with compositions close to eutectic with a deep eutectic, composed of:

– metal - non-metal. These are formed by metals of the group I of the periodic table (Ag, Au) and group VII (Fe, Ni, Co, Pd, Pt, Rh) with non-metals such as Si, Ge, P and C, the content of which in the eutectic is usually 20 to 25%. Such alloys become amorphous with the application of relatively small cooling rates: 10^5 to 10^7 K/s;

– metal - rare earth metal. These are formed by metals with normal valences (Ag, Au, Cu, Al, In, Sn) with rare earth metals (La, Ce, Nd, Y, Gd), the content of which reaches 20%; the amorphous state is formed easily;

– metals - refractory metals, e.g., Fe, Cu, Co, Ni with Ti, Zr, Nb, Ta. Amorphization takes place when cooling rates exceed 10^7 K/s.

2) Hypereutectic alloys:
– on a base of tellurium with Ag, Ga, Cu, In; they do not form a low-melting eutectic and are characterized by non-metallic bonds;
– on a base of lead and tin: Pb-Sn, Pb-Si, Pb-Ag, Pb-Au, Sn-Cu; the eutectic is situated near the low-melting component; amorphization is possible only when the cooling rate is greater than 10^8 K/s.

Amorphous alloys exhibit high strength and hardness. As an example, FeBSi alloys, prior to amorphization, have a hardness of 3500 to 5800 MPa and 7100 to 11700 MPa after amorphization, while retaining significant ductility. Although brittle under tensile loading, they allow substantial deformation - up to 50% - under compressive and bending stresses. At low temperatures, their strength drops substantially and the alloys exhibit a very good resistance to corrosion. Some alloys also exhibit special magnetic properties. Amorphous alloys without phosphorus exhibit high resistance to radiation. In some laser-amorphized alloys a crack network appears.

To date, many different alloys have been successfully vitrified by a pulse laser. Among these are FeCSn [177], CuZr, NiNb, FeBSi ($Fe_{80}B_{16}Si_4$, $Fe_{77}B_{19}Si_4$), FePSi ($Fe_{83}P_{13}Si_4$, $Fe_{79}P_{17}Si_4$), $Fe_{72}B_{14}C_{10}Si_4$, $Fe_{73}P_{12}C_{11}Si_4$ [29], FeB (e.g., in triple laser-glazed $Fe_{83}B_{17}$ alloy, a tri-zone structure was obtained: homogenous crystallites, heterogeneous crystallites and metal glass at the surface [178]), $Pd_{77}Si_{17}Cu_6$, $Fe_{74.5}Cr_{4.5}P_{7.8}C_{11}Si_{2.1}$, $Fe_{81}B_{13.5}Si_{3.5}C_2$ [50].

There are also known methods of reglazing, consisting of laser remelting of e.g., strip ready-made from metallic glass [50].

A big future is predicted for alloying of steel and cast iron with elements which enhance their tendency to amorphize (e.g., with boron or silicon). The process may be carried out in two stages. The first pass of the laser beam is for alloying and the second pass (with different parameters) for glazing [29].

Laser glazing is seldom applied but a significant development is foreseen in this area, mainly for raising resistance to tribological wear, including primarily magnetic elements, as well as components of assemblies and instruments working in conditions of severe corrosion hazard.

Densifying (healing). Densifying consists of remelting of the surface layer or a deposited coating (or coating and superficial layer - Fig. 3.40) to a certain depth in order to obtain a material of greater density which usually is associated with a decrease of porosity, but also involves the liquidation of surface defects in the form of scratches, delaminations, cracks and open pores [179]. This is accompanied by homogenization of microstructure which is important in the case of material prior subjected to plastic deformation. It is also accompanied by a change of residual stresses and, in the case of coatings, obtaining of a better metallic bond between coating and substrate than by spraying alone [180]. In densification processes, relatively low power densities and low treatment rates are applied. This allows gases present in the melted material to escape to the surface of the laser-melted pool. A usable effect of densification is an increase of hardness and improvement of surface smoothness, an enhancement of

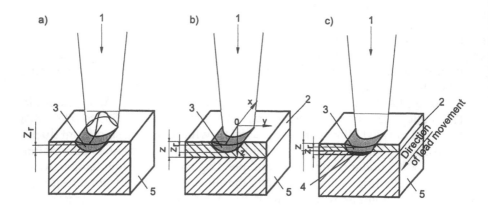

Fig. 3.40 Schematic of laser densifying of: a) superficial layer; b) surface layer of coating; c) coating and superficial layer of substrate; *1* - laser beam; *2* - thermally sprayed porous coating of thickness z; *3* - remelted zone z_r; *4* - zone of metallurgical fusion; *5* - substrate.

tribological, anti-corrosion and decorative properties. In special cases there is also an improvement of vacuum properties, e.g. degassing of the surface layer. After laser densification, the rate of material gassing is 2 to 13 times less than without laser treatment [29].

Laser densification is applied to seal surfaces of:

– Sintered steel containing carbides. As an example, in the 5XGWM* grade with a 10 to 20% titanium carbide content, such treatment, besides reducing porosity, causes increase in hardness, strength and abrasive wear resistance, in comparison with corresponding properties of sintered P/M without laser treatment [179].

– Thermally sprayed coatings [180, 181] by the arc method [182], by plasma [183–187] or by detonation [188]. Repair arc spraying of crankshafts of high compression engines by alloys such as 110MnCrCoTi8*, HNV3, 30MnCrTi5* and 10MnSi6*, laser densified by surface remelting after grinding off of the surface, allowed a 2- to 4-fold reduction of wear in comparison with journals not subjected to repair treatment [182]. Laser densification of NiCrBSi coatings (Fig. 3.41), ZrO_2 and Ti also ensures good protection of the substrate, beginning from a thickness of 10 µm of the densified coating [183]. Plasma densification is also applied to the following types of sprayed coatings: cobalt-based (CoCrWC), nickel (NiCrBSi, NiCrWC [180], NiCoCrAlY [184]), coatings of Cr, Co or alloyed NiAl, plasma sprayed on casting alloy substrates of nickel base [187] and oxide base [181], e.g., Al_2O_3, TiO_2, ZrO_2 [185], alloy CoCrWC, NiCr8Si, NiCrWC [180], metal ceramic [186], e.g., B_4C-Al (up to 50%) of up to 200 µm thickness and a hardness higher by 26 to 43% after remelting, as compared to the not remelted coating [189]. Laser densification can also be successfully applied to detonation coatings of relatively low porosity, made from heat-resistant alloys, e.g., NiCr, NiAl, NiCrAl, as well as composites of Ni, Cr, B, Si, WC-Co

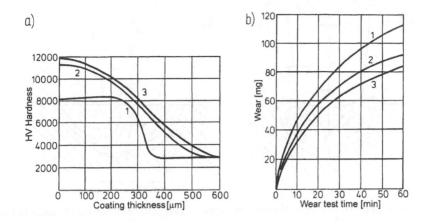

Fig. 3.41 Effect of laser remelting on properties of plasma sprayed coatings from mixture of 80% powdered GSR-3 material (composition: Ni, Cr, B, Si) and 20% powdered TiC: a) hardness distribution; b) abrasive wear resistance; 1 - plasma sprayed coating; 2 - plasma sprayed and laser remelted coating; 3 - detonation sprayed coating.

and WC-Ni [188]. For a significant improvement of corrosion resistance it is sufficient to remelt the thermally sprayed coating to a depth equal to 20 to 50% of its thickness [50].

– Electrodeposited coatings (primarily to remove scratches and cracks) [11].

Smoothing. Laser smoothing of surfaces is carried out with the use of the same range of process parameters as remelt hardening. In the microstructure of the material subjected to smoothing, the same phase and structural transformations take place as in remelt hardening. These transformations are not, however, the main technological aim of the process, but rather the reduction of surface roughness and a change in the profile of surface unevenness. They occur under the influence of hydrodynamic mixing of the molten material, due to thermocapillary forces which bring about convection. In the pool of the molten material, a high temperature gradient is formed, along with related gradients of surface tension. This causes a rapid circulation of the liquid, but only limited to a zone thinner than the entire melted layer (Fig. 3.42). For example, the rate of circulation in molten iron may even reach 150 mm/s [29]. Pressure changes within the molten pool are compensated by a change of shape of the pool surface.

The strongest effect on smoothing is that exhibited by power density. Within the range $5 \cdot 10^3$ to $5 \cdot 10^4$ W/cm^2 it is possible to obtain a surface which is smoother than after machining [29]. The recommended practice calls for low power densities and big diameters of the laser spot, in order to remelt the surface layer only to a shallow depth. This is because in such conditions convection whirlpools are broken down into a series of smaller vortexes, conducive to a smoother surface. Relatively low treatment rates also lead to the obtaining of a more favorable surface profile: lower asperities and greater asperity peak radius.

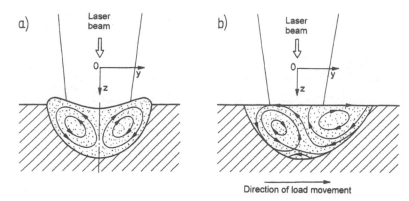

Fig. 3.42 Schematic of convection movements in laser pool: a) in section perpendicular $z0y$ to direction of load (or beam); b) in section $z0x$, parallel to direction of load (or beam). (From *Liu, J.* [169]. With permission.)

It is also possible to affect the asperity profile by the introduction into the molten pool of special additives which reduce the speed and direction of circulation. For example, if sulfur is deposited on the steel surface, the direction of circulation will be opposite to that observed after depositing carbon. An addition of a mixture of sulfur and carbon allows the obtaining of small whirlpools in the molten phase and, by the same token, high surface smoothness [29]. This method, however, causes not only changes of surface geometry parameters but also its chemical composition, thus its properties. Such a method is nothing less than **smoothing by alloying**.

Laser smoothing of steel and alloy surface may be applied as an independent treatment or it may be the finishing operation after remelt hardening, laser densification, surface remelting, etc.

3.5.2.2 Alloying

The alloying[1] of a solid depends - in accordance with the laws of diffusion - mainly on the gradient of temperature, gradient of concentration and on time of the diffusion process. Since in the action of a laser beam on the material surface, very high temperature gradients are formed in very short times (of diffusion) with concentrations of diffusing elements usually typical of traditional methods, the depth of diffusion alloying is very small. Diffusion is a slow process and in times of the order of milliseconds it is practically too small to be noticed. For example, during a laser pulse of 4 ms and a power density which does not cause remelting, the depth of diffusion is only approximately 1 μm [29]. For this reason, in laser heating, the only way of

---

[1] In an attempt to unify terminology, the term *alloying* as used here is meant to encompass all methods of introduction of alloying elements into the material being alloyed by: mechanical mixing of both materials in the powdered form, followed by their sintering (**mechanical alloying**), hydrodynamic mixing of both materials, of which at least one is in the liquid state (**remelt alloying: metallurgical or laser**); diffusion of one material into the other in the solid state (**diffusion alloying**), traditionally known as **diffusion saturation**.

introducing an alloying additive into the surface layer of the material is by remelting with hydrodynamic mixing before solidification.

Laser Surface Alloying (LSA) consists of simultaneous melting and mixing of the alloying and the alloyed (substrate) material. The action and pressure of the laser beam cause both materials to melt; a pool of molten material is formed in which intensive mixing, due to convection and gravitational movements, forms a flash at the pool surface (Fig. 3.42). At the interface between solid (substrate) and liquid (alloy), a very thin diffusion zone appears, usually not exceeding 10 μm. Only in some rare applications do the alloying components diffuse to depths of 200 to 300 μm. This takes place by diffusion by narrow canals of the molten phase along solid grain boundaries and grain blocks or, in the case of displacement of atoms by dislocations, due to local deformations.

When the action of the laser beam ceases, the alloy thus formed solidifies while the substrate material in its direct vicinity becomes self-hardened. Structure, chemical composition, and physical as well as chemical properties of the alloy are different than those of the substrate or of the alloying material. First of all, the layer of the alloy does not, in principle, exhibit the characteristic layer structure, typical of diffusion processes. Due to convection mixing of the alloy, there are no transitions from phases with a higher concentration of the alloying element to phases with lower concentration. All phases in the remelted layer are uniformly distributed along its entire depth. An exception to this is the earlier mentioned very thin diffusion zone at the interface between solid and liquid. The alloy layer is bound metallurgically with the substrate.

The alloy layer, rich in alloying components, usually exhibits a higher hardness than that of the substrate, a higher fatigue strength, better tribological and corrosion properties, but at the same time with poorer smoothness of the surface in comparison with the condition prior to alloying. These properties depend to a very high degree on the uniformity of mixing of the alloy in the molten phase, which, in turn, depends on the intensity of convection exchange of mass in that zone [193].

Depending on the method of introducing the alloying additive to the molten pool, we distinguish *remelting* and *fusion* (Fig. 3.43).

Remelting. Remelting is a two-stage process, consisting of prior deposition of the alloying material on the substrate and subsequent remelting it together with the surface layer of the substrate material (Fig. 3.43a). Usually, the thickness of the remelted surface layer is comparable with the thickness of the deposited alloying material, i.e., the mixing coefficient k_p is approximately 0.5. The process of remelting begins from the alloying coating and propagates by convection and conductivity into the surface layer of the substrate. The alloying material dissolves completely in the substrate material.

Alloying is accomplished with the employment of power densities in the range of $5 \cdot 10^4$ to 10^6 W/cm^2, which are greater than those used in hardening, and exposure times from tenth to thousandth parts of a second

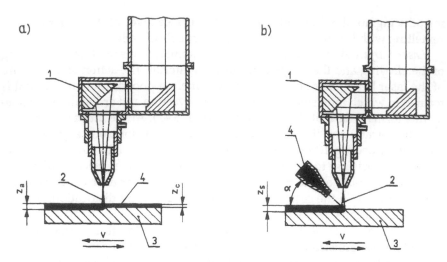

Fig. 3.43 Schematic of laser alloying: a) remelting; b) fusion; *1* - laser objective lens; *2* - laser beam; *3* - alloyed material; *4* - alloying material (remelted or fused); z_a - thickness of alloyed layer; z_c - thickness of alloying coating. (From *Thiemann, K.G., et al.* [191]. With permission.)

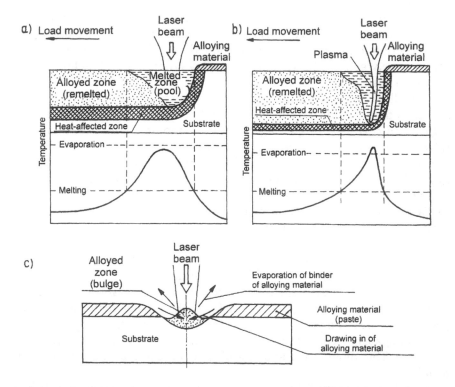

Fig. 3.44 Schematic of laser alloying: a) without plasma formation; b) with plasma formation; c) shaping of laser path in the form of a ridge. (Fig. a and b - from *Gasser, A., et al.* [192], Fig. c - from *Grigoryantz A.G., and Safonov, A.N.* [29]. With permission.)

[29, 192]. The greater the power density, the bigger the depth of remelting. High power densities may lead to the formation of plasma and vaporization of material (Fig. 3.44).

In principle, remelting is always accompanied by the occurrence of plasma and vaporization of material On the one hand plasma screens the surface from further laser heating; on the other, however, it interacts with the surface of the melted metal pool exerting pressure and causing the displacement of components of the molten material. In the pool, precisely at the site of penetration of the laser beam into the material, a conical pit is formed. The surface of this funnel is acted upon by the hydrostatic pressure of the liquid from below and by vapour pressure from above. Between the two, an unstable equilibrium is formed, constantly disturbed by, among other factors, relative movement of the beam and the treated object. The pit moves toward as yet unmelted material (in a direction opposite to that of the object relative to the beam). Behind the displaced pit, vapour pressure causes a filling in of the discontinuity. In consequence, on the molten surface there appears a characteristic waviness, similar to that which is typical of a weld seam.

Because of the above-described two-directional interaction of plasma on the molten pool, different methods of slowing down this action on the molten material are used. Among these are blowing away of the plasma cloud by a neutral gas, heated in order not to impair the energy effect. There are also methods of enhancing the action of plasma, e.g., by blowing away the plasma cloud but with the simultaneous recycling of the reflected laser radiation back to the treatment zone by a set of plane mirrors or a mirror dome. Naturally, the flow of protective gas always protects the optics of the laser head against the deposition of gases, vapors and solid particles, created during treatment.

Alloying is accomplished with the application of one or several passes of the laser beam. The alloying material is deposited on the substrate by [40, 41, 48, 53, 54]: painting, spraying of suspensions, covering by adhesive powders or pastes (containing P/M ferrous alloys of alloying metals, boron carbides, tungsten and titanium carbides and borax), thermal spraying (flame, arc, plasma and detonation), vapour deposition, electrodeposition, thin foil, plates, rods or wires, or by E.D.M. The thickness of the deposited coating ranges from several to more than 100 μm.

In the case of P/M materials, the efficiency of laser heating is greater than for solid materials, on account of the higher coefficient of absorption of laser radiation through powder, usually approximately 0.6. A relatively significant role is played by substrate surface roughness. Its growth causes an improvement of adherence of the powder mass to the substrate and thus an improvement in the passage of alloying components into the molten pool, attributed to rapid melting of asperities.

Alloying components can also be introduced to the substrate from the melt (Fig. 3.45). The alloyed part is placed in a liquid; the laser beam reaches the surface through a vapour/gas channel formed in the liquid

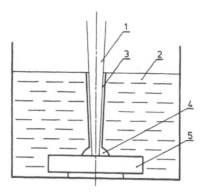

Fig. 3.45 Schematic showing remelt alloying in a liquid: *1* - laser beam; *2* - liquid containing alloying components; *3* - vapor/gas channel; *4* - hemispherical bell cap; *5* - alloyed object. (From *Grigoryantz, A.G., and Safonov, A.N.* [29]. With permission.)

and heats that surface. At the surface, the channel broadens, forming a hemispherical space, filled by vapors of the alloying liquid which reflects back the radiation initially reflected from the melted pool of the material. The so-called *hydrolaser alloying* can be applied in carburizing and nitriding [29].

Alloying from the gas phase surrounding the alloyed material is possible only if very high pressure, of the order of several tens of atmospheres, or more is ensured [29].

In all cases the parameters of the process must be selected to strictly match the type of alloying additive and the required depth of alloying. With an increase in power density or energy density in a pulse, as well as in treatment rate (feed rate of the load relative to the laser beam) and pulse duration, the dimensions of the alloying zones grow while the concentration of the alloying element drops. In turn, with a rise of thickness of the alloying element layer, the dimensions of the alloyed zones decline while the content of the alloying additives rises. Each thickness of the alloying element layer is characterized by certain critical parameters which cause the energy transmitted by the laser beam to be used only for melting the surface layer, without alloying. It should be noted that when beam paths overlap, hardness decrease in the reheated zones is significantly lower than in similar cases in remelt hardening [29].

The average thickness of remelt alloyed layers is in the range of 0.3 to 0.4 mm for pulse heating and 0.3 to 1.0 mm for continuous heating. After alloying the height of flashes is 20 to 100 μm, which usually calls for subsequent grinding.

Used most often as alloying components are [29]

1) **Non-metals**: carbon, nitrogen, silicon, and boron, while processes bear corresponding names such as:

– laser carburization, carried out with the utilization of solutions of graphite or soot in acetone, alcohol or other solvents (dusting off from surface covered by them) or in varnishes, e.g., bakelite together with activation additives, such as ammonium chloride or borax, or in liquid hydrocarbons or liquids containing carbon, e.g., hexane, acetone, toluol, carbon tetrachloride, mineral oil, etc. Carburization is applied in order to raise the hardness of plain carbon steels (4500 to 14000 MPa);

– laser nitriding: in pastes containing ammonium salts, urea $(NH_2)_2CO$, in gaseous or liquid nitrogen. Nitriding is applied to steels, as well as titanium, zirconium, hafnium or alloys of these metals, in order to increase hardness, resistance to tribological wear and to elevated temperatures;

– laser siliconizing: in pastes containing silicon powder or in liquids (e.g., in a suspension of silica gel H_2SiO_3) in order to enhance thermal, corrosion and tribological resistance of steel;

– laser boriding: in pastes constituting mixtures of boron powders, anhydrous boric acid B_2O_3, boron carbide B_4C, borax $Na_2B_4O_7 \cdot 10H_2O$, ferroboron with filler material, e.g., glue. This process is carried out in order to increase hardness and abrasive wear resistance of metals.

2) **Metals:** Co, Cr, Sn, Mn, Nb, Ni, Mo, W, Ta, V or their alloys, e.g., Cr-Mo-W, Ni-Nb. An unfavorable property of remelt alloying with metals is the formation of supersaturated solid solutions, significantly exceeding solubility in equilibrium conditions. The formation of intermetallic compounds is also possible. The utilization of metals and their alloys leads to changing of mechanical properties of ferrous, aluminum, titanium and copper alloys.

3) **Different compounds**, mainly carbides of refractory metals: TiC, NbC, VC, TaC, WC, Nb_2C, Ta_2C or alloys of carbides of these metals, deposited by thermal spraying and by electrodischarge, as well as in the form of pastes (powder + liquid glass, powder + silicate glue, etc.).

Alloying is applied to metals and alloys, mainly to steels and cast irons (Fig. 3.46 and 3.47) by single elements raising heat resistance, corrosion resistance and abrasion or erosion wear resistance. Among these are Mo, W; C, Cr, B, Mn, Ni, Co, Zn, Cd, Si, Al and composites of elements, e.g. B-C, B-Si, Co-W, Cr-Ti, Fe-Cr, C-Cr-Mn, Al-Cr-C-W and alloys, e.g., Cr_2C_3, Cr_3C_2-NiCr$_2$, WC-Co, oxides Cr_2O_3, TiO_2, B_2O_3 [29, 192, 195, 201], all allowing the obtaining of a better set of properties than by alloying with only single elements.

Alloying is most often applied to different types of steels [194-196]:

– Structural carbon, e.g., 1045 [197-200] and low alloy grades with carbon, chromium, molybdenum [198-200], P/M carbides, e.g., WC, TiC or mixtures of WC-Co [197], chromium pastes [201], boron, deposited electrolytically or in the form of paste. As an example, the microhardness of carbon (0.2% C) steel is increased by alloying from 2.5 GPa to 8.5 GPa, with a layer thickness of 0.4 mm [202].

– Tool steels: with boron [202], boron carbide or its composites with chromium (e.g., 75% B_4C + 25% Cr [208]), by different composites of carbides [209], by tungsten, tungsten carbide and titanium carbide [210], by chromium or vanadium boride [211], by vanadium carbide [212] or by Mo-Cr-B-Si-Ni composites [213].

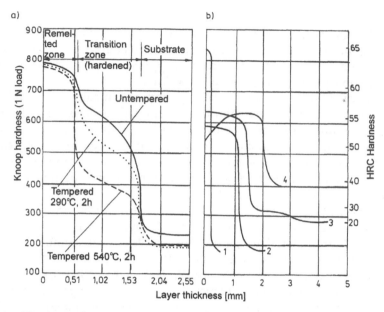

Fig. 3.46 Hardness distributions across sections of laser alloyed layers: a) gray cast iron alloyed with chromium; b) structural steels alloyed with different elements; *1* - 1018 steel, alloyed with C, Cr and Mo; *2* - 1018 steel, alloyed with Al, C, Cr and W; *3* - 4815 steel, alloyed with C, Cr; *4* - 4.320 steel, alloyed with Cr.

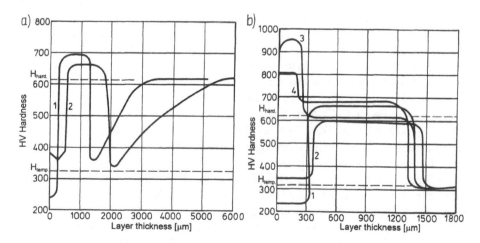

Fig. 3.47 Hardness distributions across sections of laser alloyed 3140 steel: a) hardened (after alloying); b) hardened and tempered (after alloying); alloying with: *1* - nickel; *2* - chromium; *3* - cobalt; *4* - not alloyed, hardened from the melt. (From *Artamonova I.V., et al.* [206]. With permission.)

Naturally, alloying is applied only or mainly to machine components or tools subjected to particularly severe conditions, e.g., cutting edges of stamping dies or cutting tools [207].

Remelt alloying is often applied to cast irons [214], particularly gray [215-217] and high strength [218]. These are alloyed with the use of Fe-Si powder packs [215], carbon (up to a content of 22% C) in order to enhance resistance to erosion wear[216], with boron [218], silicon, nickel and its alloys [217] and with chromium [214].

Good results have been obtained by remelt alloying of aluminum alloys [219], including Al-Si alloys [220, 221]. For example, the Al25* grade, alloyed with the application of pastes based on powders of NiCr, FeCuB, or NiCrMo, exhibits a significant increase in hardness and resistance to abrasive wear [219], in a way similar to the D16* grade, alloyed by carbides, e.g., B_4C, Cr_3C_2, B_4C+Cr, $B_4C+Cr_3C_2$, or by a composite $B_4C+Cr_2O_3+CaF_2$ [224]. Powder pack alloying of Al-Si alloys by nickel, chromium, iron, silicon and carbon clearly raises their heat resistance [221]. Similarly, alloying by Fe, Fe+B, Fe+Cu, Fe+Cu+B powders, predeposited by painting, in a mixture with zapon varnish significantly raises hardness, although the distribution of alloys in the remelted zone is not homogenous [220].

Alloying of titanium by remelting of electroplated chromium, manganese, iron or nickel coatings causes a rise in hardness of the surface layer from below 1500 MPa for the titanium substrate to 5500 to 10000 MPa for the alloyed layer [222]. The hardness of a laser hardened WT3-1* titanium alloy rises, relative to the initial hardness value by a factor of 1.1 to 1.6. This may be further enhanced by alloying with powders of Al_2O_3, FeCr, a-BN and others [223] or by borides and carbides of transition metals (Mo_2C, Mo_2B_5, WC, W_2B_5, VB_2, B_4C, $B_4C +CaF_2$) together with chromium [224].

Research is currently being conducted to study the strengthening of low-carbon overlays by alloying, e.g., by chromium, predeposited by electroplating [225].

Fusion. Fusion is a single stage process. It involves creating a pool of molten substrate material with the laser beam and the introduction into this pool of the alloying material in the form of solid particles (powder or paste) completely or partially soluble in the substrate, or in the gaseous form (see Fig. 3.43b). Fusion is accomplished only with the aid of continuous operation lasers because the alloying material may be introduced to the molten zone only while laser heating is on, and not during lapse between pulses.

The aim of fusion is the same as that of remelting, i.e., the obtaining of a surface layer in the form of an alloy or a coating with properties which are better than those of either the alloyed or the alloying material.

In the case of **powder fusion alloying**, the process of melting of both materials is simultaneous: solid particles of the alloying material are heated and may melt already at the moment of entering the site of the laser beam. Not completely melted, they drop into the pool of the simultaneously melting alloyed material.

The powder added may be a homogenous material or it may constitute a mixture of powders of several materials. The powder should be introduced in a stream of protective atmosphere in order to avoid oxidation (Fig. 3.48a). However, the gas may cause porosity of the alloyed layer.

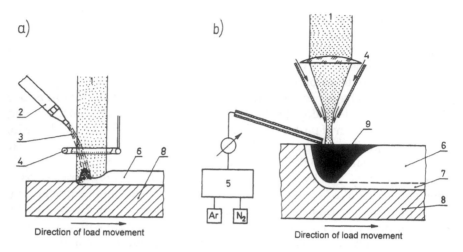

Fig. 3.48 Schematic showing the process of fusion: a) powder; b) gas: *1* - laser beam; *2* - powder container; *3* - alloying powder; *4* - blown in protective gas (e.g. argon or nitrogen); *5* - gases alloying melted pool; *6* - layer of alloyed material; *7* - heat-affected zone; *8* - alloyed material (substrate); *9* - melted pool. (Fig. a - from *Feinle, P., and Nowak, G.* [243], Fig. b - from *Gasser, A., et al.* [238]. With permission.)

Powder fusion alloying is accomplished with the use of fine-grain powders and high uniformity of grain size. The used powders are those of: silicon [228], aluminum [229], titanium carbide [230], tungsten carbide [231], tungsten carbide with cobalt [231], boron nitride [232] and stellites [233-236]. The alloyed materials are steels, especially tool steels, as well as titanium.

During alloying of titanium by silicon, intermetallic phases are formed, such as Ti_5Si_3, which raise hardness (up to 600 HV with a content up to 20% Si) [228]. Alloying of titanium with aluminum, with the formation of titanium aluminides Ti_3Al, $TiAl$ and $TiAl_3$, enhances resistance to oxidation [229].

Tool steels are alloyed by titanium carbides [230], tungsten carbides [231], boron nitrides [232] and tooling stellites [233, 234], as well as by hastellites (50%WC+50%NiCrSiB) [236]. Stellites which raise heat and corrosion resistance are used for alloying austenitic stainless steels [235] and plain carbon steels [233]. In carbon steels alloyed by stellites, tensile stresses prevail in the alloyed layer [233]. The thickness of the stellite alloyed layer is usually 0.3 to 1.0 mm [234].

In the case of **gas fusion alloying**, the alloying gas is blown into the pool of molten substrate material and, by entering into a direct or indirect chemical reaction with it, forms a coating, qualitatively different from it and constituting a chemical compound of a constituent of the substrate and a constituent of the gas (Fig. 3.48b). To this time, the predominant subject of research in this area was laser beam nitriding of titanium and its alloys with nitrogen, according to the reaction:

$$\text{Ti} + \text{''N}_2 \Leftrightarrow \text{TiN} - 336 \text{ kJ/mole} \qquad (3.17)$$

The titanium nitrided (TiN) formed features a hardness of over 2000 HV [237–239]. If the alloyed grade is TiAl6V4, the titanium nitrided layer obtained has a thickness of 50 to 500 μm [237, 238]; if the alloyed grade is titanium the thickness may reach 1 mm [239]. Some problems may, in this case, be caused by oxidation and the formation of TiO_2 oxides [239]. This method may be used in alloying of plasma sprayed coatings made of titanium and its alloys on steel.

Laser carburizing of low carbon steels by carbon formed from the decomposition of pure propane or propane mixed with neutral gases, e.g. argon, neon or helium, has also been studied. The carburized layers thus obtained exhibit thicknesses up to several millimeters [240]. Carburizing gases have also been utilized in attempts to form titanium carbide (TiC) on titanium and its alloys [29, 192].

3.5.2.3 Cladding

Cladding, also known as laser **plating** or **hardfacing**, is accomplished with process parameters similar to those in alloying and consists of melting a thick layer of the plating material and surface melting of a very thin layer of the substrate material. In hardfacing, the coefficient of mixing k_p is usually 0.1. The aim of hardfacing is not the mixing of the plating material with that of the substrate but melting of the deposited coating material or its deposition and remelting in order to obtain better resistance to erosion, corrosion, abrasion and to other service hazards than that of the substrate material. The coating material may be soluble or insoluble in the substrate [241–252]. The intermediate layer formed between the substrate and the plating is usually of a metallurgical character, in which case it causes strong bonding of the latter to the substrate [11]. In a manner similar to alloying, the plating material may be deposited onto the substrate by the following means:

– A **two-stage** process prior to laser treatment. The remelting process progresses from the top which favors the formation of defects, such as blisters and incomplete melting near the substrate. The formation of these defects is also favored by the binding material (most often: zapon varnish, epoxy resin, fats, synthetic materials, liquid glass, silicon glue, shoemaker's glue, solutions of borax and acetone, isopropyl alcohol, alcohol solutions of resin, nitrocellulose and indiarubber glue, as well as crazy glue) [27]. Such materials, added to the powders in order to ensure better adhesion to the substrate, evaporate during laser heating and periodically screen the melted pool, resulting in non-homogenous remelting of the substrate. Since this process requires the prior deposition of the plating material onto the substrate in the form of a layer of powder, foil, thin sheet or thick coating) it can, for simplicity, be called a **coating** process. Each form of plating material requires different processing parameters, mainly input power density and hardfacing rate. Parameters which are higher than optimum cause excessively deep remelting of the substrate while lower parameters may cause a droplet form of the plated layer. In the case of powdered or paste materials there usually occurs a burnout of the binding material through the sides of the plating, often causing the necessity of redeposition of the powder or paste when carrying out two or more plating operations.

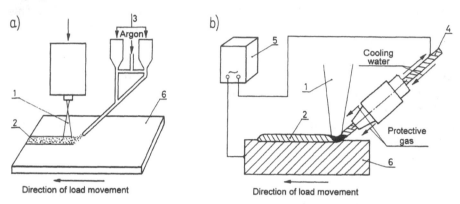

Fig. 3.49 Schematic of cladding: a) two-component, powder; b) wire: *1* - laser beam; *2* - laser path on clad material; *3* - containers with components of cladding powder; *4* - cladding wire; *5* - source of direct resistance heating current; *6* - substrate. (Fig. a - from *Abbas, G., et al.* [244], Fig. b - from *Hinse-Stern, A., et al.* [252]. With permission.)

– In a **single-stage** process, during laser treatment (Fig. 3.49), using the coating material in the form of powder or rod (strip). The powder is fed by pouring (by inertia or by vibratory mechanism) or is transported with the aid of a gas, non-neutral (e.g., air) or neutral (e.g., nitrogen, helium or argon) to the coating material.

Similarly as in fusion, it is possible to utilize active gas for the delivery of powder because an exothermic reaction with this gas may intensify the plating process. The rod is fed mechanically in a continuous manner. In the case of plating from a powder form, the process is termed **injection plating**; in the case of remelting of a rod, the term used is **overlay plating**.

A powder or a powder mixture (Fig. 3.49a) blown into the zone of the laser beam is melted and in the melted form falls onto the surface of the substrate. The plating layer therefore forms from the bottom up, as a result of solidification of the melted material at the surface of the substrate material. This type of hardfacing requires the use of a nozzle, coupled with the laser head and feeding the powder in a direction with or counter to the movement of the load. It also requires the use of powders of appropriate granulation (Fig. 3.50). The most often used powder granulation is 40 to 80 µm, while the rate of powder feed does not exceed 1 g/s. Powders should be dried before application [11]. Powder plating by laser is characterized by low energy consumption, in comparison with other

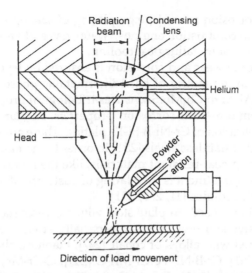

Radiation / Condensing
beam / lens

Helium

Head

Powder and argon

Direction of load movement

Fig. 3.50 Schematic of laser head coupled to powder dispenser. (From *Burakowski, T., et al.* [11], *Burakowski, T., and Straus, J.* [50]. With permission.)

laser methods. The laying of a single hardface of 1 mm thickness by the method of injection plating requires a unit energy rate of 30 to 50 J/mm², by remelting a prior deposited powder paste of 60 to 90 J/mm², while by remelting of a prior plasma sprayed coating, 180 to 350 J/mm² [29].

A rod of the plating material may be melted onto a cold substrate or one that is preheated (as in typical hardfacing) or simultaneously heated in order to avoid the formation of excessively big residual stresses and, consequently, of cracks. In the second case, the substrate may be resistance heated directly (Fig. 3.49b) [252], by arc or by another method.

Refinement of structure, dissolution of hard carbide phases and the formation of supersaturated solid solutions, which is possible to obtain in laser hardfacing, allow a significant improvement of service properties of substrate materials. First of all, there is a substantial increase in wear resistance, especially under high unit loads. Moreover, very high strength of the metallurgical bond between the plating and the substrate is obtained. It is equal to the tensile strength of the weaker of the two materials. A significant increase is also noted of resistance to the action of elevated temperatures. A serious fault of laser platings is their tendency to the formation of cracks. This can be clearly prevented by preheating the substrate to 300 to 400°C and by an appropriate choice of composition of the plating materials. Laser platings can also be smoothed by the laser beam [27].

Hardfacing processes have found application in two chief groups of materials: corrosion resistant, including those resistant at high temperatures, and heavy duty, wear resistant, mainly tooling alloys.

The unique properties of plating materials make them predestined to applications in conditions of heavy loads and high temperatures, as well as erosion and corrosion hazards, e.g., plating of sealing surfaces of valve seats and valves in combustion engines, water, gas and vapor separators, as well as components of metallurgical tooling.

Joint resistance to wear and corrosion is ensured by layers of Co-Cr-Mo-Si. The presence in the matrix of hard intermetallic phases of composition ranging from CoMoSi to Co_3Mo_2Si ensures tribological properties, while the presence of chromium - anti-corrosion properties. Similar resistance is obtained by the application of Cr-Ni-B-Si-Fe plating. There are known examples of plating austenitic stainless steels [251] in order to increase their sear resistance and of plating heat-resistant materials, like the Nimonic alloy. A typical laser hardfacing treatment is the plating of austenitic stainless steels with tungsten or cobalt carbides [11, 250, 251].

Laser hardfacing is used to plate steel with creep-resistant layers that are resistant to abrasive and erosive wear, especially at elevated temperatures. Steel may be plated with alloys of cobalt [247], titanium, alloys and/or mixtures of: Cr-Ni [251], Cr-B-Ni, Fe-Cr-Mn-C [249], C-Cr-Mn, C-CrW, Mo-Cr-CrC-Ni-Si [29], Mo-Ni [243], $TiC-Al_2O_3$-Al, TiC- Al_2O_3-B_4C-Al, aluminum, stellite [241], Hastelloy [251], carbides WC, TiC, B_4C [246], SiC [244], nitrides, including BN [245], oxides of chromium and aluminum, etc. Alloys of cobalt may be plated with alloys of nickel, yielding high temperature erosion-resistant so-called superalloys. Titanium alloys may be plated with boron nitrides [245], while Al-Si alloys with silicon. Aluminum and copper may be hardfaced with a mixture composed of 91% ZrO_2-9% Y_2O_3 or ZrO_2-CaO [242]. A mixture often used for laser rebuilding of worn surfaces is Cr-Ni-B-Fe, infrequently with the addition of C and Si. It is predicted that practical applications will be found for laser glazed SiO_2 platings, deposited on creep-resistant alloy substrates, used for heating elements, working at 900 to 1000°C in strongly oxidizing, carburizing or sulfiding environments, e.g., Incoloy 800H.

Coating composites of hard and refractory materials form mixtures of powders of these materials with powders of the plated steel or powders of iron, used as bonding. For example, in the case of the WC+Fe composite (or WC+Co, WC+NiCr), the rapid plating process, rendering diffusion transformation WC⇔Fe impossible, causes carbides to retain their hardness of approximately 11000 MPa which is equal to maximum achievable hardness of tungsten carbides present in the matrix of tool steels containing tungsten after conventional heat treatment.

Tool steels, especially those for forming tools, are laser plated by stellite in order to protect against abrasive wear. The creep resistant alloy Nimonic 8A (over 70% Ni, 20% Cr, additives of Al, Co) used for gas turbine blades, laser-plated with stellite, exhibits an almost 100-fold increase in resistance to abrasive wear. It is noteworthy that laser stellite plating is more favorable (higher hardness, finer structure) than weld stellite overlaying (by the TIG and gas methods). The Cr_2O_3 oxide, laser deposited onto austenitic steel (e.g., 321 grade), doubles its creep resistance. In the automotive industry, valve seats and flanges have been laser plated for an appreciable time.

The thickness of hardface layers is greater than that of alloyed layers and may reach several millimeters. The width of a single plated seam (with the laser beam scanning the surface with a frequency of 10 to 300 Hz, transversely to the movement of the load) may exceed 10 mm [24]. The productivity of plating is from several to over 100 mm/s [50]. The quality of plated layers, in terms of tightness, bonding to substrate and hardness, is better than that of thermally sprayed (including plasma)coatings [29, 195]. Laser hardfacing may cause a lowering of the fatigue limit for structural steels, depending on the type of plating. The more refractory the plating material, the greater the fatigue limit drop.

In laser hardfacing, the problem of surface roughness appears to be greater than in other laser alloying. Usually, surface roughness rises with the melting point of the plating material. For that reason, often mixtures are used, most often powder mixtures, containing powders with a high and a low melting point, e.g. $TiC+Al_2O_3+Al+B_4C$, although this has a negative effect on the hardness of the plating [50, 51].

3.5.3 Evaporation techniques

Laser beam evaporation techniques are divided into three groups: pure evaporation, rapid evaporation, combined with detonation hardening, and ablation cleaning.

3.5.3.1 Pure evaporation

Pure evaporation consists of utilization of only the thermal action of the laser beam for relatively slow evaporation of the material which later - due to various physical and chemical phenomena - is deposited by itself or together with another material, e.g., gas, thereby forming a coating. Methods involving these phenomena are the so-called CVD (see Part II, Chapter 5) or PVD (see Part II, Chapter 6).

3.5.3.2 Detonation hardening

This method involves the utilization of the energy of a shock wave and of the mechanical pulse to mechanically strengthen the material. A certain analogy exists here to the explosive hardening of materials by a shock wave which utilizes the energy of the explosive material or an explosive gas mixture. The latter method has been used for a long time to harden objects or their selected sites. When the laser beam is used, it is possible to locally harden metallic materials but only on condition that they are heated by an ultra-fast pulse method [253].

During the action of a laser radiation pulse on the material, the duration of which significantly exceeds relaxation times which range from 10^{-9} to 10^{-11} s, thermal energy is conducted into the material. Its heating, melting and evaporation occur in accordance with traditional laws of energy transfer [29].

When, on the other hand, the duration of the pulse is of the order of 10^{-8} to 10^{-10} s, and power density is 10^{10} to 10^{12} W/cm² and higher, the time of action of the laser pulse on the material approaches that of the time of relaxation. For this reason, it is impossible to conduct the energy away into the core of the material. A very high power density in the microzone of the surface layer causes the material to transform into the plasma state. The heated plasma expands, very high pressures are set up, similarly to the case of an explosion, and a **shock wave** may be formed. But it is formed only when the duration of the pulse is shorter than the time of propagation of the shock wave in the microzone. During this time the pressure in the surface layer of the material is very high, while in the core of the material it drops rapidly. Non-homogeneity of propagation of pressure is the cause of formation of the shock wave [29].

For the majority of solids (of approximately 1 cm thickness) the time of propagation of shock waves is approximately 10^{-5} s. In this case, laser pulses of even 10^{-6} to 10^{-7} s may bring about the formation of a shock wave. When the power density is 10^9 W/cm², the duration of the pulse is 10^{-8} s and the coefficient of absorption of the incident radiation is 0.1, pressure values for CO_2 laser radiation are $0.212 \cdot 10^5$ MPa for copper, $0.121 \cdot 10^5$ MPa for aluminum and $0.060 \cdot 10^5$ MPa for beryllium. The action of radiation from a ruby laser is twice as strong [254]. Extreme values may be significantly higher and may reach values comparable to pressures occurring during the explosion of a hydrogen bomb [8].

Besides the shock wave formed with a pulse laser of 10^{-8} s duration and 10^8 W/cm² power density, there is an additional effect on the material of the mechanical impact, formed as the result of a rapid transition of the material into the plasma state and its evaporation into the surrounding space. A strongly ionized plasma cloud causes the screening of the treated object from the laser beam, especially from medium wave infrared radiation [11]. The plasma formed expands to the surface of the treated material, a crater is formed and the plasma runs up its walls to the outside, strengthening the material by its pressure [6, 14, 44, 53, 54].

The mean evaporation rate v_{avg} is

$$v_{avg} = \sqrt{\frac{8RT_n}{\pi M}} \qquad (3.18)$$

where: M - mean particle mass of material; R - gas constant; T_n - mean temperature of material after radiation by n laser pulses.

In such a case the material is subjected to the mechanical impact of unit energy equal to [29]:

$$E = \rho z_0 v_{avg} \qquad (3.19)$$

where: z_0 - thickness of layer of evaporated material (Fig. 3.51); ρ - density of treated material.

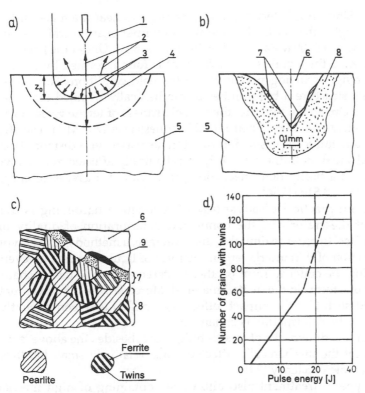

Fig. 3.51 Detonation hardening: a) schematic of processes taking place during detonation hardening; b) schematic of structure of superficial layer of low carbon steel following action by laser pulse of 35 J energy and duration of 10^{-8} s; c) microstructure of superficial layer of low carbon steel after laser hardening; d) dependence of number of grains in 1020 steel on energy of laser pulse; *1* - laser beam; *2* - evaporation; *3* - shock wave; *4* - mechanical impulse; *5* - hardened material (low carbon steel); *6* - crater; *7* - zones of remelting and heat affected; *8* - mechanically affected zone; *9* - remelted zone. (Fig. a, b and c - from *Grigoryantz, A.G., and Safonov, A.N.* [29]. With permission.)

A diagram of processes taking place in the detonation hardened material is shown in Fig. 3.51a. The combined action of the shock wave and the mechanical impact may cause changes in microstructure in the treated material, and even its cracking and even complete rupture [254].

As has already been mentioned, evaporation of a part of the material causes the formation of a conical crater. If the pulse energy is 10 to 35 J, the depth of the crater is 0.45 mm, while the ratio of diameter to depth is 1.7 to 1.9 [254].

On the side surface of the crater, in a layer of approximately 20 μm thickness, a remelted zone is formed but it does not necessarily have to be continuous (Fig. 3.51b). In the case of carbon steels it has a martensitic structure and is characterized by high microhardness, reaching 7600 MPa. Under the remelted zone there is the heat affected zone, also of approximately 20 μm thickness. Since in this technique it is possible to heat the material to temperatures higher than critical, in the sites of former pearlite grains, a martensitic structure of 6900 MPa microhardness is formed. Under the heat affected zone (HAZ), a thick zone (700 to 750 μm) is found, formed by mechanical strengthening, occurring practically in the cold condition. This zone contains in its structure a big number of mechanically twinned ferrite grains (Fig. 3.51c) Due to the twining, the microhardness of ferrite grains grows to approximately 2300 MPa. Their initial hardness was 1700 MPa. The effect of mechanically induced plastic deformation decreases with distance from the crater walls [29]. Both the size of the mechanically affected zone as well as that of the plastically deformed zone grow with an increase of energy of radiation (Fig. 3.51d) [254].

The main practical disadvantage of detonation hardening is the formation of the crater and the connected deterioration of quality of the object surface. In order to limit this disadvantage, a method is used involving the deposition of thin interlayers in the form of liquid or paste, or even thin foil causing the crater to form on these interlayers, while the mechanically affected zone forms in the treated material. Moreover, the interlayer material is selected such that the impact effect is amplified and, consequently, the effect of plastic deformation is enhanced.

Among the drawbacks of this technique are, besides the above, small dimensions of the mechanically affected zone, relatively small hardness increase and low efficiency [29].

This type of treatment also allows the obtaining of high mechanical strengthening in the case of soft materials, e.g., aluminum, or of very hard materials, including some steel grades. Although in practice, impact hardening has been applied to high strength aluminum alloys, it has, nevertheless not gone beyond the phase of experimentation [5, 29, 48, 53]. Laser-generated shock waves may also liquidate microcracks [8].

3.5.3.3 Ablation cleaning

Evaporation or sublimation of the solid under the influence of high power density laser beam radiation is called **laser ablation**. This phenomenon, discovered in the 1970s, has been utilized to clean surfaces. The mechanism of ablation cleaning is akin to that of detonation hardening, although it serves to strip coatings from substrates.

At the moment of formation of plasma, under the influence of laser radiation, the thermal energy of the plasma is, by way of convection and electron conduction, delivered to the material (both coating and substrate) which at that time are not reached by the electron beam. This is due to the fact that the beam is screened off by the plasma or there is an interruption between pulses. At the same time, due to evaporation and sublimation, the stream of energy and matter (evaporated particles of coatings) moves in the opposite direction, i.e., away from the material. The zone in which two opposite fluxes of energy meet forms the **ablation front** - a boundary of high density and plasma temperature gradients. In the zone of material movement in the direction of the substrate, a shock wave is formed as the reaction of the system to explosive evaporation of the material from the surface of the coating.

After passing through a thin coating, the shock wave rebounds from the surface of the substrate (interface), changes its direction of propagation and causes detachment (spalling) of the coating material. If, on the other hand, the coating is thick, the velocity of shock wave propagation is reduced sufficiently for it to transform into an acoustic (sound) wave, causing vibrations of the substrate which are also conducive to detachment of coating particles. The cycle is repeated until total stripping of coating material is effected. Next the shock wave is absorbed by the substrate.

The depth of the ablation front depends on the wavelength of laser radiation and is usually within the range from 0.3 μm (excimer laser) to approximately 1 μm (Nd-YAG laser). This is also the thickness range of one stripping of the coating, i.e., after one pass of the laser beam. Passes may be repeated.

For a laser of $2 \cdot 10^{-8}$ s pulse, momentary pressure is approximately 20 MPa while the velocity of explosively evaporated particles reaches 0.1 km/s, with an acceleration of 10^{10} m/s. Momentary temperature is low and does not exceed 300°C [261].

The ablation technique may be used to strip varnish coatings from metal objects, e.g., thin conductors of 0.05 to 0.3 mm are stripped of their insulating coatings [29]. It is also possible to clean works of art (paintings, sculptures) made of metal, stone, porcelain, glass, wood, cardboard, ivory, cloth or feathers of even centuries-old contaminations like patina, fats, oils, oxides, paints, varnishes, fungus, soot and solid particles adhering by other means [261].

3.5.4 Laser techniques for formation of thin and hard coatings

The laser beam allows the formation of thin and hard coatings of high hardness or with special properties, similar to or same as those obtained in typical CVD and PVD processes. It may be applied as an individual operation or as part of several other treatment cycles. Currently, it is possible to distinguish four methods of laser formation of thin and hard coatings: fusion alloying in gas, pure evaporation, pyrolytic and chemical.

3.5.4.1 Coating formation by the fusion alloying in gas method

By the utilization of fusion alloying in the gas phase, described in Section 3.5.2.2, it is possible to form coatings of greater thickness and better adhesion to the substrate than by typical CVD and PVD methods.

To this day, a known and practiced method is fusion alloying in gas coatings made from titanium and its alloys, sprayed onto any metal substrate [255].

In the latter case, after plasma spraying, in a protective atmosphere of coatings of 100 to 200 μm thickness, made from powder of 10 to 40 μm grain size, the coatings are subjected to the action of a laser beam of 10^6 W/cm^2 power density, with simultaneous blowing of nitrogen into the pool of molten titanium at 120 kPa pressure. A weakly ionized plasma, composed of nitrogen and titanium vapours is obtained. As the result of a synthesis of titanium from the substrate with atomic nitrogen from the ionized plasma, the TiN nitride is formed. The thickness of the obtained titanium nitride coating is approximately 100 μm; it is thus more than 10 times greater than that obtained by typical CVD and PVD methods. The consumption of nitrogen is 1 l/s [255].

3.5.4.2 Formation of coatings by the pure vapour deposition method

During pure evaporation of the coating material with the aid of a laser beam, and the deposition of vapours of that material (e.g., Al, Cr) on the substrate in its pure form or in the form of a compound obtained from the synthesis of this material (e.g., Ti) with the reactive gas (e.g., N$_2$), coatings are obtained which are exactly the same as those obtained by typical PVD methods, such as electron beam, resistance or arc evaporation [256]. In the case of laser evaporation, scattered debris of melted material and its contaminants may be expected. For this reason, for each type of evaporated material, the appropriate type of laser should be selected. Generally, the rule is that for vapour deposition of metals, better results are obtained

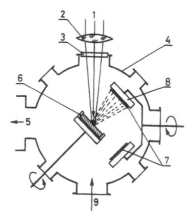

Fig. 3.52 Schematic of equipment used for evaporation PVD technique, using the laser beam to evaporate a superconductor and to vapour deposit it on a semiconductor material: *1* - laser beam (Nd-YAG laser, *l* = 1.064 μm (0.532 μm, 0.359 μm, 0.266 μm), *f* = 1 to 10 Hz, or excimer laser XeCl (λ = 0.308 μm, *f* = 5 Hz) with energy density in spot = 7.5 J/cm^2; *2* - quartz lens; *3* - laser window; *4* - vacuum tight shell; *5* - turbomolecular pump, ensuring 10^{-4} Pa in chamber; *6* - rotating target with disk of laser evaporated YBa$_2$Cu$_3$O$_7$ (which at 90 K becomes a superconductor); *7* - plates of polycrystalline ZrO$_2$ and monocrystalline SrTiO$_3$ (which at 85 K becomes a superconductor), on which vapours of YBa$_2$Cu$_3$O$_7$ condense, rotating relative to target; *8* - plate holder and heater (maximum temperature 600°C); *9* - inlet for oxygen to ensure required partial pressure. (From *Endres, G., et al.* [256]. With permission.)

with the application of Nd-YAG lasers (λ = 1.06 µm), while for non-metals, the recommended laser is the CO_2 type (λ = 10.6 µm) [29]. Naturally, exceptions to this rule do exist.

The laser beam may be utilized to vapour deposit metals (e.g., Al, W, Ti, Cr), semi-conductors (e.g., ZnS, ZnSe, GaAs) and insulators (e.g., SiO_2). The rate of deposition ranges from fractions of a nanometer to hundreds of nanometers in 1 s [29].

A significant, 100- to 1000-fold increase in deposition rate may be obtained by additional action of the laser beam on the electrically or electrolytically deposited coating, due to the increase of the temperature gradient and the consequent rise in concentration of the deposited ions [29].

Vapour deposition by laser of materials for the needs of electronics must be carried out in vacuum [257]. Fig. 3.52 shows a laboratory system for laser evaporation of a superconductor and its deposition on a semiconductor substrate [256].

Laser vapour deposition of coating materials is applied less frequently than electron beam deposition and when it is, it is for the need of electronics rather than for the machine industry.

3.5.4.3 Pyrolytic and photochemical formation of coatings

Pyrolytic formation of coatings involves the utilization of the laser beam to effect the thermal decomposition (pyrolysis) of chemical compounds, as the result of a breakdown of chemical particles of greater molecular mass into smaller particles. Pyrolysis is used to degas coal or for industrial-scale obtaining of soot from methane. The decomposed chemical compounds are gases, the components of which react chemically with the components of the substrate material. This method can be utilized in surface engineering. The condition is that pyrolysis leads to the obtaining of atoms with a high affinity to the atoms of the substrate, resulting in good adhesion of the coating to the substrate. Presently, this method has been successfully used to form coatings of TiN and the titanium carbonitride Ti(C, N), as well as their composites on pure titanium and its alloys [258]. The crystalline lattice of the substrate and the coating are similar.

Fig. 3.53 shows a diagram of the system used for pyrolytic formation of coatings. Besides the laser, its main component is the vacuum chamber into which a CO_2 laser beam is introduced through a window, permeable to laser radiation. The beam is introduced in such a way that it falls on the coated titanium or titanium alloy object placed there earlier. After evacuating the vacuum chamber, it is backfilled with ammonia (NH_3) which, under the effect of laser pyrolysis, decomposes into hydrogen and nitrogen. The latter, reacting with titanium from the substrate, forms titanium nitride, according to the reaction:

$$2NH_3 \xrightarrow{\;laser\;} 2N + 3H_2 \qquad (3.20)$$

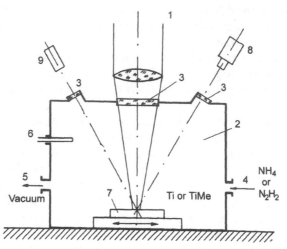

Fig. 3.53 Schematic of equipment for laser pyrolitic or chemical formation of coatings (LCVD or LACVD): *1* - laser beam; *2* - vacuum chamber; *3* - windows transmitting infrared radiation (from CO_2 laser and place of heating of coated object); *4* - inlet of reactive gas(es); *5* - vacuum pump fitting; *6* - pressure sensor; *7* - coated object; *8* - thermovision camera for observation of process of coating deposition in infrared radiation; *9* – pyrometer.

$$Ti + N \xrightarrow{\text{laser}} TiN \tag{3.21}$$

The average hardness of the titanium nitride coating, with an attractive golden appearance, is 1900 to 2000 HK and is 3 to 4.5 times higher than that of the substrate (approximately 750 to 900 HV). The coating thickness depends on the chemical composition of the substrate and on process parameters and may range from 0.3 to 30 μm [258].

The introduction of acetylene into the evacuated chamber can also yield a coating composed of the TiC titanium carbide.

It is also possible to obtain, by the pyrolytic method, double-layer coatings: after the deposition of a TiN coating of 0.2 to 3 μm, a mixture of ammonia (NH_3) and acetylene (C_2H_4) is introduced into the vacuum chamber and, by their pyrolysis, atomic nitrogen and atomic carbon are obtained. The result, obtained in the zone of action of the laser beam and with the application of appropriate process parameters, is a layer of titanium carbonitride Ti(N, C) on top of a titanium nitride TiN layer, according to reactions:

$$C_2H_4 \xrightarrow{\text{laser}} 2C + 2H_2 \tag{3.22}$$

$$Ti + N + C \xrightarrow{\text{laser}} Ti(C,N) \tag{3.23}$$

The ratio of ammonia to acetylene ranges from 2.5 to 8 with best properties of the coating obtained when this ratio is 5. The thickness of the obtained double layer coating is within the range of 0.3 to 30 μm, while the average hardness of the carbonitride coating is from 2700 to 2800 HK. The resistance of the coating to tribological wear is 9 to 11 times higher than that of the substrate, adhesion of the coating to the substrate is also very good and corrosion resistance is better than that of the substrate [258].

Photochemical coating formation involves the utilization of the laser beam to cause a chemical reaction of decomposition of gas particles and the formation of a new compound with the substrate material [29]. The photochemical reaction, occurring under the influence of laser radiation, may also be utilized to decompose gases, as well as to effect a synthesis of their components at the substrate surface. This method may be applied for laser deposition of cadmium, aluminum and their compounds [29], although in this case, the method should be classified as chemical, belonging to the LCVD group.

3.5.4.4 Formation of coatings by chemical methods (LCVD)

This group is known as *Laser-Induced Chemical Vapour Deposition* (LCVD), sometimes also called *Laser-Assisted Chemical Vapour Deposition* (LACVD). The method may be carried out with the utilization of equipment shown in Fig. 3.53. The difference between this method and the method described in Section 3.5.4.3 lies in the fact that laser decomposed components of gases react chemically with one another (but not with the substrate material) and are deposited on the substrate, forming an adhesive bond with it [259]. A mixture of reactive gases, e.g., NH_3 and SiH_4, introduced into the evacuated vacuum chamber, is decomposed to nitrogen, hydrogen and silicon and subsequently nitrogen reacts with silicon, forming silicon nitride, while hydrogen is pumped out of the chamber. The silicon nitride Si_3N_4 is deposited on the steel substrate, e.g., 1045 grade or 2Cr13* grade [258]. The deposition process proceeds according to the following reaction:

$$4NH_3 + 3SiH_4 \xrightarrow{\text{laser}} Si_3N_4 + 12H_2 \qquad (3.24)$$

and occurs above 640°C at a very fast rate, reaching 15 μm/s. The coating is of a dark gray color and is composed of α-Si_3N_4 particles, of 0.1 to 0.3 μm diameter. Its thickness may be controlled within the range of 0.2 to 40 μm. The average hardness of the coating is 2200 HK. The coating's tribological wear resistance is 2.2 to 11.75 times greater than that of the substrate and corrosion resistance is also good [258].

This method may be used to deposit other types of coatings.

It is worth mentioning that the process may be accomplished either independently or coupled with simultaneous remelt-free hardening of the substrate.

3.6 Application of laser heating in surface engineering

Laser treatment of surfaces is applied in a technological manufacturing cycle in a manner similar to electron beam treatment.

Since, like electron beam treatment, it allows the treatment of selected, small areas of material, it also allows minimization of mechanical deformation stemming from the heat effect, reducing it exclusively to the heat affected zone and the formation of residual stresses.

The range of applications of laser heating is similar to, although broader than that in electron beam heating, because heating of the load usually takes place in air which facilitates manipulation of the radiating beam. It can be used to reach elements which are otherwise difficult to access, e.g., inaccessible to an inductor in induction hardening (hardening of partially assembled rear axles of autos [5, 48] or of selected fragments of load surface.

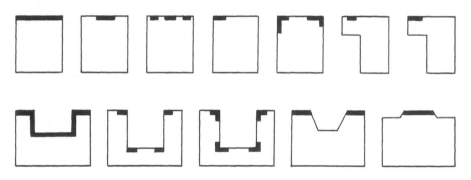

Fig. 3.54 Examples of laser hardening of flat surfaces.

Most often, the laser beam is utilized to treat long, flat surfaces (Fig. 3.54) or objects with a triangular cross-section (e.g., guide rails) [138], surfaces of rotational symmetry (rubbing surfaces of bush bearings, crankshafts, pistons, cylinders, piston rings, clamps, bearing races, etc.), specially shaped surfaces (cams, plates, clutch elements, valve seats) (Figs. 3.55 to 3.58), surfaces forming the geometry of cutting edges (cutting tools, knives, saws) or surfaces of forming tools, e.g., forging dies (Fig. 3.59).

The laser beam may be used to heat not only materials situated in air but also in those placed in other partially transmitting environments (in other gases or liquids, e.g., in water) or to heat through partially transmitting media. Best effects of laser beam transmission, naturally, are obtained in vacuum.

Lasers allow, moreover, especially in the case of pulse treatment, to deliver to the selected spot such great amounts of energy within such short a time (even of the order of billionths of a second) that the temperature of zones adjacent to the heated spot does not change

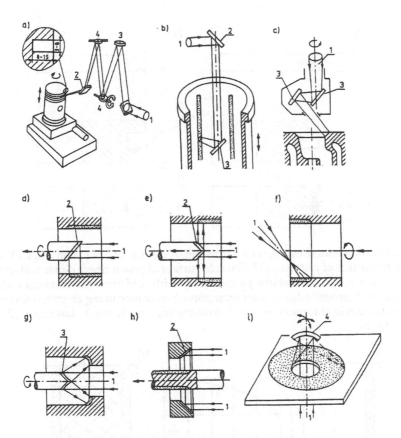

Fig. 3.55 Schematics of laser processes: a) hardening of ring grooves in pistons; b) hardening of cylinder walls of combustion engines (with simultaneous cylinder rotation, the obtained laser path is a screw-thread); c) alloying or cladding of valve seats; d - f) hardening of internal cylindrical surfaces; g) hardening of internal radius changes; h) hardening of shafts; i) hardening of flat surfaces; 1 - laser beam; 2 - plane mirror; 3 - condensing mirror (utilization of condensing mirrors allows obtaining higher power densities); 4 - oscillating mirror.

during heating. By conduction, heat is passed on to those areas only after the completion of the heating cycle which practically eliminates oxidation or burning of combustible materials during heating and creates the possibility of heating materials through absorption layers without contamination of the surface by materials belonging to those layers [6, 8]. Moreover, thanks to automation and computerization of the heating cycle, there is a possibility of precise control and rapid change of heating parameters.

Interesting technological effects may be obtained through the combination of laser treatment with other technological methods belonging to the realm of surface engineering, such as heat and thermo-chemical treatment, shot-peening (before or after laser treatment), thermal spraying, deposition by E.D.M. or detonation, with ion implantation and even with CVD and PVD methods of thin layer formation.

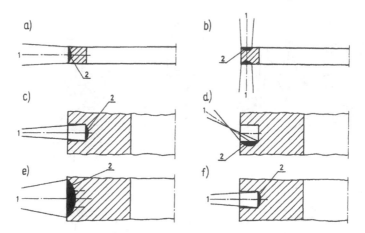

Fig. 3.56 Schematics showing laser hardening of pistons and of piston rings: a) outer diameter surface of piston ring; b) flat side surface of piston ring; c) surface of groove bottom in cast steel and cast iron piston; d) side surface of groove in cast steel and cast iron piston; e) groove edge in aluminum piston before machining groove; f) surface of groove bottom in aluminum piston after machining out groove; 1 - laser beam; 2 - site of hardening.

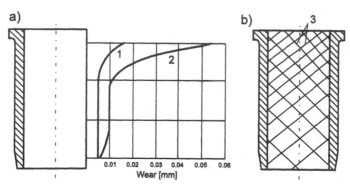

Fig. 3.57 Laser hardening of cylinder wall: a) comparison of wear; b) way of hardening causing more uniform wear; 1 - curve of wear of laser hardened cylinder wall; 2 - curve of wear of cylinder wall not hardened by laser; 3 - laser paths: hardened places or engraved groove.

Favorable effects may be obtained by combining laser heating with machining of materials otherwise difficult to machine or with their welding.

The laser beam may also be utilized for preheating of materials prior to subsequent main laser treatment (especially by low power lasers).

The advantages of laser heat treatment are similar to those of electron beam treatment, broadened by the elimination of harmful X-ray radiation, vacuum, essential in electron beam technology, as well as the necessity to demagnetize the surface. The disadvantages are also similar to those in electron beam heating but, additionally, there are strict safety rules to be

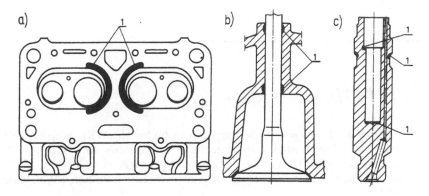

Fig. 3.58 Examples of sites hardened by laser technique (in automotive parts): a) semi-circular strips between combustion chambers in a cylinder block; b) valve seat; c) fuel nozzle seat; 1 - site of hardening. (Fig. a - from *Grigoryantz, A.G., and Safonov, A.N.* [29], Fig. b - from *Yessik, M., and Schmaltz, J.D.* [72], Fig. c - from *Funk, G., and Müller, W.* [60]. With permission.)

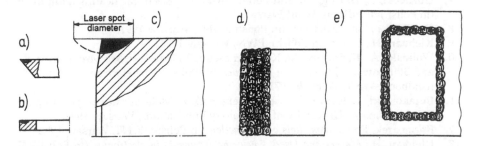

Fig. 3.59 Examples of pulsed laser hardening of tools: a) cutting tool; b) metal slitting saw; c) cutting edge of stamping die; d) working surfaces of stamping dies (before grinding); e) working surfaces of forging die (before grinding). (From *Grigoryantz, A.G., and Safonov, A.N.* [29]. With permission.)

obeyed (transmission of the laser beam must proceed along special optical guides), the necessity to increase absorption of the heated surface, short life of mirrors, complicated design of laser heaters and their high cost, ranging from several to several tens of U.S. dollars per 1 W of power [5, 29, 48, 54]. The advantages of laser treatment exceed the disadvantages. Due to laser treatment, the life of treated objects rises from several tens percent to a factor of several times. An increasing number of industrial plants apply the laser beam to enhance the surface of metallic materials. In order to facilitate the utilization of laser heaters, companies manufacturing them and research facilities develop plots and nomograms allowing optimization of treatment parameters. The number of technological lasers and the range of their application are constantly broadening.

References

1. **Stankowski, J.**: *Masers and their applications* (in Polish). WKL, Warsaw 1965.
2. **Oczoś, K.**: *Material shaping by concentrated energy fluxes* (in Polish). Publ. by Rzeszów Technical University, Rzeszów, Poland 1988.
3. **Kaczmarek, F.**: *Introduction to laser physics* (in Polish). II Edition, PWN, Warsaw 1987.
4. **Nowicki, M.**: *Lasers in electron beam technology and in material treatment* (in Polish). WNT, Warsaw 1978.
5. **Burakowski, T., and Straus, J.**: Development of laser techniques for technological needs (in Polish). *Metaloznawstwo, Obróbka Cieplna, Inżynieria Powierzchni (Metallurgy, Heat Treatment, Surface Engineering)*, No. 88, 1987, pp. 3-7.
6. **Domański, R.**: *Laser radiation - effect on solids* (in Polish). WNT, Warsaw 1990.
7. **Gozdecki, T., Hering, M., and Łobodziński, W.**: Electronic heating equipment (in Polish). II Edition. WSiP, Warsaw 1983.
8. **Dubik, A.**: *Laser application* (in Polish). WNT, Warsaw 1991.
9. **Klejman, H.**: *Lasers* (in Polish). PWN, Warsaw 1975.
10. **Woliński, A., Kaźmirowski, K., Adamowicz, T., Nowicki, M., Mroziewicz, B., and Stankowski, J.**: *Quantum electronics* (in Polish). The electronics engineer's handbook, WNT, Warsaw 1971.
11. **Burakowski, T., Roliński, E., and Wierzchoń, T.**: Metal surface engineering (in Polish). Warsaw University of Technology Publications, Warsaw 1992.
12. **Kaczmarek, F.**: *Fundamentals of laser action* (in Polish). WNT, Warsaw 1983.
13. **Klejman, H.**: *Masers and lasers - new achievements in electronics* (in Polish). II Edition, MON Publications, Warsaw 1967.
14. **Rykalin, N.N., Uglov, A.A., Zuev, I.V., and Kokora, A.N.**: *Laser and electron beam treatment of materials* (in Russian). Handbook, Publ. Masinostroenye, Moscow 1985.
15. **Trzęsowski, Z.**: High power CO_2 technological lasers (in Polish). Part I and II. *Metaloznawstwo, Obróbka Cieplna, Inżynieria Powierzchni (Metallurgy, Heat Treatment, Surface Engineering)*, No. 88, 1987, pp. 8-17.
16. **Kujawski, A.**: *Lasers* (in Polish). Warsaw Technical University Publications, Warsaw 1986.
17. **Witteman, W.J.**: *The CO_2 laser*. Springer Verlag, New York - London - Berlin 1987.
18. **Nighan, W.L. et al.**: *Gas lasers*. Academic Press, New York - London 1982.
19. **Golubev, V.S., and Lebedev, F.V.**: *Engineering principles of technological laser design. Laser technology and techniques* (in Russian), No. 2. Publ. Vissaya Skhola, Moscow 1988.
20. **Dubik, A.**: *1000 words about lasers and laser radiation* (in Polish). MON Publications, Warsaw 1989.
21. **Dudley, W.W.**: *CO_2 lasers - effects and applications*. Academic Press, New York 1976.
22. **Cruciani, D., Cantello, M., Lavona, G., Ramous, E., and Tiziani, A.**: Laser surface treatment with different absorption coatings. Report No. C3-1. XI Congreso International de Electrotermia, Malaga 1988.

23. **Zielecki, W.**: *Modification of technological and service properties of steels by the laser and electron beam* (in Polish). Ph.D. Thesis. Rzeszów Technical University, Rzeszów, Poland 1993.
24. **Przybyłowicz, J., and Przybyłowicz, K.**: Exothermic, reflecting coatings in laser treatment of steels (inPolish). Proc.: II Polish Conference on *Surface Treatments*, Kule, 13-15 Oct. 1993, pp. 117-119.
25. **Andrjachin, V.M.**: *Processes of laser welding and thermal treatment* (in Russian). Publ. Nauka, Moscow 1988.
26. *Industrial application of laser.* Edited by H. Köber. John Wiley and Sons, Chichester-New York 1984.
27. **Vedenov, A.A., and Gladish, G.G.**: *Physical processes in laser treatment of materials* (in Russian). Publ. Energoatomizdat, Moscow 1985.
28. **Grigoryantz, A.G.**: *Principles of laser treatment of materials* (in Russian). Publ. Masinostroenye, Moscow 1989.
29. **Grigoryantz, A.G., and Safonov, A.N.**: *Methods of surface treatment by laser beam. Laser technology and techniques* (in Russian). No. 3 Publ. Vissaya Shkola, Moscow 1987.
30. **Burakov, V.A., Burakova, N.M., Semenov, A.S., Burakowski, T., and Serzysko, J.**: Determination of treatment conditions of tool steels by pulsed laser relative to temperature and time (in Polish). *Metaloznawstwo, Obróbka Cieplna, Inżynieria Powierzchni (Metallurgy Heat Treatment, Surface Engineering)*, No. 101-102, 1989, pp. 58-65.
31. **Burakov, V.A., Burakova, N.M., Semenov, A.S., Burakowski, T., and Serzysko, J.**: Analytical determination of pulsed laser treatment conditions for steel (in Polish). *Przegląd Mechaniczny (Mechanical Review)*, Vol. 8, 1991. pp. 187-190.
32. **Dekumbis, R.**: Oberflächenbehandlung von Werkstoffen mit CO_2 - Hochleistunglasern. *Technische Rundschau Sulzer*, No. 3, 1986, pp. 24-28.
33. **Kusiński, J.**: Changes in structure and mechanical properties of steel caused by laser treatment (in Polish). *Transactions of AGH Mining and Metallurgy Academy - Metallurgy and Casting*, No. 132, Cracow 1989.
34. **Zimny, J.**: Properties of LBM laser treatment of refractory alloy steels (in Polish). *Transaction of Cracow Technical University*, Vol. 2, Cracow 1984.
35. **Przybyłowicz, K., Kusiński, J., Krehlik, R., and Malczyk, R.**: Life extension of cutting tools by laser treatment (in Polish). *Studies and Materials*, Vol. IV, No. 1-2, 1985, Gorzow Wlkp., Poland, pp. 418-429.
36. **Gregson, V.**: Proceedings of Conference *Laser Heat Treatment*, paper no. 15,. Laser Institute of America, Toledo, Ohio 1981.
37. **Ashby, M.F., and Easterling, K.E.**: The transformation hardening of steel surfaces by laser beams - I. Hypo-eutectoid steels. *Acta Metallurgica*, Vol. 32, No. 11, 1984, pp. 1935-1948.
38. **Geissler, E., and Bergmann, H.W.**: Calculation of temperature profiles, heating and quenching rates during laser processing. *Fachberichte Metallpraxis*, Vol. 65, No. 2, 1988, pp.119-123.
39. **Kraposkin, V.S.**: Dependence of hardening depth of steels and cast irons on conditions of laser treatment (in Russian). *Fizika i Khimia Obrabotki Materialov*, No. 6, 1988, pp.88-96.
40. **Mayorov, V.S.**: Calculation of parameters of laser hardening with scanning (in Russian). *Fizika i Khimia Obrabotki Materialov*, No. 1, 1989, pp.38-43.
41. **Parkitny, R., Winczek, J., Jabreen, H., and Thiab, S.M.**: Temperature fields in steel elements subjected to continuous and pulsed laser heating source (in Polish). Proc.: II Polish Conference on *Surface Treatment*, Kule, 13-15 October, 1993, pp. 129-135.

42. **Uglov, A.A., Smurov, I.Yu., Ignatev, M.B., Mirkin, L.I., and Krapivin, L.L.**: Calculation of melting process parameters in laser-plasma synthesis of metal nitrides in an atmosphere of nitrogen at elevated pressure (in Russian). *Fizika i Khimia Obrabotki Materialov*, No. 3, 1986, pp.18-20.

43. **Dorozkin, N.N., Vetrogon, G.I., Kukin, S.F., Dubnyakov, V.N., and Pasach, E.V.**: Calculation of dimensions of wear resistant superficial layers obtained by laser hardening of structural steel (in Russian). *Trenye i Iznos*, t. VII, No. 6, 1996, pp. 1054-1061.

44. **Rykalin, N., Uglov, A., and Kokora, A.**: *Laser machining and welding*. Pergamon Press, Oxford 1987.

45. **Gurtney, S., and Steen, W.M.**: *Source book on application of the laser in metal working*. ASM, Metals Park, Ohio, 1981, pp.195-208.

46. **Redi, D.**: *Action of strong laser radiation*. Publ. Mir, Moscow 1974.

47. **Baier, R.**: Oberflächenvergüten mittels Laserstrahlen. Alles unter Kontrolle. *Industrie Anzeiger*, No. 5, 1987, pp. 18-22.

48. **Burakowski, T.**: Techniques of producing superficial layers - metal surface engineering (in Polish). Proc.: Scientific Conference *Techniques of Producing Surface Layers*, Rzeszów, Poland, 9-10 June, 1988, pp. 5-27.

49. **Woliński, W.**: Introduction to a symposium on technological application of lasers (in Polish). *Technological Application of Lasers* - Section of Technological Fundamentals of the Machine Building Committee of the Polish Academy of Sciences, Warsaw, 22 June, 1989, pp. 1-3.

50. **Burakowski, T., and Straus, J.**: Application of lasers to heat treatment (in Polish). Proc.: Symposium on: *Technological application of lasers*, Section of Technological Fundamentals of the Machine Building Committee of the Polish Academy of Sciences, Warsaw, 22 June, 1989, pp. 4-18.

51. **Straus, J., and Burakowski, T.**: Problems and perspectives of laser heat treatment (in Polish). *Metaloznawstwo, Obróbka Cieplna, Inżynieria Powierzchni (Metallurgy, Heat Treatment, Surface Engineering)*, No. 88, 1987, pp. 36-42.

52. **Krehlik, J., and Malczyk, B.**: Improvement of service properties of cutting tools by laser treatment (in Polish). Proc.: Symposium on: *Technological Application of Lasers* - Section of Technological Fundamentals of the Machine Building Committee of the Polish Academy of Sciences, Warsaw, 22 June, 1989, pp. 19-21.

53. **Burakowski, T.**: Technological lasers and their application in metal surface engineering (in Polish). *Przegląd Mechaniczny (Mechanical Review)*, No. 11-12, 1993, pp.15-16 and 25-31.

54. **Burakowski, T.**: Lasers and their application in surface engineering (in Polish). *Mechanik (Mechanic)*, No. 5-6, 1992, pp.197-204.

55. **Abilsitov, G.A., and Safonov, A.N.**: Laser engineering development and use for material treatment. Proc.: *VII International Congress on Heat Treatment of Materials*, Moscow, 11-14 Decemeber 1990, pp.331-335.

56. *Laser electron beam and spark discharge technique of surface hardening* (in Russian). Scientific and technical progress in machine-building, Edition 9: Contemporary methods of surface hardening of machine components. International center of Scientific and Technical Information - A.A. Blagonravov Institute of Soviet Academy of Sciences, Moscow, 1989, pp. 80-204.

57. **Velichko, O.A.**: Laser hardening and cladding of industrial products (in Russian). Collection of reports on *New processes for gas-thermal and vacuum coatings*. Soviet Academy of Sciences, Kiev, 1990, pp. 17-21.

58. **Abilsitov, G.A., and Safonov, A.N.**: Modification of material surface with the laser beam (in Polish). *Metaloznawstwo, Obróbka Cieplna, Inżynieria Powierzchni (Metallurgy, Heat Treatment, Surface Engineering)*, No. 91-96, 1988, pp. 75-81.

59. **Mordike, B.L.**: Trends in the development of the application of CO_2 lasers in materials technology. *Zeitschrifft für Werkstofftechnik*, 14, 1983, pp. 221-228.

60. **Funk, G., and Müller, W.**: Temperaturgeregeltes Laserhärten in Präzisionsmengenfertigung. *Härterei-Technische Mitteilungen*, Vol. 46, No. 3, 1991, pp. 184-189.

61. **Lepski, D., and Reitzenstein, W.**: Computergestützte Prozessoptimierung bei Laser- Umwändlungshärtung von Eisenwerkstoffen. *Härterei-Technische Mitteilungen*, Vol. 46, No. 3, 1991, pp. 178-183.

62. **Katulin, V.A.**: Lasertechnologie. *Neue Hütte*, 29, No. 5, 1984, pp. 171-174.

63. **Duley, W.W.**: *Laser processing and analysis of materials*. Plenum Press, New York-London 1983.

64. **Holtom, D.P.**: Opportunities for laser treatment in the automotive industry. *Metallurgia*, Vol. 53, No. 5, 1986, pp. 183-184.

65. **Dekumbis, R.**: Oberflächenbehandlung von Werkstoffen mit CO_2 - Hochleistungslasern. *Fachberichte für Metallbearbeitung*, Vol. 63, No. 11/12, 1986, pp. 549-553.

66. **Chrissolousis, G.**: *Laser machining theory and practice*. Springer Verlag. Berlin 1991.

67. **Steen, W.M.**: *Laser material processing*. Springer Verlag, Berlin 1991.

68. **Gutman, M.B., Rubin, G.K., and Seleznyev, Yu.N.**: Laser-plasma-arc treatment of metal componenets (in Russian). *Avtomobilna Promislennost*, No. 10, 1986, pp. 32-33.

69. **Sadowski, A., and Krehlik, R.**: *Lasers in material treatment and in metrology* (in Polish). WNT, Warsaw 1973.

70. **Deriglazova, I.F., Mulchenko, B.F., Vorobev, S.S., Bogolyubova, I.V., and Sokolov, A.M.**: Laser hardening of aluminum piston grooves (in Russian). *Avtomobilna Promislennost*, No. 9, 1987, p. 25.

71. **Thiemann, K.G., Ebsen, H., Marquering, M., Vinke, T., and Haferkamp, H.**: Reparaturbeschichten von Turbinenschaufeln. *Laser-Praxis*, Oct. 1990, ISSN 0937-7069, Carl Hanser Verlag, München, pp. LS 101-106.

72. **Yessik, M., and Schmaltz, J.D.**: Laser processing at Ford. *Metal Progress*, May 1975, pp.61-66.

73. **Steen, W.M.**: Surface treatment of materials by laser beams - a review. Proc.: 2 European Conference on Laser-Metal Treatment *ECLAT '88*. 13-14 Oct. 1988, Bad Nauheim, pp. 60-64.

74. **De Dambornea, J., Vazquez, J., and Gonzalez, J.A.**: Effect of gas protection in surface treatments with a laser. *Journal of Material Science Letters*, No. 8, 1989, pp. 473-474.

75. **Gjurkowski, S.**: Laser heating of steel at ultra high rate (in Polish). *Przegląd Mechaniczny (Mechanical Review)*, No. 10, 1989, pp.32-33.

76. **Bergmann, H.W., Juckenath, B., and Lee, S.Z.**: Surface treatments with excimer lasers. Proc.: 2 European Conference on Laser-Metal Treatment *ECLAT '88*, 13-14 Oct. 1988, Bad Nauheim, pp. 106-109.

77. **Poluchin, V.P., Vermeevich, A.N., and Kryanina, M.N.**: *Application of laser treatment in high pressure processes* (in Russian). Collection of reports: Teoria i technologia metallo- i energosberegayuschikh processov obrabotki metallov davlenyem. Publ. Metallurgia, 1986, pp. 143-146.

78. **Hallouoin, M., Gerlance, M., Cottet, F., Romain, J.P., and Marty, L.**: Modifications microstructurales residueles du fer soumis a un choc laser. *Memoires et Etudes Scientifiques Revue de Metallurgie*, No. 9, September 1986, pp. 473.

79. **Woliński W.**: Photon beams in technology (in Polish). Proc.: Conference on Electron Technologies, Wrocław-Karpacz, Poland, September 1992, pp.186-190.

80. **Schneider, D., Winderlich, B., Ermich, M., and Brenner, B**.: Untersuchung des Anlassverhaltens laserhärterer Stähle mittels Ultraschall-Oberflächenwellen. *Neue Hütte*, 34, No. 3, 1989, pp. 100-105.

81. **Winderlich, B., Pollack, D., and Schneider, D**.: Untersuchungen zum Anlassverhalten des laserhärten Stahls 90SiCr5. *Neue Hütte*, 31, No. 11, 1986, pp. 418-423.

82. **Guriev, V.A., and Tesker, E.I**.: Application of laser treatment to components with stress-raisers (in Russian). *Metallovedene i Termicheskaya Obrabotka Metallov*, No. 3, 1991, pp.4-5.

83. **Przetakiewicz, W., and Napadłek, W**.: Laser hardening of low-carbon cladding (in Polish). Proc.: II Polish Conference on *Surface Treatment*, Kule,13-15 Oct. 1993, pp. 135-143.

84. **Jasiński, J., Jeziorski, L., and Pisarek, J**.: Gradiental treatment of steel with a low power CO_2 laser. Proc.: II Polish Conference on *Surface Treatment*, Kule, 13-15 Oct. 1993, pp. 361-367.

85. **Yao Shanchang**.: Laser surface transformation hardening (in Chinese). *Jinshu Rechuli (Heat Treatment of Metals)*, No. 2, 1987, pp. 39-43.

86. **An Shimin, Wang Ru, and Qi Zengfeng**: The research of transformation during heating by laser. Proc.: 4th International Congress on Heat Treatment of Materials. Berlin 1985, June 3-7, pp. 952-969.

87. **Mandziej, S., Seegers, M.C., and Godijk, J**.: Effect of laser heating on substructure of 0.4%C steel. *Materials Science and Technology*, Vol. 5, No. 4, 1989, pp. 350-355.

88. **Ciszewski, B., and Bojar, Z**.: Laser hardening of steels and cast irons (in Polish). *Bulletin of Military Technical Academy*, Year Vol. XXXVII, No. 5 (441)May 1989, pp. 99-114.

89. **Biryukov, V.P**.: Laser hardening for surface wear resistance of gray cast iron by a scanning beam (in Russian). *Trenye i Iznos*, Jul-Aug, Vol. VII, 1986, pp. 718-721.

90. Laser hardening for surface wear resistance of gray cast iron by a scanning beam (in Russian). *Trenye i Iznos*, No. 4, 1986, pp. 718-721.

91. **Kocańda, S., Natkaniec, D., and Śnieżek, L**.: Structure and microhardness of laser hardened elements made of 1045 steel (in Polish). *Bulletin of the Military Technical Academy*, Year XL, No. 2-3 (462-463), 1991, pp.59-84.

92. **Kocańda, S., Górka, A., Natkaniec, D., and Śnieżek, L**.: Structure and microhardness of laser hardened superficial layer on 1015 steel (in Polish). *Bulletin of the Military Technical Academy*, Year XXXVIII, No. 12 (448), 1989, pp.59-73.

93. **Kocańda, S., Lech-Grega, M., and Natkaniec, D**.: Residual stresses in laser hardened elements made from 1015 and 1045 steel (in Polish). Quarterly of the Mining Academy *Mechanics*, Vol. 9, book 2, 1990, pp. 29-37.

94. **Kocańda, S., Lech-Grega, M., and Natkaniec, D**.: Residual stresses in laser hardened elements made from 1015 steel (in Polish). *Bulletin of the Military Technical Academy*, Year XXXVIII, No. 12 (448), 1989, pp.75-96.

95. **Kocańda, S., and Natkaniec, D**.: Analytical description of residual stresses in steel components subjected to laser heat treatment (in Polish). *Archives of Material Science*, Vol. 12, book 3, 1991, pp. 143-163.

96. **Kusiński J**.: Residual stresses in chromium-bearing structural steels, subjected to laser heat treatment (in Polish). Proc.: Polish Conference on *Surface Treatment*, Kule, 13-15 Oct. 1993, pp. 121-129.

97. **Kocańda, S., Bogdanowicz, Z., Górka, A., and Kur, J**.: Contact fatigue strength of laser hardened model components made from 1055 steel (in Polish). *Bulletin of the Military Technical Academy*, Year XXXVII, No. 12 (436), 1988, pp.23-50.

98. **Kocańda, S., and Natkaniec, D.**: Formation and development of fatigue cracks in laser hardened 1045 steel components (in Polish). *Bulletin of the Military Technical Academy*, Year XL, No. 6 (466), 1991, pp.23-46.

99. **Kocańda, S., Natkaniec, D., and Sadowski, J.**: Microstructure of fatigue crack surfaces in laser hardened 1045 steel components (in Polish). *Bulletin of the Military Technical Academy*, Year XL, No. 7 (467), 1991, pp.21-43.

100. **Kocańda, S., and Natkaniec, D.**: Fatigue crack initiation and propagation in laser hardened, medium carbon steel. *Fatigue Fract. Engineering Materials Structure*, Vol. 15, No. 12, 1992, pp. 1237-1249.

101. **Kocańda, S., and Śnieżek, L.**: Development of fatigue cracks in laser hardened components of low carbon steel (in Polish). Proc.: *XV Symposium of Experimental Mechanics of Solids*, Jachranka, Poland, 8-10.Oct. 1992, pp. 149-152.

102. **Przetakiewicz, W., Napádłek, W., and Górka, A.**: Analysis of the effect of laser treatment on selected properties of repair layers, deposited by vibratory cladding and by concealed arc (in Polish). Proc.: *II International Conference on Effect of Technology on the State of the Superficial Layer* - WW '93. Gorzow Wlkp. - Lubniewice, Poland, 1993, pp.448-451.

103. **Stenishceva, L.N., and Seleznev, J.N.**: Laser-arc treatment of steels (in Russian). *Metallovedenye i Termicheskaya Obrabotka Metallov*, No. 1, 1989, pp. 13-15.

104. **Zenker, R., Reisse, G., and Zenker, U.**: Some aspects of laser heat treatment of steels (in Polish). *Metaloznawstwo, Obróbka Cieplna (Metallurgy and Heat Treatment)*, No. 83-84, 1986, pp. 13-16.

105. **Kremnev, L.S., Cholodnov, E.V., and Vladimirova, O.V.**: Selection of steels for laser treatment (in Russian). *Metallovedenye i Termicheskaya Obrabotka Metallov*, No. 9, 1987, pp. 49-52.

106. **Zenker, R., and Zenker, U.**: Combination heat treatment of steel - nitrocarburizing and laser hardening (in Russian). *Fizika Metallov i Metallovedenye*, Vol. 66, Edition 6, 1988, pp. 1150-1158.

107. **Zenker, R., and Zenker, U.**: Kombination Karbonitrieren/Laserstrahlhärten - eine neue Variante der Randschichtwärmebehandlung. *Neue Hütte*, Vol. 31, No. 11, 1986, pp. 407-413.

108. **Zenker, R.**: Prinzip, Ergebnisse und Anwendungsmöglichkeiten der Verfahrenskombination Gaskarbonitrieren/Hochgeschwindigkeitswärmebehandlung. Proc.: *Anorganische Schutzschichten - Oberflächenschutz von Verschleiss ASS 87*, 30 Sept. - 2 Oct. 1987, Karl-Marx Stadt, Vol. 8, pp. 116-122.

109. **Zenker, R., and Zenker, U.**: Laser beam hardening of nitrocarburized steel containing 0.5% C and 1% Cr. *Surface Engineering*, Vol. 5, No. 1, 1989, pp. 45-54.

110. **Bande, H., L'Esperance, G., Islam, M.U., and Koul, A.K.**: Laser surface hardening of AISI 01 tool steel and its microstructure. *Materials Science and Technology*, Vol. 7., No. 5, 1991, pp.452-457.

111. **Makarov, A.V., Korsunov, L.G., and Osintseva, A.L.**: Effect of tempering and friction heating on wear resistance of laser hardened U8 steel (in Russian). *Trenye i Iznos (Friction and Wear)*, Vol. 12, No. 5, 1991, pp. 870-877.

112. **Wołyński, A., and Waligóra W.**: Effect of laser treatment on abrasive wear of 1045 steel (in Polish). *Tribologia*, No. 3, 1991, pp. 61-63.

113. **Mitin, V.J., Tesker, E.I, and Guriev, V.A.**: Effect of surface pattern of laser hardening on cyclic fatigue strength of 1045 steel (in Russian). *Metallovedenye i Termicheskaya Obrabotka Metallov*, No. 10, 1988, pp. 34-36.

114. **Paul, H., and Pollack, D.**: Laserflächenhärtung von Bauteilen. *Neue Hütte*, No. 31, Vol. 11, November 1988, pp. 426-428.

115. **Hitchcox, A.L.**: Pinpoint hardening with CO_2 lasers. *Metal Progress*, April 1986, pp. 31-32.

116. **Morgner, W., and Reuter, M.**: Physikalische Eigenschaften lasergehärter Werkstoffe. *Neue Hütte*, No. 34, Vol. 4, April 4, 1987, pp. 140-142.
117. **Pergue, D., Pelletier, J.M, and Fouquet, F.**: Intéréts de traitements laser dans le cas d'un acier bas carbonne: obtentions d'état hors d'équilibre. *Memoirs et Études Scientifiques Revue de Métallurgie*, No. 3, 1986, p.1.
118. **Dorozhkin, N.N., Vetrogon, G.I., Kukin, S.F., Dubnyakov, V.N., and Pasach, E.V.**: Calculation of dimensions of wear resistant surface layers, obtained through laser hardening of structural steels (in Russian). *Trenye i Iznos (Friction and Wear)*, Vol. VII, No. 6, 1986, pp. 1054-1061.
119. **Malian, P.A.**: Engineering applications and analysis of hardening data for laser heat treated ferrous alloys. *Surface Engineering*, Vol. 2, No. 1, 1986, pp. 19-28.
120. **Hick, A.J.**: Rapid surface heat treatments - a review of laser and electron beam hardening. *Heat Treatment of Metals*, No. 1, 1983, pp. 3-11.
121. **Ding Chauxuan**: Laser heat treatment of piston rings. *Zeitschifft für Werkstofftechnik*, Vol. 14, No. 3, 1983, pp. 81-85.
122. **Lachtin, J.**: Surface hardening of corrosion-resistant steels with the utilization of a laser (in Russian). *Masinostroenye*, No. 2, 1984, pp. 124-127.
123. **Waligóra, W., and Nowicki, W.**: Investigation of the effect of laser treatment conditions on the state of residual stresses in a superficial layer of heat treated bearing steel (in Polish). Proc.: *Effect of Technology on the State of the Superficial Layer*, Gorzów (Poland), 1993, pp. 146-150.
124. **Andrzejewski, H., and Wieczyński, Z.**: Effect of basic technological parameters on the results of surface hardening by a laser beam (in Polish). *Metaloznawstwo, Obróbka Cieplna (Metallurgy and Heat Treatment)*, No. 53-54, 1981, pp. 24-28.
125. **Kwaczyński, Z., and Dzioch, R.**: Tests of hardening steel by a continuous CO_2 laser of 150 W power (in Polish). *Metaloznawstwo, Obróbka Cieplna (Metallurgy and Heat Treatment)*, No. 41, 1979, pp. 20-27.
126. **Burakov, V.A., and Zhurakovski, V.M.**: Enhancement of steel quality by doping by means of laser treatment and ultra-rapid hardening (in Polish). *Metaloznawstwo, Obróbka Cieplna (Metallurgy and Heat Treatment)*, No. 81-82, 1986, pp. 18-21.
127. **Dyachenko, V.S.**: Effect of process parameters of pulsed laser treatment on the structure and properties of high speed steel (in Russian). *Metallovedenye i Termicheskaya Obrabotka Metallov*, No. 9, 1986, pp. 11-14.
128. **Dubrovskaya, E.A., Kopetski, Ch.V., Kraposhin, V.S., and Rodin, I.V.**: Selection of parameters of laser heating of carbon steels to obtain a predetermined depth of hardening (in Russian). *Metallovedenye i Termicheskaya Obrabotka Metallov*, No. 9, 1986, pp. 132-35.
129. **Dubnyakov, V.N.**: Surface hardening of copper alloys by laser beam (in Russian). *Metallovedenye i Termicheskaya Obrabotka Metallov*, No. 9, 1987, pp. 52-54.
130. **Lakhtin, J.M., Gulyaeva, T.V., Tarasova, T.V., Syrovatkin, A.I., and Chizhmakov, M.B.**: Structure and properties of 20H13 steel after laser hardening (in Russian). *Metallovedenye i Termicheskaya Obrabotka Metallov*, No. 10, 1988, pp. 36-39.
131. **Pompe, W., Reizenstein, W., Brenner, B., and Läschau, W.**: Verbesserung der Verschleissverhaltens von Eisenwerkstoffen durch Laserbehandlung. Proc.: *Fachtagung - Anorganische Schutzschichten - Oberflächenschutz von Verschleiss, AS 87*, Karl-Marx-Stadt, 30 Sep - 2 Oct., 1987, Vol. 8, pp. 19-32.
132. **Zhiping, H.**: Research on compound layer heat treatment for steel 45 by ion nitriding and laser hardening. *Jinshu Rechuli (Heat Treatment of Metals)*, No. 5, 1990, pp. 12-16.
133. **Guangjun, Z., Quidun, Y., Yungkong, W., and Baorong, S.**: Laser transformation hardening of precision V slide way. Proc.: *3rd International Congress on Heat Treatment of Materials*, Shanghai, 7-11 Nov., 1983, pp. 81-88.

134. **Malinov, L.S., Kharlamova, E.J., Tumanova, M.V., Lisakovich, A.V., and Lokshina, E.B.**: Different treatment to obtain self-hardening superficial layers on manganese-bearing steel (in Russian). *Metallovedenye i Termicheskaya Obrabotka Metallov*, No. 3, 1991, pp. 8-10.

135. **Kochubiński, O.Yu.**: Assessment of technical possibilities of hardening with the utilization of continuous gas laser (in Russian). *Metallovedenye i Termicheskaya Obrabotka Metallov*, No.1, 1980, pp. 24-26.

136. **Koncjancic, B., and Dengel, D.**: Einige Ergebnisse der konduktiven und der Laser - Kurzzeitstahlhärtung mit hocher Leistungsdichte. *Fachberichte Hüttenpraxis Metallweiterverarbeitung*, Vol. 18, No. 12, 1980, pp. 1102-1107.

137. **Howes, M.A.H.**: Laser case hardening of steel components. Proc.: *Second International Conference on Surface Engineering*. Stratford-upon-Avon, 16-18 June, 1987, pp. 91-104.

138. **Sharp, M.C., and Parsons, G.H.**: Laser transformation hardening in practice. Proc.: *Second International Conference on Surface Engineering*. Stratford-upon-Avon, 16-18 June, 1987, pp. 83-90.

139. **Tesker, E.I., Mitin, V.Ya., Karpova, A.P., and Bondarenko, Yu.V.**: Hardening of instruments made from R6M5 steel by continuous laser beam (in Russian). *Metallovedenye i Termicheskaya Obrabotka Metallov*, No.10, 1989, pp. 18-20.

140. **Stolar, P., Suchanek, J., Honzik, O., Novakova, I., Halasek, J., and Moravec, M.**: Eigenschaften von lasergehärten Schichten auf Kohlenstoffstählen. Proc.: *Härtereitechnik 1987*, 28-30 Oct. 1987, Suhl, pp. 222-231.

141. **Frąckiewicz, H.**: Technology of laser shaping of metals: methods, problems, outlook (in Polish). Proc.: *Surface Engineering Summer School*. Kielce, Poland, 6-9 Sept. 1993, pp. 51-59.

142. **Mucha, Z.**: Application of the laser in technology (in Polish) Proc.: *Surface Engineering Summer School*. Kielce, Poland, 6-9 Sept. 1993, pp. 61-70.

143. **Kuiwu, Z., Xiaohui, Ch., Zhuxiu, H., and Baoru, S.**: The components of electromagnetic clutch with laser hardening. Proc.: *4th International Seminar of IFHTSE - Environmental and Energy Efficient Heat Treatment Technologies*, Beijing, China 15-17 Sept. 1993, pp. 164-172.

144. **Chabrol, C., Nowak, I.F., and Leveque, R.**: Traitements superficiels des aciers par laser. *Memoires et Etudes Scientifiques Revue de Metallurgie*, No. 9, Sept. 1986, pp. 484.

145. **Völmar, S., Pompe, W., and Junge, H.**: Homogene Laserstrahlhärtung mittels hochfrequenter Strahloszillation. *Neue Hütte*, No. 31, Vol. 11, Nov. 1986, pp. 414-418.

146. **Winderlich, B., Pollack, D., and Schneider, D.**: Untersuchungen zum Anlassverhalten des lasergehärten Stahls 90SiCr5. *Neue Hütte*, No. 31, Vol. 11, Nov. 1986, pp. 418-423.

147. **Kusiński, J.**: Laser hardening of medium carbon chromium-bearing steels (in Polish). Proc.: III Seminar: *Steel-mill Heat Treatment - 87* on energy and material-effective processes of manufacturing of thermally treated steel mill products. Gliwice (Poland), 1987, pp. 269-286.

148. **Dyachenko, V.S.**: Effect of pulsed laser treatment on the structure and properties of high speed steels (in Russian). *Metallovedenye i Termicheskaya Obrabotka Metallov*, No. 9, 1986, pp. 11-14.

149. **Debuigne, M., and Kerrand, E.**: Modélisation des transferts thermiques appliquée au durcissement d'acier par laser CO_2. *Memoires et Études Scientifiques Revue de Metallurgie*, No. 9, Sept. 1986, p. 471.

150. **Bergmann, H.V.**: Current status of laser surface melting of cast iron. *Surface Engineering*, Vol. 1, No. 2, 1985, pp. 137-155.

151. **Bergmann, H.V., and Mordike, B.L.**: Metallurgical considerations in the laser surface melting of iron-base alloys. *Zeitschrifft für Werkstofftechnik*, No. 14, 1983, pp. 228-237.
152. **Lian, S., Chenglao, L., Xiuling, W., Lihua, M., Daozhen, Z., and Jiajin, Z.**: The microstructure and wear resistance of laser surface processed gray cast iron. *Jinshu Rechuli Xuebao (Transactions of Metal Heat Treatment)*, No. 1, 1990, pp. 20-31.
153. **Gillner, A., Wissenbach, K., and Kreutz, E.W.**: Laser surface hardening of cast iron: processing parameters, structures, hardness. *Fachberichte für Metallbearbeitung*, Vol. 64, No. 5, 1987, pp. 487-489.
154. **Bergmann, H.W.**: Laser surface melting on cast iron containing intercooled graphite. *Zeitschrift für Werkstofftechnik*, Vol. 14, No. 7, 1983, pp. 237-241.
155. **Ivanov, J.A., Ivashov, G.P., Pikunov, A.S., and Safonov, A.N.**: Characteristics of laser hardening of steel and cast iron parts in scanning mode (in Russian). *Fizika i Khimia Obrabotki Materialov*, No. 4, 1987, pp. 50-54.
156. **Kusiński, J.**: Laser treatment of medium carbon chromium steels (in Polish). *Hutnik (Metallurgist)*, No. 7, 1988, pp. 218-225.
157. **Kusiński, J., and Krehlik, R.**: Superficial layer of tool steels (in Polish). *Mechanik (Mechanic)*, No. 3, 1981, pp. 170-171.
158. **Kusiński, J.**: Laser melting of T-1 high speed steel. *Metallurgical Transactions A*, Vol. 19A, Feb. 1988, pp. 377-382.
159. **Kusiński, J.**: Precipitation of carbides in the laser melted T-1 high speed tool steel. Proc. IV International Conference: *Carbides-Nitrides-Borides*, Poznań-Kołobrzeg, 29 Sep.-3 Oct. 1987, pp. 35-39.
160. **Kusiński, J., and Przybyłowicz, K.**: Changes in structure and chemical composition of high speed steel after laser beam heating (in Polish). Proc.: X Meeting 1981-1984, Metallurgy Committee of Polish Academy of Sciences, Kozubnik (Poland), Oct. 1984, pp. 314-318.
161. **Sokolow, K.N., and Serżysko, J.**: Surface treatment of high speed steel with laser beam remelting (in Polish). *Metaloznawstwo, Obróbka Cieplna (Metallurgy and Heat Treatment)*, No. 91-96, 1988, pp.71-75.
162. **Bekrenev, A.N., Gladys, G.G., Droyazko, S.V., and Portnov, V.V.**: Laser thermal treatment of high speed steel with surface melting (in Russian). *Fizika i Khimia Obrabotki Materialov*, No. 4, 1988, pp. 63-67.
163. **Burakov, V.A., and Burakova, N.M.**: Characteristics of structures formed during laser hardening of tool steel from the melt (in Russian). *Izvestiya Uchebnikh Zavedeni - Chornaya Metallurgia*, No. 2, 1989, pp. 92-96.
164. **Costa, A.R., Domningues, R.P., Ibanez, R.A.P., and Villar, R.M.**: Laser surface melting: a microstructural study. Proc.: *VII International Congress on Heat Treatment of Materials*, Moscow, 11-14 December 1990, pp.50-55.
165. **Yavtseva, I.L.**: Structure and properties of powder metallurgy high speed steels after laser treatment (in Russian). *Metallovedenye i Termicheskaya Obrabotka Metallov*, No. 4. 1988, pp. 48-50.
166. **Navra, V.K., Laskayeva, N.S., Kryanina, M.N., and Shupenkov, E.P.**: Laser treatment of sintered alloys TiC-steel (in Russian). *Metallovedenye i Termicheskaya Obrabotka Metallov*, No. 10. 1987, pp. 57-59.
167. **Stähli, G.**: Possibilités et limites du durcissement superficiel rapide de l'acier. *Traitement Thermique*, No. 139, 1979, pp. 64-74.
168. **Blaes, L., Bauer, P., Gonser, U., and Kern, R.**: Depth profile of a laser irradiated steel. *Zeitschrift für Metallkunde*, Vol. 79, No. 5, 1988, pp. 278-281.
169. **Liu, J.**: The solidification characteristic and the nucleation mechanism of the laser dynamic solidification structure. *Jinshu Rechuli Xuebao (Transactions of Metal Heat Treatment)*, No. 3, 1990, pp. 13-23.

170. **Grigoryants, A.G., Safonov, A.,N., Mayorov, V.S., Baskov, A.F., and Ivashov, G.P.**: Distribution of residual stresses in surfaces of steels hardened by continuous CO_2 laser (in Russian). *Metallovedenye i Termicheskaya Obrabotka Metallov*, No. 9. 1977, pp. 45-49.

171. **Komorek, Z., Bojar, Z., and Komorek, A.**: Measurement of residual stresses in the superficial layer modified by a laser beam (in Polish). Proc.: V International Symposium of the Institute of Mechanical Vehicles of the Military Technical Academy: *Perfection of Design and Methods of Vehicle Service*, Warsaw, 2-3 Dec. 1993, pp. 146-151.

172. **Bell, T., Bergmann, H.W., Lanagan, J., Morton, P.H., and Staines, A.M.**: Surface engineering of titanium with nitrogen. Proc.: 4th International Congress of Heat Treatment of Materials, Shanghai, June 3-7, 1985, Vol. 2, pp. 146-151.

173. **Wiiala, U.K., Sulonen, M.S., and Korhonen, A.S.**: Laser hardening of TiN-coated steels. *Surface and Coatings Technology*, No. 36, 1988, pp. 773-780.

174. **Ling, Z., Zhirong, Z., and Yue, X.**: Structure feature and oxidation behaviour of pure nickel after laser melting and solidifying. *Jinshu Rechuli Xuebao (Transactions of Metal Heat Treatment)*, Vol. 13, No. 2, 1992, pp. 29-32.

175. **Mordike, B.L., and Veit, S.**: Oberflächenumschmeltzen von Aluminiumlegierungen mit Laserstahl. Proc.: 2 European Conference on Laser-Metal Treatment *ELCAT '88*, 13-14 Oct. 1988, Bad Nauheim, pp. 95-96.

176. **Perque, D., Pelletier, J.M., and Fouquet, F.**: Possibilite d'amorphisation superficielle dans divers alliages metalliques par faisceau laser. *Mémoires et Études Scientifiques Revue de Metallurgie*, No. 9, 1986, pp. 482.

177. **Zhang, J.G., Zhang, X.M., Lin, Y.T., and Jun, K.**: Laser glazing of an Fe-C-Sn alloy. *Journal of Materials Science*, Vol. 23, 1988, pp. 4357-4362.

178. **Hornbogen, E., and Monstadt, H.**: Solidification behaviour and the effect of unintentional and intentional reheating on overlapping laser-glazed Fe-B alloys. Proc.: 2 European Conference on Laser-Metal Treatment *ECLAT '88*, 13-14 Oct. 1988, Bad Nauheim, pp. 81-85.

179. **Narva, V.K., Loshkaryeva, N.S., Kryanina, M.N., and Byelokonova, T.A.**: Filling in of residual porosity in sintered carbon steel by laser treatment (in Russian). *Metallovedenye i Termicheskaya Obrabotka Metallov*, No. 10. 1989, pp. 16-18.

180. **Gasser, A., Kreutz, E.W., Leibrandt, S., and Wissenbach, K.**: Verdichten von thermisch gespritzen Schichten mit CO_2 - Laser-Metallbearbeitung. Proc.: 2 European Conference on Laser-Metal Treatment *ECLAT '88*, 13-14 Oct. 1988, Bad Nauheim, pp. 81-85.

181. **Kobylańska-Szkaradek, K., and Swadźba, L.**: The influence of laser remelting treatment upon structure of oxides thermal sprayed coatings. Proc.: *Thermal Spraying Conference '93*, 3-5 March, 1993, Aachen, pp. 321-324.

182. **Novotny, S., and Kunzmann, E.**: Verbesserung der Verschleisseigenschaften thermisch gespritzen Schichten durch Nachbehandlung mittels CO_2-Laser. *Schmierungstechnik*, Vol. 19, 1988, No. 6, pp. 166-169.

183. **Matteazzi, P., Sturlese, S., Uglov, A., Pekshev, P., Smurov, I., Krivonogov, J., and Naumkin A.**: Thermal and thermochemical laser surface treatment of APS and LPPS plasma sprayed coatings. Proc.: *VII International Congress on Heat Treatment of Materials*, Moscow, 11-14 Dec. 1990, pp. 309-315.

184. **Sivakumar, R., and Mordike, B.L.**: Laser melting of plasma sprayed NiCoCrAlY coatings. *Surface Engineering*, Vol. 3, No. 4, 1987, pp. 299-305.

185. **Burchards, H.D., and Mordike, B.L.**: Laserstrahlumschmeltzen von keramischen Oberschichten. Proc.: 2 European Conference on Laser-Metal Treatment *ECLAT '88*, 13-14 Oct. 1988, Bad Nauheim, pp. 79-81.

186. **Kahrmann, W.**: Laserstrahl-Oberflächenbeschichten mit Cermets. Proc.: 2 European Conference on Laser-Metal Treatment *ECLAT '88*, 13-14 Oct. 1988, Bad Nauheim, pp. 119-121.

187. **Komorek, Z., and Bojar, Z.**: Effect of chemical composition and parameters of laser treatment of superficial layer on selected properties of protective layers of nickel-base casting alloy (in Polish). Proc.: *II International Conference Effect of Technology on State of Superficial Layer* - WW '93, Gorzów (Poland), 1993, pp. 151-154.

188. **Kovalchenko, M.S., Alfintseva, R.A., Paustovski, S.V., and Kurinnaya, T.V.**: Effect of laser treatment on protective properties on the dry plated coatings. Proc.: *VII International Congress on Heat Treatment of Materials*, Moscow, 11-14 Dec. 1990, pp. 39-48.

189. **Kovalchenko, M.S., Paustovski, A.V., Boleyko, B.M., and Zhidkov, A.B.**: Laser surface hardening of boron carbide base cermets (in Russian). *Poroshkova Metallurgia*, No. 5, 1988, pp. 77-80.

190. **Lju, J.**: Study on the characteristics of laser alloying on metal surface. *Jinshu Rechuli Xuebao (Transactions of Metal Heat Treatment)*, No. 2, 1991, pp. 49-57.

191. **Thiemann, K.G., Ebsen, H., Marquering, M., Vinke, T., and Haferkamp, H.**: Reparaturbeschichten von Turbinenschaufeln. *Laser-Praxis*, Oct. 1990, pp. 101-106.

192. **Gasser, A., Wissenbach, K., Gillner, A., and Kreutz, E.W.**: Laser surface alloying of Cr_3C_2, Cr_3C_2/NiCr and WC/Co layers on low carbon steel. *Fachbereiche für Metallbearbeitung*, Vol. 64, No. 5, 1987, pp. 480-483.

193. **Hegge, H.J., and de Hossen, J.Th.M.**: The influence of convection on the homogeneity of laser applied coatings. In: Surface Engineering Practice - Processes, Fundamentals and Applications in Corrosion and Wear. Publ. Ellis Horwood, New York-Toronto-Sydney-Tokyo-Singapore 1989, pp. 160-167.

194. **Andrzejewski, H., and Wieczyński, Z.**: Saturation of superficial layers of iron and its alloys by metallic elements with the application of laser technology (in Polish). *Metaloznawstwo, Obróbka Cieplna (Metallurgy and Heat Treatment)*, No. 46, 1980, pp. 29-34.

195. **Bei, C.A., Cerri, W.E., Mor, G.P., and Fiorini, O.A.**: Surface treatment by high power CO_2 laser: hardfacing alloy deposition. Report No. C3-2. XI International Electrothermal Congress, Malaga 1988.

196. **Chande, T., and Mazumder, J.**: Composition control in laser surface alloying. *Metallurgical Transactions*, Vol. 148, No. 6, 1983, pp. 181-190.

197. **Wang, H., Fen, Y., and Tang, Ch.**: The effect of element Ti on the modification of microstructure by laser surface alloying. *Jinshu Rechuli Xuebao (Transactions of Metal Heat Treatment)*, Vol. 14, No. 1, 1993, pp. 25-29.

198. **Radziejowska, J.**: Laser enrichment of steel superficial layer with tungsten (in Polish). *Inżynieria Materiałowa (Materials Engineering)*, No. 3, 1991, pp. 59-63.

199. **Radziejowska, J.**: Laser generation of alloy coatings (in Polish). Proc.: *III International Symposium INSYCONT: Tribological Problems of Components in Contact*, Publ. AGH, 1991, pp. 13-22.

200. **Handzel-Powierża, Z., and Radziejowska, J.**: Investigations of utilization of the laser beam for modification of steel superficial layer (in Polish). *Postępy Technologii Maszyn i Urządzeń (Advances in Technology of Machines and Equipment)*, No. 1, 1992, pp. 27-39.

201. **Govorov, I.V., Kolesnikov, J.V., and Mirkin, L.I.**: Enhancement of surface strength of carbon steel by laser deposition of chromium-bearing coatings (in Russian). *Fizika i Khimia Obrabotki Materialov*, No. 5, 1988, pp. 68-71.

202. **Mishakov, G.A., Rodionov, A.I., and Simachin, J.F.**: About mass transfer of boron in the heat affected zone under the molten pool by laser remelting of a metal (in Russian). *Fizika i Khimia Obrabotki Materialov*, No. 5, 1991, pp. 100-103.

203. **Łysenko, A.B., Kozina, N.N., Gulyayeva, T.V., Shibaev, V.V., and Glushkov, A.G.**: Structure and properties of steels after boriding with the utilization of laser heating (in Russian). *Metallovedenye i Termicheskaya Obrabotka Materialov*, No. 3, 1991, pp. 2-4.

204. **Chrissoloussis, G.**: Laser surface melting of some alloy steels. *Metals Technology*, Vol. 10, No. 6, 1983, pp. 215-223.

205. **Przybyłowicz, K., and Szyda, M.**: Effect of laser heating on the process of element fusion into the superficial layer of steel (in Polish). Proc.: Conference on *Heat Treatment in Steelmaking*, Jaszowiec (Poland), 1985, pp. 17-23.

206. **Artamonova, I.V., Nikitin, A.A., and Rizhkov, I.A.**: Effect of surface laser alloying on structure and mechanical properties of 40XN steel (in Russian). *Metallovedenye i Termicheskaya Obrabotka Materialov*, No. 10, 1989, pp. 5-7.

207. **Il'in, V.M., and Kravets, A.N.**: Extension of tool life by laser alloying (in Russian). *Vestnik Masinostroenya*, No. 1, 1987, pp. 43-45.

208. **Tomsiński, V.S., Gavrilov, V.B., and Pelenev, R.S.**: Laser heat and chemicothermal treatment of the steels Y10A and X12M. Proc.: *VII International Congress on Heat Treatment of Metals*, Moscow, 11-14 Dec. 1990, pp. 31-38.

209. **Bernstein, M.L., Kryanina, M.N., and Shchukin, V.N.**: Obtaining superficial laser-alloyed PNP-steel-carbides layers (in Russian). Publ. VUZ, *Chornaya Metallurgia*, No. 9, 1986, pp. 156.

210. **Kim, T.H., Suk, M.G., Park, B.S., and Suh, K.H.**: The formation of surface-alloyed layers on carbon tool steel with high temperature materials (W, WC, TiC) by CO_2 laser and the effect of cobalt addition. *Fachberichte Metall-Praxis*, Vol. 65, No. 6, 1988, pp. 572-582.

211. **Ebner, R., Rabitsch, K., Major, B., and Ciach, R.**: Boride laser surface modification of SW7M (AISI M2) high speed steel. Proc.: I Polish Conference on *Surface Treatment*, Kule, 13-15 Oct. 1993, pp. 101-113.

212. **Kriszt, B., Ebner, R., Major, B., and Ciach, R.**: Vanadium carbide laser surface modification of SW7M (AISI M2) high speed steel. Proc.: I Polish Conference on *Surface Treatment*, Kule, 13-15 Oct. 1993, pp. 97-105.

213. **Archipov, V.E., Birger, E.M., and Smolonskaya, T.A.**: Structure and properties of claddings obtained with the utilization of the CO_2 laser (in Russian). *Metallovedenye i Termicheskaya Obrabotka Materialov*, No. 3, 1989, pp. 25-28.

214. **Liu, J., Qiquan, L., and Zhongxing, O.**: The effect of scanning speed of laser on composition and structure in chromium surface alloying layer. *Jinshu Rechuli Xuebao (Transactions of Metal Heat Treatment)* Vol. 14, No. 3, Sept. 1993, pp. 33-37.

215. **Liu, J., and Liag, H.**: The study of laser alloying on gray cast iron surface with silicon. *Jinshu Rechuli Xuebao (Transactions of Metal Heat Treatment)*, Vol. 13, No. 4, 1992, pp. 19-24.

216. **Tomlinson, W.J., and Brandsen, A.S.**: Fabrication, microstructure and cavitation erosion resistance of a gray iron laser surface alloyed with 22% c. *Surface Engineering*, Vol. 4, No. 4, 1988, pp. 303-307.

217. **Lin, J.**: Study of solidification for laser nickel-base alloying on cast iron surface. *Jinshu Rechuli Xuebao (Transactions of Metal Heat Treatment)*, Vol. 13, No. 2, 1992, pp. 24-28.

218. **Tanako, I.A., Levchenko, A.A., Guyba, R.T., Guyba, V.A., and Sitsevaya, E.J.**: Laser boriding of high-strength cast iron (in Russian). *Fizika i Khimia Obrabotki Materialov*, No. 5, 1991, pp. 89-95.

219. **Bogolyubova, I.V., Deriglazova, I.F., and Mulchenko, B.F.**: Laser surface alloying of AL35 alloy (in Russian). *Metallovedenye i Termicheskaya Obrabotka Materialov*, No. 5, 1988, pp. 24-25.

220. **Grechin, A.N., Shlapina, I.P., Grechina, I.A., and Yegorov, N.A.**: Enhancement of wear resistance of silumins by laser treatment (in Russian). *Metallovedenye i Termicheskaya Obrabotka Materialov*, No. 3, 1989, pp. 23-24.

221. **Grechin, A.N., Shlapina, I.P., Nabutovski, L.Sh., and Grechina, I.A.**: Laser alloying of component surface from silumins (in Russian). *Metallovedenye i Termicheskaya Obrabotka Materialov*, No. 3, 1991, pp. 12-15.

222. **Bekrenev, A.N., and Morozova, E.A.**: Modification of structure and properties of superficial titanium layers by laser alloying (in Russian). *Fizika i Khimia Obrabotki Materialov*, No. 6, 1991, pp. 117-122.

223. **Lyubchenko, A.P., Satanovski, E.A., Pustovoyt, V.N., Brover, G.I., Varavka, V.N., and Katselson, E.A.**: Some characteristics of pulsed laser hardening treatment of titanium alloys (in Russian). *Fizika i Khimia Obrabotki Materialov*, No. 6, 1991, pp. 130-134.

224. **Tomsiński, V.S., Postnikov, V.S., and Peleneva, L.V.**: Laser treatment of titanium and aluminium alloys. Proc.: VII International Congress on Heat Treatment of Materials, Moscow, 11-14 Dec. 1990, pp 24-30.

225. **Napadłek, W., Przetakiewicz, W., and Górka, A.**: Laser saturation of low carbon cladding by chromium (in Polish). Proc.: V International Symposium of the Institute of Mechanical Vehicles of the Military Technical Academy, Warsaw, 2-3 Dec. 1993, pp. 229-335.

226. **Bergmann, H.W., Breme, J., and Lee, S.Z.**: Laser hardfacing by melt bath reactions. Proc.: 2 European Conference on Laser-Metal Treatment *ECLAT '88*, 13-14 Oct. 1988, Bad Nauheim, pp. 70-73.

227. **Zhang, K., Zhang, Y., Zhao, J., and Ji, H.**: Study on laser surface alloying powder. Proc.: 5th International Congress of IFHT on *Heat Treatment of Metals*, Budapest, 1986, Vol. 3, pp. 59-64.

228. **Abboud, J.H., and West, D.R.F.**: Laser surface alloying of titanium with silicon. *Surface Engineering*, Vol. 7, No. 2, 1991, pp. 159-163.

229. **Abboud, J.H., and West, D.R.F.**: Processing aspects of laser surface alloying of titanium with aluminium. *Materials Science and Technology*, Vol. 7, No. 4, 1991, pp. 353-356.

230. **Bernstein, M.L., Kryanina, M.N., and Tsukin, W.N.**: Laser surface alloying of TRIP- type steel (in Polish). *Metaloznawstwo, Obróbka Cieplna (Metallurgy and Heat Treatment)*, No. 91-96, 1988, pp. 68-71.

231. **Yongqiang, Y., and Yuhe, Y.**: Research on the procedure and porosity of laser cladding WC-Co by powder feeder. *Jinshu Rechuli Xuebao (Transactions of Metal Heat Treatment)*, Vol. 13, No. 2, 1992, pp. 33-37.

232. **Molian, P., and Rayasekhara, H.**: Laser melt injection of BN powders on tool steels. I - Microhardness and structure. *Wear*, 114, 1987, pp. 19-27.

233. **Boll, P.O., Hauert, R., and Roth, M.**: Residual stresses in laser treated surfaces. Proc.: 2 European Conference on Laser-Metal Treatment *ECLAT '88*, 13-14 Oct. 1988, Bad Nauheim, pp. 180-183.

234. **Engström, H., Hansson, C.M., Johansson, M., and Sörensen, B.**: Combined corrosion and wear resistance of laser-clad Stellite 6B. Proc.: 2 European Conference on Laser-Metal Treatment *ECLAT '88*, 13-14 Oct. 1988, Bad Nauheim, pp. 164-168.

235. **Cantello, M., Pasquini, F., Ramous, E., Tiziani, A., Giordano, L., and La Rocca, A.V.**: Cladding of austenitic stainless steel by laser. Proc.: 10th Congress UIE, 18-22 June 1984, Stockholm, paper No. 6.9.

236. **Schmidt, A.O.**: Tools and engineering materials with hard, wear-resistant infusions. *Journal of Engineering Industry*, No. 8, 1969, pp. 549-552.
237. **Burchards, H.D., and Weisheit, A.**: Gasnitrieren von Titanlegierungen mit Laserstahl. Proc.: 2 European Conference on Laser-Metal Treatment *ECLAT '88*, 13-14 Oct. 1988, Bad Nauheim, pp. 64-66.
238. **Gasser, A., Kreutz, E.W., Schwartz, M., and Wissenbach, K.**: Gaslegieren von TiAl6V4 mit CO_2 Laserstrahlung. Proc.: 2 European Conference on Laser-Metal Treatment *ECLAT '88*, 13-14 Oct. 1988, Bad Nauheim, pp. 147-150.
239. **Seierstein, M.**: Surface nitriding of titanium by laser beams. Proc.: 2 European Conference on Laser-Metal Treatment *ECLAT '88*, 13-14 Oct. 1988, Bad Nauheim, pp. 66-69.
240. **Kosyrev, F.K., Zhelenzov, N.A., and Barsuk, V.A.**: Carburizing of low carbon steels with the utilization of continuous radiation by a CO_2 laser (in Russian). *Fizika i Khimia Obrabotki Materialov*, No. 6, 1988, pp. 54-57.
241. **Corchia, M., Delgou, P., Nenci, F., Belmondo, A., Corcoruto, S., and Stabielli, W.**: Microstructural aspects of wear resistance of stellite and Colmonoy coatings by laser processing. *Wear*, Vol. 119, 1987, pp. 137-152.
242. **Koichi, T., Futoshi, U., Yoshihiro, O., and Yasuo, K.**: Ceramic coating technique using laser spray process. *Surface Engineering*, Vol. 6, No. 1, 1990, pp. 45-48.
243. **Feinle, P., and Nowak, G.**: Auftragen von molybdänhaltigen Verschleissschutsschichten mit dem CO_2-Laser. Proc.: 2 European Conference on Laser-Metal Treatment *ELCAT '88*, 13-14 Oct. 1988, Bad Nauheim, pp. 73-75.
244. **Abbas, G., Steen, W.M., and West, D.R.F.**: Lasercladding with SiC particle injection. Proc.: 2 European Conference on Laser-Metal Treatment *ECLAT '88*, 13-14 Oct. 1988, Bad Nauheim, pp. 76-78.
245. **Molian, P.A., and Hualun, L.**: Laser cladding of Ti-6Al-4V with BN for improved wear performance. *Wear*, 130, 1989, pp. 337-352.
246. **Pons, M., Gharnit, H., Galerie, A., and Caillet, M.**: Revêtement de carbure de bore sur les élaboré - s sous irradiation laser. I - Élaborations de revêtements, II - Oxydation des revetements. *Surface and Coating Technology*, No. 35, 1988, pp. 263-273, 275-285
247. **Antoszewski, B., and Cedro, L.**: Laser deposition of claddings of alloys with Laves phases (in Polish). Proc.: Polish Conference on *Surface Treatment*, Kule, 13-15 Oct. 1993, pp. 143-144.
248. **Pantalenko, F., Sieniawski, J., and Konstantinov, W.**: Producing of protective coatings with boron on titanium and on carbon steel by the laser method (in Polish). Proc.: Polish Conference on *Surface Treatment*, Kule, 13-15 Oct. 1993, pp. 173-179.
249. **Singh, J., and Mazumder, J.**: Evaluation of microstructure in laser clad Fe-Cr-Mn-C alloy. *Materials Science and Technology*, Vol. 2, July 1986, pp. 709-713.
250. **Valov, V., Gemonov, V., Ivanova, V., and Yatronov, D.**: Enhancement of heat-resistance of austenitic steel by laser deposition of coating (in Russian). *Fizika i Khimia Obrabotki Materialov*, No. 6, 1986, pp. 80-83.
251. **Watanabe, I., Kosuge, S., Ono, M., and Nakada, K.**: Surface processing with a high power CO_2 laser. Proc.: Second International Conference on Surface Engineering. Stratford-on-Avon, 16-18 June 1987, pp. 131-140.
252. **Hinse-Stern, A., Burchards, D., and Mordike, B.L.**: Laserdrahtbeschichten mit vorgewärmten Zusatzwerkstoff. *Materialwissenschaft und Werkstofftechnik*, Vol. 22, No. 11, 1991, pp. 408-412.
253. **Kovalenko, V.S.**: *Treatment of materials by pulsed laser radiation* (in Russian). Publ. Vissaya Shkola, Kiev 1977.

254. **Mirkin, L.I.**: *Physical fundamentals of material treatment by laser radiation* (in Russian). Publ. MGU, Moscow 1975.
255. **Kovalev, E.P., Malshev, D.G., Ignatev, M.B., Melekhin, I.V., Uglov, A.A., and Voloshin, V.M.**: Application of laser synthesis of titanium nitride for life extension of tribotechnical nodes (in Russian). *Vestnik Masinostroenya*, No. 8, 1988, pp. 8-10.
256. **Endres, G., Kautek, W., Roas, B., and Schultz, L.**: Präparation von Hochtemperatur-Supraleiter-Filmen durch Laserstrahlverdampfen. Proc.: 2 European Conference on Laser-Metal Treatment *ECLAT '88*, 13-14 Oct. 1988, Bad Nauheim, pp. 103-105.
257. **Woliński W., and Nowicki M.**: Laser treatment of materials (in Polish). *Transactions of the Institute of Electron Technology*, Wrocław Technical University, No. 10, 1973, pp. 51-67.
258. **Feng, Z., Guo, L., Hou, W., Han, J., Liang, Y., Tong, B., and Wnag, Y.**: Laser induced films with hardness and excellent wear - and corrosion-resistance. Proc.: 4 th International Seminar of IFHTSE: *Environmental and Energy Efficient Heat Treatment Technologies*, Beijing, China, 15-17 Sept. 1993, pp. 179-182.
259. **Esrom, H., and Wahl, G.**: Modelling of laser CVD. Proc.: *Sixth European Conference on Chemical Vapour Deposition*, Jerusalem. Ed.: R.Porat, Nahariya, 1987.
260. *Laser hardening and cladding of components*. Brochure by the Ukrainian Academy of Sciences - E.O. Paton Institute for Electro-welding.
261. **Marczak, J.**: Art renovation with the aid of laser radiation (in Polish). *Przegląd Mechaniczny (Mechanical Review)*, No. 15-16, 1997, pp. 37-40.

chapter four

Implantation techniques (ion implantation)

4.1 Development of ion implantation technology

4.1.1 Chronology of development

Ion implantation takes its roots from solid-state atomic physics. From other radiation processes it differs mainly by the character of its effect on the crystalline lattice of irradiated materials, broad range of masses of implanted ions, their non-homogenous concentration distribution with depth of implantation and, consequently, radiation defects in the very thin subsurface layer [1].

The factor triggering the development of ion implantation was the rapid development of semi-conductor technology. After W. Shockley obtained his patent, in 1954, for implantation of dopants in the form of an ion beam, it was found that the technique may be competitive to traditional doping of semi-conductors, applied on an industrial scale since 1958. Broader research and first laboratory applications of ion implantation of dopants to single crystal semi-conductor materials followed during 1964-1965, while the first industrial applications came in the late 1960s and early 1970s. From the beginning of the 1970s, **ion beam implantation** has been utilized on an industrial scale in microelectronics as the best technique of precision doping of semi-conductors (including the utilization of secondary ion implantation and ion mixing), and in the production of highly integrated circuits. It allows the introduction of a strictly defined dopant within a broad range of concentrations, to very small depth of penetration and high (approximately 1 μm) surface and volume homogeneity of dopant distribution. Of special significance is the obtaining of shallow p-n junctions at low concentration levels [2-4]. Presently, the semi-conductor industry employs more than 2000 units of ion beam implantation equipment (ion beam implanters) [5], of which many allow the obtaining of precise volume concentrations of dopants down to 0.1% [6].

It was found sometime later that non-equilibrium but controlled implantation of ions of elements into subsurface layers of solids may be utilized to enhance polycrystalline materials. It can thus be competitive to diffusion saturation of surface layers of metallic materials in order to improve their service properties. In the 1960s, research was undertaken and in the 1970s, ion beam implantation was practically applied to improve mechanical properties (tribological, fatigue, corrosion, creep resistance, microhardness and ductility) of metals and alloys, predominantly, though, of pure metals. Presently, ion beam implantation technology outside of the semiconductor industry is gaining increasing practical application in highly developed countries [7]. Requirements regarding the precision of implantation are less stringent, while those regarding the depth of implantation are greater than in ion implantation of semiconductors [8, 9].

In the second half of the 1980s, J.R. Conrad and his collaborators from the University of Wisconsin proposed **plasma ion implantation** [10]. Somewhat later, independently, research in this field was undertaken at the Australian Nuclear Science and Technology Organisation (ANTSO) [11]. Research in the area of plasma ion implantation came later and is continues to this day at various research centers of the world, without, however, broader industrial application [12]. For this reason, plasma implantation will be discussed later and only in very general terms. The main emphasis will be placed here on ion beam implantation, gaining increasing practical application worldwide.

4.1.2 General characteristic of plasma and beam implantation of ions

Ion implantation is one of the ion techniques and belongs to the group of technologies for modification of structure, i.e., crystaline, geometric and chemical, of the superficial layer of solids, with the aid of ions.

Ions may come from:

- Plasma formed in the neighborhood of (around) a treated material surface. We then speak about **plasma technology** (plasma etching, plasma sputtering, plasma deposition - see Chapter 5 and 6); **plasma ion implantation** belongs to this group;

- Ion guns. We then speak about **ion technology** (e.g., ion etching, ion sputtering). A modification of this group is the ion beam technology in which a flux of ions of lesser or greater condensation is aimed at the treated surface. **Ion beam implantation** belongs to this group.

Sometimes, plasma and ion technologies are looked upon as aliases, especially when it comes to terminology.

A common characteristic of plasma and beam implantation of ions is the imparting, by means of electric energy, of such high kinetic energy to positive ions (see Fig. 6.1) that they can penetrate into the treated material to depth of even whole micrometers. Lesser kinetic energies cause ions to be deposited on the surface of the treated material or to penetrate to only very shallow depths.

There are some basic differences between plasma and beam implantation techniques. These are

1. Plasma is formed by a set of ions, usually with an energy of less than 1 keV (in most cases from 10^{-2} to 10^3 keV) and electrons, exhibiting all the characteristics of a gas, i.e., electrical quasineutrality, temperature and pressure. The beam is formed by positive ions with an energy usually higher than 1 keV (up to several MeV).

2. Plasma and an ion beam are obtained under different partial pressures. Usually, in an ion gun the pressure is lower by a minimum of 1 to 2 orders of magnitude (higher vacuum) than in plasma. There are also differences of pressure in work chambers. These differences usually fluctuate about approx. one order of magnitude. The work chamber in a plasma unit is the plasma chamber.

3. Different possibilities of exposing the treated object to the action of ions. In plasma techniques the treatement affects simultaneously the entire surface of the object while in beam techniques this is limited to the spot where the beam falls on the object. This requires scanning of the treated surface and rotation or movement of the object in order to prevent a shadowing effect.

4. Surface temperatures of the treated objects are usually different in the two techniques and may be within the range from ambient to several hundred K. Temperatures in plasma techniques are usually higher because better results are obtained above 300°C as opposed to below 200°C for beam techniques.

4.2 Plasma source ion implantation

Plasma source ion implantation (PSII), as it is known in the U.S., and PIII or simply PI³ (Plasma Immersion Ion Implantation), as it is known in Australia [11], consist of the formation of a plasma of the working gas, the introduction into it of the implanted object and the application of a high negative alternating potential (up to approximately 100 kV) - Fig. 4.1a.

At the moment of application to the treated object of a negative potential pulse, plasma electrons in the vicinity of the object begin to be strongly repelled from it in a time which is equal to a reciprocal of the plasma electron frequency. At the same time, positive ions of the plasma, due to their inertia (they are heavy), retain their positions [10, 13]. Thus, after the repulsion of the electrons, there remains behind them a uniformy charged zone of spatial positive charge, constituting an ion shell. Due to the forces of electrical attraction, the ions are accelerated (in a time which is reciprocal to the frequency of plasma ions) in a direction perpendicular to the object's surface. They strike it with a high kinetic energy and penetrate inside, i.e., they are implanted. Finally, the dropping density of ions of the internal zone (under the ion shell) causes a corresponding drop of electron density such that the shell expands at a rate close to the speed of sound (Fig. 4.1b).

In the case of a pulse which is greater than the reciprocal of plasma ion frequency but sufficiently short to prevent the ion shell from expanding to reach the wall of the plasma implanter, the energy of ions reaching the surface is equal to the product of the ion charge and the potential applied. It is usually contained within the range of 20 to 200 keV [13]. The voltage pulse is then repeated.

The working gas is usually nitrogen, less frequently hydrogen, argon or methane. The pressure of plasma is approximately $(2 \text{ to } 3) \cdot 10^{-2}$ Pa. Plasma may be

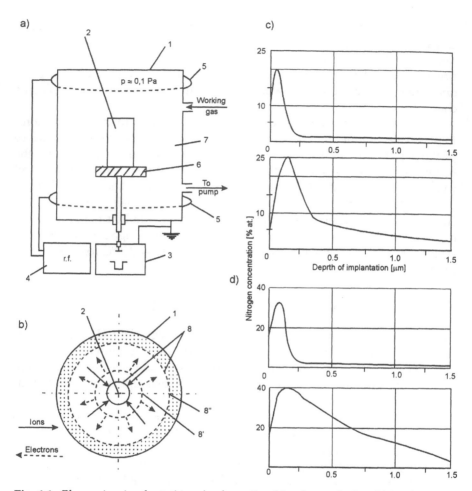

Fig. 4.1 Plasma ion implantation: a) schematic of implanter design; b) implantation model; c) typical profile of implantation of carbon steel by nitrogen; d) typical profile of implantation of stainless steel by nitrogen; *1* - work chamber, made of non-metallic material; *2* - implanted object (may be auxiliary heated) with high pulsating negative potential, up to -100 kV; *3* - source and modulator of pulses; *4* - generator of high (radio) frequency (12-14 MHz); *5* - inductor coils; *6* - base for fixturing object (may be heated and/or cooled); *7* - plasma; *8* - ion seam; *8'* - initial situation of seam edge, *8''* - subsequent situation of seam, expanding with ion acoustic speed, approximately 0.25 m/ms.

produced in a glow discharge, and its density varied by the emission of electrons from a glowing cathode, or by means of radio frequency discharges (11 to 14 MHz). Plasma density varies from 10^7 to 10^{11} ions per cm^3. Voltage generated by the pulse generator is usually from 50 to 100 kV, the duration of the pulse is from several to several hundred microseconds and time of repetition of the pulse is from several tens to several hundred μs. Ion current in plasma implanters is high - approximately 100 to 1000 mA. By comparison, in beam implanters it is only 10 to 15 mA. The power rating of plasma implanters may reach close to 1 kW [10-13].

By heating the object to a temperature of 300 to 400°C a severalfold greater depth of implantation is obtained and the implantation profile changes its character (see Section 4.3.1) (Fig. 4.1c,d). In the case of the best researched implantation of nitrogen ions (90% N_2^+ + 10% N^+) higher temperatures enable the obtaining of implantation depths exceeding 1 μm [12]. It is worthwhile mentioning that at treatment temperatures of 300 to 400°C and in times which are 10 times shorter than those of glow discharge nitriding, it is possible to obtain nitrogen diffusion to depths in excess of 1 μm. Some sources quote nitrogen diffusion in those conditions reaching depths of 100 μm [12].

Plasma implantation may be used for same applications (and similar materials) as beam implantation (see Sections 4.6 and 4.7). The implanted ion doses are similar, the most common being several × 10^{17} ions per cm^2. The results are similar to those obtained in low pressure glow discharge treatments. The time of implantation is approximately 2 to 3 h.

Because of similar pressures and same working gases as in the case glow discharge nitriding, it is possible to carry out duplex treatments in one equipment (glow discharge nitriding + implantation of nitrogen ions), as well as the application of a pulse generator in PVD units (e.g., TiN + N_2 coatings), especially those which utilize high energy ions in so-called ion assisted deposition, or in a combination with ion plating (plasma ion implantation + ion plating) [12].

4.3 Physical principles of ion beam implantation

Ions may be implanted into a solid **continuously** (long in use and well mastered) and by **pulse** (in the laboratory research phase, and insufficiently mastered) [8-10].

4.3.1 Continuous ion beam implantation

Continuous ion beam implantation involves constant introduction (implantation) of atoms of a selected element, in a condition of single or multiple ionization into a solid. This is effected due to the very high kinetic energy attained by these ions in vacuum ($6·10^{-5}$ Pa), in an electric field which accelerates the ions and forms them into a beam. The implanted ions, with energies ranging from in the teens of keV (1 eV = $1.602·10^{-19}$ J) to several tens of MeV, penetrating the solid, gradually lose their energy, due to two types of interactions: non-elastic, with electrons, and elastic, with nuclei of atoms belonging to the crystalline lattice of the host material subjected to the implantation[1] and becoming immobilized. During the initial period of their movement within the solid, the ion interacts mainly with free electrons and electrons belonging to coatings. This interaction is accompanied by the ionization of substrate atoms and the exchange of electrons between the implanted ion and substrate atoms. During this period of ion move-

---

[1] Sometimes the implanted material is termed target.

ment the main phenomenon occurring is deceleration by electrons. In the final stages of the ion movement, however, its collisions are of an elastic character and the dispersion of energy proceeds according to laws laid down by E. Rutherford.

Only an insignificant portion of the ions is expelled from the implanted material (initially dispersed), while a substantial amount causes the expulsion of atoms of the implanted material (ion sputtering). In certain conditions, an equilibrium is reached between penetration of ions into the implanted material, backscatter and sputtering. Conditions of the implantation process are selected such as to make ion penetration dominate sputtering. In some PVD processes, ion sputtering is the most utilized effect, e.g., magnetotron sputtering, with negligibly small penetration of ions [8, 9].

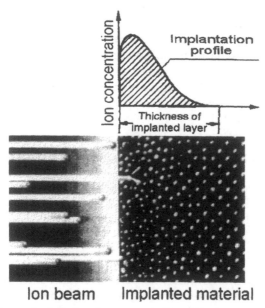

Fig. 4.2 Schematic showing the process of implantation.

As the result of ion implantation, a certain number of atoms are introduced into the subsurface zone, thus creating an implanted layer of 0.01 to 1 µm thickness (Fig. 4.2) and physico-chemical properties differing from those of the substrate. The range of penetration of the implanted ions, which could be termed **depth of penetration (implantation)**, and the distribution of implanted ions in the host material depend on their kinetic energy, atomic number, angle of incidence and on the properties of the host material, such as atomic number and mass of atoms forming it, as well as microstructure [3, 6]. Penetration depth increases rapidly with the rise of ion energy; at approximately 0.1 MeV it does not exceed 0.1 µm.

The range of penetration and distribution of implanted ions depend on the type of the host material. In the case of an amorphous body, these are

random variables with a given density of probability distribution, and therefore their distribution is of character similar to that of a Gaussian curve (Fig.
4.3) which can be described by the equation:

$$N(x) = N_0 \exp\left[-\frac{(R_p - x)^2}{2\Delta R_p^2} \right]$$

(4.1)

where: $N(x)$ - concentration of ions at distance x from surface [ions/cm³];
N_0 - concentration of ions at $x = R_p$ [ions/cm³]; R_p - effective range of penetration by an ion, dependent on real range R_c (projection of R_c on the x- axis
[μm]; R_c - real range of penetration, i.e., distance covered by the ion in the
implanted material - from the material surface to the place where it reaches
the state of immobility [μm]; ΔR_p- standard deviation of the R_p value [μm].

The Gaussian distribution curve characterizes the following:
– the location of maximum concentration R_p, the value of which depends predominantly on the energy of ions and their atomic mass m_1;
– the scatter ΔR_p, the relative value of which, $\Delta R_p/R_p$, depends predominantly on the ratio of atomic masses of implanted ions m_1 and atoms of
the substrate m_2.

In the case of ion implantation into a crystalline body, their range of
penetration, number and distribution depend predominantly on the orientation of structure relative to the direction of ion beam incidence (see
Fig. 4.3a). If the ions move within a substrate material along given crystallographic directions, e.g., [110] or [111], there occurs correlated interaction
of ions from the beam with atoms from the crystalline lattice and the range of
penetration of the ions rises by an order of magnitude or more which is
accompanied by a change in their concentration distribution. This phenomenon is known as **tunneling** (or channeling) of ions in the structure of the
substrate material. It is accompanied by a decrease in the number of defects
caused by the ions, while their range strongly depends on its crystallographic
orientation and on surface condition, its temperature, as well as on the direction and number of ions introduced [8]. The distribution of ion concentration
comprises several ranges:
– The first - corresponding to the distribution introduced by the non-
tunneling portion of the ion beam and containing more than 20% of the
implanted ions. The values of R_p and DR_p ascribed to it are like those for
an amorphous body.
– Second (and possibly subsequent) - corresponding to ions penetrating far deeper, thanks to the tunneling effect. The concentration of implanted ions usually decreases monotonically, often in a manner similar
to that of an exponential function.

The boundary of ion penetration is the maximum range R_{max}, determined by phenomena which dampen the tunnel effect, e.g., thermal vibrations of the crystalline lattice. In some crystals the damping of the tunnel

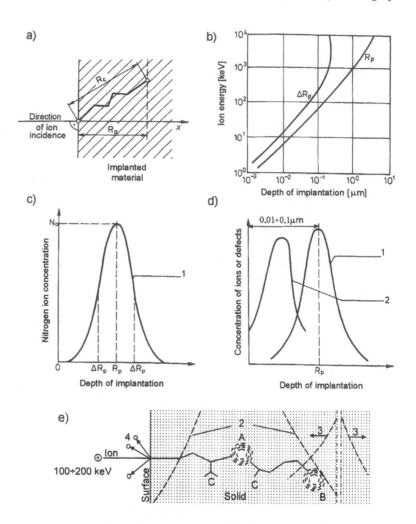

Fig. 4.3 Depth of implantation: a) schematic portraying the range of ion implantation; b) dependence of ion implantation range R_p and standard deviation ΔR_p on energy of implantation of nitrogen ions into iron; c) Gaussian implantation profile for nitrogen ion implantation in iron; d) distribution of concentration of ions implanted in an amorphous body and distribution of defects caused by them; e) defect formation in substrate C by an incident ion (several thousand atom translocations in the lattice) and the formation of sputter cascades A and B (several hundred vacancies and several hundred interstitial atoms); 1 - implantation profile; 2 - defects profile; 3 - implanted atoms diffusion; 4 - sputtered atoms.

effect is insignificant and then there may occur an additional maximum of concentration of implanted ions in the vicinity of R_{max}.

The curve which determines the distribution of implanted ions at different depths in the host material is known as the **implantation profile**. Maximum concentration of implanted ions, especially those of light elements, is not at the surface of the host material but at a distance of some tenths of a micrometer from it, due to backscatter and the non-elastic character of interaction of the ions with electrons of the host material. Implanted ions, colliding with atoms of the host material, causes their displacement which, in turn, causes the formation of radiation defects. Since the energy of the ion (several dozen keV) is several thousand times greater than the energy of atom bonds in the lattice (in metals it is approximately 25 eV), one ion is capable of knocking even several thousand atoms out of their nodular positions. In this way, the implanted ion generates along its path a strongly defected zone, known as a **cascade**, of several to several tens nanometers, which propagates laterally to the direction of the ion movement, due to the secondary interaction of atoms knocked out from their positions. The number of defects exceeds the number of implanted ions by a factor of 2 to 3 and usually is so big that there is a defect saturation [14]. Because the probability of defect formation depends on the cross-section of nuclear deceleration, the distribution profile of atom displacement of the implanted material (radiation defects) is similar to the profile of implantation, with the defect maximum, however, always occurring closer to the surface [5, 15] (Fig. 4.3d). When the ion dose exceeds 10^{14} ions per cm^2, separate disturbed zones superimpose and the enhanced density of point defects may cause the formation of amorphous zones, dislocations and microporosity, through the coagulation of point defects. The boundary value of the dose necessary to amorphize the substrate decreases with a rise in the mass of ions and with a drop in substrate temperature. The formation of microblisters may occur during implantation of metals by ions of inert gases [1].

The implanted ions may be located at dislocation boundaries, take up substitution positions or form inclusions of a new phase [14]. Many defects formed during the implantation suffer atrophy already at room temperature, while the distribution of the remaining ones is similar to that of an alloying additive. The number and range of atom displacement in the implanted material depend strongly on the **ion dose**, i.e., on the number of implanted ions per unit surface [3, 6, 9].

In the majority of cases of practical utilization of ion implantation, it is important to obtain in the surface layer of a concentration of the alloying additive from several to several tens per cent. The corresponding doses should be contained within the range of 10^{16} to 10^{18} ions per cm^2. Such doses are delivered within tens and hundreds of seconds [1].

For small doses, the implantation profile corresponds well to the theoretical Gaussian distribution (see Fig. 4.3c and Fig. 4.4 - curve 1').

An increase of the ion dose causes an increase in the number and range of atom displacements which are connected with significant defects in the crystalline lattice. Deepest penetrating are those ions, the incidence direction of which is in agreement with the direction of "empty channels" in the lattice (stemming from the spatial distribution of nodes) and with the crystallographic axis of the implanted material. This is the tunneling effect, mentioned earlier. Partially decelerated ions also migrate athermically

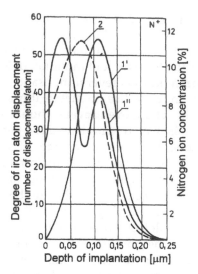

Fig. 4.4 Nitrogen ion concentration profiles of 100 keV energy, implanted into pure iron with a dose of $1\cdot10^{17}$ N$^+$/cm^2 (1' - at room temperature; 1" at 200şC) and distribution of displacement concentration (at room temperature) of iron atoms (curve 2) caused by the implanted nitrogen ions. (Curves 1' and 2 - typical curves, curve 1" - from *Iwaki, M.* [20]. With permission.)

along the crystalline lattice. There also occurs the effect of abnormal diffusion, consisting of a rapid migration of implanted ions along defects caused by the implantation. The rate of that diffusion strongly depends on temperature [3, 9, 14-18].

The theoretical correlation between concentration and dose of implanted ions Q_i is described by the equation:

$$N(x) = \frac{1}{2\pi} \frac{Q_i}{\Delta R_p} \exp\left[-\frac{(R_p - x)^2}{2\Delta R_p^2} \right]$$

(4.2)

which for $x = R_p$ assumes the form [3]:

$$N_0 = \frac{1}{2\pi} \frac{Q_i}{\Delta R_p}$$

(4.3)

From the above equations it follows that with a rise in the dose of ions, their concentration in the implanted material increases and that the character of the implantation profile may change.

In reality, the distribution concentration curve $N(x)$, for greater Q_i doses, deviates from the Gaussian distribution (Fig. 4.5) and is not symmetrical relative to the maximum value N_0 (for $x = R_p$) (see Fig. 4.4, curve 1").

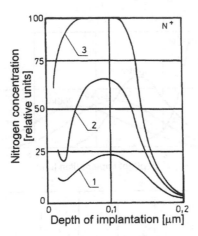

Fig. 4.5 Dependence of profile for nitrogen ions, implanted at 75 keV energy into pure iron on ion dose: *1* - dose: $3 \cdot 10^{17}$ N⁺/cm²; *2* - dose: $6 \cdot 10^{17}$ N⁺/cm²; *3* - dose: $1 \cdot 10^{18}$ N⁺/cm². (From *Iwaki, M.* [20]. With permission.)

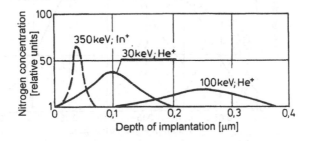

Fig. 4.6 Dependence of profile and depth of implantation on atomic number of ions implanted into gold, and on ion energy. (From *Deicher, M. et al.* [21]. With permission from Elsevier Science.)

Usually, deviations are observed at the surface in the direction of increased concentration, which is the result of ion etching of the implanted material, or under the surface of the material (for $x > 2\Delta R_p$), where ions may occur due to random tunneling (Fig. 4.6).

Besides the above, ion mixing of materials in the deceleration zone, as well as possible diffusion processes, causes deviations from the Gaussian distribution curve [14].

The depth of penetration of ions into the solid is relatively small - it exceeds 1 μm only exceptionally and drops rapidly with a rise of ion mass (Fig. 4.6). For ions of the same elements it rises with a rise of the accelerating voltage (Fig. 4.7a) and ion energy (Fig. 4.7b). In order to obtain an implantation profile without a sharp maximum peak it is possible to raise the energy of the ions during the implantation process. The final profile will be an algebraic sum of implantation profiles obtained at different energy levels (Fig. 4.7b). Another method is to increase the ion dose [8].

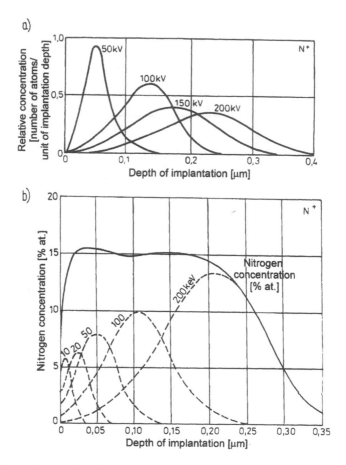

Fig. 4.7 Dependence of profile and depth of implantation of nitrogen ions into steel on: a) accelerating voltage; b) ion energy. (Fig. a - from *Hulett, D.M., and Taylor, M.A.* [19], Fig. b - from *Fromson, R.E., and Kossowsky, R.* [22]. With permission.)

The achievement of greater implantation depths and higher concentrations by raising the energy and dose of the implanted ions is counteracted by ion sputtering from the surface, causing etching of a portion of the material. One ion is capable of knocking out ten to twenty atoms from the surface. This effect occurs most strongly when the implanted ions are those of heavy elements. In that case, the implantation profile exhibits maximum ion concentration at the surface of the implanted material [8].

From the point of view of formation of chemical bonds between the atoms from the substrate and those implanted, ion implantation is a non-equilibrium thermodynamical process. Due to the thorough intermixing of atoms of the host material and implanted atoms, diffusion phenomena may occur 10^4 (and more) times faster than in normal metallurgical processes involving melting, thus yielding metastable phases, impossible to obtain by traditional methods [1].

Since implantation involves the introduction of ions to the host material but basically without arise in volume, ion implantation is accompanied by the formation of compressive stresses and a local rise in the surface temperature of the implanted material. The bombarding ion may bring about, in the cascade zone, local temperature of approximately 1000°C in a time shorter than 10^{-11} s. Heating up of the implanted material depends predominantly on energy and dose of ions and is described by the density of power supplied. At power density of 10 kW/m² the surface of the material heats up to approximately 100°C, while at 100 kW/m², it heats up to 350 to 500°C within several minutes. At the maximum achievable power density of 6000 kW/m², the material may melt or even vaporize [8, 9, 24, 25]. Usually, the implantation process is conducted in such a way as not to allow the temperature of the implanted material to exceed 200°C, thus minimizing or eliminating changes of properties and deformation of the implanted material [14].

4.3.2 Pulse ion beam implantation

On account of the non-stationary character of interaction of the pulse ion beam with the solid, due to the sufficiently short ion pulse (ns, μs) of very high energy, the beam causes melting of the thin surface layer of the solid (which is not observed with the application of the continuous beam) and the introduction of a foreign component (beam ions) into the molten liquid.

In the case of metals and alloys, modification of surface properties takes place as the result of the coexistence of three processes [26]:

– thermal processes: melting, recrystallization, rapid cooling (at rates of 10^7 to 10^{11} K/s);

– stress processes, caused by the propagation of stresses formed by the shock wave (ablative or dilatation) connected with extremely rapid vaporization of a portion of the molten material;

– physico-chemical processes, connected with the supply of a foreign component to the molten zone in the form of ions or material coating the substrate, which are sputtered out by the ions of the beam, combined with thermal processes; these are, naturally, alloying processes.

The relative participation of these processes depends on the parameters of ion beams, their mean energy and thermal properties of the material subjected to implantation. From the scientific point of view, these processes have not been thoroughly researched to date.

4.4 Ion beam implantation equipment

To carry out ion beam implantation, special ion accelerators are used, called **ion beam implanters** [27].

Ion implanters, in the most general sense, may be divided into two groups:

– with continuous ion beam, traditionally called simply ion implanters,

– with pulse ion beam, which are versions of high voltage ion diodes or ionotrons.

The first group has been used for ion implantation since the advent of implantation technology, while the second group is in existence since the 1980s and used only for laboratory purposes [9].

4.4.1 Continuous ion beam implanters

The main component systems of the implanter are (Fig. 4.8): the ion source, separator, focusing and accelerating system, deflecting system vacuum system and the working (implantation) chamber [28–35].

Ion source. This is the most significant functional element of the implanter, serving to produce the initially formed beam of positive ions of a given type. It comprises a discharge chamber where ionization takes place of gas, vapour or gaseous compounds of solids, and an extractor, used for extracting ions from the ionization zone, their initial formation into a beam and directing it to the focusing-accelerating system. Properties of the source are determined by technological possibilities and the effectiveness of the implanter [28–35].

Depending on the mechanism of ionization of the ion-forming substance, ion sources are divided into three groups [33–37]:

1. With **discharge in the gas phase**, often referred to as **plasma implanters**:

– **with extraction of ions from the discharge plasma**: spark (very seldom used in implanters), with capillary arc discharge (also seldom used);

– **with low voltage arc discharge in magnetic field or without a magnetic field**; the best known and used in implanters are the Bernas, Nielsen and Freeman sources. The Bernas source operates with an arc discharge at several tens to several hundred volts and a current of several tens amperes in a gas under pressure of approximately 1.33 Pa. It operates in a magnetic field generated by an electromagnet with controlled induction up to several hundred Gauss, perpendicular to the direction of ion extraction. The field serves to limit the escape of ions from the central portion of the discharge chamber and to enhance the effectiveness of ionization by extending the path of moving electrons (Fig. 4.9a). Discharge in the source takes place between the hot cathode and the anode (anti-cathode). In order to avoid condensation of the ionized material, the discharge chamber may be heated by a resistance heater and screened by foil, e.g., tungsten or molybdenum. At the same time it must be intensively water cooled. The ionized material can be a pure element or a chemical compound. It is supplied to the discharge chamber in the form of vapour under appropriate pressure in quantities ranging from fractions of a milligram to several hundred milligrams, depending on the dose of implanted ions and on the type of chemical substance used for the formation of ions of the given element. Vapours of the implanted elements or chemical compounds are formed in separate heating chambers and supplied in a controlled manner to the discharge cham-

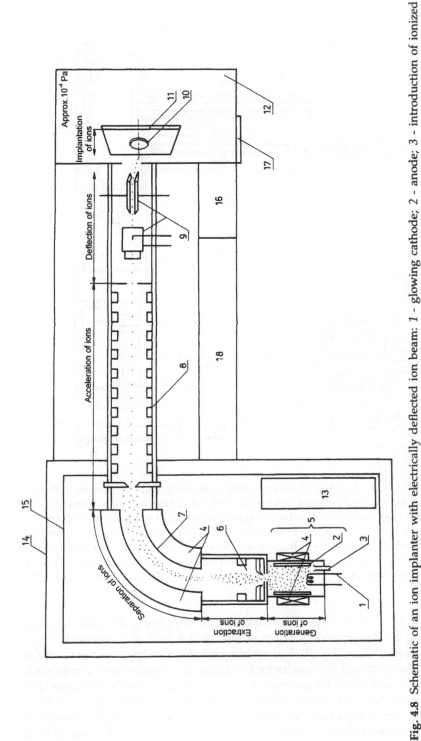

Fig. 4.8 Schematic of an ion implanter with electrically deflected ion beam: *1* - glowing cathode; *2* - anode; *3* - introduction of ionized medium; *4* - magnets; *5* - ion source; *6* - extraction chamber; *7* - ion separator; *8* - accelerating tube; *9* - deflecting system; *10* - implanted material (load); *11* - stage for fixturing load; *12* - work chamber; *13* - power supply; *14* - protective screen (safety); *15* - high voltage zone; *16* - control console; *17* - viewing port; *18* - pump system.

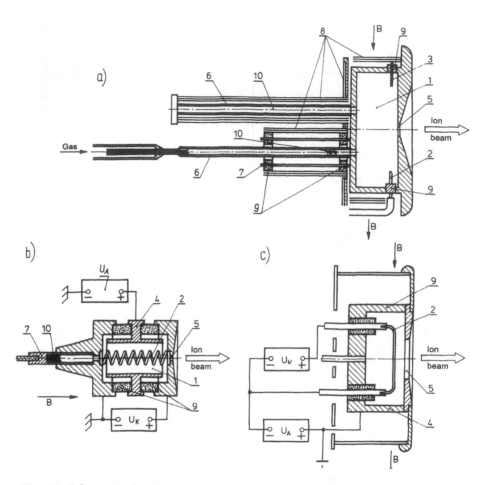

Fig. 4.9 Schematic showing ion sources with low voltage arc discharge: a) Bernas type - in magnetic field perpendicular to direction of ion extraction; b) Nielsen type - in magnetic field parallel to direction of ion extraction; c) Freeman type - in magnetic field perpendicular to direction of ion extraction; *1* - discharge chamber; *2* - cathode; *3* - anticathode; *4* - anode; *5* - extraction aperture; *6* - reservoirs; *7* - heater; *8* - reflector; *9* - insulators; *10* - primary material; *B* - direction of induction of magnetic field. (Fig. a - from *Drwięga, D., et al.* [28], Fig. b and c - from *Mączka, D., et al.* [31]. With permission.)

ber. Usually there is also the possibility of supplying an auxiliary gas, sustaining the discharge(e.g.,argon or other noble gas), or a chemically reactive gas (e.g., PCl_3, ClF_3, Cl_2, F_2), enabling the formation of halogens of the desired elements and the conduction of a chemical synthesis of compounds in the ion source, in order to obtain a material which is easily ionizable [28]. High temperature of reservoirs and of the discharge chamber (up to 1600°C) allows the obtaining of ions of the majority of elements at high intensity of extracted currents (up to several mA) and with 2-, 3- and 4-fold ionization of the elements which, in turn, allows high energy of ions without the need to raise the extraction voltage [28–35].

In the Nielsen source (Fig. 4.9b) the cathode, powered by direct current in the form of a coil, supplies electrons to the discharge zone and the anode is a graphite cylinder. The source serves to produce ions of both gaseous and solid elements.

The above-described ion sources belong to the most versatile and allow the production of the majority of ions within the range of atomic masses from 1 to 240. Their characteristic feature is the hot cathode and the heater which enables vaporization of solid materials. Such sources are used mostly in laboratory implantators [30].

In industrial implantators there are usually applied specialized sources of ions of one element, often with a cold cathode which, although featuring long service life, allow the obtaining of relatively small currents of the ion beam and can operate with gases only [31]. An example of an ionization source, especially for the generation of strong fluxes of nitrogen ions, is the J.H. Freeman slot source (Fig. 4.9c) in which the cathode, made of tungsten wire, is located very close to the extraction opening in the shape of a slot. This, in conjunction with the strong cathode glow current (up to approximately 100 A) and the effect of the magnetic field, perpendicular to the direction of ion extraction, allows the obtaining of maximum plasma density opposite the extraction slot [31];

– **with electron oscillation** (so-called F.M. Penning sources), also referred to as **sources with cold or hot cathode**. The best known are G. Sidenius sources and their different modifications. Fig. 4.10 shows the principle of operation of such a source. The ion emitter here is plasma of low pressure discharge, ignited in the electrode system, in which the cathode has the shape of a hollow cylinder (either a wire coil [Fig. 4.10b] or solid [Fig. 4.10c]), forming a niche. The glow of the hot hollow cathode is from direct current and the cathode is heated to a temperature at which a high electron emission current is obtained (Fig. 4.10b). When appropriate pressure conditions are met, arc discharge is activated in the discharge chamber and plasma is formed, screened from the hot cathode by a bipolar layer of spatial electrical charge. Practically the entire used interelectrode voltage is situated in this layer, and in this zone electrons, emitted by the cathode, achieve sufficient energy for the ionization of the gas. In the discharge chamber there may or may not exist a magnetic field. In the second case, the magnetic field created by the flow of current through the cathode may be compensated by an external field. In the absence of a magnetic field, electrons move perpendicularly to the axis of the source and after passing through plasma they are decelerated by the electric field in the layer situated opposite the spot where they are emitted by the cathode. The electrons are slowed down, their direction is reversed and they are accelerated in the direction of the discharge plasma; next, decelerated at the cathode, and thus the cycle repeats itself. Electrons oscillate inside the hollow cathode. Maximum ionization occurs near the

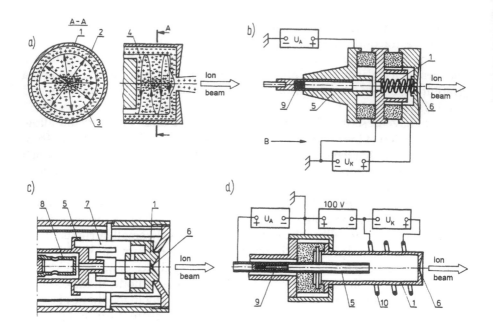

Fig. 4.10 Hollow Cathode Ion Source (HCIS): a) principle of operation; b) Sidenius ion source with hot cathode (tungsten); c) ion source with cold cathode (graphite); d) ion source with hot cathode heated indirectly; *1* - hollow cathode; *2* - bipolar layer; *3* - plasma zone; *4* - electron paths; *5* - anode; *6* - extraction aperture; *7* - insulator; *8* - reservoir; *9* - evaporator; *10* - tungsten coil. (Fig. a, b and d - from *Mączka, D., et al.* [31]. With permission.)

source. By the application of an appropriate extraction voltage, ions are extracted from the source in the form of a beam of high intensity [31]. The hot cathode source was demonstrated for the first time by G. Sidenius [29]. It allows the obtaining of dense plasma in the discharge zone with a relatively high efficiency of the ionization process [37].

In the cold cathode source (Fig. 4.10c) ionization of gas or of vapours from the vaporizer is obtained with the utilization of electron bombardment, emitted by the cold cathode under the influence of the applied voltage. Usually, the beginning of operation of the source is initiated by discharge in argon which is later replaced by a different material, designated for ionization [29].

Fig. 4.10d shows an ion source with a cylindrical cathode, heated directly through bombardment by electrons emitted by a glowing tungsten filament. After heating of the cylindrical cathode to the appropriate temperature, electrons emitted from the internal surface of the cylinder are accelerated in the electrical field between the anode and cathode and they ionize the substance designated for the implantation process. This source may also operate as thermoemissive, which is of special importance in the generation of ions of elements with a low ionization potential [31];

– **very high frequency** (seldom used in implanters);

– **duoplasmotrons**, i.e., sources with extraction of ions from plasma which diffused from the discharge chamber to the zone of beam formation. Rather seldom used in implanters; characterized by high values of ion currents.

2. **Thermoemissive** - with thermal emission of ions from the surfaces of solids (Fig. 4.11). This source operates on the principle of utilization of ionization of atoms colliding with the surfaces of metals. The hollow tungsten cylinder with the extraction orifice of 0.2 mm diameter, together with rhenium foil is heated to approximately 2500 to 3000°C by electrons emitted by glowing external cathodes. Atoms of the substance designated for implantation pass from the vaporizer to the inside of the hollow cylinder where, after making contact with the surface of the rhenium foil they undergo thermoionization. Such sources allow the obtaining of ions not only of single atom elements but also of molecules [31].

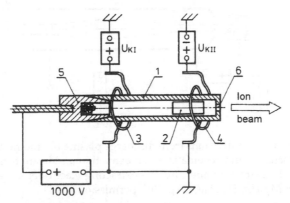

Fig. 4.11 Schematic of thermoemissive ion source: *1* - tungsten cylinder; *2* - rhenium foil; *3* - cathode I; *4* - cathode II; *5* - evaporator; *6* - extraction aperture. (From *Mączka, D., et al.* [31]. With permission.)

3. **Field** - with surface ionization in which the difference between the work of exit of electrons from the metal and the ionization potential of elements for ionization is utilized.

4. **With bombardment by accelerated electrons** (not used in implanters).

Extraction systems. These systems are built of one, two or three flat, cylindrical or conico-cylindrical electrodes, the first of which serves to extract ions and the remaining two to the formation of the beam (Fig. 4.12). To the electrodes, voltages are applied of several tens kV. Control of the extraction system involves either a change of value of the applied voltage, or change of location of electrodes relative to ion source [8, 9, 28, 30–37].

In the case of single electrode extraction systems, the extraction and acceleration of ions occur with the aid of a conico-cylindrical electrode, placed near the extraction orifice with a negative potential relative to

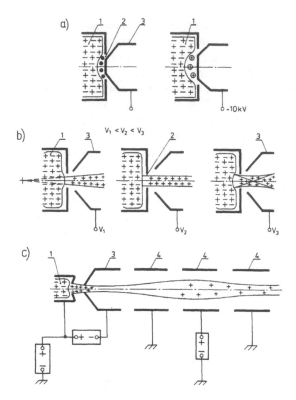

Fig. 4.12 Extraction of positive ions from discharge plasma (a), methods of formation of plasma meniscus (b) and schematic of ion extraction and beam formation system (c): *1* - plasma; *2* - extraction aperture; *3* - extracting electrode; *4* - electrodes; *V* - potential. (From *Mączka, D., et al.* [31]. With permission.)

that of the ion source (Fig. 4.12a). Depending on the value of this potential and on the discharge parameters, plasma situated in the source may flow or pass into the source interior. Its boundary surface assumes the shape of a convex or concave meniscus (Fig. 4.12b) [31].

Focusing-accelerating system. This system comprises a high voltage power supply and a single optical system, generating an appropriate electrical field in which ions are accelerated and formed into a beam of round cross-section (usually of 2 to 10 μm diameter) or rectangular. In the simplest case, the focusing system comprises three electrostatic lenses (Fig. 4.12c). Usually, the single optical system, also termed the acceleration pipe, contains several to between ten and twenty beam stops of increasing potential. The rule is that the lower the accelerating voltage, the less stops. In a system with 40 kV, there is only one beam stop. The selection of the voltage depends on the desired depth of ion penetration and on the type of ions used. In semiconductor implantation technology, accelerating voltages used are in the range of 5 to 300 kV; in non-semiconductor implantation, the current range is 50 to 100 kV for industrial implanters and up to 500 kV in laboratory units [9, 28].

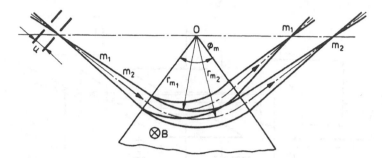

Fig. 4.13 Principle of action of sector magnetic field on a divergent beam of monoenergy ions: m_1, m_2 - ion masses; Φ_m - angle of divergence of magnetic field; U - potential accelerating ions; r_{m1}, r_{m2} - curvature radii of ion paths with masses m_1, and m_2; B - vector of induction of magnetic field.

Mass separators. Separators are used for precise selection of ions in the beam, based on an analysis of mass of the beam ions. Separators allow through only ions with a given e/m (charge to mass) ratio, utilizing the effect of the homogenous magnetic field, limited by two planes (so-called magnetic lens). This field affects the beam of single energy ions but of different masses in two different ways: it focuses the divergent beam of ions of same masses but the sites of focusing depend on masses (Fig. 4.13) in accordance with the equation [31]:

$$r_m = \frac{mv}{eB} \tag{4.4}$$

where: r_m - radius of curvature of ion path; v - ion velocity; B - magnetic induction.

Thanks to this, the beam passed through the separator is free from contaminations in the form of ions or particles of a mass other than desired. The source of such contaminations may be incompletely purified gases or vapours, desorbed particles of gas from the ion source electrodes, atoms sputtered out of the electrode surface and ions from vapours of an ionized compound, supplying the implanted ions. For example, if the substrate of Cr ions is CrCl, the Cl ions are separated and arrested in the separator, while only Cr ions are allowed to pass through, similarly to Na ions, obtained from the NaCO compound [22].

Most often used separators are electromagnetic, as well as separators operating on the principle of crossed electrical and magnetic fields [16].

Scanning systems. To scan the treated material with an ion beam in order to ensure the required homogeneity of the implantation process, the following systems are used:

1. System for deflecting the ion beam in the x-z axis (scanner):

a) mechanical - in the form of a rotating and possibly laterally moving shield with openings, modulating the ion beam mechanically [39];

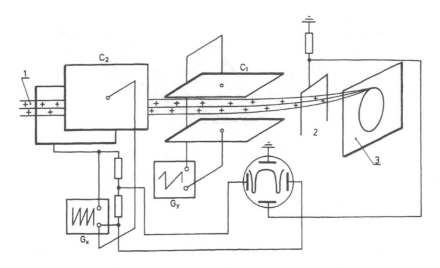

Fig. 4.14 Schematic of scanning system (collector) of Polish-built UNIMAS-79 implanter: *1* - ion beam; *2* - probe; *3* - implanted material. (From *Mączka, D. et al.* [31]. With permission.)

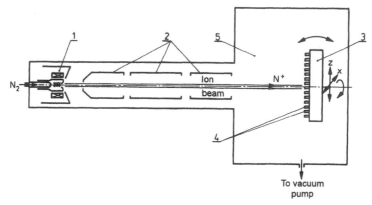

Fig. 4.15 Simplified schematic of ion implanter (without separator and beam deflector) with fixed beam and movable stage (in *z-x* plane): *1* - discharge chamber with ion source and extracting electrode; *2* - three-electrode beam focusing system; *3* - movable stage; *4* - treated load; *5* - work chamber.

b) electrical - in the form of two generators of saw-tooth shaped voltage, producing voltage with an amplitude dependent on maximum ion energy and on size of implanted surface, with a frequency of deflection voltage from 1 to 100 kHz (Fig. 4.14).

2. System for mechanical feed in the *x-z* plane of the stage with the implanted material and a fixed ion beam (Fig. 4.15) with possible masking off of beam.

3. System for mechanical deflection of the ion source together with the extraction and acceleration system, usually in one direction, with masking off of ion beam.

The last two systems usually find application in industrial implanters for the implantation of gases and metals [33-35].

Vacuum systems. Systems of vacuum pumps, valves and vacuum gauges serve to obtain vacuum in the zone of extraction and acceleration within the working chamber. High vacuum is of special significance in systems with low working voltages, while soft vacuum in the single optical system may cause losses of beam current even up to 90% and deteriorate resolution [8, 9, 28, 30].

Work chamber. This chamber is used for placing of the treated load. It should be equipped with mechanical systems for fixing and moving (in two directions) or rotating the load (of special importance for complex shapes) in such a way that the treated elements do not mutually screen off the beam incidence line (Fig. 4.16). The chamber is, moreover, equipped with loading/unloading systems which are usually cooled. The cooling is either forced by a water jacket or water canals, or it may be natural, through radiators which give off heat. Laboratory implanters are equipped with more systems than industrial implanters, e.g., heaters, goniometer, refrigerators, etc. [40].

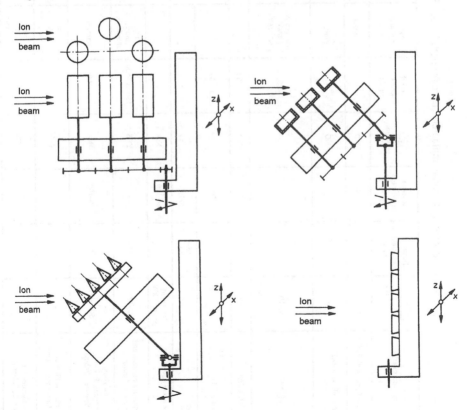

Fig. 4.16 Examples of providing movement to load by means of movable z - x stage. (From *Podgórski, A. et al.* [38]. With permission.)

Table 4.1

Technical data relating to some implanters (Data from [8, 17, 33, 34, 42] and various other sources)

Brand (manufacturer)	Ion energy [keV]	Beam current intensity [mA]	Accelerating voltage [kV]	Mass separator	Load temperature [°C]	Type of implanted ions	Maximum zone of implantation [mm]	Load mass [kg]	Remarks
PIMENTO (ARE, Harwell, U.K.)	10–100	5	100	NO	-	nitrogen	50–80	-	First industrial implanter. Working chamber 20 times smaller than the high current unit
High current implanter (ARE, Harwell, U.K.)	60	10–25	-	NO	-	nitrogen	1000×20	1500	Used for implantation of large size components. Diameter of working chamber: 2.5 m; length: 2.5 m. Pressure: $1.3×10^{-5}$ Pa.
ZYMET Z 100(USA)	50–100	5	-	NO	<200	nitrogen N$^+$ (40%) + + N2$^+$ (60%)	20×20 or φ250	-	First serial-produced implanter for tooling and machine components. Maximum load size: 300 mm. Homogeneity and reproducibility of implantation: ±10%. Typical dose: $3×10^{17}$ ions/cm^2. Duration of implantation for a single component: 30 min.; for a multi-component load: 90 min.
ZYMET Z 200(USA)	100	10		NO	ca. 200				For dynamic ion mixing
TECVAC 221(U.K.)			90		ca. 100		10×16.5		For implantation of primary ions and dynamic ion mixing. Maximum load size: 1 m; duration of implantation: 40 min.
Westinghouse (USA, 1983)		0.5–2.5	100	-	-	-			For implantation of tooling
Fraunhofer Institute (Germany, 1980)	Ion beam power:1 kW	1	60–100	NO	-	-	35×35	35	-
Riken (Institute for Physical and Chemical Research, Japan)	20–200	13	-	YES	–150 to +500	titanium, aluminum	15×24	-	
IMTE (Institute for Basic Technical Problems, Polish Academy of Sciences, Poland, 1986)	50–100	1	60–100	NO	20–600	nitrogen	15x15	-	Implanter designed by Institute for Electronic Materials Technology
WiD-63 (Institute of Physics, Marie Curie University, Poland)	80			YES	-	varied 1–200 atomic mass			Radiation homogeneity: 5%
UNIMAS-79 (Institute for Nuclear Research, Poland)	330			YES	up to 600	varied 1–200 atomic mass	φ50		Radiation homogeneity: 3%

Single and multibeam implanters. Ion implanters constitute a design solution of mass spectrophotometers, from which they differ by values of ion currents and by ion energy. In the ion implanters, currents are more than a million times stronger and range from 1 to 100 µA [31].

The majority of ion implanters are single beam units (see Table 4.1) Double beam implanters, allowing simultaneous direction of two ion beams from two independent sources, are used very seldom. Even more seldom used are three-beam implanters, while only exceptionally - and mainly for laboratory purposes - four-beam implanters.

In the world today there are between 3000 and 4000 ion implanters, laboratory and industrial units. They are used mainly for semiconductor implantation, less often for other types of implantation. Usually, laboratory implanters are more versatile, therefore of a more complex design. Industrial implanters are usually built as simpler versions, often without ion separators, in most cases designated only for the implantation of one type of ions.

4.4.2 Pulse ion implanters

Pulse ion implanters comprise high voltage ion diodes and ionotrons.

Ion diodes for implantation purposes, mainly for the semi-conductor industry, came into use in the later 1970s [29, 31]. In the early 1980s, Russian physicists from the Institute of Nuclear Physics in Tomsk began tests aimed at utilization of pulse ion beams, produced with the aid of high voltage ion diodes with magnetic insulation (Table 4.2).

Table 4.2
Technical data related to pulse ion implanters (From *Piekoszewski, J., and Langner, J.* [45]. With permission.)

Type	Ion energy [keV]	Pulse duration [µs]	Energy density [J/cm^2]	Ion dose [ions/cm^2]	Input energy [kJ]	Type of ions
Ion diode(Russia)	200-600	0.05-0.1	1-8	10^{12}-10^{14}		Hydrogen +carbon
RPI-15(Poland)	5-30	0.3-1.0	1-6	10^{12}-10^{15}		Deuterium or nitrogen
Ionotron-86(Poland)	5-20	1	2-5.6	10^{13}-10^{15}	30	Hydrogen or nitrogen

A pulse ion beam obtained with the use of parameters given in Table 4.2 may fill the role of a carrier of energy with deep penetration - up to 35 to 45 µm in steel and up to 150 to 180 µm in aluminum and copper [45]. This allows the remelting of the substrate or of prior deposited layers but due to the low ion doses it is not suitable for doping or alloying of steels or alloys.

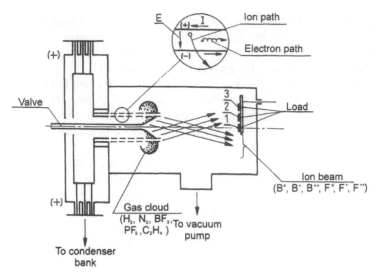

Fig. 4.17 Schematic of ionotron design. (From *Piekoszewski, J. et al.* [26]. With permission.)

Ionotrons belong to ion generators of the rod gun type. The ionotron is a pulse plasma accelerator, developed for the needs of ion fusion. The ions are produced in a strong current electric (plasma) discharge in a gas under partial pressure between transparent cylindrical electrodes. Acceleration of the ions takes place with strong choking of the electron current flow by its own magnetic field (Fig. 4.17). The generated pulsed ion-plasma beams, with currents of the order of several tens kiloamperes and momentary powers reaching gigawatts, enable the remelting of the subsurface layer to a depth of approximately 1 µm, with the simultaneous introduction of the doping component [26, 45]. Low energy of the ions (5 to 20 keV) allows the doping of semiconductor materials but is rather too low for alloying of steel. Initial results indicate that in some cases, after the remelting of the surface, its microhardness rises but its roughness deteriorates. Microcraters - so-called "orange peel" - may be formed [46]. It seems that the development of ionotrons, aiming at combining implantation with strong heating of the surface, but without the need to remelt it, may also enable alloying of metals and alloys, sufficient for technological applications [46]. This can be achieved through giving the ions energies of the order of tens and even hundreds eV, extension of pulse duration to approximately 10 to 100 µs and an increase of the ion dose to 10^{16} to 10^{18} cm^{-2}.

4.5 Ion beam implantation techniques

The fundamental advantage of ion implantation technology is the possibility of implanting[1] of any material, metal or non-metallic, by any chosen components - ions of gases or solids. Besides, the implanted material may be

---
[1] Implanting of semiconductor materials is referred to as *doping*, while implanting of metals - *ion alloying* [1].

either not coated, as in the overwhelming majority of cases, or prior coated, usually by a layer of metal, less often by a combination of layers or coated during the implantation by atoms or ions, in most cases of metallic elements. This last case has, so far, found the least number of applications but it appears to be very promising. Limitations of the level of implantation, due to sputtering of the surface by the ion beam, may be significantly minimized by the application of different modes of the implantation process, allowing the achievement of a series of additional advantages [15, 47].

Depending on the energy of the ions, on their dose and on the rate of implantation, as well as the absence or presence of layers or atoms of foreign metals and on the number of beams used in the process, all versions of the implantation process may be divided into four groups (Fig. 4.18) [20, 48].

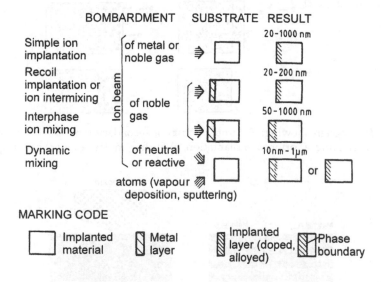

Fig. 4.18 Diagram showing different types of implantation.

1. **Simple Ion Implantation**. The oldest, best known and best researched technique of implantation, used for alloying of metallic materials with no prior coating. It allows the alloying of materials up to between 10 and 20% of the implanted elements, depending on the type of implanted material and on the ions [38, 48]. No measurable dimensional or roughness changes of the surface of implanted material take place.

2. **Recoil Implantation**. Implantation through recoil by a prior deposited layer. It consists of bombarding a thin layer, deposited by electrolysis, PVD or CVD techniques, on the surface of the implanted metalic material by ions of noble gases. Due to collisions with ions, the atoms of that layer attain significant energy and penetrate into the implanted material

(Fig. 4.19A) [38, 49]. This technique allows the level of implantation to exceed 50% but requires the use of implanters with higher working parameters than those used for simple ion implantation. The energy of the ions depends on the thickness of the deposited layer and is usually well over 150 keV (for simple ion implantation 60 keV is sufficient), while the ion current is from approximately 10 μA to approximately 100 mA. The typical duration of the implantation processs ranges from 10 to 100 s/cm² of surface [48].

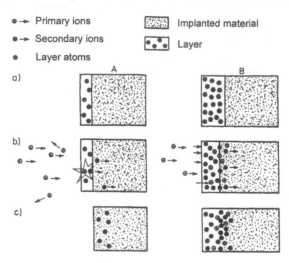

Fig. 4.19 Diagram showing: *A* - implantation of secondary ions; *B* - interphase ion beam intermixing; a) prior to implantation; b) during implantation; c) post implantation.

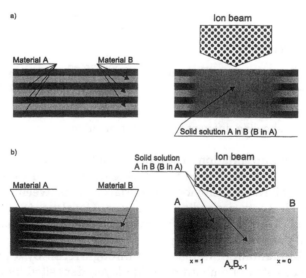

Fig. 4.20 Examples of interphase ion beam intermixing (also true for dynamic intermixing) for two different multi-layer structures: a) with fixed thickness; b) with variable thickness.

Since a change of process parameters may be accompanied not only by recoil but also simple implantation, a process carried out in this version is termed *Ion Mixing* (or *Ion Beam Mixing*). Both versions cause no change in the geometry of the implanted material [8, 9, 25, 44, 51, 52].

3. **Ion Beam Intermixing** - constitutes a version of ion beam mixing, used for greater layer thicknesses or for a composite of layers (A and B), deposited on the substrate predominantly in order to improve the bonding between it and the layer [53]. By appropriate selection of process parameters, intermixing of phase boundaries is obtained (see Fig. 4.20). This includes interfaces: substrate - layer (and A - B). The result is that the deposited layer is characterized by gentle transition into the coated material and by very good adhesion (Fig. 4.19B) [44]. There is a change in the dimensions of the coated material. This technique is used, e.g. to improve the adhesive bonding of a TiN layer, obtained by any PVD technique, deposited on tool material, stainless steels, etc. [54]. The technique may be used at higher temperatures of the implanted material in order to enhance diffusion.

4. **Dynamic Mixing** [20] (known also as *Dynamic Recoil Mixing* [56] or *Ion Enhanced Position Mixing* [49, 57]) - is the latest technique, showing definite signs of promise, constituting a version of ion mixing in which the mixing process takes place either during subjecting of the substrate to simultaneous implantation by two or more ion beams, or during deposition onto the substrate of a coating by evaporation (thermal - Fig. 4.21,

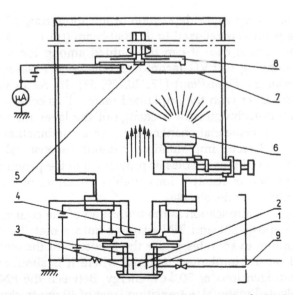

Fig. 4.21 Schematic of laboratory equipment for dynamic ion beam intermixing: *1* - ion emitter; *2* - ion source; *3* - magnet; *4* - system of focusing-accelerating electrodes; *5* - implanted and coated material; *6* - source of vapors of metal vaporized by ion beam; *7* - layer thickness gage; *8* - fixture for implanted and coated material; *9* - inlet for ionized gas in ion source. (From *Wolf, G.K.* [48]. With permission.)

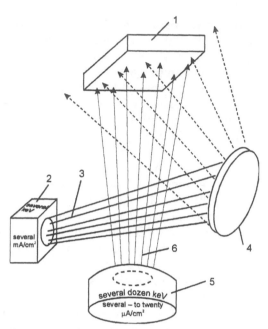

several keV

several mA/cm²

several dozen keV
several – to twenty
µA/cm²

Fig. 4.22 Diagram showing method of dynamic ion beam intermixing, utilizing ion sputtering: *1* - surface to be vapor deposited or coated; *2* - source of low energy ions sputtering target material; *3* - plasma; *4* - target made from sputtered material; *5* - source of high energy ions; *6* - ion beam.

electron, or laser), by vapor deposition or ion sputtering (Fig. 4.22). In the first case the process is referred to as dual-beam (tri- or quad-beam) deposition. In the second case it can be counted among ion beam enhanced processes belonging to PVD techniques (*Ion Beam Assisted Deposition or Ion Beam Enhanced Deposition*) [17, 20, 29, 58]. Its result is either better bonding of the layer with the implanted material, as compared to conventional adhesion techniques or a change of the layer microstructure (accompanied by dimensional changes) or even a combination of simple ion implantation and recoil implantation (without dimensional changes). Implantation of neutral ions enables improvement in the penetration of layers, while implantation of reactive ions enables obtaining of different surface layers, e.g., nitride, oxide, etc. [48].

An example of application is the utilization of the equipment shown in Fig. 4.21 to deposit BN and TiN coatings onto a metallic substrate. The boron or titanium evaporates under the influence of the electron flux and is deposited on the implanted material where it is simultaneously bombarded by nitrogen ions of 30 keV energy. Between the BN or TiN layer and the implanted material, a transition layer of 40 nm thickness is formed, comprising ions of nitrogen, implanted into the coated material [48]. Special mention should be made of the fact that the bonding obtained by this technique is 10 to 45 times stronger than a conventional adhesion connection [59]. Besides, the thickness of the deposited layer is not limited by physical factors [58]. The method may be utilized not only for implantation of metal ions but also for insulators and also for the production of very thin, continuous layers (e.g., gold of 4 nm thickness), impossible to obtain by any other method without a loss in layer continuity [59].

Fig. 4.23 shows diagrams of realization of ion implantation processes.

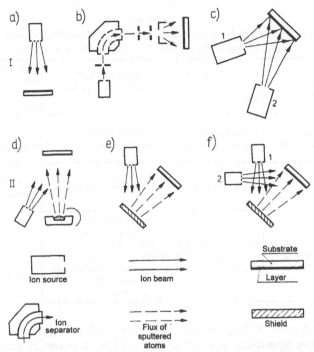

Fig. 4.23 Diagrams showing how ion implantation processes are accomplished: *I* - primary or secondary ion implantation; a) Direct Ion Beam Implantation; b) Ion Beam Implantation with utilization of mass separator; c) Dual Ion Beam Implantation; *II* - Ion Beam Mixing; ion beam assisted processes of coating deposition by PVD methods; d) Ion Beam Assisted Evaporation; e) Ion Beam Deposition by Sputtering; f) Dual Ion Beam Deposition; *1* - first ion source; *2* - second ion source.

4.6 Modification of properties of implanted materials

The utilization of ion implantation to the modification of properties of implanted materials was made possible thanks to the advent of high current efficiency implanters, exceeding 1 mA. For several gases, e.g., nitrogen, this efficiency may exceed 15 mA and even reach 300 mA [4]. Implantation allows the selective introduction of one or more elements into the surface layer of a metal (either uncoated or coated by a layer of a different metal) regardless of thermodynamic equilibrium [60]. This causes the obtaining of non-equilibrium structures [15], amorphous structures [16, 61, 62], supersaturated solutions and metastable compounds and phases [60, 63, 64]. Strong compressive stresses prevail in them. At the same time, implanted layers feature significant defects (Table 4.3) [65–67]. The introduction of foreign atoms and the formation of fine phases limit dislocation movements, thereby increasing strength [68, 69]. The initial structure can usually be obtained by recrystallization [10, 14].

Table 4.3
Effect of implanted elements in the metallic matrix, depending on temperature range of matrix (From *Wolf, C.K.* [70]. With permission.)

Temperature range of implanted material	Implanted element, soluble in implanted material	Implanted material, insoluble in implanted material	Implanted element reacting with implanted material (reactive implantation)
Very low temperatures (down to liquid helium)	Strongly defected structure (amorphous)	Strongly defected structure (amorphous)	Strongly defected structure (amorphous)
Low temperatures (down to ambient)	Stable solid solutions	Metastable solid solutions	Formation of phases (creation of compounds)
High temperatures (up to slightly below melting point)	Partial recovery. Behavior in accordance with phase diagrams. Full recovery	Partial recovery. Formation of clusters (agglomerations characterized by higher concentration of elements or intermetallic compounds)	Partial recovery. Stabilization or decomposition of compounds. Full recovery

Bombardment by high energy ions may lead to fragmentation of inclusions and precipitations present in the implanted material and to a refinement and homogenization of the structure of that material [38, 65].

Atrophy of grain boundaries [71] and dislocations [38] is observed. An increase in the dose of implanted ions may lead to an atrophy of the initial structure of the material [38]. In consequence, it is possible to obtain new, hitherto unknown versions of surface layers of high hardness, high resistance to abrasive wear and to corrosion, with better mechanical properties and with enhanced fatigue strength [63], while at the same time featuring better ductility.

Table 4.4. shows the effect of ion implantation on the properties of materials. Changes in properties marked by bold font are of greatest significance to the machine building industry.

Appropriate changes of properties of surface layers on metallic materials are obtained by the selection of the type or types of implanted elements, their dose and energy, as well as temperature of the implanted surface [72, 73]. For each of the implanting elements there is an optimum dose, the value of which varies, depending on the desired properties (e.g. tribological, fatigue, corrosion). To improve tribological properties by nitrogen ion implantation, the optimum dose is $2 \cdot 10^{17}$ ions/cm² (a dose of $3.5 \cdot 10^{17}$ ions/cm² corresponds to a surface concentration of 20 at% [11]).

The same dose in carbon ion implantation is $1 \cdot 10^{17}$ ions/cm², while for argon ions it is approximately $6 \cdot 10^{16}$ ions/cm² [4].

Table 4.4

Effect of ion implantation on material properties

Mechanical	Chemical	Electromagnetic
Wear	**Corrosion**	Superconductivity
Friction	Oxidation	Photoconductivity
Hardness	Catalysis	Resistivity
Fatigue	Electrochemical	Magnetic properties
Plasticity		Reflectivity
Ductility		Dielectric properties
Adhesion		

Metals and alloys may be implanted by non-metallic elements (e.g., nitrogen, carbon, silicon, boron, oxygen, argon, deuterium, neon) or by metals (e.g., titanium, aluminum, chromium, copper, tellurium, gold, silver, molybdenum, zirconium. yttrium, tantalum, platinum, tin) [4]. It is also possible to implant radioactive ions (e.g., the production of cryptanates for the need of tribological research) [20, 58, 72].

4.6.1 Tribological properties of implanted materials

Implantation increases wear resistance of metallic materials. The majority of reports devoted to the positive effect of implantation on tribological properties pertains to different grades of steel [1, 51, 74]. Several works pertain to implantation of titanium and its alloys. After implantation they are characterized by a lesser tendency to galling, less wear and changes of the friction coefficient [1, 33-35].

Table 4.5

Implanting ions improving tribological wear resistance

Implanted material	Ions
Be alloys	B
Cu alloys	B; N; P
Ti alloys	N; C; B
Zr alloys	C; N; Cr + C
High alloy steels	Ti + C; Ta
Low alloy steels	N
Stainless steels	N
Tool steels	N
Bearing steel	Ti; Ti + C
Superalloys	Y; C; N
Cobalt, coated with tungsten carbide	N; Co; B; C

An increase of tribological wear resistance of metallic materials, ceramics, sintered carbides and synthetic materials may be obtained through implantation of surfaces ions [75] of: N, Ti, C, B, Ta, Mo, Y, Cr, O, P, Zr, Ti + C, Y + N, Ti + Ni (Table 4.5) [74]. In most cases, however, in alloying of metallic materials lighter ions (nitrogen, carbon) and ions of neutral gases are utilized. By alloying stainless steels with carbon, boron, nitrogen, titanium or a combination of titanium with one of the above metalloids, it is possible to achieve an approximately 100% rise in wear resistance [28].

Enhancement of wear resistance is usually connected with the following factors:

1. Rise in hardness occurring due to the introduction of ions of foreign elements (usually ions of light elements, e.g., nitrogen, carbon, boron or noble gases) and hardening of the surface zone due to the formation of significant compressive stresses, blocking of dislocation movement [67] or the formation of hard inclusions (most common are nitrides, carbides and borides) in state of fine dispersion [42, 76-79]. These hard inclusions ensure good load-carrying capacity while the surrounding soft matrix has vibration-damping properties [1].

2. Greater ductility of metallic surface through implantation of heavy metal ions (e.g., Sn, Mo), which cause smoothing of the friction surface without spalling, as well as formation of a solid lubricant layer through implantation of those ions which improves lubrication properties (e.g., Sn, Mo, S, Mo + S, N + Ca, N + Mo) or actually forming a layer of lubricant in the process of wear (e.g., Cr). Another possibility is the formation of thin layers of soft oxides of tin or yttrium at the surface, decreasing forces of friction and protecting deeper-lying layers against rapid wear [1].

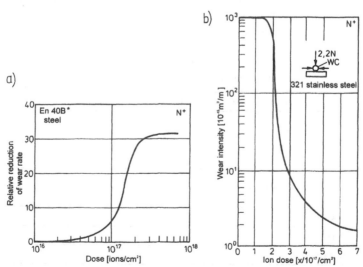

Fig. 4.24 Wear rate vs. nitrogen ion dose: a) relative reduction of wear intensity of 40B* grade steel, implanted by ions of 50 keV energy; b) wear intensity of 321 stainless steel. (Fig. a - from *Dearnaley, G.* [17]. With permission. Fig. b - from *Dimigen, H., et al.* [40]. With permission from Elsevier Science.)

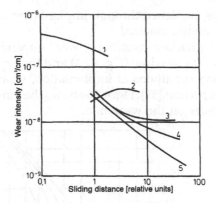

Fig. 4.25 Effect of sliding distance on wear intensity of En40B* nitriding steel for a pin-on-disk rubbing pair: *1* - not implanted; *2* - implanted with dose of $2 \cdot 10^{17}$ N$^+$/cm^2; *3* - implanted with dose of $2 \cdot 10^{17}$ C$^+$/cm^2; *4* - implanted with dose of $2.6 \cdot 10^{17}$ N$^+$/cm^2; *5* - implanted with dose of $2 \cdot 10^{17}$ B$^+$/cm^2. (*From Hulett, D.M., and Taylor, M.A.* [19]. With permission.)

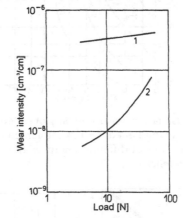

Fig. 4.26 Effect of load force on wear intensity of En40B* nitriding steel for a pin-on-disk rubbing pair [19]: *1* - not implanted; *2* - implanted with nitrogen ions of 50 keV energy and dose of $2 \cdot 10^{17}$ N$^+$/cm^2. (*From Hulett, D.M., and Taylor, M.A.* [19]. With permission.)

From among the different elements which are implanted, the best researched effect on tribological properties is that of nitrogen [5, 16, 17, 19, 22, 40, 43, 50, 51, 67, 71, 72, 81–91]. Implantation by nitrogen improves wear resistance of practically all metallic materials [40, 50, 71, 76, 85, 92-104]. Some research results negate this generalization, e.g., implantation of nitrogen ions into Nitralloy 135M caused a rise in abrasive wear by 37% [23]. The friction coefficient may be raised or decreased by several to several tens percent, through the implantation of ions of different elements [14].

In the process of friction there occurs in steel microspalling of asperity peaks in implanted materials. The load-bearing surface increases and the roughness valleys, hardened through implantation, exhibit higher wear resistance than the matrix material.

Given the same friction conditions, wear depends on conditions of implantation, i.e., on the ion dose (Fig. 4.24) and the type of ions (Fig. 4.25).

Given the same conditions of implantation, wear depends on the type and conditions of friction [39]. Fig. 4.26 shows the effect of the load acting on the rubbing surfaces on the wear rate.

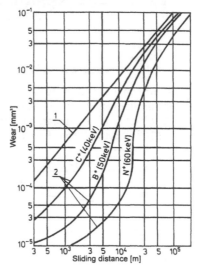

Fig. 4.27 Effect of sliding distance on wear intensity of sintered carbides for a pin-on-disk rubbing pair: *1* - disk not implanted; *2* - disk implanted with dose of $3 \cdot 10^{17}$ cm^{-2} of carbon, boron and nitrogen ions of different energies. (From *Richter, E.* [105]. With permission from Elsevier Science.)

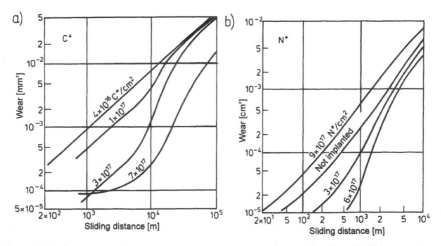

Fig. 4.28 Effect of sliding distance on wear intensity of: a) hard HG-110 plates, implanted by different doses of carbon ions of 50 keV energy; b) C100W1˙ steel, implanted by different doses of nitrogen ions of 50 keV energy. (Fig. a - from *Barnavon, Th., et al.* [50], Fig. b - from *Hochmuth, K. et al.* [18]. With permission from Elsevier Science.)

Wear of the implanted material is not uniform throughout the entire friction process. Usually, at the beginning of the process it is less and subsequently rises. Fig. 4.27 shows the correlation between the amount of wear on the sliding distance for the same material implanted by ions of different elements. Fig. 4.28 shows the same correlation for two different materials, implanted with different ion doses.

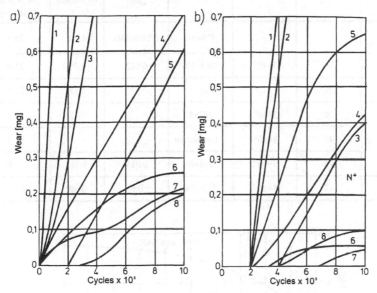

Fig. 4.29 Effect of number of impacts on impact wear of different metals and alloys: a) not implanted; b) implanted by nitrogen ions of 100 keV energy, dose of $4 \cdot 10^{17}$ N^+/cm^2, beam current of 5 μA and beam spot diameter of 2.5 to 3 mm, in partial vacuum of 1.3 Pa: 1, 2, 3 - soft materials, 2024 Al alloy (1), 7075 Al alloy (2) and Cu (3) with hardnesses of 29, 72 and 76 HRF correspondingly; 4, 5, 6, 7 - medium-soft materials: 4 - 90%Cu-10%Sn bronze - 80 HRF; 5 - 70%Cu-30%Zn cartridge bronze - 98 HRF; 6 - 82%Cu-18%Al ambralloy bronze - 98 HRF; 7 - sintered iron 5%Cu-1%C - 91 HRF; 8 - hard material - 1018 steel - 28 HRC. (From *Shih, K.K.* [88]. With permission from Elsevier Science.)

Fig. 4.29 shows the effect of ion implantation on the impact wear resistance (20 to 140 impacts per s) of different materials [88]. A similarly positive effect is exhibited ion implantation on impact wear resistance of high speed steel [90].

Table 4.6. shows the (mostly positive) effect of implantation on the tribological properties of phosphor bronze, using abrasive wear on a plate-rod rubbing pair [106].

Lesser abrasive wear of implanted materials than that of materials which are not implanted is, to a great extent, the result of a lower coefficient of friction, which becomes apparent in later stages of the wear process (Fig. 4.30). The initial coefficient of friction of implanted layers may sometimes have a higher value than that of not implanted material (see Table 4.6), although these values later switch around [107, 108].

Table 4.6

Effect of implantation on microhardness and tribological properties of PB102* phosphor bronze (From *Saritas, S., et al.* [106]. With permission from Elsevier Science.)

Implanted ions	Ion energy[keV]	Ion dose [ions/cm²]	Micro-hardness HV	Implantation depth [μm]	Initial coefficient of friction	Wear after 500 m sliding distance (under 10 N load) [×10⁻⁴cm³]
No implantation	-	-	178±10	0.01(oxide)	0.12	16.5
B⁺	40	5×10¹⁷	284±10	0.012	0.35	11.0
C⁺	20	1×10¹⁷	183±7		0.18	16.8
N⁺	20	1×10¹⁷	207±4		0.15	15.5
	40	1×10¹⁷	not measured		-	15.5
	40	5×10¹⁷	218±6	0.015	0.16	13.0
P⁺	40	5×10¹⁷	269±10	0.9	0.30	11.6

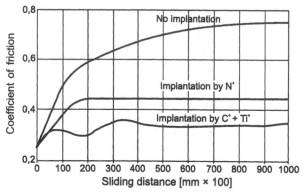

Fig. 4.30 Effect of sliding distance in pin-on-disk rubbing pair on coefficient of friction of stainless steel, not implanted and implanted by C⁺ + Ti⁺ and N⁺ ions.

A low coefficient of friction is exhibited by steel implanted with tin or molybdenum, and subsequently by sulfur (Table 4.7). Implantation with nitrogen of TiN coatings increases their wear resistance [109], and it has a similar effect on TiB coatings [110]. Alloying with molybdenum or cobalt of coatings implanted by nitrogen causes improvement of lubricity [4].

The depth to which the properties of the implanted material are modified, which may be likened to the thickness of the implanted layer, is not great. In an extreme case it reaches 1 μm. On steels it usually does not exceed 0.2 to 0.3 μm. However, despite a really thin initial thickness, it practically averages values greater by a factor one to two orders of magnitude during the process of abrasive wear, due to

Table 4.7

Implanted ions improving friction properties

Implanted material	Ions
Ti alloys	Sn; Ag
High alloy steels	Sn; Ag; Au; Mo + S
Low alloy steels	Sn
Stainless steels	C + Ti

– stimulated radiation diffusion which causes significant (up to several orders of magnitude) increase of the coefficient of diffusion of implanted atoms in the matrix material, mainly as the result of defects in the latter. With ion energy of 200 keV and a dose of 10^{18} ions/cm², a significant concentration of radiation defects is obtained [1];

– local heating of surface of implanted material in contact with another rubbing material; the components of the matrix material and their compounds increase their mobility [1];

As the result of both effects forced migration occurs of implanted atoms into the material (Fig. 4.31) because in the implantation process they are introduced with a significant surplus relative to the quantity required to harden the surface. For example, the surface concentration of nitrogen implanted into steel reaches 26% at. and thus is greatly in excess of equilibrium solubility of nitrogen in steel. The dislocation lattice and its decorating nitrogen (so-called A.H. Cotrell atmosphere) are subjected to forced

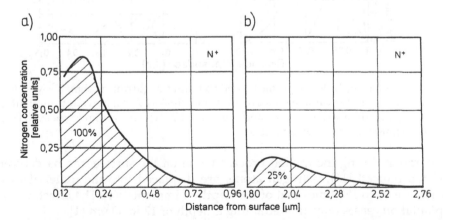

Fig. 4.31 Nitrogen concentration profile before (a) and after (b) wear test of 432 stainless steel, implanted by nitrogen ions (ion dose of $4 \cdot 10^{17}$ N⁺/cm², beam current: 0.5 µA; beam cross-section: 100×100 µm). (From *Dimigen, H., et al.* [40]. With permission from Elsevier Science.)

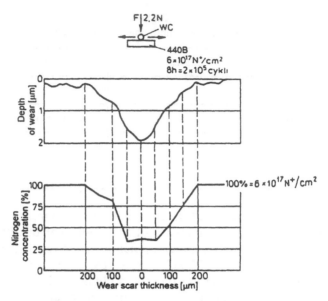

Fig. 4.32 Migration of nitrogen into 440B steel during abrasive wear test. (From *Wolf, G.K.,* [70]. With permission.)

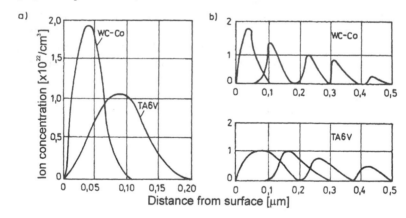

Fig. 4.33 Initial implantation profile (a) and its migration (b) during abrasive wear test of WC-Co (tungsten carbide - 8% Co) alloy and Ti-6Al-4V titanium alloy, implanted by nitrogen ions of 40 keV energy and dose of $1 \cdot 10^{17}$ N$^+$/cm^2. (From *Fayeulle, S.* [71]. With permission from Elsevier Science.)

migration during the friction process while on the surface, the hard layer is reproduced [14]. Nitrogen atoms are displaced into the material to a distance 10 to 150 times that of the depth of implantation. The effect of implantation ceases only after attaining a depth of 12 to 30 μm [4].

The effect of nitrogen migration, later also termed **auto-** or **quasi-implantation** of nitrogen, was observed for the first time in 1978 by G. Dearnaley and N.E.W. Hartley. They concluded that at a depth 100 times greater than the initial depth of implantation, after the friction process there still remained approx. 40% of the implanted ions. Other researchers have confirmed that the effect of forced migration of nitrogen dominates in abrasive wear of ferritic steels [40, 70, 71]. Fig. 4.32 shows such migration of nitrogen into steel grade 440B, while Fig. 4.33 shows the shift in the implantation profile in WC-Co and TA6V̊ alloys.

4.6.2 Strength properties of implanted materials

Ion implantation, especially of nitrogen ions, but also of C, Ba, Mn, Ni, Ti, Ta, W, Re (Table 4.8), causes a rise of fatigue strength [102, 111-113]. This is attributed to the presence of radiation defects and the formation of significant compressive stresses, caused by a fine dispersion of hard phase precipitations [38], hardening of structure, blocking of dislocation movement which usually occurs after implantation and, finally, a certain "smoothing" of the surface. In the macro scale, ion implantation may be looked upon as a lightly smoothing treatment.

Table 4.8
Implanted ions improving fatigue properties

Implanted material	Ions
Ti alloys	N; C; Ba
High alloy steels	N; Mn; C; B; Ni
Low alloy steels	N; Ti

The positive effect of implantation on low cycle fatigue strength is connected mainly with the formation of radiation defects increasing homogeneity of deformations (decreasing slip planes) and decreasing structure defects upon formation of new phases [1].

High cycle fatigue strength is more affected by residual stresses forming in the implanted layer. Compressive stresses increase the value of fatigue limit while tensile stresses lower it [1, 114].

Implantation causes a rise in fatigue strength by several tens percent (Fig. 4.34). Titanium alloys, alloyed by nitrogen, carbon and barium, cause a rise of fatigue strength by 10 to 20%. It should be added that nitrogen and carbon, besides saturating the solid solution, form finely dispersed nitrides and carbides which strengthen the structure of the implanted layer. Barium effectively impedes the migration of oxygen to the surface layer.

The combination of implantation with earlier thermo-chemical treatment or by glass beading usually also causes an increase of the fatigue limit [119]. For example, in grade 30HGSNA* steel the fatigue limit, which prior to implantation is approximately 500 MPa, rises to approximately 760 MPa after glass beading and after subsequent nitrogen ion implantation at 100 keV

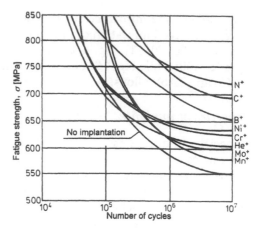

Fig. 4.34 Increase in fatigue strength of 30HGSNA' steel due to implantation by ions of different elements, of 40 keV energy. (From *Vasileva, E.V., et al.* [116]. With permission.)

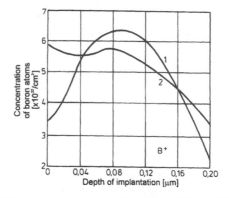

Fig. 4.35 Profile of boron ion implantation (50 keV, $1 \cdot 10^{18}$ B$^+$/cm^2) into nickel: *1* - prior to fatigue test; *2* - after fatigue test (20100 cycles). (From *Hochmuth, K., et al.* [117]. With permission from Elsevier Science.)

energy to approximately 850 MPa, with no scatter observed during fatigue tests [28]. A rise in the fatigue limit has also been observed in carbon steels, hardened after thermo-chemical treatment and subsequently subjected to nitrogen ion implantation. The same has been found in the 12H2I4ASz' grade after carburizing and subsequent nitrogen ion implantation [28].

Due to fatigue loads, with the rise of temperature and changes in stress concentration, the implantation profile is flattened as the result of migration of ions from sites with highest concentration (Fig. 4.35).

Ion implantation also causes a rise of static strength, e.g., resistance to brittle cracking (Fig. 4.36).

In some implantation processes one can observe the phenomenon of post-implantation aging which either improves or causes deterioration of strength properties [38]. For example, implantation of 1018 grade steel with nitrogen by a dose of $2 \cdot 10^{17}$ ions/cm² yielded an insignificant increase of the fatigue limit value, while after natural or artificial aging the fatigue limit of samples subjected to rotational bending rose by a factor of 10.

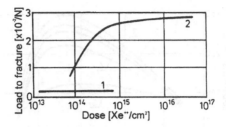

Fig. 4.36 Effect of xenon ion dose of 600 keV energy on brittle fracture of graphite: *1* - not implanted graphite; *2* - implanted graphite. (From *Hirvonen, I.P., et al.* [118]. With permission from Elsevier Science.)

Resistance to brittle cracking of ceramics and sintered carbides is enhanced by the implantation of Zr, Cr and Ti.

4.6.3 Hardness and adhesion of implanted materials

Ion implantation of the majority of metallic, ceramic and synthetic materials by N, P, Co, Y, Cr, Ti, Mo, Zr, Nb and Ta causes an increase of hardness of the implanted metallic layer (Table 4.9) and of the coating deposited on the substrate prior to or during the implantation process [119, 120, 121].

Implantation causes a rise of microhardness, due to the formation of strong compressive stresses and hard microinclusions of nitrides, carbides

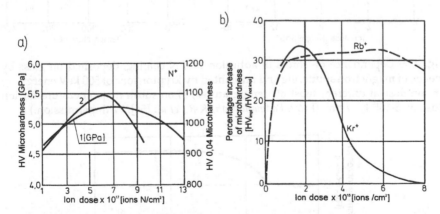

Fig. 4.37 Variations of microhardness vs. ion dose: a) steel implanted by nitrogen ions of different energy levels: *1* - E51100 steel, implanted with ions of 60 keV energy; *2* - 100W1˙ steel, implanted by ions of 50 keV energy; b) TiN coating, deposited on stainless steel by magnetron sputtering, implanted by ions of 500 keV energy. (Curve *1* - from *Kolitsh, A., and Richter, E.* [121], curve *2* - from *Hochmuth, K., et al.* [18], Fig. b - from *Padmanabhan, K.R., et al.* [54]. With permission.)

and borides in a manner which depends on the type of ions and on their dose (Fig. 4.37), as well as on the temperature of the implanted material (Fig. 4.38).

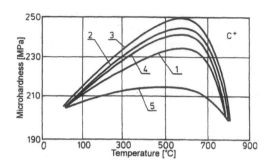

Fig. 4.38 Effect of steel temperature on microhardness of C100W1˙ steel, implanted by carbon ions of different doses: *1* - dose of $1 \cdot 10^{17}$ C$^+$/cm^2 ; *2* - $3 \cdot 10^{17}$ C$^+$/cm^2 ; *3* - $6 \cdot 10^{17}$ N$^+$/cm^2; *4* - $1 \cdot 10^{17}$ C$^+$/cm^2; *5* - $3 \cdot 10^{17}$ N$^+$/cm^2. (From *Hochmuth, K., et al.* [18]. With permission.)

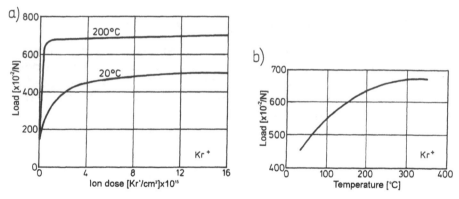

Fig. 4.39 Dependence of forces of adhesion of TiN coating, deposited on glass by means of magnetron sputtering and implanted by krypton ions of 500 keV energy: a) on ion dose at different substrate temperatures; b) on substrate temperature for ion dose of $5 \cdot 10^{15}$ Kr$^+$/cm^2. (From *Padmanabhan, K.R., et al.* [54]. With permission.)

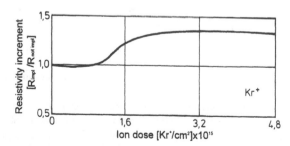

Fig. 4.40 Changes in resistivity of Ni layer, deposited on glass by magnetron sputtering, vs. dose of implanted krypton ions of 500 keV energy. (From *Padmanabhan, K.R., et al.* [54]. With permission.)

In the case of TiN coatings, deposited on the substrate material by different PVD techniques or by electroplating, implantation causes a rise in adhesion, again in a manner dependent on the ion dose (Fig. 4.39a) and on the substrate temperature (Fig. 4.39b). Implantation also causes changes of electrical properties of TiN coatings (Fig. 4.40) [54].

Table 4.9
Implanted ions raising hardness

Implanted materiel	Ions
Al alloys	N
Be alloys	B
Ti alloys	N; C; B
Zr alloys	C; N
Cu alloys	B; C; N; P
High alloy steels	Ti + C
Low alloy steels	N
High speed steels	N; B
Cobalt coated by tungsten carbide	N; Co
Sintered ceramic	Y; N; Zr; Cr

4.6.4 Corrosion resistance of implanted materials

Corrosion resistance of metallic materials is improved by implantation of ions, mainly the following elements: N, Cr, Al, Ta, Y, Sn (Table 4.10), but

Table 4.10
Implanted ions improving corrosion resistance

Implanted material	Ions
Al alloys	Mo
Cu alloys	Cr; Al
Zr alloys	Cr; Sn
High alloy steels	Cr; Ta; Y
Low alloy steels	Cr; Ta
Superalloys	N; C; Y; Ce
Copper	N
Surgical alloys	N

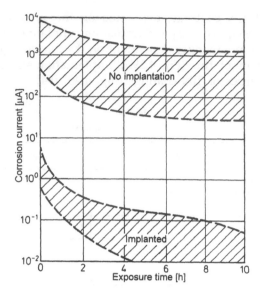

Fig. 4.41 Corrosion current vs. time of exposure in a bath of 0.9% NaCl (pH=7) for an implanted and not implanted Ti-6Al-4V surgical alloy; the value of 5μA·h corresponds to a material loss of 1 nm. (From *Wolf, G.K.* [48]. With permission.)

also of Ar, He, Xe, Cu, Ni, Mo. The majority of investigations carried out to date was devoted to the effect of implantation on atmospheric corrosion. Other topics of research were corrosion in acids [70], in body fluids (Fig. 4.41) [57, 123, 124] and in water [125].

Ion implantation leads to the improvement of corrosion resistance of metallic materials in oxidizing atmospheres through [38]:

– formation in the surface layer of compounds with new physico-chemical properties, e.g., corrosion resistance of titanium implanted by palladium rises more than 1000 times in comparison with not implanted material [126, 127];

– formation - with appropriate selection of ions and their dose - of amorphous properties which also feature enhanced corrosion resistance, similarly to amorphous materials obtained by splat cooling [128]. Amorphous properties may be obtained by implantation of pure metals, e.g., copper by ions of tantalum or tungsten and of iron by titanium. Amorphous structures formed by implantation also feature high stability, e.g., copper implanted by tungsten does not lose its amorphous properties at 600°C in 1.5 h [127];

– blockage of channels of easy oxygen diffusion (in the case of implanting of atoms greater than those of the lattice, e.g., of barium, strontium, calcium, rubidium or cesium into titanium) [38];

– formation of compact oxide layers, like Al_2O_3 (by e.g., implantation of Al ions into copper) or Cr_2O_3, SnO_2, SiO_2, and $YCrO_3$ (by e.g., implantation of yttrium into chromium-bearing steel) and $CaTiO_2$, which constitute a barrier for oxidation processes [129];

– plastization of brittle oxide layers which prevents the formation of microcracks exposing the implanted surface, modification of defect distribution and conductivity by oxides;

– implantation of ions which act catalytically, e.g., of platinum and ions of metals belonging to the same group, which slow down oxygen migration;

– implantation of ions which slow down cathodic processes, e.g., Pb;

– modification of electrical conductivity of oxides [38].

In practice, ion implantation is sporadically applied to improve corrosion resistance of metals.

4.6.5 Other properties of implanted materials

Catalytic properties of metallic materials and ceramics are improved by the implantation of Pt, Mo and Pd ions.

Hydrogen embrittlement in steels is diminished by the implantation of Pt and Pd ions.

Nitride formation in steels and aluminum is facilitated by the implantation of Ti and Mo ions.

Optical properties of glasses and of synthetic materials is modified by the implantation of Nb, Ti, Mo, Zr and Y ions.

4.7 Application of implantation technology

In surface engineering of metals, the method applied on an industrial scale is mainly primary ion beam implantation with recoil ion beam implantation trailing behind [130]. Currently the beginnings of a rapid advance of industrial application of ion mixing, as well as of research in the field of applications for plasma ion implantation are observed [42].

Of the techniques of primary ion beam implantation, implantation of nitrogen ions has been predominant in industry worldwide [43, 95, 130], especially applied to cutting and forming tools, less often to machine components. This technique is responsible for a rise of tool life by 2 to 10 times (Tables 4.11 and 4.12). The life increase obtained depends not only on the type of implanted material and implanting ions, but also, to a great extent, on the mating material of the rubbing pair or of the object treated by the implanted tool [6, 22, 55, 59, 84, 88, 131-134].

Nitrogen ion implantation has a lot in common in the utilitarian sense with gas and/or plasma nitriding, which both also significantly increase the service life of machine parts and tools. Fig. 4.42 shows volumetric wear of stainless steel, nitrided and ion implanted. It can be seen that right until the moment of complete wear of the nitrogen ion implanted layer, the material exhibits a better behavior than the nitrided one [81]. Interesting effects may also be obtained by the implantation by nitrogen ions of prior nitrided layers.

Table 4.11
Examples of application of ion implantation technology taken from different sources
(Data from [22, 42, 47, 65, 105] and various other sources.)

Implanted tools or components		Implanting material (ions)	Life increase factor
Type	Material		
1	2	3	4	5

	Type	Material	Implanting material (ions)	Life increase factor
	2	3	4	5
Forming tools	drawing dies for copper wire	WC-Co	carbon	5
	shears	tool steel	nitrogen	2-4
	shears	WC-Co	nitrogen	5
	shears for plastics	90%Mn 8%V; diamond	nitrogen	2-4
	dies and stamps	steel, WC,WC-Co	nitrogen	2-5
	anvils for riveting of noble metals	D3	nitrogen	2-5
	rollers for aluminum and copper rolling	alloy steels	nitrogen	3-6
	tools for extrusion of aluminum cans and tubes	D3	nitrogen	3-5
	die cast tools	steel	nitrogen	3-5
Cutting tools	thread taps	tool steel	nitrogen	8-10
	thread dies	tool steel	nitrogen	3-4
	dentist's drills	WC-Co	nitrogen	2-3
	drills for metals	tool steel, hard sintered P/M	nitrogen	0.2-6
	drills for printed circuits	high speed steel, sintered P/M inserts	nitrogen	4
	drills for graphite	WC	-	6
	hobs	high speed steel (C100W4)*	nitrogen	2-3
	thread cutters	high speed steel	nitrogen	5
	circular cutters	high speed steel	nitrogen	11
Forming tools	components of injectors for plastics (nozzles, dies, inserts, worm gears)	tool steel	nitrogen	2-10
	fuel injectors	tool steel	nitrogen	100
	precision aero bearings	M50, 440C or 52100	nitrogen	better resistance to pitting and corrosion
	beryllium alloy bearings	beryllium alloys	boron	3-5
	ball bearings	4210 steel	chromium	3× less corrosion in sea water

Table 4.11 continued

1	2	3	4	5
Forming tools	ball bearings	M50	tantalum	reduced wear and corrosion
	components of extruders of fiberglass	tool steel	titanium	wear significantly reduced
	turbine blades	Ni alloys	yttrium	high resistance to oxidation
	vapor valves	steel	tin	friction reduced 10×
	pump components	17-4 PH	titanium + carbon	reduced wear
Other	artificial hip joints	titanium alloys Ti-6Al-4V	- nitrogen	100 400

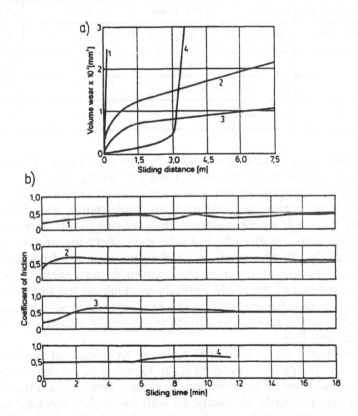

Fig. 4.42 Volume loss due to wear (a) and coefficient of friction (b) vs. sliding distance and sliding time of steel cylinder (4400 stainless steel, hardened and tempered to 58 HRC) rubbing against disk made of 15-5PH steel, aged for 4 h at 550°C to 40 HRC: *1* - no treatment; *2* - gas nitrided (ammonia, 540°C, 44 h); *3* - ion nitrided (40% N, 40% Ar, 20% H, 540°C, 22 h); *4* - implanted by nitrogen ions with dose of $2 \cdot 10^{17}$ N^+/cm^2 at 100°C (process parameters: $6.6 \cdot 10^{-3}$ Pa, 30 mA, 100 keV). (From *Cohen, A., and Rosen, A.*[81]. With permission from Elsevier Science.)

Table 4.12

Practical application of ion implantation technology (From [18, 22, 59, 65, 83].)

Application	Implanted material	Treatment (dose and type of ions)	Extension of service life	Reference
Paper shears	1%C, 1.6%Cr steel	$8 \cdot 10^{17}$ N+/cm^2	×2	[22]
Drills for plastics	high speed steel	$8 \cdot 10^{17}$ N+/cm^2	×4 ×5	[18] [22]
Shears for cutting latex	WC - 6%Co	$8 \cdot 10^{17}$ N+/cm^2	×12	[22]
Die inserts	1%Ni, 1%Cr steel	$4 \cdot 10^{17}$ Co+/cm^2	30%	[22]
Dies and punches	12%Cr, 2%C steel	$4 \cdot 10^{17}$ N+/cm^2	adhesive wear significantly reduced	[22]
Dies for copper rods	WC - 6%Co	$5 \cdot 10^{17}$ C+/cm^2	×5	[22, 83]
Drawing dies for steel wire	WC - 6%Co	$5 \cdot 10^{17}$ C+/cm^2	×0.3-3	[22, 59, 65]
Drawing dies for copper wire	WC - 6%Co	$5 \cdot 10^{17}$ N+/cm^2	×5	[18, 65]
Injector nozzles	D3	$5 \cdot 10^{17}$ N+/cm^2	×5	[18, 65]
Threading dies	M2 high speed steel	$8 \cdot 10^{17}$ N+/cm^2	×5	[59]
Thread taps for phenolic resins	M2 high speed steel	$8 \cdot 10^{17}$ N+/cm^2	×12	[59]

Some interesting results of complex changes of tribological and anti-corrosion properties in various steels used for tooling, implanted by N+ and NO+ ions (with optimum ion doses), are given in Table 4.13 [133].

It should be emphasized that prior to implantation the surface requires very thorough washing because the presence of oils, dust or other contaminants may significantly and adversely affect the implanted metallic material. Ion implantation, as a technique, is so sensitive that even oil vapors or residual atoms from the vacuum pump system, as well as those sputtered out of the shield partially masking off the implanted object, all contaminate the material by the implantation of contaminants in it [123]. After implantation, the surface is especially sensitive to the action of humidity [38].

Also developed on an industrial scale are techniques of implantation of metals by zinc, silicon and carbon [130]. Expected in the near future are practical applications of implantation by boron, titanium, aluminum, yttrium and other metallic elements, as well as implantation by several combinations of different elements (e.g., Ti + C, N + O, Mo + S, Cr + C). After successful laboratory tests, also expected is a broader application of combined technique of ion implantation with coating deposition by PVD and CVD techniques [57, 129]. The possibility of diverse applications of the technique of ion mixing is shown in Table 4.14, and dynamical mixing in Table 4.15 [42]. Besides other applications [52], ion mixing presently

Table 4.13

Steels used for tooling and their properties after implantation (From *Iwaki, M., et al.* [133]. With permission from Elsevier Science.)

Type of steel	AISI grade	Chemical composition [%]									Heat treatment	Knoop hardness (under 2N load)	Changes		
		C	Si	Mn	Ni	Cr	Mo	V	Co	other			hardness	wear	anodic current
1	2	3	4	5	6	7	8	9	10	11	12	13	14	15	16
Tool steels	H13	0.32-0.42	≤1.5	≤1.5	≤0.25	4.5-5.5	1.0-1.5	0.4	-	0.18S	PH				
	P21	0.15	0.3	1.5	3.0	0.8	0.3	-	-	1.0Cu, 1.0Al	PH	430-460			
	17-4PH	0.07	-	-	4.5	17.0	-	-	-	3.0Cu, 0.3Nb	PH				
Alloy steels	D2	1.5	-	-	-	12.0	1.0	0.4	-		QT				
	A2	1.0	0.2	0.6	-	5.3	1.1	0.2	-		QT	630-690			
	H13	0.35-0.42	0.8-1.2	0.3-0.5	-	4.8-5.5	1.2-1.6	0.5-1.1	-	0.13S	QT				
	M1	≤0.03	≤0.1	≤0.1	17.0-19.0	-	4.5	-	7.0÷8.5	0.3-0.5 Ti0.05-0.15Al	SA	520-700			
	M3	≤0.03	≤0.1	≤0.1	17.0-19.0	-	4.5-5.5	-	3.5-9.5	1.2-1.8 Ti0.05-0.15Al	SA				
	M4	≤0.03	≤0.1	≤0.1	17.0-19.0	-	3.5-4.5	-	12.0-13.0	1.2-1.8 Ti0.05-0.15Al	SA				
Stainless steels	420	0.38				13.6					QT	580-600			
	420					13.6					QT				

Notation: PH - prehardening, QT - quench and temper, SA - solution annealing and aging

■ - at $2.5 \cdot 10^{17} NO^+/cm^2$; □ - at $2 \cdot 10^{17} N^+/cm^2$

constitutes the only possibility of obtaining uniformity within a broad range of alloys, in the solid solution, of systems which feature limited solubility in the solid and liquid state [1, 128, 136, 137].

Table 4.14
Possibilities of application of ion mixing

Application	Substrate material	Elements in mixture, magnetron sputtered, ion plated	Ion beam
Wear resistance	Ti-6Al-4V alloy	Sn	N⁺
Oxidation resistance	Superalloys, steels	Y	Ar⁺
Surface catalysis	Carbon	Pt	Ar⁺
Corrosion resistance	Steel titanium iron	CrPd; PtAl; Cr	Ar⁺; Kr⁺ Ar⁺ Ar⁺; Xe⁺
Resistance to surface tarnishing	Copper	Al; Cr	Ne⁺
Improvement of adhesion	Al2O3, quartz, ceramic, plastics	Al; Cu; Au	He⁺; Ne⁺

Table 4.15
Possibilities of application of dynamic mixing

Application	Substrate material	Vapour deposited elements	Ion beam
Obtaining of superhard regular boron nitride	Steel	B	N⁺
Strongly adherent hard layers (TiN, HfN)	Steel	Ti; Hf	N⁺
Strongly adherent metallic layers of high density and small porosity	Any material	Al; Cu; Au	Ne⁺
Anti-corrosion coatings	Steels	Cr; Ta	Ne⁺; He⁺
Preparation of substrate for PVD coatings	Any material	Ti	N⁺

In perspective, one can visualize a combination of implantation with diffusion, made possible by additional heating of the load [25].

4.8 Advantages and disadvantages of ion implantation techniques

Advantages of ion implantation are the following:
– The already mentioned potential possibility of implantation by any chosen element of any material in a short time (of the order of 10 to 100 s/cm² of surface) and at any temperature (but not exceeding 600°C), although very

limited in practical applications; furthermore, the possibility of introducing combinations of alloying additives.

– Possibility of obtaining concentrations of alloying additives which exceed their solubility in the alloyed material (usually approximately 20%, up to over 50% maximum).

– Ease of electrical process control, as well as possibility of precise control of concentration and distribution of alloying additives by programming the dose and energy of ions, with the possibility of monitoring.

– Possibility of conducting the process at low temperatures (usually below 200°C), allowing its application independently of classical heat treatment of finished components, without changes of shape and dimensions. This technique allows the retention of tool tolerances down to several nm without the need for subsequent finishing operations [39].

– Independence of technique of the effect of adhesion.

– Low electrical energy consumption.

– Cleanliness of process (vacuum) and non-pollution of the environment.

– Material economy.

Among the disadvantages are

- The inherent beam characteristic of the process (it is possible to implant only surfaces in the line of beam operation; best results are obtained on surfaces which are perpendicular to the ion beam axis). This does not apply to plasma ion implantation.

- Small depth of implantation (maximum up to 1 μm) in the case of beam implantation and greater in the case of plasma ion implantation, which increases, however, during service.

- Impossibility of implantation of loads with complex geometrical shapes and walls of deep holes (this does not apply to plasma ion implantation).

- Very high cost of ion beam implanters, of the order of $150,000 to 200,000 U.S. As an example, the price of a the big "Tecvac 221" implanter is £200,000 (pound sterling).

- High operating costs of implantation, from $0.03 to $0.10 per cm^2 [19] to $0.4 to $0.65 per cm^2 [105] of implanted surface.

- Necessity of very thorough cleaning of load surface prior to implantation.

- Differences between technological possibilities of implantation. As an example, it is relatively easy to implant metallic materials with nitrogen, boron, carbon, aluminum, tin, cerium or silicon but difficult to implant with titanium, palladium, yttrium and very difficult with platinum.

- Requirement of very highly trained and qualified personnel to operate implanters.

- Necessity to use radiological shields, protecting the operator from X-ray radiation which occurs during the operation of ion beam implanters.

- Only small automation of the implantation process: to this day single chamber implanters are being built. It is possible, however, to build a three-chamber implanter (Fig. 4.43), in a manner similar to vacuum chamber furnaces, in which the work chamber would be separated by vacuum locks from the first chamber (initial pump-down to vacuum) and from the final chamber (backfill to atmospheric pressure).

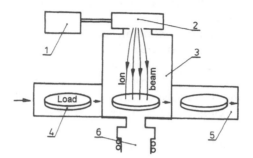

Fig. 4.43 Schematic of a three-chamber ion implanter: *1* - power supply system; *2* - ion source; *3* - work chamber; *4* - loading chamber; *5* - unloading chamber; *6* - pumping system. (From *Dearnaley, G.* [55]. With permission from Elsevier Science.)

There is limited possibility of implantation of big size loads; the biggest industrial ion implanter allows implantation of loads up to 1.5×1.5 m or maximum length up to 2.5 m [5].

References

1. *Scientific and technical progress in machine-building.* Series 9: Contemporary methods of hardening machine component surfaces (in Russian). International Center for Scientific and Technical Information; A.A. Blagonravov Institute for Machine-building, USSR Academy of Sciences, Moscow, 1989.
2. **Hałas, A., Szreter, M., Marks, J., and Pytkowski, S.**: The role of vacuum technology in the development of electron techniques (in Polish). Proc.: *1st Conference on Electron Technology*, Wrocław (Poland), Sept. 1982, pp. 57-72.
3. **Rosiński, W.**: *Selected applications of ion implantation in science and technology* (in Polish). Publ. Ossolineum, Wrocław (Poland) 1978.
4. **Rosiński, W.**: Application of the electron beam to modify properties of a solid (in Polish). Proc.: *1st Conference on Electron Technology*, Wrocław, Sept. 1982, pp. 57-72.
5. **Savage, J.E.**: Adding wear resistance via ion implantation. *Metal Progress*, Nov. 1984, pp. 41-44.
6. Joint report, edited by **J.K. Hirvonen**: *Ion implantation*. Academic Press, New York-London-Toronto-Sydney-San Francisco, 1980.
7. **Townsend, P.D., Kelly, I.C., and Hartley, N.E.W.**: *Ion implantation sputtering and their applications*. Academic Press, New York, 1976.

8. **Burakowski, T.**: Possibilities of application of ion implantation in metal surface engineering (in Polish). *Przegląd Mechaniczny (Mechanical Review)*, 1988, Part I, No. 16, pp. 5-11; Part II, No. 17, pp. 15-32.

9. **Burakowski, T.**: Ion implantation and possibilities of its implementation for modification of metal superficial layer properties (in Polish). *Trybologia*, 1989, No. 5, pp. 4-12.

10. **Conrad, J.R., Radtke, L., Dodd, R.A., Worzala, F.J., and Tran, N.C.**: Plasma source ion-implantation technique for surface modifications of materials. *J. Appl. Phys.*, 62, 1987, pp. 4591-4596.

11. **Tyrkiel, E. (g.ed.), and Dearnaley, P. (c.ed.)**: *A guide to surface engineering terminology*. The Institute of Materials - IFHF, London 1995.

12. **Hutchings, R., Kenny, M.J., Miller, D.R., and Yeung, W.Y.**: Plasma immersion ion implantation at elevated temperatures. In: *Surface Engineering Processes and Applications*. Technomic Publishing Co., Inc, Lancaster, Basel 1995, pp. 187-200.

13. **Ławrynowicz, Z.**: Mechanism and technology of plasma implantation. *Metaloznawstwo, Obróbka Cieplna, Inżynieria Powierzchni (Metallurgy, Heat Treatment, Surface Engineering)*, No. 115-117, 1992, pp. 50-58.

14. **Gawlik, G., Jagielski, J., and Podgórski, A.**: Modification of the superficial layer of metals and alloys by ion implantation (in Polish). *Prace Instytutu Lotnictwa (Transactions of the Institute of Aeronautics)*, Vol. 2-3, No. 121-122, 1990, pp. 74-91.

15. **Fnidiki, A., and Eymery, I.P.**: Sur la transition ordere-disordere crée par implantation ionique dans Fe-Co 50% at., *Scripta Metallurgica*, Vol. 19, No. 3, March 1985, pp. 329-332.

16. **Villain, I.P., Poola, O., and Moine, P.**: Amorphisation par implantation ionique et resistance f l'usure des aliages intermétalliques Ni-Ti. *Memoires et Études Scientifiques Revue de Métallurgie*, No. 9, Sep. 1986, p. 483.

17. **Dearnaley, G.**: Practical applications of ion implantation. *Journal of Metals*, No. 9, 1982, pp. 18-28.

18. **Hohmuth, K., Koltisch, A., Rauschenbach, B., and Richter, E.**: Beeiflüssung mechanischer Eigenschaften durch Ionimplantation. *Neue Hütte*, Vol. 29, No. 5, 1984, pp. 174-178.

19. **Hulett, D.M., and Taylor, M.A.**: Ion nitriding and ion implantation - a comparison. *Metal Progress*, Aug. 1985, pp. 18-21.

20. **Iwaki, M.**: Surface modification of steels by ion implantation. *Journal of the Iron and Steel Institute of Japan*, Vol. 71, No. 15, Nov. 1985, pp. 1734-1741.

21. **Deicher, M., Grubel, G., Recknagel, E., Reiner, W., and Wichert, Th.**: Helium vacancy interaction in helium implanted gold. *Materials Science and Engineering*, No. 69, 1985, pp. 57-62.

22. **Fromson, R.E., and Kossovsky, R.**: Ion implantation process and its results for tools. *Industrial Heating*, No. 9, 1984, pp. 26-28.

23. **Zielecki, W.**: *Modification of technological and service properties of steel by the electron and laser beam* (in Polish). Ph.D. Thesis, Rzeszów Technical University, Rzeszów (Poland), 1993.

24. **Burakowski, T.**: Ion implantation (in Polish). Proc.: Conference on *Technology of Superficial Layer Formation on Metals*, Rzeszów (Poland), 9-10 June 1988, pp. 106-113.

25. **Burakowski, T.**: Trends in development of heat treating (in Polish). Section for Fundamentals of Technology of the Machine-Building Committee of the Polish Academy of Sciences, Institute of Precision Mechanics, Warsaw, March 1988, pp. 78-84.

26. **Piekoszewski, J., Werner, Z., Langner, J., Jakubowski, L., Pochrybniak, C., and Harasiewicz, A.**: Pulse implantation doping - concentration profiles and surface morphology. *Nuclear Instruments and Methods in Physics Resaerch,* 209/210, 1983, pp. 477-482.

27. International Conference on Ion implantation Equipment and Techniques. Jeffersonville, 23-27 July, 1984. *Nucl. Instrum. Methods Phys. Res.,* Section B, 1985, Vol. 6 (1/2), pp. 1-93.

28. **Drwięga, M., Lipińska, E., Lazarski, S., and Wierba, M.**: IFJ Ion implanter and its utilization in ion engineering (in Polish). Report No. 1656/AP of the Institute of Nuclear Physics, Cracow, Sept. 1993.

29. **Drwięga, M., Lipińska, E., Lazarski, S., and Wierba, M.**: Adaptation of the IFJ ion implanter for the Ion Beam Assisted Deposition dual beam technique (in Polish). Report No. 1660/AP of the Institute of Nuclear Physics, Cracow, Jan. 1994.

30. **Marakhtanov, M.K.**: Industrial ion beam equipment (in Russian). *Baumann MVTU,* Moscow 1976.

31. **Mączka, D., Wasiak, A., and Partyka, J.**: Ion implantation at the Department of Nuclear Physics at the Marie Curie University (in Polish). *Prace Instytutu Lotnictwa (Transactions of the Institute of Aeronautics),* No. 121-122, Vol. 2-3, 1990, pp. 51-73.

32. **Simonov, V.V., and Solovieva, T.E.**: Equipment for ion implantation (in Russian). *Zarubiezhnaya Radioelektronika,* 1982, No. 9, pp. 47-69.

33. **Burakowski, T.**: Implantation of ions into metals (in Polish). *Prace Instytutu Lotnictwa (Transactions of the Institute of Aeronautics),* No. 121-122, Vol. 2-3, 1990, pp. 5-50.

34. **Burakowski, T.**: Ion implantation and possibilities of its utilization in metal surface engineering (in Polish). *Metaloznawstwo, Obróbka Cieplna, Inżynieria Powierzchni (Metallurgy, Heat Treatment, Surface Engineering)* No. 91-96, 1988, pp. 29-50.

35. **Ryssel, H., and Ruge, I.**: *Ionenimplantation.* Ed. G. Teubner. Stuttgart 1978.

36. **Cnad, J.R., Dodd, R.A., and Worzala, F.J.**: Plasma source ion implantation: a new, post-effective, non-line-of-sight technique for ion implantation of materials. *Surface and Coating Technology,* 1988, Vol. 36, No. 3-4, pp. 927-937.

37. **Grabovich, M.D.**: *Plasma ion sources* (in Russian). Publ. Naukova Dumka, Kiev 1964.

38. **Podgórski, A., Gawlik, G., and Jagielski, J.**: Development of initial design for a laboratory implanter of nitrogen ions for metal treatment (in Polish). Report on I and II stage of Project No. CPBR 02-13. Institute for Technology of Electronic Materials, Warsaw 1986.

39. **Frey, H., and Kienel, G.**: *Dünnschicht Technologien.* VDI Verlag, Düsseldorf 1987.

40. **Dimigen, H., Kobs, K., Leutenecker, R., Ryssel, H., and Eichinger, P.**: Wear resistance of nitrogen-implanted steels. *Materials Science and Engineering,* 69, 1985, pp. 181-190.

41. **Leutenecker, R., Ryssel, H., and Spohrle, H.P.**: 60 keV implanter for metals. Proc.: *Int. Conf. Modif. Met.,* Heidelberg 1984.

42. Brochures by: AERE Harwell, Beam Alloy Corp., Hawker-Siddley, Ion Beam Materials Processing Corp., Ion Surface Technology, Surface Alloys Corp., Surface Modification Spire Corp., Tecvac Ltd., Vardict Ion Technology, Westinghouse Electric Co., Zyment Inc.

43. **Brunel, M., Chabrol, C., Jaffrezic, H., Marest, G., Moncoffre, N., and Tosset, G.**: Amélioration de la résistance f l'usure par implantation d'azôte dans les aciers: Évolution des phases form et en fonction de la fluence et de la température. *Memoires et Études Scientifiques Revue de Métallurgie,* No. 9, Sept. 1986, pp. 485.

44. **Drwięga, M., and Lipińska, E.**: Application of ion beams for modification of the superficial layer on solids, with special emphasis on the IBAD technique, as carried out at the Institute of Nuclear Physics (in Polish). Report No. 1583/AP, Institute of Nuclear Physics, Cracow 1992.

45. **Piekoszewski, J., and Langner, J.**: High intensity pulsed ion beams in material processing: equipment and applications. *Nuclear Instruments and Methods in Physics Research*. B53, 1991, pp. 148-160.

46. **Piekoszewski, J., Waliś, L., Langner, J., Werner, Z., Białoskórski, J., Nowicki, L., Kopcewicz, M., and Grabias, A.**: Alloying of austenitic stainless steel with nitrogen using high intensity pulsed beams of nitrogen plasma. *Nuclear Instruments and Methods in Physics Research*. B114, 1996, pp. 263-268.

47. **Ehinger, M., Böhm, P., and Lauffer, H.J.**: Ionimplantieren und Anwendungspotentiale ionenimplantierter Werkzeuge. *Werkstattstechnik*, No. 77, 1978, pp. 475-478.

48. **Wolf, G.K.**: Die Anwendung von Ionenstrahlen zur Veränderung von Metalloberflächen. *Metalloberfläche*, 40, 1986, 3, pp. 101-105.

49. **Kaut, R.A., and Sartwell, B.D.**: Ion beam enhancement of vapour deposited coatings. *Journal of Vacuum Science and Technology*, A, 1985, Vol. 3, No. 6, pp. 2675-2676.

50. **Barnavon, Th., Jaffrezic, H., Marest, G., Moncoffre, N., Tousset, J., and Fayeulle, S.**: Influence of temperature of nitrogen implanted steel and iron. *Materials Science and Engineering*, No. 69, 1985, pp. 531-537.

51. **Dearnaley, G.**: Adhesive and abrasive wear mechanisms in ion implanted metal. *Nucl. Instrum. Methods Phys. Res.*, sect. B, 1985, Vol. 7/8 (pt. 1) pp. 158-165.

52. **Deutschmann, A.H., and Partyka, R.J.**: Practical applications of ion beam mixing: a new surface treatment technique. *Industrial Heating*, 1988, Vol. 55, No. 2, pp. 30-31.

53. **Van Rossum, M., Cheng, Y., Nicolet, N., and Johnson, W.L.**: Correlation between cohesive energy and mixing rate in ion mixing of metallic bilayers. *Appl. Phys. Lett*, 1985, Vol. 46, No. 6, pp. 610-612.

54. **Padmanabhan, K.R., Hsieh, Y.F., Chevalier, J., and Sorensen, G.**: Modifications to the microhardness, adhesion and resistivity of sputtered TiN films by ion implantation. *Journal of Vacuum Science and Technology*. A, Vol. 1, No. 2, Part 1, April-June 1983, pp. 279-283.

55. **Dearnaley G.**: Application of ion implantation in metals. *Thin Solid Films*, 107, 1985, pp. 315-326.

56. **Ziemann, P.**: Amorphisation of metallic systems by ion beams. *Materials Science and Engineering*, 69, 1985, pp. 95-103.

57. **Sioshansi P.**: Ion beam modification of materials for industry. *Thin Solid Films*, 118, 1984, pp. 61-71.

58. **Junqua, N., Pimpert, S., and Delaforce, J.**: Le mixage dynamique: appareilage et premiers resultats. *Memoires et Études Scientifiques Revue de Métallurgie*, No. 9, Sep. 1986, p. 486.

59. **Armini, A.**: Formation of new surface alloys by ion implantation technology. *Industrial Heating*, January 1986, pp. 17-19.

60. **Guseva, M.I.**: Ion implantation in metals (in Russian). *Poverchnost, Fizika, Khimia, Mechanika*, 1982, No. 4, pp. 27-50.

61. **Knapp, J.A., Folistaedt, D.M., and Doyle, B.L.**: Characterization of amorphous surface layers in Fe implanted with Ti and C. *Nuclear Instruments and Methods*, 87, 8, 1985, pp. 38-43.

62. **Chupyatova, L.P.**: Alloying of the superficial layer of iron with aluminum and nickel by ion mixing (in Russian). *Izv. Chornaya Metallurgia*, 1988, No. 9, pp. 155-158.

63. **Dearnaley, G., Freeman, J.H., Nelson, R.S., and Stephen, S.**: *Ion implantation*. North Holland Publ., Amsterdam 1973.

64. **Hall, B.O.**: Surface hardening in ion-implanted metals. *Journal of Nuclear Materials*, 1983, Vol. 116, pp. 123-126.

65. **Greggi, I. jr.**: The observations of defect structures in nitrogen ion implanted WC-Co. *Scripta Metallurgica*, Vol. 17, No. 6, June 1983, pp. 765-768.

66. **Guseva, M.I., Martynenko, Ju.V., and Ribalko, B.F.**: Surface effects during radiation (in Russian). *Voprosy Atomnoy Nauki i Tekhniki*, Moscow, Ed. 4(18), pp. 35-48.

67. **Hall, I.W.**: Microstructural effects of ion implantation on molybdenum. *Metallography*, Vol. 15, No. 2, May 1982, pp. 105-120.

68. International Conference on Surface Modification of Metals by Ion Beams, Heidelberg, T-21, Sep. 1984, *Materials and Science Engineering*, 1985, Vol. 69(1), pp. 155-252.

69. **Martan, J., and Olszewska-Mateja, B.**: Ion implantation into metals (in Polish). *Elektronika*, No. 3, 1985, p.27.

70. **Wolf, C.K.**: Ionenbeschuss als Methode in der Korrosionsforschung und - verhutung. *Werkstoffe und Korrosion*, 30, 1979, pp. 853-857.

71. **Fayeulle, S.**: Tribological behaviour of nitrogen-implanted materials. *Wear*, Vol. 107, No. 1, 1986, pp. 61-70.

72. **Goode, P.D., and Baumvol, I.J.R.**: The influemce of ion implantation parameters on the surface modification of steels. *Nuclear Instruments and Methods*, 189, 1981, pp. 161-168.

73. **Zdanowski, J.**: Ion technology (in Polish). Proc.: *Conference on Electron Technology*, Wrocław (Poland), Sep. 1982, pp. 159-174.

74. **Kustas, F.M., and Mistra, M.S.**: Application of ion implantation to improve the wear resistance of 52100 bearing steel. *Thin Solid Films*, 1984, Vol. 122, No. 4, pp. 279-286.

75. **Pursche, G., Rosert, R., Resch, H., Steinhäuser, S., Fabian, D., Ebersbach, G., Richter, E., Zenker, R., Zenker, U., and Büge, B.**: Verschleissschutz durch Oberflächenschichten - eine Betrachtung zum Entwicklungsstand und zu den Entwicklungstendenzen. *Schmierungstechnik*, No. 6, 1986, pp. 168-173.

76. **Dienel, G., Jörren, C., and Richter, E.**: Zum Verschleissverhalten ionenimplantierer Stahoberflächen. *Schmierungstechnik*, No. 16, 1985, pp. 118-119.

77. **Madakson, P.B., and Dearnaley, G.**: The role of ion implantation on the tribological properties of steel. *Materials and Science Engineering*, No. 69, 1985, pp. 155-160.

78. **Madakson, P.B., and Smith, A.A.**: Friction and wear of ion implanted aluminium. *Nuclear Instruments and Methods*, No. 209, 1983, pp. 983-988.

79. **Nazarenko, P.V., Popovich, M.N., Pilyavski, V.S., and Labuzhets, W.F.**: Effect of ion bombardment by He^+ on the formation of ultra-dispersed structure of material during friction in motion (in Russian). Lectures by the Soviet Academy of Sciences, Series A, 1986, No. 2, pp. 72-75.

80. **Potter, D.J., Ahmed, M., and Lomond, S.**: Metallurgical surfaces produced by ion implantation. *Journal of Metals*, No. 35, 1983, pp. 17-22.

81. **Cohen, A., and Rosen, A.**: The influence of nitriding process on the dry wear resistance of 15-5 PH stainless steel. *Wear*, Vol. 108, No. 2, 1986, pp. 107-158.

82. **Dillich, S.A., and Singer, I.L.**: Effect of Ti implantation on the friction and surface chemistry of Co-Cr-W alloy. *Thin Solid Films*, 108, 1983, pp. 219-227.

83. **Erck, R.A., Fenske, G.R., Erdemir, A., and Nichols F.A.**: Applications of ion beam processes in tribology. Proc.: *5th International Congress on Tribology - EUROTRIB: Friction-Wear-Lubrication*, Helsinki, June 1989, Vol. 2, pp. 102-109.

84. **Gerva, A., and Wiesner, L.**: *Oberflächenvergütung von Werkstoffen durch Ionenimplantationstechniken. Studie der Arbeitsgemeinschaft Ionenstrahltechnik*, Odenthal, 1987.

85. **Pivin, J.C.**: Dry friction of Ti and TiAl implanted with N, C, B, or N+O ions. Proc.: *5th International Congress on Tribology - EUROTRIB: Friction-Wear-Lubrication*, Helsinki, June 1989, Vol. 1, pp. 271-276.

86. **Hirvonen, J.K.**: Nitrogen ion implantation for wear applications. *Journal Vacuum Science Technology*, A, 1985, Vol. 3, No. 6, pp. 2691-2692.

87. **Dorn, E.**: Microstructure of tool steels, implanted by nitrogen (in Polish). *Prace Instytutu Lotnictwa (Transactions of the Aeronautics Institute)*, No. 121-122 (Vol. 2-3/1990), pp. 92-95.

88. **Shih, K.K.**: Effect of nitrogen implantation on impact wear. *Wear*, Vol. 105, 1985, No. 4, pp. 341-347.

89. **Sommer, T.I., Hale, E.B., Burris, K.W., and Kohser, R.A.**: Characterization of wear modes in ion-implanted steel from Auger measurements. *Materials Science and Engineering*, Vol. 69, No. 1, Feb. 1985, pp. 149-154.

90. **Wu Bing Shu**: The study of improvement of impact wear resistance of high speed steel by nitrogen ion (N^+). *Jinshu Rechuli (Heat Treatment of Metals)*, No. 7, 1987, pp. 24-28.

91. **Zielecki, W., and Perłowski, R.**: Investigations of wear of 20HN3A steel, implanted by nitrogen ions (in Polish). Transactions of the Rzeszów Technical University, No. 82, *Mechanika*, Vol. 28, 1991, pp. 301-307.

92. **Dos Santos, C.A., Behar, M., and Baumvol, I.J.R.**:Surface modifications and the mechanical properties of carbon steels implanted with nitrogen. *Journal of Physics D: Applied Physics*, 17, 1984, pp. 551-562.

93. **Ecer, G.M., Wood, S., Boes, D., and Schreurs, J.**: Friction and wear properties of nitrided and N^+ implanted 17-4PH stainless steel. *Wear*, 89, 1983, pp. 201-214.

94. **Rauschenbach, B., and Kolitsch, A.**: Formation of compounds by nitrogen ion implantation in iron. *Phys. Stat. Solids*, 80, 1983, pp. 211-222.

95. **Rauschenbach, B., Kolitsch, A., and Richter, E.**: Formation of AlN by nitrogen ion implantation. *Thin Solid Films*, No. 109, 1983, pp. 37-45.

96. **Singer, I.L., and Jeffries, R.A.**: Effects of implantation energy and carbon concentration on the friction and wear of Ti-implanted steel.*Appl. Phys. Lett.*, No. 43, 1983, pp. 925-927.

97. **Singer, I.L., and Jeffries, R.A.**: Surface chemistry and friction behaviour of Ti-implanted 52100 steel. *Vacuum Science Technology*, A, Vol. 1, No. 2, Part 1, Apr-June 1983, p. 317.

98. **Suchanek, J., Bakula, J., and Cerny, F.**: Tribological characteristics of steels implanted by nitrogen (in Polish).*Metaloznawstwo, Obróbka Cieplna, Inżynieria Powierzchni (Metallurgy, Heat Treatment, Surface Engineering)*, 1989, No. 101, pp. 23-27.

99. **Takodum, J., Pivin, J.C., and Shaumont, J.**: Friction and wear of amorphous Ni-B, Ni-P films obtained by ion implantation into nickel.*Journal of Material Science*, 1985, Vol. 20, No. 4, pp. 1480-1493.

100. **Tousset, I., Mancoffre, N., Barnavon, Th., Teyseulle, S., Treheus, D., Guivaldeng, M., and Roblet, A.**: Amelioration de la résistance f l'usure par implantation ionique. *Mémoires et Études Scientifiques Revue de Métallurgie*, No. 9, 1983, p.469.

101. **Vasileva, E.V., and Savicheva, S.M.**: Enhancement of wear resistance of steel by ion implantation (in Russian).*Metallovedenye i Termicheskaya Obrabotka Materialov*, No. 1, 1987, pp. 59-62.

102. **Vasileva, E.V., and Savicheva, S.M.**: Effect of nitrogen and carbon ion implantation on the strength of bearing steel (in Russian). *Fizika i Khimia Obrabotki Materialov*, No. 1, 1989, pp. 80-82.

103. **Wiesner, L., and Gerve, A.**: Theoretical concept for the prediction of the extension of the lifetime of ion-implanted tools from short time measurements.*Materials Science and Engineering*, Vol. 69, No. 1, Feb. 1985, pp. 227-231.

104. **Zhai, C.F., De, L.L. and Zhong, Z.X.**: Modification of tribological characteristics of metals of N implantation. *Nuclear Instruments and Methods*, 209, 10,1983, pp. 881-887.

105. **Richter, E.**: Verschleissshutz durch Ionenimplantation. *Schmierungstechnik*, No. 11, 1985, pp. 324-326.

106. **Saritas, S., Procter, R.P.M., Ashworth, V., and Grant, W.A.**: The effect of ion implantation on the friction and wear behaviour of a phosphor bronze. *Wear*, Vol. 82, No. 2, 1982, pp. 233-255.

107. **Ohmae, N.**: Recent work on tribology of ion plated thin films. *Journal Vacuum Science and Technology*, 1976, Vol. 4, pp. 82-87.

108. **Sioshansi, P., and Au, J.J.**: Improvement of sliding wear for bearing grade steel implanted with Ti and C. *Materials Science and Engineering*, No. 69, 1985, pp. 161-166.

109. **Wanka, K., Kimura, T., and Shimonura, J.**: TiN film formation on 1% C roll steel by nitrogen ion beam implanted vapour deposition and the resulting improvement of the wear resistance. *Tetsuo-Hagane*, 1988, Vol. 74, No. 11, pp. 2177-2184.

110. **Yust, C.S., McHargue, C., and Harris, L.A.**: Friction and wear of ion-implanted TiB$_2$. *Materials Science and Engineering*, Dec. 1988, T.A. 105/106, pp. 489-496.

111. **Łabędź J.**: The effect of ion implantation on the fatigue life of self-aligning ball bearings. Proc.: *5th International Congress on Tribology - EUROTRIB: Friction-Wear-Lubrication*, Helsinki, June 1989, Vol. 2, pp. 359-365.

112. **Vasileva, E.V., and Savitseva, S.M.**: Structure and properties of 12H2N4A steel - implanted by nitrogen and carbon atoms (in Russian). *Materialovedenye i Termicheskaya Obrabotka Materialov*, 1988, No. 10, pp. 43-45.

113. **Vladimirov, B.G., Guseva, M.I., Ivanov, S.M., Terentiev, V.F., Fedorov, A.V., and Stepanchikov, V.A.**: Enhancement of fatigue strength of metals and alloys by ion implantation (in Russian). *Prochnost: Fizika, Khimia, Mechanika*, 1982, No. 7, pp. 139-147.

114. **Tarkowski, P.**: Change of the distribution of stresses in a contact zone after implantation with nitrogen ions. Proc.: *5th International Congress on Tribology - EUROTRIB: Friction-Wear-Lubrication*, Helsinki, June 1989, Vol. 1, pp. 388-393.

115. **Łunarski, J., and Zielecki, W.**: Assessment of fatigue strength of turbine blade alloys after implantation and shot-peening (in Polish). *Zeszyty Naukowe Politechniki Rzeszowskiej* (Transactions of the Rzeszów Technical University), No. 46, Mechanika, Vol. 17, pp.107-118.

116. **Vasileva, E.V., Fedorov, A.V., Savicheva, S.M., and Sapozhenikov, A.G.**: Effect of ion implantation on the properties of components made of 30HGSA steel (in Russian). *Vestnik Mashinostroenya*, No. 1, 1986, pp. 13-15.

117. **Hohmuth, K., Richter, E., Rauschenbach, B., and Blochowitz, C.**: Fatigue and wear of metalloid-ion-implanted metals. *Materials Science and Engineering*, No. 69, 1985, pp. 191-201.

118. **Hirvonen, I.P., Stone, D., Nastasi, M., and Hannula, S.P.**: Hardening of graphite surface by ion beam amorphisation. *Scripta Metallurgica*, Vol. 20, No. 5, May 1986, pp. 649-652.

119. **Follstaedt, D.M., Knapp, J.A., and Pope, L.E.**: Effects of ion implantation on the microstructure and surface mechanical properties of Fe alloys implanted with Ti. *Appl. Phys. Lett.*, 1984, Vol. 45, No. 5, pp. 529-531.

120. **Kaletta, D.**: The mechanical behaviour of ion implanted vanadium and Ti-V alloys investigated by means of microhardness techniques. *Journal of Nuclear Materials*, 104, 19882, pp. 907-912.

121. **Kolitsh, A., and Richter, E.:** Change of microhardness on ion implanted tungsten carbide. *Crystall Research Technology*,18, 1983, K 5-7.
122. **Hohmuth, K., and Rauschenbach, B.:** High-fluence implantation of nitrogen ion into titanium. *Materials Science and Engineering*, No. 69, 1985, pp. 489-499.
123. **Singer, I.L.:** Carbonization of steel surfaces during implantation of Ti ions at high fluence. *Journal Vacuum Science and Technology*, A, 1983, pp. 419-422.
124. **Williams, I.J., and Buchanar, R.A.:** Ion implantation of surgical Ti-6Al-4V alloy. *Materials Science and Engineering*, Vol. 69, No. 1, Feb. 1985, pp. 237-246.
125. **Hicks, R.D., and Robinson, E.P.A.:** The aqueous corrosion behaviours of an ion implanted 12% chromium steel. *Corrosion Science*, 1984, Vol. 24, No. 10, pp. 885-900.
126. **Garodetski, A.E., Guseva, M.I., and Tomasov, N.D.:** Formation of corrosion-protective layers on titanium by means of ion implantation of palladium (in Russian). *Prochnost: Fizika, Khimia, Mechanika*, 1982, No. 3, pp. 83-88.
127. **Kopetski, C.V., and Vyatkin, A.F.:** About certain trends of development of contemporary metallurgy (in Russian). *Vestnik AN SSSR*, 1982, No. 1, pp. 47-56.
128. **Bhattacharaya, R.S., Rai, A.K., Raffoul, C.N., and Pronko, P.P.:** Corrosion behaviours of amorphous Ni based alloy coatings fabricated by ion beam mixing. *Journal Vacuum Science and Technology*, A, 1985, Vol. 3, No. 6, pp. 2680-2683.
129. **Dearnaley, G., Goode, R.D., and Minter, F.J.:** Ion implantation and ion assisted coating of metals. *Journal Vacuum Science and Technology*, 1985, Vol. 3, No. 6, pp. 2684-2690.
130. **Sartwell, B.D., Baldwin, D.A., and Singer, I.L.:** Summary Abstract: Effects of pressure and temperature on implantation induced carburization of steels. *Journal Vacuum Science and Technology*, A, Sec. Series, Vol. 3, No. 3, Part 1, May-June 1985, pp. 589-590.
131. **Baumvol, I.J.R.:** Ion implantation metallurgy. *Ion Implantation Science and Technology*, Edited by: J.F. Ziegler, Academic Press 1984.
132. **Smidt, F., and Sartwell, B.D.:** Manufacturing technology program to develop a production ion implantation facility for processing bearing and tools. *Nuclear Instruments and Methods*, 86, 1985, pp. 70-77.
133. **Iwaki, M., Fujihana, T., and Okitaka, K.:** Mechanical and electrochemical properties of various tool steels implanted with nitrogen or NO ions. *Materials Science and Engineering*, No. 69, 1985, pp. 211-217.
134. **Korycki, J.:** An assessment of tribological properties of the implanted layer (in Polish). *Prace Instytutu Lotnictwa (Transactions of the Aeronautics Institute)*, No. 121-122 (Vol. 2-3), 1990, pp. 100-118.
135. **Scharf, W.:** Charged particle accelerators. Applications in science and technology (in Polish). PWN, Warsaw, 1989.
136. **Kaczmarek, J.:** Investigations of properties of ion implanted tools (in Polish). *Mechanik*, No. 11/12, 1990, pp. 385-386.
137. **Podgórski, A., Jagielski, J., and Gawlik, G.:** Perfection of ion implantation method in order to obtain optimum properties of the technical surface layer (in Polish). *Mechanik*, No. 11/12, 1990, pp. 393-396.

chapter five

Glow discharge methods and CVD technology

5.1 Conception and development of glow discharge methods

The term "glow discharge treatment" encompasses such processes as nitriding, carbonitriding, sulfonitriding and carburizing, carried out in the presence of glow discharge, as well as boriding, siliciding and *Plasma Assisted Chemical Vapour Deposition* (PACVD) methods. The latter involve deposition with the participation of a chemical reaction and in conditions of electrical activation of the gaseous environment. All of these processes are aimed at obtaining hard surface layers, e.g., carbides, nitrides, borides, and oxides of transition metals. The only difference between these methods are process parameters, due to different gas mixtures used, and design features of some sub-assemblies of equipment used to carry out these processes.

The best known and commonly used on an industrial scale is the process of glow discharge nitriding which, despite attempts at utilizing it in the closing years of World War II to improve the life of artillery barrels by a method proposed by B. Berghaus in the late 1930s [1, 2], found proper industrial application only in the 1970s. It was only at that time that research concerning, among others, the stabilization of the glow discharge with the aid of high intensity current created a basis for industrial application of the process. In 1972, the German company Klöckner Ionon GmbH launched the sale of industrial equipment for glow discharge nitriding [3].

Presently, such equipment is manufactured in many countries, among others in Germany, U.S., Poland, Japan, Russia, France and Bulgaria, and has found broad application in industry [4].

5.2 Physico-chemical basis of glow discharge process treatment

Thermo-chemical treatment in a gaseous environment with the utilization of the effect of glow discharge involves the following.

Treated objects (cathode) are placed in the working chamber whose walls, as well as appropriately designed screens, constitute the anode. The reactive gas, e.g. in glow discharge nitriding - ammonia (NH_3) or a mixture of hydrogen and nitrogen ($H_2 + N_2$), or BCl_3 vapors in a mixture with hydrogen in the process of boriding, are introduced into the working chamber at pressures in the range of 1 to 13 hPa into a so-called dynamic vacuum, i.e., with continuous flow of a given gas mixture through the chamber. A potential difference is applied between the cathode and the anode, ranging from 400 to 1800 V, depending on the chemical composition of the gas mixture. Such conditions enable the maintenance of abnormal glow discharge whose characteristic feature is the rise of current intensity with the rise of voltage and the existence of a cathode potential drop in which elementary processes responsible for the formation of the surface layer take place [5-7]. A change in the voltage of the glow discharge in the given range causes a change in the current intensity and, in consequence, heating of the treated objects to desired treatment temperatures. The treated object is heated by radiation from glow discharge regions close to the cathode and by bombardment by gas ions [9]. The make-up of the active particles is quite complicated. Their type depends on the chemical composition of the gas mixture, on the pressure inside the working chamber, on voltage, and on the material of the treated object. However, the knowledge of the local gas composition in the discharge zone close to the cathode is not essential to control the process. This is effected by the selection of the gas mixture, pressure and discharge current density. A significant role is played in this method by the gas environment in which the process is carried out and which affects, among others, the kinetics of layer formation, its structure and properties [7-9].

The problem of selection of the appropriate reactive mixture and, connected with it, control of structure and phase composition of the layers produced, is one of the most important in thermo-chemical treatment utilizing glow discharge.

5.2.1 Glow discharge

Electrical discharge in gases may be of a variable character, depending on the working conditions of the system, among others, pressure and type of gas, intensity and frequency of the electric field, on electrode material and on their temperature. Hence, there is a variety of types of discharges and some difficulties in their unequivocal systemization [5, 10-12].

The main parameters of each glow discharge are voltage, current intensity and gas pressure. Depending on the combination of these parameters it is possible to distinguish e.g., the following types of discharges: glow, corona, spark and arc [5, 6].

Glow technology belongs to a group of plasma methods which utilize plasma that is non-equilibrium, low temperature and non-isothermal, formed as the result of continuous drawing of energy from the electric field. In glow methods, one type of electrical discharge in gas is utilized, i.e., the glow discharge, which occurs in gases within the range of pressure from 10^{-3} to 13 hPa (see Fig. 5.1). Due to the fact that glow discharge is accompanied by a

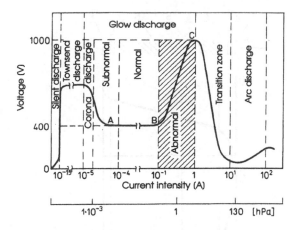

Fig. 5.1 Voltage-current characteristic of glow discharge in argon.

characteristic glowing of gas close to the cathode (so-called *cathodic glow*), these methods have come to be known as **glow** methods, sometimes also called **plasma** or **ion** methods (e.g. *plasma nitriding, ion nitriding or nitriding by glow discharge*).

The current-voltage characteristic curve for electric discharge in low pressure gas is shown in Fig. 5.1. Near point A, a glow appears around the cathode. A characteristic feature of interval AB of the curve is that it is almost totally independent of current intensity. The discharge is stable and is referred to as **normal glow discharge**. In this zone the current density is constant and the glow covers only a portion of the cathode. As the current intensity is increased, the glow covers an increasing portion of the cathode surface but without any change in voltage, until the entire cathode is covered by the glow. Interval BC corresponds to unstable, so-called **abnormal glow discharge**, utilized in thermo-chemical treatment. Its characteristic feature is that voltage rises with a rise of the current intensity. Beyond point C begins the range of arc discharge. In the space between electrodes it is possible to observe several zones of varied extent and glow intensity [5, 6, 14]. Glow discharge in the space between electrodes is not homogenous, as shown by the model in Fig. 5.2. The example shown here is one of discharge in the pressure range of 1.33 to 13.3 hPa, in a quartz tube with flat electrodes, taking into account potential distribution and the intensity of the electric field [5].

For processes carried out in conditions of glow discharge in a constant electric field, the most important zone is the one between the cathode and the Faraday dark space [9, 12, 13]. Positive ions in the Crooks dark space and the negative glow regions, being accelerated within the cathode potential drop zone, strike the cathode, sputtering electrons out of its surface. These, in turn, being accelerated in the opposite direction, increase their energy, causing first excitation of particles (cathode glow) and next their ionization. As the result of these processes, both at the cathode surface as well as in the surrounding gas phase, active particles may be created, responsible for the formation of the surface layer.

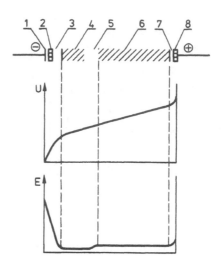

Fig. 5.2 Distribution of light effects in a glow discharge, potential (*U*) and electric field intensity (*E*) between electrodes: *1* - Aston dark space; *2* - cathodic glow; *3* - cathodic dark space; *4* - negative glow; *5* - Faraday dark space; *6* - positive aurora; *7* - anodic dark space; *8* - anodic glow.

In conditions of glow discharge, a dark zone is formed at the cathode in which a high potential drop occurs. At the edge of that zone negative glow begins. It has been determined that under 26 hPa pressure it begins at a distance of 1 mm from the cathode [5, 14]. In its further portion this negative glow changes to Faraday's dark zone and this, in turn, into the positive column. Within Faraday's dark zone and in the positively charged column concentrations of electrons and ions are the same. If the distance between electrodes varies, the only zone to undergo change is the length of the positively charged column, while the zones at the cathode remain unchanged [5, 6, 14].

The carriers of electric current in glow discharge are electrons and positive ions. At the moment of application of potential, these particles are accelerated. During bombardment by ions the cathode heats up which causes the gas surrounding the cathode to heat up also, thus reducing its density. This causes a change in the voltage-current characteristic of the discharge, in particular of the width of the cathodic drop zone [15] which also depends on the pressure inside the working chamber. In the case of nitriding at pressures of 1.33 hPa to approximately 9 hPa, this width ranges from 4.4 to 0.8 mm [17], while the mean energy of nitrogen ions (N^+) under pressure of 8 hPa is approxmately $9.6 \cdot 10^{-18}$ J [15]. The highest intensity of the electric field, as well as the greatest potential drop between electrodes - as can be seen from Fig. 5.2 - occur in zones 1 to 3 of the glow discharge space. From the point of view of activation of chemical processes in glow treatment, these zones are of greatest significance [9, 13].

The course of chemical reactions is also affected by the non-uniform distribution of density of electric charge carriers, thus by the non-uniform distribution of current density. Where the greatest density of electric charge carriers occurs, among others of electrons, activation of chemical processes is most intensive. Consequently, in conditions of glow discharge treatment, in the working chamber there may come about a stabilization of a very non-uniform distribution of density of active particles. For example, the reaction of dissociation of nitrogen particles occurs with greatest efficiency in the cathodic potential drop zone [13, 18], where there is a sufficiently big number of electrons with energies higher than the energy of dissociation of nitrogen particles ($1.52 \cdot 10^{-18}$ J), as well as the energy of nitrogen ionization ($2.5 \cdot 10^{-18}$ J). The degree of ionization depends on the energy of electrons and attains its maximum value of approximately 0.45 when the energy of electrons is about $3.2 \cdot 10^{-17}$ J [17]. Plasma created in conditions of glow discharge contains electrons whose mean energy is within the range $(1 \text{ to } 10) \cdot 10^{-17}$ J, temperature is 10^4 to 10^5 K and electron gas density is 10^{15} to 10^{18} electrons/m^3 [6, 11]. The temperature of the electrons T_e is significantly higher than the temperature of the gas particles T_g and their ratio is $T_e / T_g = 10$ to 100. It is, therefore, possible to carry out chemical reactions already at temperatures close to those of the environment, while at the same time free electrons (or some ions) have an energy sufficient to detach covalent bonds and to initiate the process of polymerization [12]. In conditions of glow discharge nitriding in an atmosphere of nitrogen at 8 hPa pressure, there occur first of all ions of N^+ and N_2^+, with the N^+ ions making up approximately 50% of all ions [15, 21]. Moreover, the density of positive ions formed in the cathodic potential drop zone is approximately 10^9 to 10^{11} ions/cm^3 at a temperature of 450 to 550°C, while that of electrically neutral particles is 10^{16} to 10^{17} per cm^3 [17, 20]. The mean energy of N^+ nitrogen atoms reaching the cathode when the total pressure inside the working chamber is approximately 8 hPa is $9.6 \cdot 10^{-18}$ J [15]. Glow discharge leads to the formation of a low temperature, non-isothermal plasma, characterized by the existence of ions, electrons, excited particles, atoms and gas particles which causes a substantial rise in the energy of the system. Fig. 5.3 shows, by way of example, the spatial distribution between the anode and cathode of some active particles formed in conditions of glow discharge in a mixture of vapours of $TiCl_4 + H_2 + N_2$ [23]. It should be emphasized that ions carry only about 10% of the total energy of the system [21]. In typical conditions of glow discharge ionization does not exceed several percent [10]. Nevertheless, ions play a major role in the formation of the superficial layer, especially on account of activation of the surface of the treated object, by the utilization of the cathode sputtering effect [9]. Table 5.1 shows examples of results of investigations concerning identification of ions reaching the cathode in the process of glow discharge nitriding. The investigations were carried out with the aid of mass spectrometer, coupled with the working chamber of an glow discharge furnace. This table presents the percentage distribution of ions reaching the cathode vs. the composition of the gas mixture, when the pressure in the working chamber is 8 hPa [15].

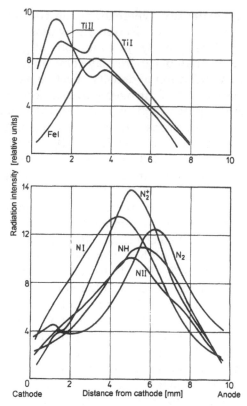

Fig. 5.3 Intensity distribution of radiation of active particles (*I* - atoms, *II* - ions Ti⁺ and N⁺) in the interelectrode space, in a mixture of $TiCl_4 + H_2 + N_2$, in conditions of glow discharge. (From *Wierzchoń, T., et al.* [23]. With permission.)

Table 5.1
Percentage distribution of ions reaching the cathode, depending on composition
of N_2+H_2+Ar gas mixture under 8 hPa pressure (From *Hudis, M.* [15]. With permission.)

Type of ions	Gas mixtures of various compositions - partial pressures [hPa]								
	N_2	Ar	H_2	N_2	Ar	H_2	N_2	Ar	H_2
	1.6	0	6.4	1.6	5.3	1.1	1.6	6.4	0
H^+		86.6			78.6			0	
N^+		0.05			1.8			48.6	
NH^+		0.16			43			0	
NH_2^+		1.62			3.4			0	
NH_3^+		6.8			7.0			0	
NH_4^+		2.8			2.9			0	
N_2^+		0.02			0.1			18.3	
$N_2H_2^+$		1.04			0.64			0	
Ar^+		0			1.2			32.7	

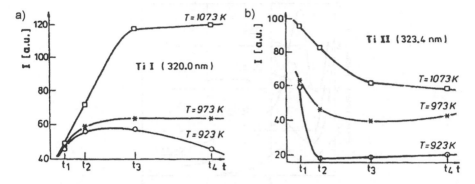

Fig. 5.4 Variation of intensity of radiation of atoms (T I) (a) and ions (T II) (b) of titanium in plasma of $H_2 + N_2 + TiCl_4$ (5%) with duration of glow discharge at different cathode temperatures, for times of $t_1=3$, $t_2=15$, $t_3=45$, $t_4=90$ min. (From *Kułakowska, B., et al.* [16]. With permission.)

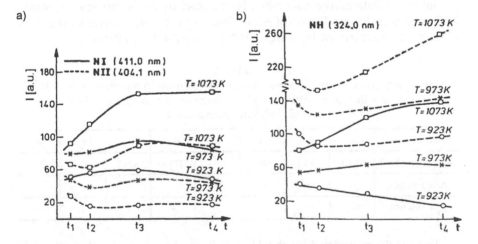

Fig. 5.5 Variation of intensity of radiation of atoms (N I) and ions (N II) of nitrogen (a) and NH radicals (b) in plasma of $H_2 + N_2 + TiCl_4$ (5%) with duration of glow discharge at different cathode temperatures, for times of $t_1=3$, $t_2=15$, $t_3=45$, $t_4=90$ min. (From *Kułakowska, B., et al.* [16]. With permission.) Times same as in Fig. 5.4.

Figs. 5.4 and 5.5 present, again as an example, the change in intensity of radiation (concentration) of atoms (TI) and ions (TII) of titanium, as well as ions (NII) and atoms (TI) of nitrogen and NH radicals in a glow discharge in an atmosphere composed of $TiCl_4 + N_2 + H_2$ during the process of formation of a titanium nitride layer for different cathode temperatures [16]. Fig. 5.6 shows changes in the intensity of radiation of boron ions vs. distance from cathode and duration of glow discharge in an atmosphere of $BCl_3 + H_2$ under 7 hPa pressure and current density of 10 mA/cm². The investigations were carried out by the emission spectroscopy method [19].

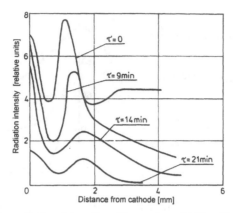

Fig. 5.6 Variation in radiation intensity of boron ions (B II) with distance from cathode and time τ of glow discharge in an atmosphere of $H_2 + 10\%$ BCl_3, under pressure of 7 hPa and current density of 10 mA/cm². (From *Szczepańska, H., et al.* [19]. With permission.)

Table 5.2 lists active particles (determined by emission spectroscopy) [19] occurring in conditions of glow discharge in zones close to the cathode, in gas mixtures of $H_2 + BCl_3$ (10%) and $Ar + BCl_3$ (10%).

Table 5.2
Active particles occurring in zones close to the cathode in mixtures of $H_2 + BCl_3$ and $Ar + BCl_3$ in conditions of glow discharge (From *Szczepańska, H., et al.* [19]. With permission.)

Gas mixture	B I	B II	B III	B_2	BCl	Cl II	Fe II	H_B
Wavelength λ [nm]	208.9	345.1	206.7	328.3	272.2	490.2	315.4	486.1
Ar + 10% BCl_3	+	+	+	+	+	+	+	-
$H_2 + 10\%$ BCl_3	+	+	-	-	+	+	-	+

The spectrum emitted by the $H_2 + BCl_3$ mixture was significantly deficient in lines and emission bands. Only the presence of atomic boron and its ions, BCl particles and atomic hydrogen was determined [19]. This is due to the participation of hydrogen in reactions of reduction of boron chloride, while the comparison of active particles occurring in the cathode zone in mixtures of $H_2 + BCl_3$ and $Ar + BCl_3$ is proof of the more effective course taken by cathode sputtering in the case of the $Ar + BCl_3$ mixture during the initial stage of boride layer formation [9, 19, 34]. Due to the high drop of cathode potential surrounding the cathode (treated object) uniformly, ions formed as the result of collisions between electrons and gas particles are accelerated in the direction of the cathode. They attain a certain kinetic energy, dependent on the voltage applied and on pressure. In the case of a glow discharge process under 800 V potential and pressure of approximately 4 hPa, energies of ions formed in a nitrogen atmosphere do not exceed $3.2 \cdot 10^{-17}$ J [26].

Hence, in conditions of glow discharge in a nitriding process there is an available source of nitrogen ions which in the cathodic space exhibit a substantial kinetic energy. This is of big significance to the kinetics of surface layer formation in glow discharge processes insofar as a short time suffices for a high concentration of the element constituting the layer to be formed at the interface. It becomes the source of an enhanced concentration gradient and, by the same token, of enhanced diffusion rate. Moreover, the bombardment of the cathode by accelerated ions has another aspect. This is the possibility of obtaining the effect of cathode sputtering which plays a significant role from the point of view of metal surface activation [9, 22, 24-27]. It also enables some influence on dimensional changes of the treated object during the process, as demonstrated by glow discharge nitriding [10].

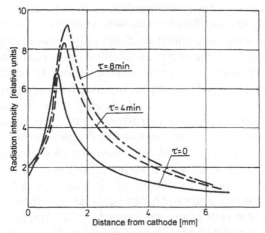

Fig. 5.7 Variation in radiation intensity of atomic hydrogen ($\lambda = 486.1$ nm) with distance from cathode and time τ of glow discharge in an atmosphere of $H_2 + 10\%$ BCl_3, under pressure of 7 hPa and current density of 10 mA/cm². (From *Szczepańska, H., et al.* [19]. With permission.)

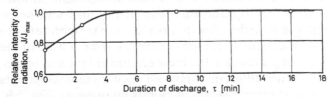

Fig. 5.8 Dependence of relative radiation intensity of atomic hydrogen ($\lambda = 486.1$ nm) on time of discharge in a mixture of $H_2 + 10\%$ BCl_3, under pressure of 7 hPa and current density of 10 mA/cm². (From *Szczepańska, H., et al.* [19]. With permission.)

The cathode (treated object), bombarded by ions, is heated as the result of collisions between ions and its surface. This method of heating is beneficial from the point of view of chemical reactions at the solid/gas interface and diffusion processes because the highest temperature is at the site of the reaction.

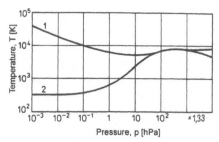

Fig. 5.9 Dependence of temperature of electrons (1) and gas particles (2) on pressure in conditions of glow discharge. (From *Zdanowski, J.* [6]. With permission.)

Charged particles forming the plasma are in an electric field which constantly accelerates them, maintaining a constant level of kinetic energy. The mean kinetic energy attained in such way may assume values which significantly exceed the mean kinetic energy of electrically neutral particles. This creates non-equilibrium thermodynamic conditions, resulting in the possibility of obtaining non-equilibrium concentrations of reaction products, i.e. higher than those corresponding to equilibrium for a given temperature [9, 13]. Thus, upon heating hydrogen in a resistance furnace to a temperature of approx. 800 K, only trace amounts of atomic hydrogen are obtained [28]. The introduction of glow discharge allows higher-than-equilibrium concentrations of hydrogen atoms at significantly lower temperatures. Figs. 5.7 and 5.8 show the changes in intensity of radiation of atomic hydrogen with distance from the cathode, for different times of glow discharge in a mixture of H_2 + BCl_3 [9, 19]. It should be emphasized that electric activation of chemical reactions is particularly significant in conditions of glow discharge [13, 28]. The low temperature plasma formed is composed of different active particles: ions, nascent atoms, excited atoms, electrons and gas particles at different temperatures. The components of the plasma have their kinetic energies and their mean values determine the temperatures of ions, electrons and gas particles. During glow discharge, electrons and ions absorb same amounts of energy but electrons lose significantly less energy than ions on account of collisions with gas particles. For that reason, the temperature of electrons (T_e) - as already mentioned - is higher than the temperature of the remaining components of plasma, i.e. the temperature of ions and of gas particles (T_g). The temperature difference between the various components, characteristic of non-isothermal plasma, depends on pressure (Fig. 5.9) because the higher the pressure the greater the frequency of collisions of electrons with ions or with gas particles. Thus, the amount of energy passed to gas particles by electrons rises and the energies of the particular gas components are equalized which may, finally, cause the formation of isothermal plasma [6]. This is a typically non-equilibrium system and the big difference in glow discharge plasma between the high temperature of electrons and relatively low temperature of gas particles creates conditions favorable for chemical reactions to proceed.

In conditions of electric activation, the state of dynamic equilibrium is the result solely of the given system of rates of elementary processes, taking place under the influence of the electric field. For that reason, both the rate of chemical transformations, as well as the final state to which these chemical reactions are directed, are, as a rule, different to those where the process is activated only thermally. This can be explained by the fact concentrations of reaction products may be greater than the equilibrium concentration corresponding to a given temperature, thus, they may be higher than equilibrium which has been demonstrated in the process of nitrogen oxide (NO) synthesis [13]. In conditions of only thermal process activation, the distribution of particle energy is in accordance with the Maxwell-Boltzmann distribution, in other words, the same as in conditions of thermal equilibrium. The system then tries to reach thermodynamic equilibrium in which the concentrations of reagents result directly from the equilibrium of chemical reactions for given process conditions. The fact that under the influence of the electric field, charge carriers, notably electrons, may attain an energy that is significantly higher than the mean energy of gas particles and that they may pass on some of that energy to them during collisions causes more active particles to appear in the system (ions, atoms, excited particles), all with higher energy than that corresponding to equilibrium at the given temperature. They all exhibit high chemical activity. The presence in the system of a significant number of chemically active particles with high energy explains the fact that in electrical discharges, chemical reactions may proceed at lower temperatures and at high rates. Electric activation of chemical reactions is particularly significant in conditions of glow discharge [28, 29] and the system, despite high rates of reactions, does not reach the state of thermodynamic equilibrium.

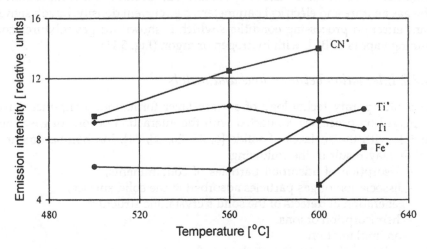

Fig. 5.10 Variation of the emission intensity of selected spectral lines vs. cathode temperature. (From *Sobiecki, J.R., et al.* [16]. With permission.)

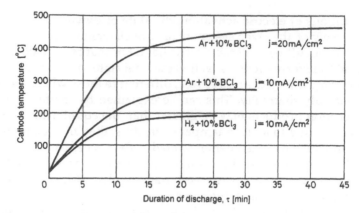

Fig. 5.11 Dependence of cathode temperature on time of glow discharge and on composition of gas mixture, under 7 hPa pressure, for different current densities. (From *Szczepańska, H., et al.* [19]. With permission.)

The type of ions and active particles reaching the cathode is also influenced by processing conditions, as well as by the type of treated material. For example, the presence of titanium in the steel cathode causes a rise of concentration of nitrogen ions and atoms in the glow discharge plasma [20, 31, 42]. Fig. 5.10 shows, by way of example, the variation in intensity of selected active particles in the plasma of tetraisopropoxytitanium $(Ti(OC_3H_7)_4)$, hydrogen and nitrogen with substrate temperature. It follows from this representation that with a rise of process temperature rises the number of CN radicals which are responsible for the process of carbonitriding, taking place at temperatures above 560°C. Above that temperature cathode sputtering also takes place which is manifest by the presence of Fe in the emission spectrum [32]. Besides, the composition of the gas mixture and electrical parameters, e.g., current density, have a significant effect on processing conditions which is shown for gas mixtures containing vapors of BCl_3 with hydrogen or argon (Fig. 5.11).

5.2.2 Interaction between ions and metals

In contemporary technology of surface layer formation, a significant role is played by processes connected with the interaction of ions of gases and metals with the surfaces of solids [6, 25, 26, 33, 34]. Ion bombardment of solids may result in the following:
- desorption of adsorbed particles of contaminants,
- dissociation of gas particles adsorbed at the solid surface,
- generation of defects of the solid's crystalline structure,
- chemisorption of ions,
- ion implantation,
- emission of electrons and photons from the cathode.

In glow discharge treatment, ion sputtering plays an important role in the formation of surface layers.

5.2.2.1 Ion sputtering

The material of the cathode bombarded by a stream of ions may be sputtered in accordance with one of the following processes [4, 6]:

1. **Chemical sputtering** where gas ions form volatile chemical compounds with the material of the sputtered cathode. These compounds are easily detached and subsequently condense on surfaces which are close. In this process, kinetic energy of ions plays a secondary role in comparison with their chemical activity.

2. **Reactive sputtering** where gas ions bombarding the electrode form chemical compounds with sputtered particles, e.g. oxides, nitrides. In this process, chemical activity of ions is of no lesser importance than their kinetic energy.

3. **Physical sputtering**, most often called **cathode sputtering**, where the detachment of particles from the bombarded electrode takes place due to imparting of impetus to them by ions.

Threshold sputtering energy E_t depends predominantly on the type of ions and of sputtered material

$$E_t \gamma_{max} \geq H \tag{5.1}$$

where: γ - coefficient of accommodation, describing the maximum amount of energy which may be passed by an ion to an atom;

$$\gamma = \frac{4 M_1 M_2}{(M_1 + M_2)} \tag{5.2}$$

where: M_1, M_2 - atomic masses of ion and of sputtered particle, H - heat of sublimation of sputtered material.

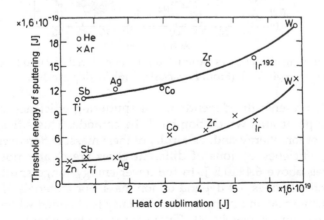

Fig. 5.12 Threshold energy of sputtering vs. heat of sublimation for different materials sputtered by ions of helium and argon. (From *Zdanowski, J.* [6]. With permission.)

Threshold energy of sputtering of the majority of gas-metal systems is within the range of (5 to 52)×10⁻¹⁹ J [26]. Fig. 5.12 shows the dependence of threshold energy of sputtering on the heat of sublimation of different materials sputtered by ions of helium and argon. There exists a strict correlation between the threshold energy and the heat of sublimation of the sputtered material which varies in sequence in the particular periods of the periodic table [6].

Sputtering efficiency S is described by a correlation which represents the mean number of sputtered atoms and ions N_a per one bombarding ion N_j.

$$S = \frac{N_a}{N_j}$$ (5.3)

Sputtering efficiency is a function of such factors as: type, energy and angle of incidence of ions bombarding the cathode, as well as crystalline structure and temperature of the cathode material.

Fig. 5.13 shows, by way of example, the dependence of sputtering efficiency on atomic number of the cathode material which is bombarded by neon ions. It follows from it that the cathode material has a decisive influence on sputtering efficiency. When the outer electron shell of the sputtered material atom is more filled, it sputters easier, i.e. sputtering efficiency is higher.

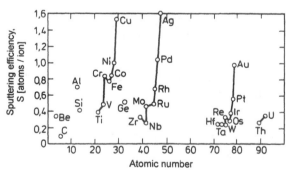

Fig. 5.13 Sputtering efficiency vs. atomic number of material of cathode, bombarded by neon ions with 6.4·10⁻¹⁷ J. (From *Zdanowski, J.* [6]. With permission.)

Fig. 5.14 presents the dependence of sputtering efficiency on the energy and type of ions which bombard the cathode. This efficiency rises with a rise of ion energy and, as a rule, of their mass. Differences between sputtering efficiency by ions of different gases are clearly noticeable at energy levels above 6.4×10⁻¹⁷ J. In the lower energy range sputtering efficiency practically does not depend on ion mass. For materials of polycrystalline structure maximum sputtering efficiency is obtained when the angle of incidence is approximately. 60°. Temperature, on the other hand, has only an insignificant effect [6, 26].

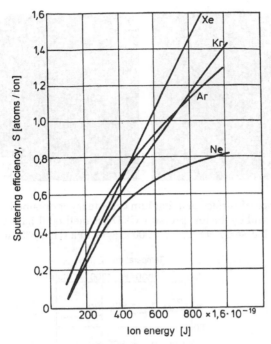

Fig. 5.14 Sputtering efficiency of iron by ions of xenon, krypton, argon and neon with different energies. (From *Plesivtzev, N.V.,* [26]. With permission.)

5.2.2.2 The role of ion sputtering in glow discharge treatments

Ion bombardment of a metal surface is an effective method of its cleaning and activation [9, 25, 26]. It creates the possibility of removal of surface layers of oxides and adsorbed gases. It has an influence on surface development by revealing grain boundaries, i.e., by etching the structure, as well as on structural changes of the superficial layer of the metal, e.g., by the formation of great numbers of displaced atoms, vacancies and dislocations [22, 25]. Bombarding ions may be adsorbed at the surface and react chemically with the metal. Hence, for processes of cathode sputtering usually neutral gases are used because they feature low heat of binding with the metal surface and they are easily desorbable (see Table 5.3) [35].

Table 5.3

Heat of adsorption of some gases at the surface of tungsten (From *Groszkowski, J.* [35]. With permission.)

Metal	Tungsten					
Gas	Ar	He	H_2	N_2	N	O_2
Heat of adsorption [kJ/mole]	7.94	8.36	191.8	354.41	646.4	809.0
Type of adsorption	physical		chemical			

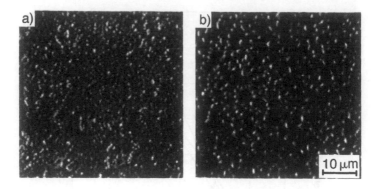

Fig. 5.15 Appearance of surface of a titanium carbide layer obtained in conditions of glow discharge (a) and by the low pressure CVD gas method (b). Process parameters are the same in both cases: $T = 830°C$, $t = 300$ s, $p = 7$ hPa ($H_2 + 8\%$ $TiCl_4$).

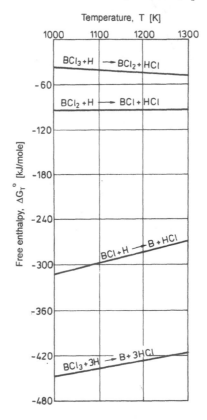

Fig. 5.16 Changes in free enthalpy ($\Delta G°_T$) of the reaction of boron chloride reduction by atomic hydrogen vs. temperature.

The effect of cathode sputtering plays a significant role in glow discharge treatments, on account of its activation of the superficial layer of the treated objects. This is demonstrated by e.g., a 10-fold increase in the density of structural defects in a molybdenum single crystal, after heating to 1500°C in hydrogen in conditions of glow discharge, with the following process parameters: potential - 700 V, current density - 0.20 A/cm², working chamber pressure - 26 hPa. The maximum number of defects of the crystalline structure was found in the superficial layer of 3 to 10 μm thickness [22]. This fact has a significant impact on the intensification of surface processes, i.e., on the effects of chemisorption and diffusion and, due to the elevated energy of the superficial layer, on the nucleation of a new phase. As an example, Fig. 5.15 shows the results of comparative investigations of the formation of titanium carbide on 9% C carbon steel in short-cycle PACVD processes, carried out under glow discharge conditions, as well as by LPCVD (*Low Pressure Chemical Vapour Deposition*)under 4 hPa pressure in an atmosphere composed of $TiCl_4 + H_2$ [9]. It follows from the plot that diffusion of carbon from the steel matrix to the reaction site is faster under glow discharge conditions and that these conditions result in the formation of a finer structure, due to a greater number of nucleation centers. Significant development of the treated surface due to cathode sputtering, electric activation of the gas environment (presence of active particles forming the layer), as well as significant effect of atomic hydrogen on the reaction of chloride reduction (Fig. 5.16) causes an iron boride FeB and Fe2B layer of 20 μm thickness to be formed in short boriding processes (tens of minutes) (Fig. 5.17). It follows from that figure that the borides once formed exhibit an orientation perpendicular to the surface and occur uniformly across the entire surface, creating a compact layer.

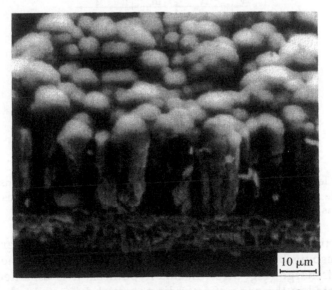

Fig. 5.17 Appearance of surface and fracture of a boride layer formed on Armco iron during heating in conditions of glow discharge in a mixture of vapors of $H_2 + 10\%$ BCl_3 to a temperature of 800°C (in 1200 s) and holding at that temperature for 300 s under 4 hPa pressure.

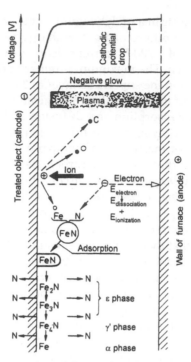

Fig. 5.18 Schematic presentation of the mechanism of formation of a surface layer in a glow discharge nitriding process. (From *Keller, K.,* [36]. With permission.)

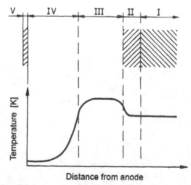

Fig. 5.19 Temperature distribution inside a work chamber in conditions of glow discharge: *I* - load; *II* - surface layer being formed; *III* - zone of heated gas; *IV* - zone of cooled gas; *V* - cooled wall of the glow discharge furnace chamber.

Reactive sputtering is, according to Keller and Edenhofer [3, 36], the effect responsible for the formation of nitrided layers in the process of glow discharge (Fig. 5.18). In this hypothesis it is assumed that ions accelerated in the zone of cathodic potential drop strike the surface of the cathode with a relatively high kinetic energy which causes "knocking out" of atoms from the surface, electron emission and heating of objects. Additional heating is provided by thermal radiation from near-cathode discharge zones - see Fig. 5.19 [28]. At the same time there occurs the process of partial return to the cathode of the products of sputtering after the reaction with atomic nitrogen in the zone of cathodic drop. According to Keller, these are both iron nitrides, as well as carbides and carbonitrides (carbon supplied by matrix). Iron nitrides, rich in nitrogen, upon deposition on the cathode undergo a breakdown into nitrides of lesser nitrogen content, due to constant bombardment of the surface by ions and interaction by the matrix. According to this hypothesis, ion sputtering and sputtered iron atoms which combine with active particles of nitrogen in the potential drop zone are responsible for the transport of nitrogen to the treated surface.

5.2.3 Chemisorption in glow discharge treatments

The process of chemisorption [4, 37–39] is an effect which plays a significant role in thermo-chemical treatment carried out under conditions of glow discharge. Its course is greatly influenced by ion sputtering and by the presence of active particles of elements which form the superficial layer, e.g., atoms and ions of nitrogen in the process of nitriding, lower chlorides of titanium, $TiCl_2$ and $TiCl_3$, as well as atoms and ions of titanium in the process of generating TiN or TiC layers by the PACVD method. Substantial defects of the crystalline structure of the superficial layer, removal from the surface of contaminants, e.g., adsorbed gases, water vapour or oxides, cause a rise of the energy of the layer, thus intensification of adsorption processes (at elevated temperatures also of chemisorption), and, in turn, of diffusion and nucleation of the new phase. This is of great significance to the kinetics of layer formation in glow discharge processes because within a short time, a high concentration of particles forming the layer collects at the interface, providing a source for enhanced gradient, hence faster diffusion. This can be said of processes such as glow discharge nitriding or boriding, or the acceleration of chemical reactions which guarantee the formation of layers of titanium nitride (TiN) or titanium carbide (TiC) by PACVD methods. The activation of the gas environment which causes accelerated reaction of boron chloride by atomic hydrogen in the process of glow discharge boriding (Fig. 5.16), as well as activation of the superficial layer of the treated load during heating of parts to the treatment temperature (Fig. 5.20) allow the obtaining of a high concentration of boron at the surface. This, in turn, spells favorable conditions for diffusion and the formation of the boride layer (Fig. 5.17) provided that the appropriate chemical composition of the gas mixture is used, i.e., H_2 + max. 10% BCl_3. When gas atmospheres containing more than 10% by volume of BCl_3 vapors are used, a brittle, porous sublayer is formed in the external zone of the surface layer. This sublayer contains significant amounts of boron, chorine and iron, which indicates chemisorption of products of chemical reactions (see Fig. 5.17), e.g., hydrogen chloride, and disturbs the process of formation of the boride layer [9].

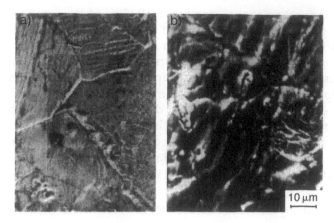

Fig. 5.20 Appearance of sample surfaces made of Armco iron, after heating in conditions of glow discharge in hydrogen (a) and argon (b) to a temperature of $T = 800°C$ in 900 s, under 3 hPa pressure.

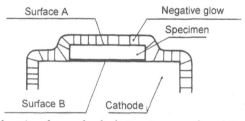

Fig. 5.21 Schematic showing the method of situating samples of titanium on the cathode in a glow discharge nitriding process. (From *Roliński, E.,* [20]. With permission.)

Besides, as was shown in Fig. 5.17, during the initial stages of formation of the titanium carbide layer, a very fine-grained titanium carbide is obtained which indicates a greater number of crystallization nuclei of the TiC phase than in the LPCVD method. This means a greater number of centers of chemisorption of active titanium particles at the surface, as well as faster diffusion of carbon to the reaction site (surface), on account of the elevated energy of the superficial layer, caused by the defected crystalline structure and cleaning of the load surface due to cathode sputtering [7]. In glow discharge nitriding, cathode sputtering intensifies chemisorption of atomic nitrogen which takes place at temperatures from above 200°C [40, 41]. Chemisorption plays a fundamental role in nitriding of titanium and its alloys [20, 42, 45]. Titanium clearly enhances dissociation of nitrogen particles [20, 31, 42] and the formation in the nitrogen plasma of N^+ ions and atoms of nitrogen. Moreover, chemisorption of nitrogen in titanium is a non-activated process, while titanium features high chemical affinity to nitrogen [58]. In glow discharge nitriding of titanium and its alloys, the fundamental role in the formation of the layer is played by chemisorption and diffusion of nitrogen [20, 43-45]. In order to determine this effect, an experiment was carried out in which titanium samples were placed on a cathode and nitrided (Fig. 5.21).

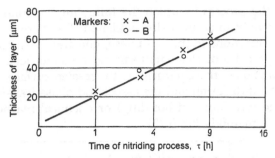

Fig. 5.22 Thickness of nitrided layer on surfaces *A* and *B* of a titanium sample, vs. time of nitriding; process temperature $T = 900°C$, $p = 3$ hPa. (From *Roliński, E.*, [20]. With permission.)

Surface A (external) was subjected to bombardment by ions and neutral particles formed in the plasma during glow discharge, while on surface B only adsorption processes took place because glow discharge does not penetrate such a narrow gap. Obtained were nitrided layers of the same phase composition and thickness on both surfaces A and B [20]. Fig. 5.22 shows thicknesses of the nitrided layer on surfaces A and B of the titanium samples vs. time of nitriding. The significant role of chemisorption is also illustrated by investigations concerning the formation of titanium nitride layers in a chamber with the so-called hot anode, which yielded surface layers both on the cathode, as well as on the anode at temperatures of the order of 550°C [46].

On transition metals, chemisorption of most gases, e.g., of nitrogen on titanium, is a non-activated process. In the case of the nitrogen-iron system, however, a relatively small activation energy of approximately 80 kJ/mole is needed [3, 20]. In the presence of atomic nitrogen, however, under glow discharge conditions, chemisorption proceeds freely without the need for activation energy. Chemisorbed particles may be deformed at the surface and dissociation chemisorption may take place [38], with the formation of free atoms and radicals. For these reasons, chemisorbed particles are chemically more active [4]. During chemisorption, when additional energy appears, a chemical reaction may proceed. As an example, in the process of formation of titanium carbide layers, the adsorbed lower titanium chlorides ($TiCl_2$ and $TiCl_3$) may react with carbon from the steel matrix, in accordance with the reactions:

$$TiCl_{2(g)} + 2H + C_{(s)} \rightarrow TiC_{(s)} + 2HCl_{(g)} \quad \Delta G°_{1000\,K} = -373 \text{ kJ/mole} \quad (5.4)$$

$$TiCl_{3(g)} + 3H + C_{(s)} \rightarrow TiC_{(s)} + 3HCl_{(g)} \quad \Delta G°_{1000\,K} = -450 \text{ kJ/mole} \quad (5.5)$$

where: $\Delta G°_{1000\,K}$ - molar free enthalpy of the reaction under standard pressure at a temperature of 1000 K.

Recapitulating, the effect of chemisorption plays a basic role in glow discharge treatments because, intensified to a large extent by ion sputtering, it affects the process in which the following separate stages, in thermodynamic equilibrium, can be distinguished. These are

1) chemical reactions in the gas phase, constituting a condition of supply of active particles of elements forming the superficial layer,

2) chemisorption of these particles at the treated surface,

3) processes of diffusion and the associated phase transformations (e.g., glow discharge nitriding and boriding) or chemical reactions at the load surface (PACVD methods, Fig. 5.23).

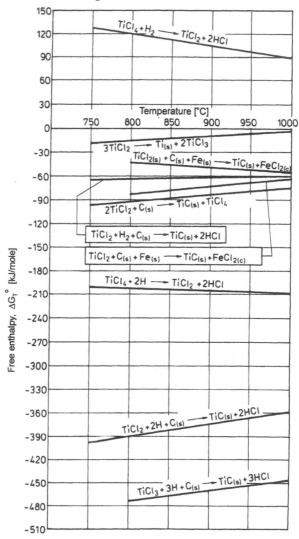

Fig. 5.23 Changes in free enthalpy ($\Delta G°_T$) of a chemical reactions determining the formation of a titanium carbide layer under glow discharge conditions.

It follows then that the formation of the superficial layer may be influenced primarily by the direction of chemical reactions in the gas atmosphere. This depends on the chemical composition and flow rate of the gas mixture, as well as electrical parameters. It is also influenced by the appropriate preparation of new load surface from the point of view of chemisorption, which has a direct effect on diffusion processes and on the course of chancel reactions which are the essential condition of formation of superficial layers in PACVD methods.

5.3 Glow discharge furnaces

Units for carrying out thermo-chemical treatment of the time described in preceding sections is called the **glow discharge furnace** and are utilized for such diffusion processes as e.g., nitriding, carbonitriding or boriding, as well as for PACVD methods. They are composed of: a working chamber, a voltage generator, a system for metering reactive gases, a vacuum system and a control-measurement system. They may differ by the shape of the working chamber, by the power of the voltage generator, design of current feed-throughs, method of load fixturing, and by the method of metering of reactive gases, in particular of chlorides and metal-organic compounds. In the case of furnaces used for PACVD methods, such furnaces may differ by the design of the working chamber and by the method of elimination of harmful components of the gas atmosphere exiting the working chamber, such as vapors of BCl_3, HCl, $TiCl_2$, $TiCl_3$, etc.

In all types of glow discharge furnaces, dynamic vacuum is used. This allows the establishment of a dynamic equilibrium between the basic stages of the process, i.e. chemical reactions in the gas atmosphere, with the aid of technological parameters such as temperature, partial pressure and chemical composition of the gas mixture. The chemical reactions are essential to guarantee a supply of active particles forming the layer through chemisorption, diffusion and the resultant phase transformations or chemical reactions of adsorbed active particles. The appropriate selection of the gas mixture and treatment conditions creates a possibility of process control, in particular of microstructure and phase composition of the layer being formed, i.e. of its properties.

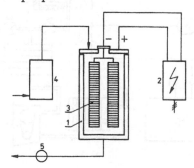

Fig. 5.24 Schematic of equipment used for glow discharge nitriding; *1* - furnace; *2* - direct current supply; *3* - load; *4* - system for metering the mixture of reactive gases; *5* - vacuum system.

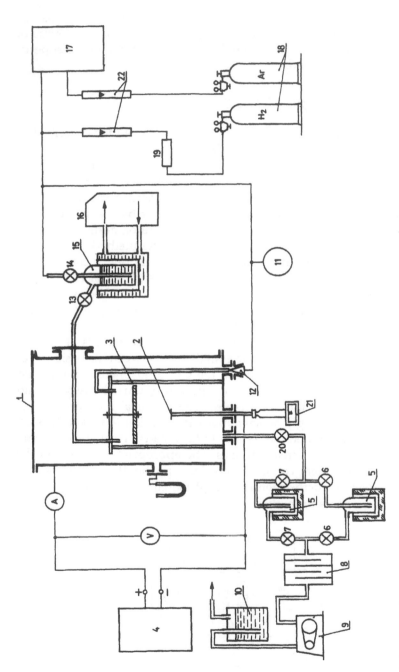

Fig. 5.25 Schematic of equipment for carrying out glow discharge processes in chloride-vased atmospheres, containing vapors of metal-organic compounds: *1* - work chamber; *2* - cathode; *3* - internal screens; *4* - direct current supply; *5* - refrigerating system; *6, 7, 20* - vacuum valves; *8* - "mechanical" filter; *9* - vacuum pump; *10* - water seal; *11,14* - metering and cut-off valves; *15* - reservoir, e.g. with boron chloride or Ti(OC$_3$H$_7$)$_4$; *16* - thermostat; *17, 19* - gas purifiers; *18* - gas cylinders; *21* - temperature measuring device; *22* - flow meters.

In glow discharge diffusion processes, especially in those accomplished by PACVD methods, two types of working chambers are used:
- **cold wall** (cooled anode - chamber wall) in which the load (cathode) is heated by glow discharge;
- **hot wall**, i.e., with auxiliary heating of the chamber (retort) walls which allows more favorable conditions of gas flow and the utilization of load polarization other than cathodic, as well as conduction of thermo-chemical treatments under reduced pressure (LPCVD methods) [33, 47, 48].

Diagrams of glow discharge furnaces for nitriding and its modifications, e.g., sulfo-nitriding, oxy-nitriding, oxy-carbonitriding and boriding, as well as diagrams of versatile furnaces for diffusion treatments and PACVD and LPCVD processes are shown in Figs. 5.24 to 5.27.

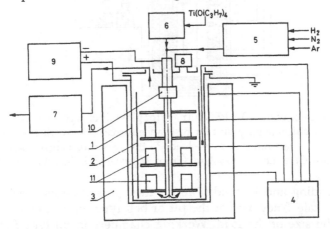

Fig. 5.26 Schematic of a versatile unit for glow discharge diffusion processes by the PACVD and LPCVD methods: *1* - work chamber; *2* - internal screens; *3* - resistance heated retort furnace; *4* - system for temperature stabilization and measurement; *5* - gas metering system; *6* - metering unit for the layer forming element, e.g. $TiCl_4$, $Ti(OC_3H_7)_4$; *7* - vacuum system; *8* - pressure gauge; *9* - voltage supply; *10* - current feed-through; *11* - load.

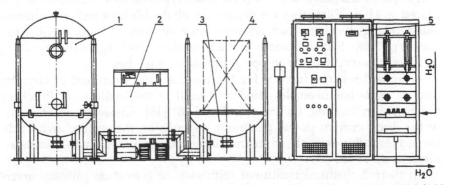

Fig. 5.27 Schematic of a stand for glow discharge nitriding with a JONIMP-500/900 bell-type furnace [48]: *1* - glow discharge furnace; *2* - vacuum pump system; *3* - furnace hearth; *4* - load; *5* - supply and control cabinets. (From *Trojanowski, J.*, [48]. With permission.)

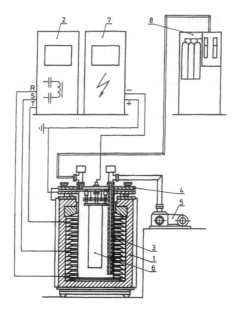

Fig. 5.28 Schematic of the JON-PEG furnace for glow discharge nitriding: *1* - resistance furnace; *2* - electrical supply unit for resistance furnace; *3* - work chamber (vacuum retort); *4* - lid; *5* - vacuum system; *6* - load; *7* - thyristor direct current supply unit; *8* - reactive atmosphere metering system. (From *Trojanowski, J.,* [48]. With permission.)

For nitriding and its modifications these are furnaces used on a wide industrial scale. They are of the pit or bell type with dimensions dependent on the size of the load. Working chambers in pit-type furnaces are designated for slender and log parts which are usually nitrided in the suspended position, e.g., crankshafts, injection mold screws, cylinders. On account of their big heights (up to 6 m) they are installed in pits. Modular design makes possible their extension as the need arises. Chambers with bigger usable diameters are designed as bell type and in this case, the load is placed on a base. Often they are designed with double base and an exchangeable working chamber which allows a more effective utilization of the furnace (Fig. 5.28). Such a furnace is used for glow discharge nitriding of big and heavy parts placed in an upright position, e.g., dies [48].

A glow discharge furnace for thermo-chemical treatment of complex shaped loads has been designed and built at the Institute of Precision Mechanics in Warsaw, Poland (see Fig. 5.28) [48]. The selection of power of the direct current power generator depends on the surface area of the treated load. For example, in the nitriding process, power supply units are rated at up to 150 kW.

In thermo-chemical treatment with electric activation pulsed current power supply may also be used, where changes of voltage (frequency) constitute a new parameter, independent of discharge power and of other process parameters, e.g., substrate temperature, treatment time, pressure and composition of gas mixture [49-51]. Schematics of furnaces powered

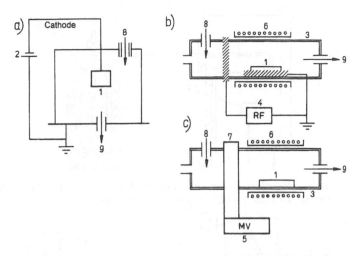

Fig. 5.29 Schematic representation of work chambers for carrying out PACVD processes with the application of: a) direct current voltage; b) pulsed radio frequency voltage; c) pulsed microwave frequency voltage [12]: *1* - load; *2* - direct current supply; *3* - quartz tube; *4* - radio frequency generator; *5* - microwave generator; *6* - load heating; *7* - microwave hollow cathode resonator; *8* - reactive gas inlet; *9* - vacuum pump. (From *Tyczkowski, J.*, [12]. With permission.)

by pulsed current with radio and microwave frequency are shown in Fig. 5.29. The application of activation by pulsed current with frequencies within the range of 10 to 50 kHz and 13.56 MHz requires design changes of the equipment and the application of appropriate power units [12, 51].

The vacuum system of glow discharge furnaces is built around vacuum pumps which, at a pressure of 1013 hPa have a pumping rate of 15 to 90 m³/h and is selected in such a way, relative to the working chamber, as to assure its pumping down within 10 to 30 min. An appropriate set of choking valves and cut-off valves enables stabilization of the vacuum in the working chamber within the range of 0.5 to 13 hPa, with required flows of reactive gases and air backfill of the chamber after the finished process.

The gas metering system comprises several reduction valves, cut-off valves and needle valves, as well as flow-meters and is adapted for precise generation of gas atmospheres applied in treatment cycles. In the case of metering of vapours of compounds, e.g., BCl_3, $TiCl_4$ or $Ti(OC_3H_7)_4$, special designs of metering devices are used. Such devices assure the generation of the reactive gas within the working chamber by either suction or by the flow-through method [9].

Fig. 5.30 shows, by way of example, a schematic of the method of generation, in the working chamber, of a gas mixture from chloride atmospheres, utilized also in PACVD methods in which the choice of the process, as well as of size, shape and number of treated parts, determines the type and geometry of the working chamber, whether hot or cold wall. In the first type of working chambers, i.e., with a hot anode, the walls of the chamber are surrounded by heating elements. Hence, the entire chamber

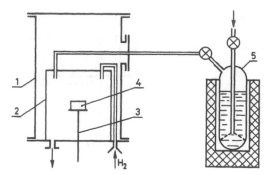

Fig. 5.30 Schematic of work chamber and inlet for BCl_3 vapours: *1* - work chamber; *2* - anode; *3* - cathode; *4* - load; *5* - BCl_3 vapour metering device.

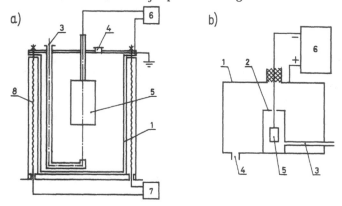

Fig. 5.31 Schematics of a hot-wall (a) and a cold wall (b) work chamber for carrying out PACVD processes: *1* - work chamber (anode); *2* - internal screen; *3* - inlet for reactive gas; *4* - outlet for gases from work chamber; *5* - cathode (load); *6* - voltage supply; *7* - system for temperature stabilization and electrical supply for resistance furnace; *8* - resistance furnace.

is heated which allows the assumption that both the chamber walls and the load are maintained at the same temperature [52]. Despite undoubted advantages of this type of chamber (Figs. 5.28 and 5.31), mainly due to favorable geometry of gas circulation, there always exists the risk of contamination of the deposited layer by components of chamber wall material on which chemical reactions occur with the gas atmosphere, in a similar way as on the load surface [52, 53]. In a cold-wall chamber, only the load is heated, hence the load temperature is considerably higher than that of the chamber (Fig. 5.31). This means that the process of layer formation occurs exclusively on the load surface, provided there is ample mixing and circulation of the gas atmosphere. Due to the temperature gradients existing in the chamber, there are unfavorable convection flows which lower the effectiveness of the deposition process [40]. During circulation of a gas mixture around a load surface, a transition layer is formed between the main stream and the surface. In this transition layer (but also

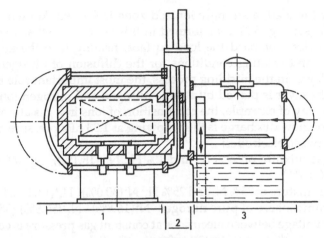

Fig. 5.32 Schematic of furnace for CARBO-JONIMP glow discharge carbonitriding and carburizing: *1* - resistance/glow-discharge heating chamber; *2* - lock; *3* - quench vestibule. (From *Kowalski, S., et al.* [54]. With permission.)

in the main stream) chemical reactions take place which have a significant effect on the rate of formation and properties of surface layers. For that reason, selection of the chemical composition of the atmosphere, its homogenization and supply to direct contact with the treated load surface is one of the fundamental problems of PACVD methods.

Special mention should be made of the chamber design for processes such as glow discharge carburizing and carbonitriding, developed by the Institute of Precision Mechanics in Warsaw [54]. It comprises two chambers, separated by a vacuum-tight baffle. The first chamber is used for carrying out the process of nitrocarburizing (atmosphere composed of H_2 + N_2 + CH_4) or carburizing (H_2 + CH_4) under glow discharge, while the second is for cooling of load (Fig. 5.32). The cooling chamber is equipped with a transport mechanism and a quench tank.

5.4 Glow discharge applications

5.4.1 Glow discharge nitriding

Glow discharge nitriding is a method used in thermo-chemical treatment, yielding diffusion cases of varied structure, featuring high hardness, very good fatigue properties, good wear resistance and resistance to corrosion in some environments.

In the conventional gas nitriding method, nitrogen available for diffusion is obtained as a result of dissociation of ammonia which flows continuously over the load surface, heated to process temperature. In glow discharge nitriding, active nitrogen particles are obtained by ionization of the reactive gas (nitrogen or a mixture of nitrogen and hydrogen), brought about by the effect of glow discharge. During the flow of current around treated elements of the load, a strongly ionized zone is formed, known as the cathodic glow (see Fig. 5.2). Ions formed in this zone, as well as high energy neutral particles, bombard the load surface, heating it to the appropriate temperature and creating conditions for the diffusion of nitrogen into the superficial layer. In the nitriding process, the most important role in the formation of the layer is played by atomic nitrogen and nitrogen ions, i.e., N^+ and N_2^+. By way of example, the composition of ions in a gas atmosphere in conditions of glow discharge is the following at 1.33 hPa pressure and 800 V potential between electrodes:

- for a gas mixture of 99% N_2 and 1% H_2: N^+ (15.1%), N_2^+ (40.0%), H_2^+ (5.1%), H^+ (11.7%),

  - for a gas mixture of 75% N_2 and 25% H_2: N^+(30.0%), H_2^+(13.0%), H^+(17.0%),

  - for an atmosphere of pure nitrogen: N^+(16.8%), N_2^+(55.5%) [55].

A rise of voltage between electrodes at constant gas pressure is conductive to an increase in the concentration of active particles, e.g., a rise of voltage from 400 to 100 V ensures a 10-fold increase in the concentration of atomic nitrogen [55]. One of the advantages of glow discharge nitriding is the possibility of optimization of layer structure to obtain certain usable properties through a change of process parameters.

In order to obtain desired results of glow discharge nitriding, four basic process parameters are controlled:

- chemical composition of reactive gas (from 5 to 80% H_2 - rest N_2),

- pressure in the working chamber within the range of 1 to 13 hPa,

- duration of process within the range of 3 to 16 h, depending on the type of substrate material and desired layer thickness,

- temperature of load surface, dependent primarily on type of load and its material.

Glow discharge nitriding features the following advantages in comparison with the conventional method of gas nitriding in an ammonia atmosphere:

1. Possibility of obtaining, in a controlled manner, of four basic types of nitrided layer structures, both on plain carbon, as well as on alloy steels (with an alloy content of up to 10%). These are only diffusion zone, diffusion zone and the Fe_4N (γ') nitride, diffusion zone with the $Fe_{2-3}(C_xN_y)$ (ε phase) carbonitride, and finally, a diffusion zone with a surface compound zone, composed of the γ' and ε phases (Fig. 5.33). This forms the basis for selection of the type of nitrided structure for given service conditions of a component;

2. Possibility of treatment of components with complex shapes, e.g., crankshafts;

3. Reduction of process time, thanks to faster heating of load to treatment temperature and activation of the environment, as well as the treated surface of the load;

4. Possibility of control of dimensional growth of components subjected to treatment;

5. Significant economy of electric energy because only the load is subjected to heating. This means that heat resistant retorts and ceramic shields

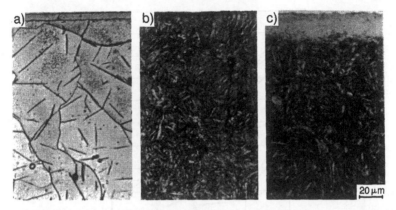

Fig. 5.33 Microstructures of nitrided layers: a) γ ' + diffusion zone on Armco iron; b) diffusion layer without compound zone on N135M steel; c) $\varepsilon + \gamma$ ' + diffusion zone on 33H3MF grade steel (0.33%C, 0.4%Mn, 2.4%Cr, 0.37%Mo, 0.2%Ni).

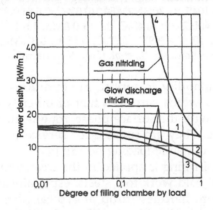

Fig. 5.34 Comparison of power density in glow discharge and gas nitriding processes at 500°C: *1* - glow discharge nitriding without insulating elements; *2* - glow discharge nitriding with one insulating screen; *3* - glow discharge nitriding with two insulating screens; *4* - gas nitriding. (From *Kowalski, S., et al.* [54]. With permission.)

are not needed. When the degree of filling of chamber by load is approximately 35%, energy consumption is only 30 to 40% that in gas nitriding (Fig. 5.34) [10].

6. Elimination of the need to use ammonia as the reactive atmosphere, although some companies, e.g., Klöckner Ionon, recommend this gas also for glow discharge processes.

Glow discharge nitriding, carried out in a broad range of temperatures (400 to 590°C), enables the obtaining of different degrees of hardness of superficial layers, depending on processing conditions and on the chemical composition of the treated steel. Fig. 5.35 shows examples of hardness profiles in nitrided layers on different steel grades [54]. The superficial layers obtained feature high wear resistance and fatigue strength, as well as better ductility in comparison with layers nitrided in the gas process. For example, nitrided layers without the compound zone feature very good fa-

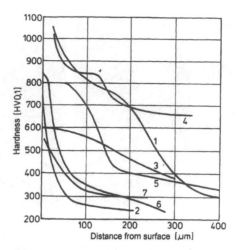

Fig. 5.35 Hardness distributions in glow discharge nitrided layers for different steel grades: *1* - N135M; *2* - 1045; *3* - 30HN2MFA* (0.3%C, 0.6 to 0.9%Cr, 2 to 2.5%Ni, 0.15 to 0.3%V); *4* - M2; *5* - 40H2MF* (0.4%C, 0.5 to 0.8Mn, 1.6 to 1.9%Cr, max. 0.3%Ni, 0.3%Mo, 0.2%V); *6* - 4140; *7* - N9E* (0.9%C, 0.15 to 0,3%Mn, 0.15 to 03%Si, max. 0.15%Cr, 0.2%Ni). (From *Kowalski, S., et al.* [54]. With permission.)

tigue properties, suitability to service in conditions of strong dynamic loading and ductility which surpasses that of other types of nitrided layers. Nitrided layers with the surface phase zone feature low ductility but high wear resistance. They are suitable for service in conditions of wear by friction when no dynamic loading occurs [54].

Glow discharge nitriding is applied in the treatment of structural alloy steels, hot work and cold work tool steels, especially those which do not suffer significant core hardness drop at typical nitriding temperatures above 500şC, of high speed steels, e.g., M2 and 9%W grades, as well as of steels with special properties, such as acid-resistant, heat resistant and creep-resistant. It can also be used in the treatment of titanium and molybdenum and their alloys. Hence, glow discharge nitriding is finding broad application in extending the service life of machine components and tools, e.g., injection mold screws and cylinders, dies, gears, punches, injection nozzles, ultrasonic wave-guides, hobs, etc. For example, the life of glow discharge nitrided hobs was extended significantly (from 3- to 10-fold, depending on steel grade) in comparison with that of same components, nitrided by other means [54, 55].

Glow discharge, ensuring homogeneity of physico-chemical conditions at the load surface, as well as the possibility of controlling layer structure, also allows the treatment of components with big dimensions and high requirements regarding dimensional tolerance, such as crankshafts. The heat treatment of these components is encumbered with some problems, on account of the type of loading and the tendency to undergo dimensional changes caused by geometry and residual stresses induced by the manufacturing process.

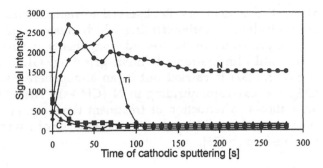

Fig. 5.36 Examples of distribution profiles for nitrogen, titanium, carbon and oxygen in a Ti(NCO) layer of 3 μm thickness, obtained on a prior glow discharge nitrided 1H18N9T* steel (0.1%C, 18%Cr, 9%Ni, 0.5%Ti).

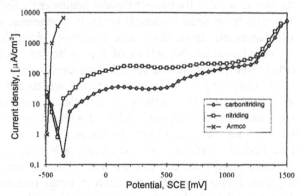

Fig. 5.37 Anodic polarization curves in 0.5 m NaCl solution of surface layers on Armco iron, formed in glow discharge nitriding and carbonitriding at 580°C.

Glow discharge nitriding also finds application in the treatment of hot work tools, among others, forging dies, punches, die-casting dies and plastic molds. Very ductile monophase layers of 6 μm thickness produced at the surface accommodate dynamic loading occurring during forging, enhance sliding properties and reduce tendency to checking as the result of thermal fatigue [59].

Glow discharge nitriding is also used on a broad assortment of cutting tools, among others on milling cutters, broaches, reamers and thread taps. The process is, for those applications, carried out at 450 to 500°C in times up to 30 min. Their core hardness (62 to 65 HRC) does not change, while surface hardness rises to 1100 to 1300 HV1. The hard nitrided layer significantly raises wear resistance without causing spalling of the cutting edge, thanks to good ductility.

Glow discharge nitriding is finding increasing application in industry, and research currently carried on is aimed at developing modifications of the process, e.g., by introducing carbonitriding, oxy-carbonitriding [56, 57], obtaining of composite layers, e.g., a nitride layer topped by a zone composed of titanium nitride [23] or of Ti(NCO) titanium-oxy-carbonitride (Fig. 5.36).

In industrial practice, a tendency is observed to replace carburizing by glow discharge nitriding or carbonitriding which creates a possibility of enhancing service properties of the treated components, ensuring dimensional accuracy and elimination of finish grinding. Glow discharge carbonitriding (Fig. 5.37) is carried out in an atmosphere composed of $H_2 + N_2 + CH_4$, and oxy-carbonitriding in N_2+CH_4+air. The utilization of glow discharge allows a reduction of treatment time, e.g. at 800şC it is possible to obtain in a time of 1 h a carbonitrided layer of 1 mm thickness and good wear resistance [56].

5.4.2 Glow discharge boriding

Formation of boride layers in a gaseous environment with the utilization of glow discharge is a new thermo-chemical treatment method, still carried out on laboratory or semi-technical scale. Nonetheless, specific properties of boride layers on steels, their high hardness, significant resistance to wear, good resistance to the action of a number of chemical agents, notably to acids [58-60], constitute a basis for research aimed at developing new methods of obtaining such layers, methods which are economically justifiable and at the same time creating possibilities of control of structure, hence, of properties. Currently known and used methods, e.g., powder pack, salt bath, paste and gas, are not fully controllable.

The most favorable conditions for obtaining a homogenous diffusion layer are those associated with the gas method. Gaseous boriding features such advantages as: lower temperatures than in other treatment methods, possibility of forming boride layers on components with complex shapes and easier process control [61]. The gas atmospheres used in such a process, however, pose hazards on account of their toxicity and strong corrosive action against the walls of the working chamber (BCl_3) or their high degree of explosiveness (B_2H_6). These difficulties may be eliminated by a reduction of the proportion of reactive gas in the gas mixture with H_2 or H_2 with Ar, and its more efficient utilization. Such conditions can be created by lowering the pressure in the working chamber and the introduction of glow discharge [9, 34, 62-64]. This allows shortening of process time and lowering of treatment temperature, as well as reduction of reactive gas consumption. The pressure applied here is within the range of 1 to 13 hPa, while the content of BCl_3 vapours is from 2 to 10% by volume [8]. Fig. 5.38 shows examples of microstructure of boride layers on different metallic substrates, while Fig. 5.39 shows an example of microstructure and distribution of boron and iron in a biphase $FeB+Fe_2B$ boride layer on 1045 steel. Boride layers are biphase, as a rule, and comprise the borides FeB and Fe_2B. Their microhardness oscillates within the range of 1500 to 2400 HV0.1, depending on the phase composition and their thickness reaches 200 µm after 6 h of treatment at 850°C. The desired goal is to obtain monophase layers comprising only the more ductile Fe_2B phase. The FeB phase, being harder and more brittle, spalls during service and acts as abrasive medium, thus accelerating the process of wear [9].

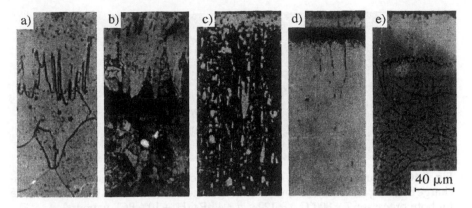

Fig. 5.38 Microstructures of borided layers obtained on different materials: a) Armco iron ($T = 800°C$, $\tau = 3$ h); b) 1045 steel ($T = 850°C$, $\tau = 3$ h); c) NC10* (1.5 to 1.8%C, 0.2 to 0.5%Mn, 11.0 to 13.0%Cr, 0.2 to 0.5%Si) steel ($T = 800°C$, $\tau = 1$ h); d)1H18N9T steel ($T = 800°C$, $\tau = 2.5$ h); e) nickel ($T = 850°C$, $\tau = 2$ h).

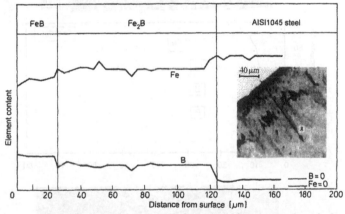

Fig. 5.39 Microstructure and distribution of iron and boron in a boride layer formed on 1045 steel. Process parameters: $T = 800°C$, $t = 4$ h, $p = 4$ hPa ($H_2 + 10\%$ BCl_3).

The presence of atomic hydrogen in processes of glow discharge boriding plays a decisive role in the initiation of reaction of reduction of boron chloride which is essential for obtaining active boron atoms and ions, responsible for the formation of boride layers [34, 66].

Cleaning of the surface and its significant development, due to cathodic sputtering taking place during load heating, in conjunction with giving chemical reactions a direction in the gas atmosphere, ensure high concentration of boron at the surface, hence, favorable conditions for diffusion and for the formation of boride layers. This is indicated by results of comparative investigations of formation of boride layers by the glow discharge method, as well as with the application of the same process parameters but with lower pressure and without glow discharge (Fig. 5.40). They show that the initial stage of formation of the boride layer is faster in the glow dis-

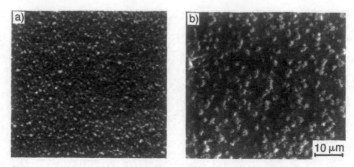

Fig. 5.40 Appearance of surface of boride layers obtained on 1045 steel by glow discharge boriding (a) and by the low pressure gas method (b). Process parameters same for both processes: $T = 800°C$, $\tau = 120$ s, $p = 4$ hPa ($H_2 + 10\%$ BCl_3 vapours).

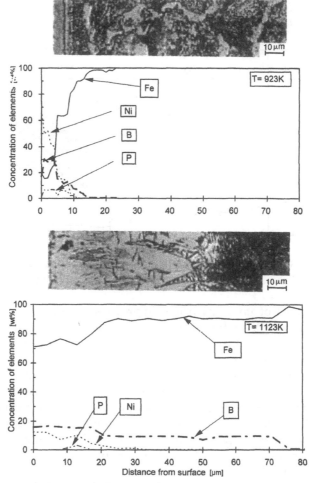

Fig. 5.41 Microstructure and distribution of Ni, Fe, B and P in multicomponent boride layers, obtained on AISI1045 grade steel at various temperatures: 923, 1023 and 1123 K.

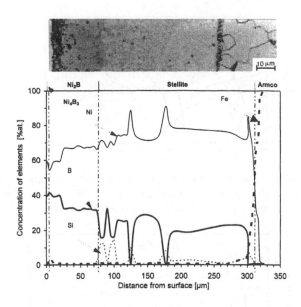

Fig. 5.42 Microstructure of composite borided layers on low carbon steel (0.18%C), obtained by the combination of plasma spraying of a nickel-base alloy PMNi35 (C+B+Si - 8%, Co - max 1%, Fe - 3-5%, Cr - 5-8%, Ni - balance) with the process of glow discharge boriding. Shown also is distribution of elements in borided layers.

charge process (Fig. 5.17). Electric activation of the gaseous medium, significant development of the treated surface, thus intensification of the process of chemisorption of boron cause the FeB boride to be formed first, as a rule, in an atmosphere composed of vapors of BCl_3 and H_2 [9]. Thus, depending on the amount of boron supplied, FeB or Fe_2B borides may be formed. Due to the effect of cathodic sputtering, borides being formed in the initial stages are fine grained and have a developed surface. This favors chemisorption of boride and its diffusion into the core of the treated steel. After the formation of a compact layer of FeB and Fe2B borides, boron may diffuse from the surface in the direction of the core of the treated component along boride grain boundaries [8, 76]. The formation of a boride layer on the steel surface is based predominantly on the mechanism of reactive diffusion [9, 65]. Borides formed during the early stages of the boriding process are clearly oriented perpendicular to the surface and constitute a compact, tight layer (Fig. 5.17), facilitating further diffusion of boron along grain boundaries. Research carried out in the field of glow discharge boriding is aimed at implementation of this process in the industry, especially as part of the formation of composite and multi-component layers. Figs. 5.41 and 5.42 show examples of multi-component and composite boride layers, formed by the combination of processes of electroless nickel plating or plasma spraying of a nickel alloy with the process of glow discharge boriding [67, 68]. These are layers featuring good corrosion resistance (Figs. 5.43 and 5.44), as well as good wear resistance (Fig. 5.45).

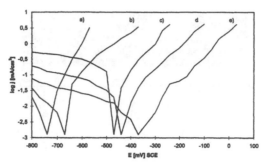

Fig. 5.43 Polarization curve in 0.5 M NaCl solution for AISI1045 steel specimens (a), borided at T = 923 K (b), nickel plated and borided at 1123 K (c), nickel plated and borided at 1023 K (d) and nickel plated and borided (e) at 923 K.

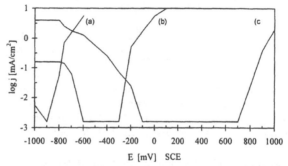

Fig. 5.44 Polarization curves for Armco iron (a) and for PMNi35 stellite before (b) and after plasma boriding at 850°C (c). (From *Bieliński, P., et al.* [68]. With permission.)

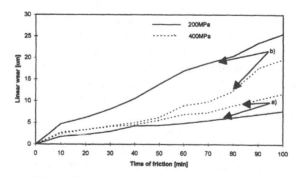

Fig. 5.45 Linear wear of borided layers obtained on AISI1045 steel with and without a nickel coating, under loads of 200 and 400 MPa, vs. sliding time: a) nickel plated and borided at 1023 K; b) without nickel plating - borided at 1023 K.

5.5 CVD methods

Chemical Vapour Deposition (CVD) methods, i.e., deposition of a layer from the gas phase with the participation of a chemical reaction, have been widely used in the industry worldwide since the late 1960s. Their aim is to produce anti-abrasion and anti-corrosion layers [69-74]. These are methods which make possible the formation of such surface layers as e.g. titanium carbide (TiC) and titanium nitride (TiN), aluminum oxide (Al_2O_3), silicon nitride (Si_3N_4), as well as multi-component layers like Ti(C,N), Ti(NCO) or composite like TiC + TiN, nitrided layer + TiN or Ti(NCO), or boride layer + TiB$_2$ or Fe$_2$B +(Ni,Fe)B + TiB$_2$.

CVD methods constitute a continuation of powder pack and salt bath methods of formation of surface layers, differing by phase structure, with a thickness up to 15 μm. In the traditional form, these are **unassisted** CVD processes, carried out under atmospheric pressure (*Atmospheric Pressure* CVD or APCVD) and *Low Pressure* CVD or LPCVD.

These are high temperature processes, thus their utilization is primarily in the case of treatment of such materials as sintered carbides or such components where in service the only important aspect is wear by friction, without major dynamic loading. In such conditions, these layers assure a significant increase in the service life of treated components [4, 72]. Table 5.4 shows data concerning the extension of service life of some components, made from different materials and treated by different CVD methods [72, 75]. A common feature of all CVD methods is the supply of the element constituting the layer usually in the form of a halogen, e.g., $TiCl_4$ in the case of TiC, TiN or Ti(C,N) layers, $SiCl_4$ in the case of Si_3N_4 layers and a mixture of halogens, e.g., $TiCl_4$ + BCl_3 in the case of TiB$_2$ layers.

Table 5.4

Examples of application of CVD methods and extension of life of treated components

Glow discharge treated components	Material	Type of layer	Application	Life extension factor after treatment
Upsetting machine punches	HSS	TiC, Ti(C,N)	low carbon steels	2.6 ×
Deep extrusion rams	NC10	TiC	stainless steel	5 ×
Rollers	NC10	Ti(C,N)	cold rolled steel	2.7 ×
Small roller	N9	TiC	thin aluminum sheet	4.7 ×
Pitch circle for external grinding of thread tap centers	NC6	TiN	abrasive material	5 ×
Masoneylen valves in coal liquefying plants	1H18N9T		flow of suspension of coal dust in oil at up to 300ₛC	10 ×

The second component of the layer may come from the substrate, e.g., nitrogen or carbon in the case of TiN, TiC and Ti(C,N) layers [23] or from the atmosphere, e.g., oxygen in the case of oxide layers [76, 77]. Traditional CVD processes require the use of high temperatures, necessary for the occurrence of chemical reactions which guarantee the formation of the layers (Fig. 5.46) which limits the scope of their application. This is because in the case of components exposed to dynamic loading in service, or

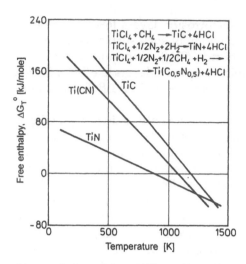

Fig. 5.46 Changes of free enthalpy of the reaction of layer formation vs. temperature, for TiC, TiN and Ti(C,N) layers. (From *Stock, H.R., et. al.* [73]. With permission.)

of components made from high speed steel, heat treatment after the CVD process is essential. This, in turn, creates the hazard of dimensional changes. Therefore, the direction to follow in the development of these methods is lowering of process temperature. This may be accomplished by the following means:

- selection of appropriate gas atmospheres and the utilization of metal-organic compounds. Such a process is described by the acronym MOCVD (*Metal-Organic* CVD). It has already found practical application in the electronics industry [78, 79].

- electrical activation of the gas environment with the aid of glow discharge [23, 80-84, 88], by high frequency currents [77, 85-87] or by the introduction of laser technology or the electron beam in an appropriate gas environment. The latter yield good effects on account of significant concentration of energy and high rate of heating with shallow penetration of the heat flux. By selecting e.g., an appropriate energy density and rate of displacement of the laser beam, it is possible to melt the thin superficial layer without affecting the structure of the substrate, thus enabling the formation of TiN or TiC layers on titanium and its alloys (the so-called *Laser* CVD) [90]. These are the so-called assisted or activated methods. If the activation of the gas environment is by electric means, the method is termed PACVD. Table 5.5 lists CVD methods which are currently being researched and developed.

Table 5.6 shows the differences between APCVD, LPCVD and PACVD methods carried out in conditions of glow discharge, as exemplified by the titanium nitride layer. The lowering of temperature significantly broadens the scope of CVD applications, among others, encompassing treatment of tools and components made from high speed steel, after prior heat treatment.

Table 5.5
CVD method designations

APCVD	Atmospheric Pressure Chemical Vapour Deposition
LPCVD	Low Pressure Chemical Vapour Deposition
DCPACVD	Direct Current Plasma Assisted Chemical Vapour Deposition
RFPCVD	Radio Frequency Plasma Chemical Vapour Deposition
MPCVD	Microwave Chemical Vapour Deposition
HFCVD	Hot Filament Chemical Vapour Deposition
EACVD	Electron Assisted Chemical Vapour Deposition
PhACVD	Photon Assisted Chemical Vapour Deposition
LCVD	Laser Chemical Vapour Deposition
MOCVD	Metal-Organic Chemical Vapour Deposition
PAMOCVD	Plasma Assisted Metal-Organic Chemical Vapour Deposition

Table 5.6
General characteristics of selected CVD methods in the process of titanium
nitride layer formation

General designation of method	APCVD	LPCVD	PACVD	PACVD
Method of heating	resistance heating of working chamber	resistance heating or so-called indirect heating with the utilization of glow discharge	heating by glow discharge or by glow discharge with the so-called hot anode	heating by glow discharge or by glow discharge with the so-called hot anode
Process temperature	1170-1220 K	1150 K	770-820 K	720-790 K
Pressure inside working chamber	atmospheric	10-500 hPa	3-13 hPa	3-10 hPa
Gas atmospheres	$TiCl_4+H_2+N_2$	$TiCl_4+H_2+N_2$	$TiCl_4+H_2+N_2$	$Ti(OC_3H_7)_4 + H_2+N_2$
Type of layer	Ti(C,N), TiN	TiC, Ti(C,N), TiN	TiN, composite layer: nitrided + TiN	layers of the type: Ti(OCN) or composite: nitrided + Ti(OCN)

A successful future and broad application is predicted for the PACVD method, carried out in conditions of glow discharge, without prior treatment (Fig. 5.47) or following surface treatment (Fig. 5.48) with the application of gas atmospheres containing metal-organic compounds, e.g. vapors of tetrapropyloxititanium - $Ti(OC_3H_7)_4$ [8, 73, 82] or $Ti[N(CH_2CH_3)_2]_4$ or $Ti[N(CH_3)_2]_4$ [84]. This process allows the formation of multi-component layers of the Ti(NCO), Ti(CN) and composite layers like nitrided layer topped by a layer of titanium nitride or titanium oxycarbonitride - Ti(NCO), as well as Ti(NCO) + TiN [8, 83]. Composite layers may be formed with the utilization of single stage processes, i.e. after the finished glow

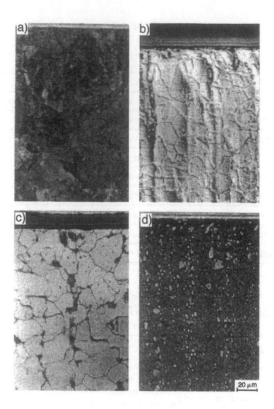

Fig. 5.47 Microstructures of surface layers, obtained by the PACVD method in conditions of glow discharge: a) TiC layer on NC6* grade steel (1.3-1.45%C, 0.4-0.7%Mn, 1.3-1.65%Cr, 0.1-0.25%V); b) Ti(NCO) layer on 1H18N9T steel; c) SiC layer on AISI1010 steel; d) TiN layer on M2 steel.

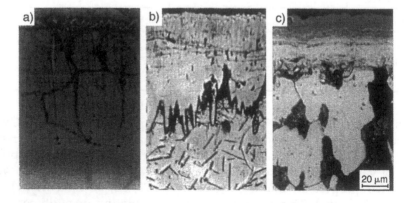

Fig. 5.48 Microstructures of composite layers obtained on different steel grades: a) nitrided layer + TiN on 1H18N9T* steel (0.1%C, 18%Cr, 9%Ni,0.5%Ti), b) nitrided layer + Fe$_2$B layer on Armco iron, c) (Fe,Ni)B layer + TiB$_2$ on AISI1045 steel.

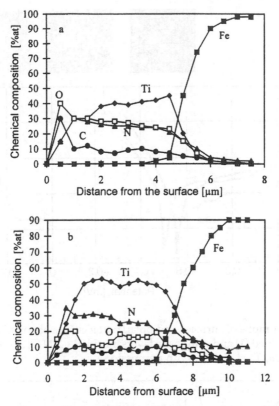

Fig. 5.49 Concentration profiles of nitrogen, titanium, carbon and iron in Ti(NCO) layers, obtained on Armco iron at 560°C (a) and 520°C (b) in an atmosphere composed of vapors of $Ti(OC_3H_7)_4$ - H_2 - N_2.

discharge nitriding process in an atmosphere of N_2+H_2, the working chamber is filled with vapours of $TiCl_4$ or $Ti(OC_3H_7)_4$ in a mixture with hydrogen and nitrogen, and by changing process parameters the Ti(NCO) type layer is formed in a controlled process. Similarly to all glow discharge treatments, by changing process parameters, and appropriate preparation of the treated surface (by e.g., cathodic sputtering) it is possible to control layer microstructure, their chemical composition and thickness (Figs. 5.49 and 5.50).

Such layers feature good adhesion to the substrate because nitrogen (from initial nitriding) and carbon from the matrix actively participate in the formation of titanium nitride layers or of Ti(CN), Ti(NCO) or TiC layers [7, 8, 23]. They all feature high surface hardness within 1600 to 2400 HV0.05 for Ti(NCO) layers, depending on carbon and oxygen content, approximately 200 HV0.05 for TiN layers and 3000 to 4000 HV0.5 for TiC layers [7].

Usable properties of these layers may be shaped by their microstructure. Better corrosion resistance and wear resistance are exhibited by layers with a fine grain structure which, as has been stressed, may be formed by the appropriate selection of chemical composition of the reactive atmosphere and the utilization of cathodic sputtering (Figs. 5.50 and 5.51).

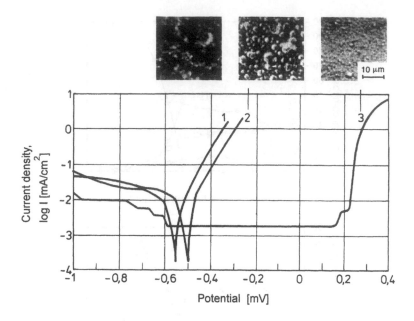

Fig. 5.50 Micrographs of surfaces and polarization curves obtained in 0.5 m solution of NaCl for nitrided (*1*) and composite layers: nitrided + TiN (*2, 3*) on 1H18N9T* steel. (From *Wierzchoń, T., et al.* [23]. With permission.)

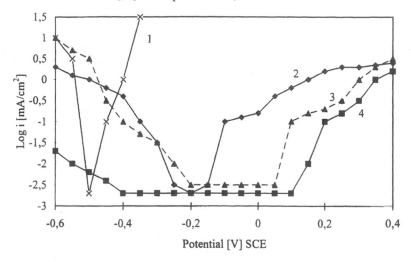

Fig. 5.51 Polarization curves for nitrided layers (1), Ti(NCO) layers (2) and composite layers, i.e. nitrided + Ti(NCO) (3) on 1H18N9T steel in comparison with the corrosion resistance of the same steel with a polished surface (4).

Figs. 5.52 and 5.53 show results of wear resistance tests, carried out by the "three rollers - cone" method (modified "four ball" wear test), on different superficial layers, formed with the aid of glow discharge [89].

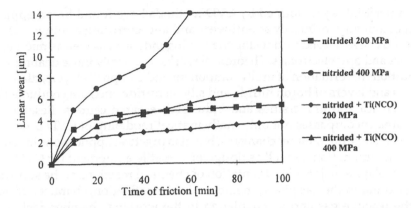

Fig. 5.52 Linear wear of nitrided layers and of composite layers, i.e. nitrided + Ti(NCO), obtained on 1H18N9T steel, vs. sliding time, under loads of 200 and 400 MPa.

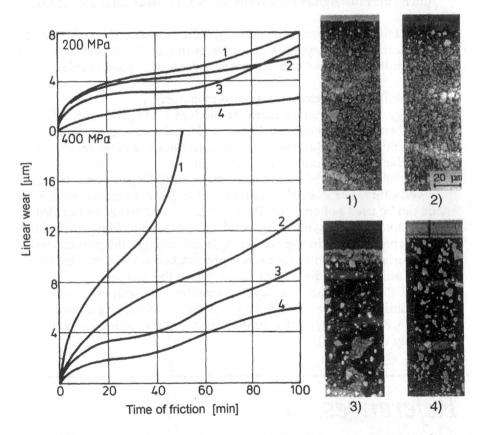

Fig. 5.53 Wear resistance of nitrided layers (1), Ti(NCO) layers (2), composite layers, i.e. nitrided + Ti(NCO) (3) and nitrided +TiN (4) on M2 steel, vs. sliding time, under loads of 200 and 400 MPa. (From *Wierzchoń, T., et al.* [8]. With permission from Elsevier Science.)

Superficial layers formed by CVD methods have found broad application in industry, primarily as anti-wear and anti-corrosion coatings. These methods have helped to develop the tooling industry, as well as microelectronics and optoelectronics. Theoretical works (especially those dealing with conditions of formation of multi-component and composite layers, diamond layers and layers of boron nitride and silicon nitride, etc.) are conducted on a broad scale in universities and industrial research centers. Research in this area encompasses not only mathematical modeling of CVD processes and investigating their mechanisms, but also practical applications in which process automation, as well as design of versatile equipment for CVD processes, plays a major role. Orientation of chemical reactions at the substrate surface and in the gas phase, methods of their activation, character of flow of the reactive gas and its circulation in the working chamber, including control of gas flow rate distribution, all constitute the basis for optimization of the CVD process.

Among the most significant advantages of CVD treatments are the following:

- possibility of formation of heterogeneous superficial layers with desired structure and properties at temperatures of 400şC and above (including multi-component and composite layers) with good usable properties;

- possibility of process control and automation;

- energy and material economy of modern CVD processes, utilizing, among others, the effect of glow discharge, for reduced energy and gas consumption through shortening of process time, lowering of process temperature and the heating of only treated surfaces;

- relatively low capital cost of equipment for CVD processes, as well as their versatility. For example, equipment for glow discharge with a hot anode can be used not only for PACVD and LPCVD processes but also for diffusion treatment, e.g., nitriding, boriding and carbonitriding. Such equipment is capable of producing composite layers through the combination of various surface treatments in one single process, besides featuring better efficiency of utilization of the working space inside the chamber;

- possibility of production of layers on components with complex shapes;

- possibility of utilization of gas atmospheres containing organic compounds [8, 84, 88, 91, 92], thus eliminating halogen-bearing atmospheres, commonly used in CVD.

References

1. **Berghaus, B.**: Process for surface treatment of metallic elements. German patent DRP 668 639 (1932).
2. **Berghaus, B.**: Vacuum furnace, heated by glow discharge. German patent DRP 851 540 (1939).

3. **Edenhofer, B.**: Physical and metallurgical aspects of ion nitriding. *Heat Treatment of Metals*, 1, 23, 1974.
4. **Burakowski, T., Roliński, E., and Wierzchoń, T.**: *Metal surface engineering* (in Polish). Warsaw University of Technology Publications, Warsaw 1992.
5. **Weston, G.F.**: *Cold cathode glow discharge tubes*. Academic Press, London 1968.
6. **Zdanowski, J.**: *Electrical discharge in gases* (in Polish). Wroclaw University of Technology Publications, Wroclaw 1975.
7. **Wierzchoń, T., and Karpinski, T.**: Grundlagen des Plasmanitrierens. *Härterei-Technische Mitteilungen*, 42, 3, 1987.
8. **Wierzchoń, T., and Sobiecki, J.R.**: PACVD method by glow discharge for depositing surface layers in $H_2 + N_2 + Ti(OC_3H_7)_4$ atmosphere. *Vacuum*, Vol. 44, No. 10, 975, 1993.
9. **Wierzchoń T.**: *Formation of iron boride layer on steel under glow discharge conditions* (in Polish). Warsaw University of Technology Publications, Vol. 101, Warsaw 1986.
10. **Marciniak, A.**: *Processes of heating and nitriding of a cathode in conditions of glow discharge* (in Polish). Ph.D. dissertation, Warsaw University of Technology Publications, 1983.
11. **Szymański, A., et al.**: *The chemistry of low temperature plasma* (in Polish). WNT, Warsaw 1983.
12. **Tyczkowski, J.**: *Thin films of polymers* (in Polish). WNT, Warsaw 1990.
13. **Szmidt-Szalowski, K.**: *Processes of NO synthesis in flow-through reactor in conditions of glow discharge*. Warsaw University of Technology Publications. Chemistry, Vol. 16, 1975.
14. **Roliński, E.**: *Effects occurring at the interface between gas and solid in the process of glow discharge nitriding* (in Polish). Ph.D. dissertation, Warsaw University of Technology Publications, 1978.
15. **Hudis, M.**: Study of ion-nitriding. *Journal of Applied Physics*, 44, No. 4, p.1489, 1973.
16. **Kułakowska, B., Żyrnicki, W., Badowski, N. et al.**: Spectroscopic studies of N_2 - H_2 - $TiCl_4$ glow discharge used for TiN deposition, *Surface Engineering*, ed. F. Broszeit et al., Vol. 1, DGM Informationsgesellschaft, Verlag Oberursel, 91, 1989.
17. **Xu, B., and Zhang, Y.**: Collision dissociation model in ion nitriding. Proc.: *5th International Congress on Heat Treatment of Materials*, Budapest, 1056, 1986.
18. **Kaufman, F.**: *Chemical reaction in electrical discharge*. Ed. B.D. Blanstein, American Chemical Society, Vol. 3, Washington 1969.
19. **Szczepańska, H. et al.**: Investigations of zones of glow discharge near the cathode in a mixture of vapours of $BCl_3 + H_2$ and $BCl_3 + Ar$ (in Polish). Report by the Institute of Flow-through Machines, No. 169/83, Gdansk-Wrzeszcz 1983.
20. **Roliński, E.**: *Ion nitriding of titanium and its alloys* (in Polish). Warsaw University of Technology Publications, Mechanics, Vol. 112, 1988.
21. **Teer, D.G.**: The energies of ions and neutrals in ion plating. *Journal of Applied Physics*, No. 9, 187, 1976
22. **Babad-Zakhryapin, A.A., and Lagutkin, M.J.**: Structural defects in glow-discharge treated materials (in Russian). *Metallovedenye i Termicheskaya Obrabotka Materialov*, No. 7, 70, 1976.
23. **Wierzchoń, T., Michalski, J. et al.**: Formation and properties of composite layers on stainless steel. *Journal of Materials Science*, No. 27, 771, 1992.
24. **Prokoskin, D.A., Arzamasov, W.N., and Ryabchenko, E.V.**: *Obtaining of coatings on metals in glow discharge* (in Russian). Naukova Dumka, Vol. 3, Kiev 1970.
25. **Kaminsky, K.**: *Atomic and ionic phenomena on metal surface*. Springer-Verlag, Berlin-New York 1965.

26. **Plesivtzev, N.V.**: *Cathodic sputtering* (in Russian). Atomizdat, Moscow 1968.
27. **Carter, G., and Colligon, J.S.**: *Ion bombardment of solids*. H.E.B., London 1969.
28. **Ryabchenko Je, W.**: Application of glow discharge for diffusion alloying of metals (in Russian). *Struktura i Svoistva Zaroprochnykh Splavov*, No. 228, 1971.
29. **Predvoditelev, A.S., Kobozev, N.I. et al.**: *Chemistry and physics of low temperature plasma* (in Russian). Moscow University Publications, Moscow 1971.
30. **Pfander, E.**: Grundlagen der Plasmatechnik. *Elektrowärme International*, Vol. 45, B. 3/4, 130, 1987.
31. **Arni, R.**: Nitriding of titanium and its alloy by N_2, NH_3 or mixtures of $N_2 + H_2$ in a d.c. arc plasma at low pressures < 10 torr. NASA Technical Memorandum 83803, November 1984.
32. **Sobiecki, J.R., Wierzchoń, T., Kułakowska, B. et al.**: Effect of temperature on the formation of surface layers in plasma-assisted chemical vapour deposition process carried out in an atmosphere containing $Ti(OC_3H_7)_4$ vapours. *Journal of Mat. Sci. Letters*, 15, 159, 1996.
33. **Wierzchoń, T., Michalski, J. et al.**: Method and equipment for production of surface diffusion layers on metals (in Polish). Polish patent no. 147847, C23C8/06, 1989.
34. **Wierzchoń, T.**: The role of glow discharge in the formation of boride layers on steel in the plasma boriding process. In: *Advances in Low-Temperature Plasma Chemistry, Technology, Applications*. ed. H.V. Boenig, Vol. 2, Technomic Publishing Co., Inc., Lancaster-Besel, USA, 1988.
35. **Groszkowski, J.**: *Vacuum technology* (in Polish). WNT, Warsaw 1972.
36. *Keller K.*: Schichtaufbau Glimmennitrierter Eisenwerkstoffe. *Härterei-Technische Mitteilungen*, No. 26, 2, 120, 1971.
37. **Ościk, J.**: *Adsorption* (in Polish). PWN, Warsaw 1983.
38. **Thompkins, F.C.**: *Chemisorption of gases on metals*. Academic Press, New York 1978.
39. **Hayward, D.O., and Trapnell, B.M.**: *Chemisorption*. Butterworths, London 1964.
40. **Lachtin, Ja. M., and Kogan, Ja. D.**: *Nitriding of steel* (in Russian). Publ. Masinostroenye, Moscow 1976.
41. **Tibbets, G.**: Role of nitrogen atoms in ion nitriding. *Journal of Applied Physics*, 45, 11, p. 5072, 1974.
42. **Avni, R., and Spalvins, T.**: Nitriding mechanisms in Ar-N_2, Ar-N_2-H_2 and Ar-NH_3 mixtures in D.C. glow discharge at low pressures less than 10 Tr. *Materials Science and Engineering*, 95, 237, 1987.
43. **Rie, K.T., and Lampe, T.**: Plasmanitrieren und Titanlegierungen. *Metal*, 37, 1003, 1983.
44. **Rie, K.T., and Lampe, T.**: Thermochemical surface treatment of titanium and titanium alloy Ti-6Al-4V by low energy nitrogen bombardment. *Materials Science and Engineering*, 69, 473, 1985.
45. **Wierzchoń, T., Fleszar, A., and Krupa, D.**: Properties of surface layers on titanium alloys produced by plasma carbooxynitriding process. Proc.: *2nd International Conference on Carburizing and Nitriding with Atmospheres*, Ed. Grosch et al., ASM International, Cleveland, p. 341, 1995.
46. **Michalski, J., and Wierzchoń, T.**: CVD of TiN layers in various glow discharge regions. *Journal of Materials Science Letters*, No. 10, 506, 1991.
47. **Boenig, H.V.**: Reactor and equipment designs in low-temperature plasma technology, In *Advances in low temperature plasma chemistry, technology, applications* ed. H.V. Boenig, Vol. 2. Technomic Publishing Co. Inc., Lancaster-Basel, 89, 1988.
48. **Trojanowski, J.**: Technology and equipment for ion nitriding (in Polish). Proceedings: Conference on *Techniques of producing surface layers on metals*, p.70, Rzeszów, 1988.

49. **Rie, K.T., Detjen, K., and Eisenberg, S.:** TiC layers on steel by pulsed DC-plasma CVD. Proc.: *1st International Conference on Plasma Surface Engineering.* Garmisch-Partenkirchen, 133, 1988.

50. **Eisenberg, S.:** Plasma CVD. *Material - Wissenschaft und Werkstofftechnik,* 20, 429, 1989.

51. **Wierzchoń, T., Rudnicki, J., Hering, M. et al.:** Formation and properties of nitrided layers produced in pulsed plasma at a frequency between 10 to 60 kHz. *Vacuum.* Vol. 48, No. 6, 499, 1997.

52. **Carlsson, J.O.:** Processes in interfacial zones during CVD: aspects of kinetics, mechanisms. *Thin Solid Films,* 130, 261, 1985.

53. **Michalski, J.:** *Conditions of formation of superficial layers in processes of chemical deposition from the TiCl₄ + H₂ and TiCl₄ + H₂ + N₂ atmospheres* (in Polish). Ph.D. dissertation, Warsaw University of Technology, 1989.

54. **Kowalski, S., Łataś, Z., Rogalski, Z. et al.:** Heat treatment technology (in Polish). Vol. 3, Publ. ZODOK SIMP, Warsaw, 1987.

55. **Buchkov, A., and Toshkov, V.:** *Ionno azotirovanye* (Ion nitriding) (in Bulgarian). Publ. Technika. Sofia 1985.

56. **Pessel, W.Z.:** Plasmawärmebehandlung von Stahl in Nitrier- und Oxinitrieratmosphäre sowie in einer Mischung von Luft und Endgas. *Härterei-Technische Mitteilungen,* No. 46, 2, 81, 1991.

57. **Lampe, T.:** *Plasmawärmebehandlung von Eisenwekstoffen in Stickstoff und Kohlenstoffhaltigen Gasgemischen.* VDI-Verlag, Düsseldorf 1985.

58. **Voroshnin, L.G., and Liakhovich, L.S.:** *Boriding of steel* (in Russian). Publ. Metallurgia, Moscow, 1978.

59. **Graf von Matuschka, A.:** *Borieren.* Carl Hansen Verlag, Munich-Vienna 1972.

60. **Dubinin, G.N., and Kogan, Ja.D.:** *Progressive methods of thermo-chemical treatment* (in Russian). Publ. Masinostroenye, Moscow 1979.

61. **Hegewalt, F., Singheiser, L., and Türk, M.:** Gasborieren. *Härterei-Technische Mitteilungen,* Vol. 39, 1, 7, 1984.

62. **Gifford, F.E.:** Plasma silicon nitriding and iron boriding. *Journal of Vacuum Science and Technology,* Vol. 10, 1, 85, 1973.

63. **Casadesus, P., and Gantois, M.:** Über das Borieren von Eisenlegierungen mittels Iononbeschuss mit Diboran. *Härterei-Technische Mitteilungen,* Vol. 33, 4, 202, 1978.

64. **Hunger, H.-J., and Trute, G. et al.:** Plasmaaktiviertes Gasborieren Bortifluorid. *Härterei-Technische Mitteilungen,* Vol. 52, 1, 39, 1997.

65. **Kastner, B., Przybyłowicz, K., and Grygoruk, J.:** Untersuchungen von Boridschichten auf Stählen. *Härterei-Technische Mitteilungen,* Vol. 34, 4, 173, 1979.

66. **Wierzchoń, T., Pokrasen, S., and Karpiński, T.,** Erzeugung und Bedeutung der Gasatmosphären bei der thermochemischen Behandlung in Plasma einer Glimmentladung. *Härterei-Technische Mitteilungen,* 4, 189, 1981.

67. **Wierzchoń, T., and Bielinski, P.:** Properties of composite borided layers on steel produced by combined surface treatments. *Materials and Manufacturing Processes,* Vol. 10, No. 2, 309, 1995.

68. **Bielinski, P., Sikorski, K., and Wierzchoń, T.:** Effect of glow discharge boriding on the distribution of elements within nickel stellite layers and the properties of these layers. *Journal of Mat. Sci. Letters* 15, 1335, 1996.

69. **Vossen, I.L., and Kern, V.:** *Thin films processes.* Academic Press, New York, London 1978.

70. **Stjernberg, K.G., Gas, H., and Hintermann, H.E.:** The rate of chemical vapour deposition. *Thin Solid Films,* 40, 81, 1977.

71. **Kortmann, W.:** Vergleichende Betrachtungen der gebräuchlichsten Oberflächenbehandlungsverfahren. *Fachbereite Hüttenpraxis Metallverarbeitung,* Vol. 24, No. 9, 734, 1986.

72. **Akamatsu, K., Kamei, K., and Ikenaga, M.**: Applications of chemical vapour deposition to some tool steels. Proc.: *3rd International Congress on Heat Treatment of Metals*, Shanghai, 9, 47, 1983.
73. **Stock, H.R., and Mayr, P.**: Hartstoffbeschichtungen mit dem Plasma-CVD-Verfahren. *Härterei-Technische Mitteilungen*, 41, 3, 45, 1986.
74. **Broszeit, E., and Gabriel, H.M.**: Beschichten nach den CVD-Verfahren. *Zeitschrifft fur Werkstofftechnik*, 11, 30, 1980.
75. **Wierzchoń, T.**: CVD methods - direction of development (in Polish). Proc.: *Conference on Surface Treatments*, Kule-Czestochowa, 214, 1996.
76. **Hess, D.W.**: Plasma-enhanced CVD: Oxides, nitrides, transition metals and transition metal silicides. *Journal of Vacuum Science and Technology*, A2, 2, 223, 1984.
77. **Michalski, J., Łunarska, E. et al.**: Wear and corrosion properties of TiN layers deposited on nitrided high speed steel. *Surface and Coatings Technology*, 72, 189, 1995.
78. **Razubayev, G.A., Gribov, V.G., and Domarchev, G.A.**: *Metal-organic compounds in electronics* (in Russian). Publ. Nauka, Moscow, 1972.
79. **Razubayev, G.A.**: *Application of metal-organic compounds to obtain non-organic coatings and materials* (in Russian). Publ. Nauka, Moscow, 1991.
80. **Deutschman, L. et al.**: Some new hard coatings prepared by plasma enhanced CVD using organometallics. *Proc. of I.S.P.C.*, Bochum, 2, 4, 1991.
81. **Rie, K.-T., Lampe, T., and Eisenberg, S.**: Abscheidung von Titannitridschichten mittels Plasma-CVD. *Härterei-Technische Mitteilungen*, 42, 3, 153, 1987.
82. **Arai, T., and Fujita, H.**: Plasma-assisted CVD of TiN and TiC on steel. Proc.: *6th International Conference on Ion and Assisted Techniques*, Brighton, U.K., 196, 1987.
83. **Wierzchoń, T., and Sobiecki, J.R.**: Properties of surface layers produced from metalorganic titanium compounds under glow discharge conditions. *Journal de Physique IV*, Vol. 5, C5-699, 1995.
84. **Rie, K.-T., and Gebauer, A.**: Plasma assisted CVD of hard coatings with metallo-organic compounds. *Materials and Science Engineering*, A1, 39, 61, 1991.
85. **Doering, E.**: Deposition technology of insulating films. *Insulating Films on Semiconductors*. Ed. M.. Schultz, G. Pensl, Springer Verlag, 1981.
86. **Bunshah, R.F.**: Overview of coating technologies for large scale metallurgical, optical and electronic applications. *Journal of Vacuum Science and Technology*, B2, 4, 789, 1984.
87. **Matsumoto, D., and Kanzaki, Y.**: Nitriding of titanium in a radiofrequency discharge. *Journal of Less Common Metals*, No. 84, 157, 1982.
88. **Wierzchoń, T., Sobiecki, J.R., and Kurzydlowski, K.**: Properties of surface layers produced from metalloorganic compounds. In Thin Films. Ed. G. Hecht, F. Richter, J.Hahn, DGM-Verlag, 195, 1994.
89. PN-83/H-04302, 1983. Polish Standard Specification. Strength testing of metals. Friction test in three-rollers-cone system.
90. **Boman, M., and Carlsson, D.**: Laser-assisted chemical vapour deposition of hard and refractory binary compounds. *Surface and Coatings Technology*, 49, 221, 1991.
91. **Jones, A.C., Houlton, D.J. et al.**: A new route to the deposition of Al_2O_3 by MOCVD. *Journal de Physique IV*. Vol. 5, C5-557, 1995.
92. **Bastianini, A., Battiston, G.A. et al.**: Chemical vapour deposition of ZrO_2 thin films using $Zr(NE^+_2)_4$ as precursor. *Journal de Physique IV*, Vol. 5, C5-525, 1995.

chapter six

Vacuum deposition by physical techniques (PVD)

6.1 Development of PVD techniques

Beginnings of PVD techniques, similarly to electron beam and ion implantation techniques, date back to Torricelli's famous experiment with an inverted glass test tube filled with mercury, in which he established the existence of vacuum. Later, the first technical instrument which made vacuum obtainable - the vacuum pump - was invented by von Guericke (Table 6.1). These inventions formed the foundations for the advent and development of vacuum technology without which progress in modern surface engineering would have been impossible.

Table 6.1
Chronology of development of vacuum deposition of coatings
(From *Burakowski, T., et al.* [1]. With permission.)

Year	Name of discoverer or inventor	Invention, discovery or introduction of term*
1643	E. Torricelli	Vacuum
1650	O. von Guericke	Vacuum pump
1834	M. Faraday	Ions*
1852	W.R. Groove	Diode sputtering
1857	M. Faraday	Thermal vapor deposition (exploding wire technique)
1887	R. Narwold	Thermal vacuum vapor deposition
1896	J.J. Thomson	Electrons
1898	W. Crookes	Ionization*
1907	F. Soddy	Reactive vapor deposition
1909	M. Knudsen	Distribution of emission of vapor deposited material
1909	J. Stark	Theory of sputtering
1910	A. Hull	Sputtering threshold energy
1928	I. Langmuir	Plasma*
1928	A.I. Shalnikov, N.N. Syemyonov	Deposition from molecular beams
1932	J. K. Roberton, C.W. Chapp	Deposition from high frequency plasma
1934	F.M. Penning	Magnetron

Table 6.1 continued

Year	Name of discoverer or inventor	Invention, discovery or introduction of term*
1937	B. Berghaus	Vapor deposition in vacuum with polarization of substrate
1950	M. von Ardenne	Deposition by ion beam
1951	L. Holland	Evaporation by ion beam
1958	W. Wroe	Evaporation by electric arc
1964	D.M. Mattox	Ion beam plating*
1968	J.R. Morley	Hollow cathode evaporation
1972	T. Takagi, I. Yamada	Deposition from ionized clusters
1977	E. Moll	Deposition by thermo-ionic arc
1977	M. and A. Sokołowski	Pulse-plasma deposition

Significant development in research, although still very far from practical application, took place in the 19th century. Groove discovered the phenomenon of **ion sputtering**, while Faraday, that of **thermal deposition** from the vapour phase.

The phenomenon of material sputtering with the help of ions accelerated in an electric field, observed by Groove in 1852, was utilized only 25 years later in the manufacture of reflecting layers of mirrors, as competitive with regard to chemical techniques. In the 1920s it was utilized on an industrial scale in the manufacture of gold coatings [1].

The phenomenon observed by Faraday in 1857 - that of deposition on a glass substrate of metal vapors from a "burnt" (by resistance heating) metal wire in a neutral atmosphere - led to the development of the **thermal vapour deposition in vacuum** which, being simpler and more economical, replaced Groove's technique from mirror production. One hundred years after its discovery has been widely implemented to deposit pure metals onto lamp reflector elements, on mirrors, for decorative purposes (e.g., coating of wrist-watch elements), semi-conductor production and for making replicas in metallography [2]. Vapour deposition in vacuum in its classical form did not prove suitable for extending the life of components and tooling, nevertheless constituted the basis for the development of techniques later termed PVD. This development consisted mainly of intensification of material evaporation processes (resistance, electron, arc and laser), of techniques of gas and vapour ionization, as well as of reagent activation by, among others, magnetron sputtering, substrate polarization, utilization of glow discharge and the application of high frequency [1-4].

The first practical application of coatings deposited on cutting tools by PVD techniques took place in the 1960s when cutting tools were coated by titanium nitride for the first time in U.S. industry [3, 5, 6].

6.2 PVD techniques

6.2.1 General characteristic

Today, there are several tens of versions and modifications of PVD (*Physical Vapour Deposition*) techniques. They all have one factor in common, and that is that they are based on the utilization of different physical phenomena which take place at pressures reduced to 10^2 to 10^{-5} Pa. These phenomena, occurring with different intensities in different techniques, are the following:

– **Obtaining of vapours** of metals and alloys (through erosion of the vapour source, due to evaporation or sputtering) which may form the substrates for a possible later chemical reaction;

– **Electrical ionization** of gases supplied and of metal vapours obtained (the higher the degree of ionization, the better);

– **Crystallization** from the obtained plasma of metal or compound. Crystallization centers may be formed in existing clusters of different gaseous phases) or on a relatively cold substrate;

– **Condensation** of components of the plasma (particles, atoms, ions) on a relatively cold substrate;

– Possible **formation of a chemical compound** of the substrates in the vicinity of or on the surface of the treated object;

– Possible **physical** (sometimes also chemical) **assistance** of processes occurring.

The mentioned stages of the process of physical deposition occur with different intensity in different versions, while some may not occur at all.

In almost all PVD techniques the coating deposited on the substrate is formed from a flux of plasma, directed electrically at a relatively cold substrate. For this reason, techniques of depositing coatings from plasma (with the utilization of ions) are referred to as **plasma assisted deposition** (*PAPVD*), **plasma enhanced deposition** (*PEPVD*) or **ion assisted deposition** (*IAPVD*). Table 6.2 lists elementary processes occurring with different intensity in different ion deposition techniques [5].

The interaction of an ion with a solid depends on the energy of the ion (Fig. 6.1) [6, 7]. The most favorable range of ion energy is from several to several tens electronvolts. This is the range of the same order as that of the energy of ion bonds at the coating surface and does not exceed the threshold energy of sputtering. With such energies desorption occurs of contaminant atoms, including residual gases. Weakly bonded atoms are displaced from interstitial positions, surface defects are formed, centers of nucleation (condensation) of high density are formed. Increased surface atom mobility and surface chemical activity occur. All these effects lead to obtaining coatings with good physical properties and good adhesion to substrate even at low temperature [28]. Further rise of ion energy leads to knocking out of particles of deposited coating and substrate (surface sputtering) and further rise to their implantation into the load surface (see Chapter 4).

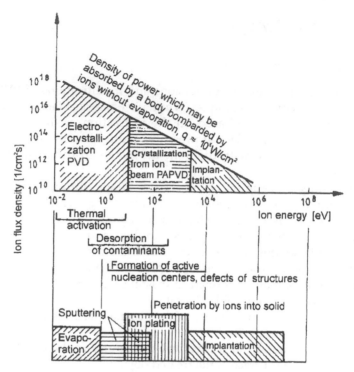

Fig. 6.1 Types of interaction between ion and solid, depending on ion energy. (From *Michalski, A.* [6]. With permission.)

Metal vapours or compounds condensing from plasma are deposited on a cold or heated (up to 200 to 600°C) substrate which allows coating of a quenched and tempered substrate without causing a hardness drop during the deposition process (due to rise in temperature), but at the same time leads to the formation of thin coatings, on account of rather weak bonding of coating to substrate. The connection is of an adhesive character (in more strict terms - adhesive-diffusion) and is weaker if the coated surface is less pure. Usually, the substrate is made of such materials which allow the obtaining of hard and very hard coatings, up to 4000 HV [1, 3, 6].

In an overwhelming majority of PVD techniques, there occur chemical reactions between plasma components or between components of plasma and the substrate (so-called reactive PVD techniques). Similarly, in the majority of CVD techniques, physical phenomena, e.g., reduced pressure and plasma, are utilized. The difference between them becomes ever smaller. It would therefore be in order to assume one common name to cover the majority of both CVD and PVD techniques, i.e., PCVD (Physical-Chemical Vapour Deposition), treating as exceptions pure PVD techniques, i.e., vacuum vapour deposition or pure CVD techniques, e.g., those carried out under atmospheric pressure, and according them one common discussion. This approach was, however, abandoned in this book in favor of the traditionally accepted divisions between the two groups.

Table 6.2
Elementary processes occurring in ion deposition (From *Frey, H.,
and Kienel, G.* [5]. With permission.)

Surface preparation (cleaning) process — Ar⁺ ions 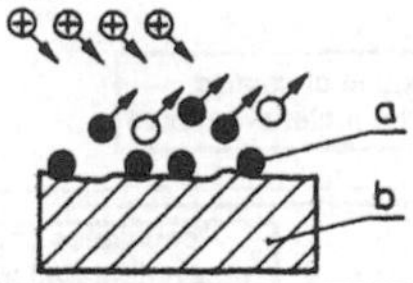 / a / b	Intensive cleaning of surface and sputtering of atoms of contaminating elements and of substrate
Ar⁺ ions 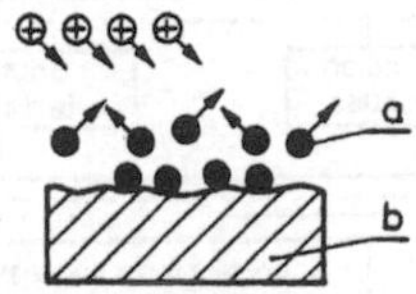 / a / b	Increase of desorption, especially by heating of substrate with ion flux
Ar⁺ ions 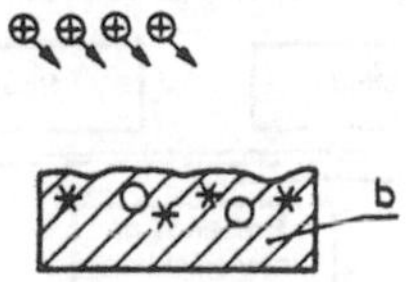 / b	Activation of substrate surface desorption and formation of defects in substrate
Coating process — Ar⁺ ions / coated material partially ionized 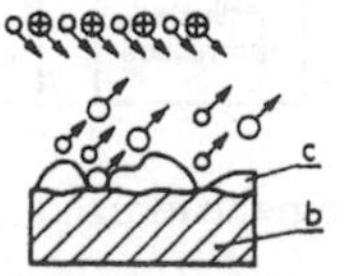 / c / b	Secondary sputtering of substrate and layer material; increase in coating is accompanied by increase of sputtering of substrate material
Ar⁺ ions / coated material partially ionized 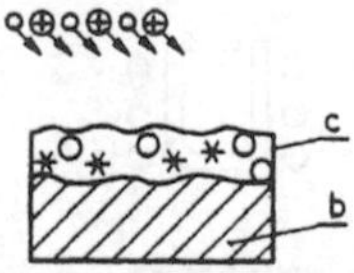 / c / b	Activation of layer and surface of substrate; formation of defects

Note: a - contaminant layer, b - substrate, c - layer obtained in process.

6.2.2 Classification of PVD techniques

Existing PVD techniques differ from one another by the following (Fig. 6.2) [1-13]:

1. Location of zone of obtaining and ionization of substrate vapours, i.e., of deposited material (separate or shared zones);

2. Technique of obtaining and ionization of substrate vapours by [13]:

– **thermal evaporation** of substrate not melted (Fig. 6.3a) or melted (Fig. 6.3b, c, d, e) by resistance, induction, electron beam (most frequent approximately 50% of all applications), arc or laser;

– **thermal sublimation,** i.e. transition of substrate from solid directly to vapour, in a continuous or pulsed arc discharge (approximately 25% of all applications);

– **sputtering** of metal or compound in the solid state (approximately 25% of all applications) by the following means:

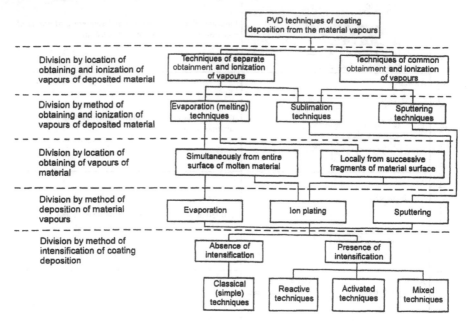

Fig. 6.2 Division of PVD techniques of coating deposition from the ionized gas phase.

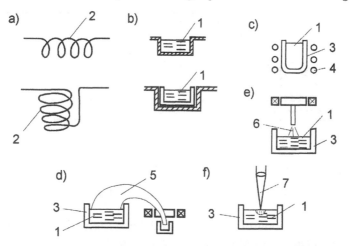

Fig. 6.3 Schematics of evaporators utilizing different methods of heating: a) direct resistance heating by wire; b) direct resistance heating by strip; c) induction heating; d) electron heating; e) arc heating; f) laser heating; *1* - evaporating material in the form of wire; *2* - molten evaporating material; *3* - metal or ceramic pot; *4* - induction coil; *5* - electron beam; *6* - arc; *7* - laser beam.

– *ion* (*cathodic* or *anodic*) - sputtering of the negative electrode (cathode) or positive electrode (anode) under the influence of bombardment by ions of opposite sign,

– *magnetron* - sputtering of the electrode in an abnormal glow discharge in crossed electric and magnetic fields in order to enhance concentration of glow discharge plasma and its localization in the direct neighborhood of the magnetron;

3. Situation of zone of obtaining of substrate vapours through evaporation which may be

– **simultaneous** from the entire surface of the molten substrate in the evaporator,

– **local** from consecutive fragments of substrate surface in the solid state;

4. Technique of depositing of metal vapour on the substrate [13]:

– **Evaporation - E**. Deposition of non-ionized (classical technique) or only insignificantly ionized (tenths of a percent) metal vapours, e.g., Al, Cr, B, Si, Ni, obtained by classical thermal techniques through evaporation. These are assisted techniques. Usually, insignificant ionization of metal or compound vapour occurs in a different zone to that of evaporation. Sometimes, deposition of insignificantly ionized vapours is counted as ion plating. The technique of evaporation is identified with deposition of condensing vapours on a substrate;

– **Ion plating - IP**. Deposition of vapours of a metal or compound, obtained by evaporation or thermal sublimation, ionized to a degree greater than in assisted techniques. Usually, ionization of metal or compound vapours occurs in the evaporation zone. There is a wide variety of ion plating techniques. In some cases, especially in the latest developments, sputtering is identified with ion plating, all the more so because sputtering characterizes the technique of obtaining and not deposition of metal or compound vapours;

– **Sputtering - S**. A modification of ion plating. The deposition of strongly ionized vapours of metal or compound, obtained by sputtering of a metal electrode (so-called shield) by ions of a neutral gas (mostly argon). In principle, the sputtering does not pertain to the technique of deposition on the substrate which is analogous to the technique of obtaining metal vapors;

5. Absence (**Simple - S**) or presence of intensification of process of layer deposition through [13]:

– utilization of reactive gases (e.g., N_2, hydrocarbons, O_2, NH_3) enabling, by way of chemical reaction with metal vapors the obtaining of an appropriate hard compound (e.g., TiN, VC, Al_2O_3) on the coated surface (so-called **Reactive - R** techniques) [12]. In principle, reactive techniques are of a physicochemical nature;

– activation of process of ionization of gas and metal vapours with the aid of additional physical processes: glow discharge, fixed (polarized electrodes or substrate) or variable electric and magnetic fields, additional sources of electron emission, heating of substrate in order to obtain diffusion (so-called **Activated - A** techniques) or through a combination of the two above techniques (so-called **mixed** or **reactive-activated** techniques) [12].

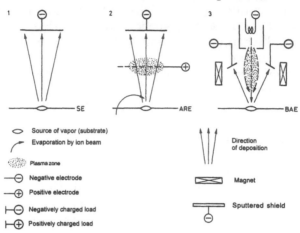

Fig. 6.4 Schematic diagrams showing versions of evaporation: *1* - classical (simple); *2* - activated reactive; *3* - activated by additional electrode. Version most frequently used - ARE. Notation used here pertains also to Figs. 6.5 and 6.6.

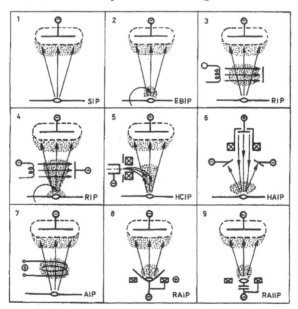

Fig. 6.5 Schematic diagrams showing versions of ion vapour deposition (*ion plating*): *1* - classical (simple); *2* - classical with melting of metal by electron flux; *3* - activated by additional flux of electrons; *4* - with melting of metal by electron flux, vapors activated by additional electron flux; *5* - activated by arc with hollow cathode; *6* - activated by hollow anode; *7* - activated by additional glow discharge; *8* - with high current continuous arc discharge; *9* - with high current pulsed arc discharge. Technique most frequently used - *RIP*.

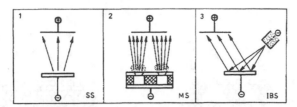

Fig. 6.6 Schematic diagrams showing versions of sputtering: *1* - classical (simple); *2* - activated by magnetic field (magnetron); *3* - activated by ion flux. Technique most frequently used - *MS*.

The classification proposed above enables (see Fig. 6.2) the naming of these techniques in accordance with their characteristic features, e.g., **ARE**, **SIP**, with a possible addition of the technique of activation, e.g., additional electrode (**Bias - B**), **Hollow Cathode - HC**, **Hollow Anode - HA**, **Magnetron - M**. In some cases the name has, additionally, a symbol denoting technique of evaporation, e.g., **Electron Beam - EB** or **Laser Beam - LB**.

Fig. 6.4 shows simple diagrams of the most important techniques of evaporation, Fig. 6.5 of vapour deposition and Fig. 6.6 of sputtering.

Lately, with increasing frequency, due to the convergence of processes of deposition in ion plating and ion sputtering, deposition techniques are divided into two groups:

– **Classical deposition** of atoms from metal vapours in vacuum (or in an atmosphere of non-ionized gas) on a clean, cold substrate. This is typical vapour deposition. The deposition process is usually slow because the evaporation surface of the metal in the evaporating chamber is several times smaller than the surface of the coated load. Metal vapours in the form of neutral atoms have low energy upon reaching the substrate. They cannot knock atoms out from the substrate and only condense (settle) on it by sublimation. Their mobility, especially on substrates at low temperatures, is low. In effect, they form coatings of low density and weak adhesion with a big amount of contaminants (Fig. 6.7a) [14, 15]. Presently, classical deposition is utilized very seldom, mainly for vapour deposition of mirror reflecting surfaces.

– **Ion deposition of substrates in vacuum** (most often reactive) on a clean surface, cold or preheated to several hundred degrees C. Ion deposition encompasses several techniques which can all share a common name of **ion plating** [16], proposed by D.M. Mattox. *The classical definition of ion plating encompasses all processes of layer deposition in which the substrate surface, and next, the surface of the deposited coating are subjected to constant bombardment by a stream of ions with an energy sufficient to cause sputtering.* Ion bombardment yields very good density, tightness and adhesion of the coating to the substrate, mainly due to removal of atoms of contaminants, heating of the substrate or even shallow implantation (Fig. 6.7b), as well as a favorable distribution of residual stresses in the vicinity of the substrate-coating interface. This facilitates deposition of relatively thick coatings, up to several micrometers. In the conventional ion plating system

a)

b)

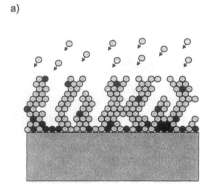
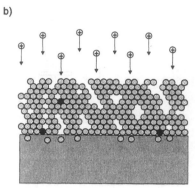

○ - Layer atom
● - Impurity atom
⊕ - Ion

Fig. 6.7 Surface processes occurring during coating deposition: a) classical vapour deposition; b) ion plating.

the deposited material was evaporated from an evaporator in the form of a pot, by an electron beam and subsequently ionized. In non-conventional systems, the deposited material may be evaporated by any preferred technique and sputtered either by ions or by a magnetron, as long as it is subjected to ionization and, usually, chemical reactions during the process or later [14-21].

In reactive ion plating (being Physical-Chemical Vapour Deposition - PCVD), the coating is formed as the result of:

– evaporation of the coating material from the evaporator or sputtering from a target by any preferred technique,

– ionization of gases and vapours of the coating material by any preferred method,

– occurrence of chemical reactions between atoms and ions of the coating material and atoms of the reactive gas during their movement in the direction of the substrate or on the substrate,

– collisions of particles of coating material with particles of gas, leading to so-called gas dispersion and to a rise of the energy of particles of the coating material,

– condensation of particles of the deposited material to the form of crystallization nuclei, negatively charged, due to the presence of plasma (even when the substrate is polarized),

– gradual ordering of the crystalline lattice, as the result of an increase of mobility of atoms of the deposited coating material, which is due to the energy passed on to them by the bombarding ions,

– removal of atoms of the gas built into the deposited coating, both during direct interactions between ions and built-in atoms, as well as due to heating of the substrate.

The mechanism of formation of crystallization nuclei in ion plating is, of course, different from that in classical vapour deposition in vacuum or in an atmosphere of non-ionized gas. High energy particles, i.e. ions and atoms, neutralized in the discharge reaction, are implanted shallow in the substrate. Particles of lower energy, i.e., neutral atoms which did not receive energy directly from the electric field during any phase of its movement in the direction of the substrate, are subject to the same mechanisms of nucleation as in vapour deposition. However, in the case of ion plating, the stream of vapours moving through the plasma zone receives energy through collisions with plasma particles. As a result, the mean energy of deposited particles rises with a rise in the number of collisions, i.e., with a rise in the distance between particle source and substrate, and this enhances adhesion of the layer to the substrate. Due to fairly uniform ion bombardment of the entire substrate, as well as uniform distribution of surface defects generated by this bombardment, the distribution of crystallization nuclei is also more uniform, with smaller nuclei (10^{-2} μm) occurring more densely. A columnar structure is formed on these nuclei and a continuous coating is obtained already at thicknesses as small as $1.3 \cdot 10^{-2}$ μm, i.e., smaller that in the case of vapour deposition [13, 16-18, 21].

Great variety of modifications of PVD techniques and a lack of uniform terminology are responsible for the fact that these techniques are named differently by different authors.

Later in this chapter, the most important PVD techniques will be discussed, using original terminology.

6.2.3 Discussion of more important PVD techniques

6.2.3.1 Techniques utilizing simultaneous evaporation of substrate from entire liquid surface

This group includes those techniques which utilize the vapours of the deposited material (substrate), heated in the evaporator until it melts. Evaporation occurs from the entire liquid surface. The means of heating does not change the principle of the method itself, although sometimes has an effect on the design of the equipment and its service parameters. Most often, electron beam heating is used, less often resistance heating (due to low effectiveness of vapors and difficulty in application to materials with a high melting point) and sometimes induction heating [1].

Diagrams of the more important techniques are shown in Fig. 6.8.

Activated Reactive Evaporation - *ARE*. This classical technique, described by R.F. Bunshah [17, 18], utilizes the electron beam and was first applied in 1963 by D.M. Mattox [16] to evaporate material (Fig. 6.8a). The surface of the molten metal serves two purposes: as a source of vapours and an emitter of electrons. Metal vapours levitating above the molten metal surface are ionized by low energy electrons, emitted also by that surface which serves as a thermal cathode. Into the thus formed plasma a

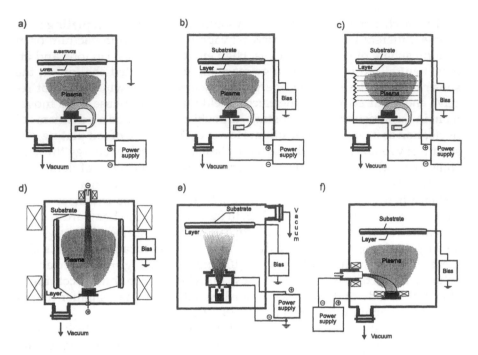

Fig. 6.8 Schematic diagrams showing techniques which utilize simultaneous evaporation of substrate from the entire surface of the liquid: a) Activated Reactive Evaporation (*ARE*); b) Bias Activated Reactive Evaporation (*BARE*) by additional electrode; c) Bias Activated Reactive Evaporation (*BARE*) by additional electron emitter; d) Thermoionic Arc activated reactive Evaporation (*TAE*); e) activated evaporation by hot Hollow Cathode Discharge (*HCD*); f) reactive deposition from Ionized Cluster Beam (*ICB*).

reactive gas is introduced which by chemical reaction with ionized metal vapors, is deposited in the form of a compound on the surface of the load, placed opposite the evaporator and having the positive bias of the plasma (anode). An improvement of this technique is *BARE*.

Bias Activated Reactive Evaporation - *BARE*. In 1937 B. Berghaus observed that the application of a negative potential to the substrate causes acceleration of ions participating in crystallization and, by the same token, improves adhesion of the deposited coating [19]. In the 1950s, this method was further improved by M. Auwärter [20]. As compared to the *ARE* technique, this one features the utilization of discoveries mentioned above and the application of negatively polarized substrates (usually 50 to 1550 V). There are numerous known modifications, differing not only by the value of polarization voltage but also by the existence of additional systems which increase the degree of ionization of vapours. This can be accomplished by: a magnetic field forming a column of plasma and raising the degree of ionization of the working gas (usually argon) [13], an additional ionizing electrode, positively polarized (Fig. 6.8b) or additional electron emitters (Fig. 6.8c), situated in many patterns between the evaporator and the substrate and under different voltage [18-20]. These techniques allow an increase in effectiveness of ionization by over 50%. This technique is offered in equipment manufactured by Balzers and by the Institute of Electron Technology from Wrocław Technical University in Poland [21].

Thermo-ionic Arc Evaporation - *TAE*. This technique was developed in 1977 by E. Moll, working with the Balzers company, and used to depositing TiN coatings [29]. A pot with the metal, constituting the anode, is heated by an electron beam. The electrons are emitted from a thermal cathode. Voltage between the two electrodes is approximately 50 V, while the beam current is approximately 100 A. Ions emitted by the anode are trapped in a magnetic trap, formed by solenoids wound around a vacuum chamber, and are deposited on the surface of the load (Fig. 6.8d) [22].

Hot Hollow Cathode Discharge - *HCD*. In 1968, J.R. Morley proposed the utilization of magnetic deflection of a beam of electrons emitted by a resistance heated hollow cathode in the presence of a neutral gas (e.g., argon) introduced into its interior and the melting, with their aid, of a metal anode, in a water-cooled pot (40 V, 400 A). The evaporated metal is partially ionized during collisions with beam electrons, and reacting with a reactive gas, supplied through additional heads, forms a chemical compound, deposited on the negatively biased (approximately 100 V) load [17, 23-25]. This technique features a high degree of plasma ionization (10-50%). This technique was used on a large scale in equipment marketed by the Japanese company Ulvac, while in Europe, it was utilized in "Tina" equipment by the once East German manufacturer - VEB Hochvakuum Dresden (Fig. 6.8e).

Ionized Cluster Beam Deposition - *ICB* or *ICBD*. This technique was developed in 1972 by I. Yamada and T. Takagi from Kyoto University [26, 27] from which come several designs of equipment. It involves melting (by induction or resistance) of a metal inside a pot, adiabatic decompression of the evaporated metal during its flow through a head to a high vacuum zone $(133.3 \cdot 10^{-6}$ Pa), resulting in the partial formation of a beam of atom clusters, i.e., conglomerations of 500 to 2000 intercombined atoms. After leaving the pot, these clusters are partially ionized in the ionizer by a lateral electron flux. Usually, up to 40% of the clusters are ionized. Positively charged clusters are then accelerated by voltage of approximately 10 kV to a supersonic velocity and directed toward the load. The load is bombarded with clusters that are ionized (with energy of several eV per atom) and non-ionized (approximately 0.1 eV per atom), as well as single atoms and ions. The current density at the load surface is a value resultant from the geometry of the source and electrical parameters and varies from fractions to tens of μA per cm². Usually, reactive gas is introduced into the chamber and then the working pressure in the chamber is higher by 1 to 2 orders of magnitude than in deposition without the gas and a chemical compound is formed at the load surface (Fig. 6.8f). At the moment of striking the load surface the cluster is broken and liberated atoms gain, among others, a transverse component of momentum, conducive to a rise in the density of packing of the coating material. The basic advantage of this technique is the high rate of deposition, ranging from fraction of to several nm per s, which can be attributed to the ratio of cluster mass to the charge, which is greater by several hundred to several thousand times with respect to corresponding values for ions of the given element [28].

6.2.3.2 Techniques utilizing local evaporation

In this group of techniques, the vapour source as a whole has temperature which is too low for thermal evaporation. Evaporation takes place locally, from small zones (usually changing their position on the surface of the source) of several μm to several mm² area, and temperature of several thousand degrees, evolved as the result of a strong-current electric arc, pulse discharge or subjection to the action of a laser beam.

Arc Evaporation or Cathode Spot Arc Evaporation - AE. This technique was developed in the early 1970's at the Physico-Technical Institute in Kharkov and by way of license and sub-license purchase has been broadly propagated by US companies like Multi-Arc and Vec-Tec System, as well as Plasma und Vakuum Technik from Germany [29-37].

Depending on the size and designation of the equipment, the vapour deposition contains from 1 to 12 sources with cathodes made from the evaporated material. At the surface of the cathode a high current, low pressure arc discharge is generated. The current intensity is 35 to 100 A and current density 10^6 to 10^8 A/cm², and the power is usually several kilowatts. The discharge takes place between the thick, water cooled target and the ring anode which is also water cooled. The main discharge is initiated by an auxiliary anode. The discharge has no fixed spatial character and is localized within the zone of so-called cathode spots which, due to sublimation, constitute a source of highly ionized material vapours. The degree of ionization of the plasma flux is 30 to 100% and depends on the type of evaporated material. The direction, size and rate of displacement (reaching 100 m/s) of cathode spots of diameter reaching 100 μm are all controlled with the help of electrostatic screens or electromagnetic systems (Fig. 6.9). The occurrence of multiple ions, their high kinetic energy

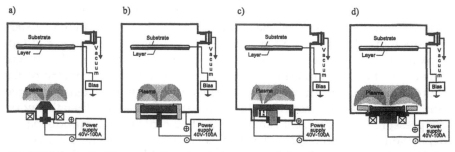

Fig. 6.9 Schematic diagrams showing techniques of electric arc evaporation with localization of electron spot: a) electrostatically; b) electromagnetically; c) electromagnetically with movable magnetic system; d) electrostatic- electromagnetically.

(10 to 100 eV), the possibility of ionic cleaning of the substrate and of making cathodes of different materials in one equipment, combined with the possibility of evaporation in a mixture of reactive gases, all render AE the most often utilized technique [36-38]. One disadvantage of this method is the presence in metallic plasma of drops of evaporated material and their participation in the formation of the coating which is limited by appropriate cathode design, controlled by the movement of cathode spots and plasma filtration [39].

Pulsed Plasma Method - *PPM*. This technique was developed by the M. and A. Sokołowski husband and wife team at the Institute for Material Engineering of Warsaw University of Technology in the 1970s [40]. It consists of evaporation from the solid phase of an electrode, made from coating material and placed centrally in a plasma generator. Evaporation is accomplished as the result of a strong current (100 kA) pulse discharge of a series of condensers of 1 to 10 kV voltage [41, 42]. At the moment of discharge, a current layer is formed which is displaced in the direction of the outlet from the plasma generator, driven by a magnetomotive force, collects the gas ahead of it (may be reactive) and causes ablation of consecutive ring-shaped fragments of the central electrode. By controlling the shape of the current layer it is possible to influence evaporation of the central electrode and the transportation of consecutive packages of plasma (and its decomposition) in the direction of the load. The time of crystallization from ionized portions of metallic vapours (plasma packages) and time of heating of the substrate by plasma at a temperature of approximately 2000 K does not exceed 100 μs when the rate of substrate temperature rise is approximately 10^7 K/s and cooling rate approximately 10^5 K/s, and the interval between two successive pulses is approximately 5 s. These phenomena may be controlled with the aid of its own (Fig. 6.10a, b) [43, 44] or external magnetic field (Fig. 6.10c) [45, 46]. Own magnetic field may be made dependent on the application of ferromagnetic materials for external electrodes of the pulse generator (surrounding the central electrode) which causes lamination the plasma flux in the generator zone [47]. In industrial units, more than one plasma generator may be utilized. Such equipment is especially well suited for coating of big loads in the form of tooling.

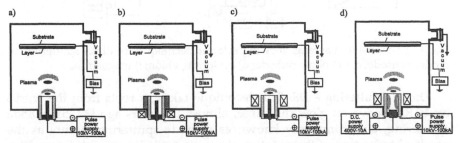

Fig. 6.10 Schematic diagrams showing techniques of pulsed-plasma evaporation: a) with non-standardized own magnetic field; b) with standardized own magnetic field; c) with magnetic field situated externally relative to the generator; d) with external magnetic field and additional power supply to generator by direct current.

Laser Beam Evaporation - *LBE*. This technique was developed in the early 1980's and involves evaporation of material by a pulsed laser beam, focused on the surface of the material. Similarly, material may also be evaporated by a pulsed electron beam. The vapours of the material are ionized in the zone of the laser spot and the generated ions are extracted in the direction of the negatively biased substrate. This technique has not yet widely reached the phase of industrial application. It gives the possibility of obtaining submicron coatings of practically any chosen composition: ceramic oxide materials, metals, biomaterials, diamond-like carbon, semiconductor superlattices. From 1988 laser beam evaporation techniques are named: *Pulsed Laser Deposition - PLD* techniques [48-50].

6.2.3.3 Techniques utilizing direct sputtering

In these techniques the material constituting the chemical substrate for the coating, in this case called the target, is sputtered by ions of gas, generated in the zone between the plasma and the load. The sputtered atoms pass through the plasma zone where they are ionized and, possibly reacting with ions and atoms of the reactive gas, are deposited in the form of a chemical compound on the load (Figs. 6.11 and 6.12) [1].

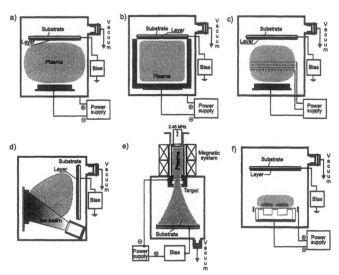

Fig. 6.11 Schematic diagrams showing selected techniques of direct sputtering: a) diode; b) triode; c) in hollow cathode; d) cyclotron; e) ion; f) magnetron.

Diode Sputtering - *DS*. This technique takes its roots from the works by W.R. Groove on glow discharge, almost 150 years ago [32, 51]. Diode sputtering, also commonly known as cathode sputtering, occurs as the result of sputtering of the negative electrode (cathode - target) by positive ions of gas, due to the application of high voltage between the electrodes, separated by gas at 1 to 10 Pa pressure (Fig. 6.11a). The load always forms the positive electrode. We distinguish direct current sputtering and alternating current, radio frequency diode sputtering (*RFDS*) [52, 53]. Presently, diode sputtering is carried out as a reactive process.

Triode Sputtering - *TS*. This technique consists of the introduction into the system of a third, auxiliary electrode, usually in the form of a thermocathode. This is aimed at forming inside the working chamber of two-zones: ion generation (situated near the cathode) and cathode sputtering. From the first zone ions are extracted in the direction of the cathode to the second zone, in order to sputter the material of the cathode. Atoms (and possibly ions) of the sputtered material are ionized while passing through the plasma zone and as the result of a chemical reaction with the reactive gas, are deposited on the load (Fig. 6.11b). Triode sputtering may be generated by direct or alternating current [54, 55].

Hollow Cathode Sputtering - *HCS*. In this technique, the cathode target takes the form of a big, cylindrical cavity, which also constitutes a big part of the working chamber of the unit (Fig. 6.11c). Similarly to the F.M. Penning cylindrical cathode, this design forces electron oscillations in the working volume of the cathode, thus allowing the obtaining of a higher degree of plasma ionization. The majority of units is supplied by high frequency alternating current and only some units (those employed in initial ion cleaning) are supplied by direct current [56].

Electron Cyclotron Resonance Sputtering - *ECRS*. A fundamental characteristic of this technique is gradual acceleration of ionizing electrons in a portion of the chamber, with the aid of an alternating electric field of constant frequency, in a magnetic field of constant intensity. This takes place until cyclotron resonance is reached where the frequency of the variable component of electron velocity is equal to the frequency of excitation (Fig. 6.11d). Such conditions are assured by the appropriate selection of the value of induction of the magnetic field and of the frequency of the high frequency generator. In such conditions, a change in parameters (power, pressure) allows control of the degree of ionization of the gas. This technique is one of the newest and is successfully applied in deposition of diamond coatings [57, 58].

Ion Sputtering - *IS*, Ion Beam Sputter Deposition or simply **Sputter Deposition**. The classical form of this technique consists of depositing a coating on the load by sputtering the material of the target by an ion beam generated by an ion source of any design and the reaction of sputtered atoms with ions from the beam and by ionized atoms (Fig. 6.11e). Modifications of the technique consist of additional introduction of reactive gas, the application of two sources of ions (sputtering of the target and ionizing sputtered atoms). The second ion beam may possibly react chemically with the sputtered material [28]. The ion beam may be employed to sputter any material with precise control of composition of the deposited coating [59-61].

Magnetron Sputtering - *MS*. The beginnings of this technique date back to 1936 when Penning, in an effort to increase plasma concentration of glow discharge, proposed the application of a transverse magnetic field

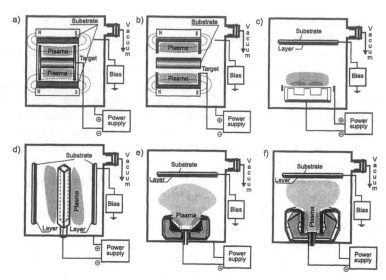

Fig. 6.12 Schematic diagrams showing techniques of magnetron sputtering with different designs of magnetrons: a) cylindrical rod; b) cylindrical hollow; c) flat; d) linear; e) conical; f) sputter gun.

[59] for better sputtering of the coaxial electrode. The main feature of this technique, and one that most significantly affects its productivity, is the magnetron, i.e., a functional group within the unit which allows glow discharge in crossed electric and magnetic fields (Fig. 6.11f). The prototype of the first magnetron was a mercury electrode lamp, first used by Kisayev and Pashkova in 1959 [60, 61] and adapted in a flat form by J.S. Chapin in 1974 to deposit coatings [62, 63]. In the second half of the 1970s, cylindrical magnetrons became popular (rod and hollow) [64-68]. Later, conical magnetrons were developed [64-71], as well as magnetron sputtering guns (Fig. 6.12). There are over 1000 units in service which utilize the principle of magnetron sputtering and employing 1 to 6 magnetron sputtering heads, including single and double twin sets (*Gemini*) of direct current flat magnetrons.

In all types of magnetron units, the constant magnetic field generated by magnets serves to localize the glow discharge on the surface of the water-cooled sputtered targets, which are made from one or several materials (alloys or chemical compounds) constituting substrates of the compound deposited on the load. Electrical discharge takes place in a mixture of a neutral and a reactive gas, continuously supplied to chamber to replenish continuous consumption. The neutral gas, e.g., argon, by diluting the reactive gas, e.g., nitrogen, allows control of the stoichiometric composition of the coating. When the nitrogen content is high, e.g., approximately 10% molar, and the target is made of titanium, titanium nitride is also formed on the target, thereby reducing the effectiveness of sputtering. The surface being coated is usually polarized by a high negative bias (approximately -500 V), which allows ionic cleaning of the load surface prior to the process of coating deposition and ensures the participation of high energy ions in the deposition process. There are known designs with radiant heating of the load prior to deposition in order to improve the connection between coating and substrate. The shape and size of deposition zones, as well as spatial shaping of plasma, depend on the power supplied to the magnetrons, the intensity of the magnetic field and on gas pressure. The technique of magnetron sputtering is one of the most broadly used techniques (over 20% of all applications) - besides arc sputtering (approximately 25% of all applications) - mainly on account of the high rate of target sputtering (1 to 2 orders of magnitude higher than in cathode sputtering) and the reduced range of operating pressures [1, 67, 72-77].

6.2.3.4 Techniques utilizing deposition from ion beams

In this group of techniques, the deposited material, constituting the substrate of the coating, is initially evaporated or sputtered in any preferred way and next ionized, usually outside of the deposition zone. Ions of the material are formed into a flux of low energy, lower than that required for implantation. In the vicinity of the surface or on the load surface, chemical reactions take place between ions and atoms of the reactive gas supplied to the chamber and ions of the beam material, as a result of which, a coating crystallizes on the surface of the load (Fig. 6.13). The potential of the load is negative [1].

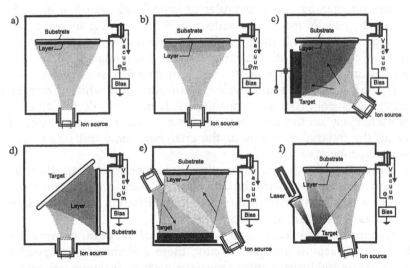

Fig. 6.13 Schematic diagrams showing techniques of deposition from ion beams: a) simple deposition; b) self-mixing; c) ion mixing with sputtering of target with negative potential and of substrate by ion beam; d) ion mixing with sputtering of target by ion beam; e) ion mixing with cathode sputtering; f) ion mixing with thermal (laser) evaporation.

Ion Beam Deposition - *IBD*. This technique consists of direct aiming of the low energy ion beam at the load being coated and of depositing the coating in this way on its surface (Fig. 6.13a). It is characterized by simple control of the deposition process, as well as possibility of control of structure and chemical composition [28, 78-83].

Ion Mixing - *IM*. This method differs from the implantation technique of ion mixing (see Section 4.5) by lower energy of the ion beam. An interesting version of this technique constitutes simultaneous sputtering of the substrate (load surface) and deposition of the coating. As the result of so-called self-mixing, an intermediate coating, strongly adhering to the substrate, is formed at the surface of the load (Fig. 6.13d) [84]. In the majority of techniques, a low energy ion beam and a flux of evaporated or sputtered material (Fig. 6.13 c,d,e) or evaporated material (Fig. 6.13f) react chemically with each other or with the substrate material and crystallize on its surface [85].

6.3 Equipment for coating deposition by PVD techniques

All equipment used for coating deposition by PVD techniques, which could be termed **vapour depositors** (**evaporative** - resistance, electron, laser, arc or pulsed plasma, or **sputtering** - diode, triode, cathode, ion, magnetron and cyclotron) regardless of the technique employed, comprise the following basic functional elements of design:

– **vacuum chamber**, of rectangular or cylindrical shape or a combination of both, usually made of stainless steel and serving to place deposition heads together with their auxiliary components, as well as elements used for fixturing and displacement of the load relative to the heads. Often, the internal surface of the vacuum chamber is covered with removable (after several work cycles) aluminum foil which protects the chamber walls from coating deposition. It is not possible to deposit particles exclusively on the load surface; to a lesser or greater extent they cover all the internal elements of the chamber. Some units have several vacuum chambers;

– **deposition heads** (correspondingly: **evaporative** or **sputtering**) for formation and direction, with the utilization of electric and magnetic fields, of ions or atoms into the ionization and crystallization zones. The latter is situated near or at the load surface;

– **systems for formation and sustaining of vacuum**, comprising oil and diffusion vacuum pumps. Usually, these systems are equipped with vacuum valves and instruments to measure vacuum (vacuum gauges);

– **systems for supply of reactive gases** (cylinders, valves, pressure and flow gages);

– **electrical** and possibly **magnetic systems** supplying the heads and auxiliary electrodes and polarizing the electrodes and load;

– **auxiliary components**, e.g., for preheating of the load or for water cooling of the radiator elements;

– **systems for fixturing and displacement** (sliding, rotation) of the load, comprising one or many elements, relative to the deposition heads. From the design point of view, these systems feature a varied degree of complication, dependent on the type of technique employed and on the size (mass may range from several grams to several hundred kilograms) and on the number of pieces in the load (from one to several hundred, e.g., 600 twist drills of 3 to 8 mm diameter). Such systems range from the simplest sliding or rotational stages to complicated planetary systems, equipped with individual, strip or jaw-type fixturing grips. Their job is always to effect such spatial positioning of the load relative to the head or heads, that regardless of the direction of particles deposited on the load, maximum uniformity of coverage is ensured;

– **control systems**, usually computerized, for controlling the process of coating deposition. Besides the computer, they comprise the optical load observation system, systems for measurement of parameters, of plasma, degree of ionization, of the coating process.

Usually the vacuum chamber, together with its equipment, constitutes a separate design sub-assembly. Supply and control systems constitute separate sub-assemblies (power supply cabinet, control console). Often, vapor depositors, together with systems for load cleaning, constitute complete production lines.

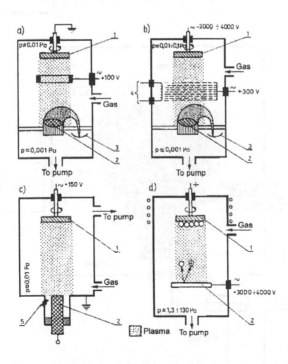

Fig. 6.14 Schematic diagrams showing designs of vapour depositors for some PVD techniques: a) Activated Reactive Evaporation (*ARE*); b) Reactive Ion Plating (*RIP*); c) Reactive Arc Ion Plating (*RAIP*); d) Simple Sputtering; 1 - coated object; 2 - coating metal; 3 - electron gun; 4 - glowing cathode; 5 - sparking electrode. (From *Michalski, A.* [6]. With permission.)

a)

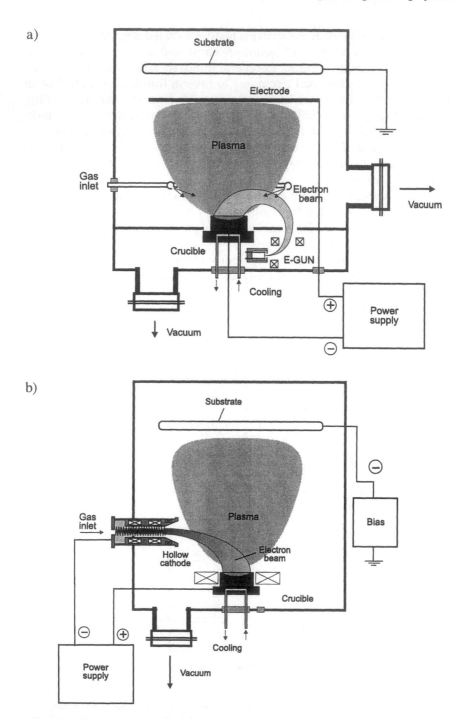

Fig. 6.15 Schematic diagrams showing designs of depositors for most frequently used PVD techniques: a) Bias Activated Reactive Evaporation (*BARE*); b) Hollow Cathode Discharge (*HCD*); c) Arc Evaporation (*AE*); d) Magnetron Sputtering (*MS*).

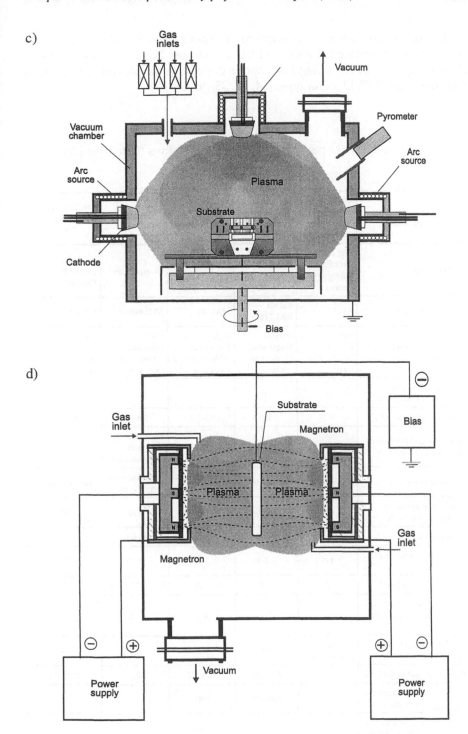

Fig. 6.15 continued

Fig. 6.14 shows schematic diagrams of vapour depositor design for some of the more interesting solutions in PVD techniques. Fig. 6.15 shows more in-depth diagrams of the design of vapour depositors employed for the most frequently used PVD techniques [14, 15, 35, 56, 66, 87].

Table 6.3
Design and service parameters of vacuum depositors (Data from [87] and various other sources.)

Original designation of technique		BARE - Bias Activated Reactive Evaporation	TAE - Thermoionic Arc Evaporation	AE - Arc Evaporation		HCD - Hollow Discharge Cathode
Manufacturer		Wroclaw Technical University, Poland	BALZERS, Liechtenstein	Multi-Arc, USA	Ulvac, Japan	Hoch-Vakuum-former East Germany
Type of equipment		AT-1	Balinit BAI 730	MAV 40	ATC 400	IPB 45
Technique of generating vapours		thermal evaporation by electron beam	thermal evaporation by electron beam	high temperature sublimation by electric arc	high temperature sublimation by electric arc	thermal evaporation by resistance-heated hollow cathode
Number of sources (deposition heads)		1	1	4	4	1
Deposition pressure [Pa]		0.5	0.1-0.4	0.4-0.8	0.4-0.8	0.4-0.8
Load bias (polarization of substrate or target) [V]			1000	500	450 (cleaning) 150 (deposition)	
Substrate preheating		-	electron beam	bombardment by Ti ions	bombardment by Ti ions	radiation heating
Substrate temperature during deposition [°C]		450	450	350-500	450	500
Substrate rotation		yes	yes	yes	optional	yes
Typical layer thickness [μm]		1-5	1.5-4	3-6	3-5	2-6
Process duration [min]		150	140	120	70	140
Working chamber volume [m³] or linear dimensions [m]		-	0.1	0.2	0.102	0.072
Energy consumption per cycle [kVAh]		source power: 6 kW	100	50	120	90
Equipment cost in [mln £]		-	0.5-0.66	0.25-0.3	0.4	0.25-0.3
Maximum load	twist drills 6 mm dia	-	700-800	600	-	-
	milling cutters 100×100 mm	-	-	40-60	-	10
Comments		-	load mass: 600 kg	load mass: 600 kg	-	-

Table 6.3 continued

		TRIP - Thermo-ionically Assisted Triode Ion Plating	KIB - Kondensatsiya veschestva v usloviyakh Ionnoy Bombardirovki		RIP - Reaktywna Impulsowo-Plazmowa, or PPM-Pulsed Plasma Method	
Original designation of technique						
Manufacturer	Hoch-Vakuum former East Germany	Tecvac Great Britain	former USSR	former USSR	Warsaw Univeristy of Technology, Warsaw, Poland	
Type of equipment	Tina 900H	IP35L	Pusk-83	Bulat NNW-6.6-12		
Technique of generating vapours	thermal evaporation by resistance-heated hollow cathode	thermal evaporation by electron beam	high temperature sublimation by electric arc	high temperature sublimation by electric arc	high temperature sublimation by pulsed arc discharges	
Number of sources (deposition heads)	1	1	1	3	1	3
Deposition pressure [Pa]	$2.6 \cdot 10^{-4}$	0.4-0.8	10^{-5}	10^{-3}	5-50	5-50
Load bias (polarization of substrate or target) [V]		400	1000	100-500	n/a	n/a
Substrate preheating		Bombardment by Ar ions	Bombardment by Ti ions	Bombardment by Ti ions	n/a	n/a
Substrate temperature during deposition [°C]	-	350-500	400	-	400	400
Substrate rotation	yes	yes	yes	yes	yes	yes
Typical layer thickness [μm]	-	2-4	3-3.5	5-40 μm/h	5	5
Process duration [min]	-	140	20-30	-	80	200
Working chamber volume [m³] or linear dimensions [m]	0.9 dia x 0.9	0.14	0.3×0.4	0.6 dia×0.6	0.04	0.35
Energy consumption per cycle [kVAh]	35 kVA (chamber) + 27 kVA (cabinet)	50	30 kW		4	50
Equipment cost [mln £]	0.25	0.15-0.17	-	0.25		0.10
Maximum load — twist drills 6 mm dia	-	1000	-	-	-	-
Maximum load — milling cutters 100×100 mm	-	40			2	25
Comments:	-	load mass: 300 kg	2 chambers operating alternately	-	-	-

Table 6.3 continued

Original designation of technique	MS - Magneton Sputtering			RFDS - Radio Frequency Diode Sputtering	DS - Diode Sputtering
Manufacturer	Leybold Heraeus, Germany	Leybold Heraeus, Germany	NPO "Avtoprompo-kritye" - former USSR	Dowty - Great Britain	TI - Abar - Great Britain
Type of equipment	ZV 1200	Z 700P2/2	Mars-650	DSC 91	Glo - Tine 24-36
Technique of generating vapours	magnetron sputtering by direct current	magnetron suttering by direct current	magnetron sputtering (6 heads simultaneously sputtering up to 3 different materials)	magnetron sputtering with radio frequency	magnetron sputtering by direct current
Number of sources (deposition heads)	2		6		1
Deposition pressure [Pa]	2.5		0.13-0.65		1.5-2.8
Load bias (polarization of substrate or target) [V]	500-600	300-1000		2000	
Substrate preheating	bombardment by Ar ions		n/a	discharge energy	radiation heating
Substrate temperature during deposition [°C]	300-500	50-300	200	max.250	450-500
Substrate rotation	optional	no	rotation of cage with load	yes	
Typical layer thickness [μm]	2-3	0.3-0.5	3-10	0.2-3	2-4
Process duration [min]	60		20-90		480
Working chamber volume [m³] or linear dimensions [m]	0.41	0.15	0.5 (0.7 dia × 0.84)	0.04	0.17
Energy consumption per cycle [kVAh]		40	85	-	10-20
Equipment cost [mln £]	0.4	-	0.2	-	0.275
Maximum load — twist drills 6 mm dia	600	800		4200	1200
Maximum load — milling cutters 100×100 mm					
Comments	3 working chambers		cycle time: 20 min	2 working chambers and air chamber	

Vapour depositors for PVD techniques employing different physical parameters (load temperature - 30 to 600°C, vacuum usually - 0.1 to 130 Pa, particle energy - 0.01 to 1000 eV and accelerating voltage from several hundred to several thousand V) yield varied results. The rate of deposition of the coating varies from 0.01 to 75 μm/min, while the uniformity of deposition and adherence of the coating to the substrate vary from low to very high.

Table 6.3 shows design and service parameters of the most popular vapour depositors.

6.4 Coatings deposited by PVD techniques

6.4.1 Coating material

Coatings deposited by PVD techniques should meet the following requirements:

– not impair mechanical properties of the substrate (and the entire product);

– improve tribological, decorative and anti-corrosion properties of the product which may be exposed to different external hazards;

– compressive residual stresses to prevail in the coating;

– bonds between coating and substrate, in most cases adhesive, to be strong and the force of adhesion to compensate residual stresses in the coating.

Not all types of materials are suitable for deposition of coating by PVD techniques. As opposed to many other techniques of coating deposition, those coatings which are deposited by PVD techniques are only in exceptional cases composed of pure evaporated material (e.g. aluminum, gold or copper). In the overwhelming majority of cases, deposited substrates are constituted by transition metals belonging to group IVb, Vb and VIb of the periodic table (most common are Ti, V, Zr, Cr, Ta, Mo, W, Nb and Hf), and reactive gases (usually nitrogen and oxygen). They may also be vapours (of e.g., sulfur, boron or silicon) or elements obtained from different chemical compounds (e.g., carbon from the dissociation of methane or acetylene) which combine to form nitrides, sulfides, carbides, oxides, borides or their combinations (Table 6.4) [86-93]. Often, PVD techniques utilize neutral gases (mainly argon) which usually do not constitute components of the coatings, although Ar may be built into a TiN coating. Compounds forming the coating are usually very hard, rather brittle, refractory and usually resistant to corrosion [92] and to tribological wear. In literature, such compounds are referred to as **hard**; sometimes they are also termed **ceramic**. Generally, they are characterized by a significant variability of chemical composition which results in changes in the type of chemical bonds and morphology [89-93].

Strong structural defects are probably the cause of the majority of excellent properties of hard coating materials [89]. The compounds forming them are non-stoichiometric, their chemical composition has a broad range of variation and the concentration of defects reaches 50% [92]. Double carbides and nitrides of transient metals, in the majority of cases, dissolve fully in each other in the solid phase, while the properties of multi-component compounds thus formed attain extreme values [89, 93].

Table 6.4
More important substrates and coatings deposited from them (From *Hebda-Dutkiewicz, E.* [90]. With permission.)

Substrates		Coating
Evaporated or sputtered material	Reactive gas	
Ti	N_2	Ti_2N; Ti_2N + TiN; TiN
Zr	N_2	ZrN
Ti	C_2H_2	TiC
Zr	C_2H_2	ZrC
Hf	C_2H_2	HfC
Ta	C_2H_2	TaC, Ta_2C
Nb	C_2H_2	NbC
V	C_2H_2	VC
Ti	O_2	TiO_2; suboxides of Ti
Ti + V	C_2H_2	VC; TiC
Ti + Ni	C_2H_2	TiC; Ni
Y	O_2	Y_2O_3
Al	O_2	Al_2O_3
V	O_2	VO_2; suboxides of V
Be	O_2	BeO
Si	O_2	SiO_2
In	O_2	In_2O_3
In + Sn	O_2	oxides of In + Sn
Cu, Mo	H_2S	$Cu_xMo_6S_x$
Ti	H_2S	TiS; Ti_2S_3; TiS_2
Nb + Ge		Nb_3Ge
Ti + V	C_2H_2	VC - TiC
Ti + Ni	C_2H_2	TiC - Ni

Atoms of substrates of metals and non-metals combine to form particles, the stability of which is determined by the bond energy, and thus form the coating.

Three types of bonds of hard coating materials (occurring not only in the coatings discussed here) are distinguished [89, 90, 93, 95-97]:

– **Metallic (M)** which do not have directional character, and occur in crystals of metals containing conducting electrons (electron gas). Positively charged ions, i.e., atomic kernels, situated in lattice nodes, are in equilibrium with electron gas which fills the lattice space and comprises free electrons, common to all metallic ions. In the case of transient metals, there occur additional bond forces, due to interaction between external electron shells. Elements belonging to the transient groups, due to incompletely filled electron shells, have a high bonding energy, e.g., 4.29 eV for Fe. In coatings containing such elements, these bonds are formed by borides, carbides and nitrides of transition metals (Table 6.5).

– **Covalent** or homeopolar (C), occurring between atoms of non-metals, formed by pairing of two electrons, each coming from one of the two partners. The electron pair thus formed is shared by both atoms. The en-

Table 6.5
Properties of coating materials with metallic bonding (Data from [5, 88, 90, 94, 102]
and various other sources.)

Type of coating		Density [g/cm³]	Hardness [HV]	Melting point[°C]	Thermal conductivity [J/(cm·s·K)]	Coeff. of linear exp. [10-6/K]	Young's Modulus [kN/mm²]	Resistivity [μW×cm]
Nitrides	TiN	5.40	2100-2400	2950	0.289	9.35-10.1	256-590	18-25
	VN	6.11	1560	2050-2177	0.113	8.1-9.2	460	85
	ZrN	7.32	1600-1900	2980	0.109	7.9	510	7-21
	NbN	8.43	1400	2200-2300	0.0374	10.1	480	58
	TaN	-	1300	2090	0.096	5.0	-	128
	HfN	-	2000	2700	0.113	6.9	-	28
	CrN	6.12	1100	1050	-	2.3	400	640
Carbides	TiC	4.93	2800-3800	3070-3180	0.172-0.35	7.61-8.6	460-470	51
	VC	5.41	2800-2900	2650-2830	0.043	6.5-7.3	430	60
	HfC	-	2700	3890	0.063	6.73-7.2	359	37
	ZrC	6.63	2600	3445-3530	0.205	6.93-7.4	355-400	42
	NbC	7.78	1800-2400	3480-3610	0.142	6.84-7.2	345-580	19-35
	WC	15.72	2000-2400	2730-2776	0.293	3.8-6.2	600-720	17
	W$_2$C	-	2000-2500	-	-	-	-	-
	TaC	14.48	1550-1800	3780-3985	0.22	6.61-7.1	291-560	15-20
	Cr$_3$C$_2$	6.68	1500-2150	1810-1890	0.188	10.3-11.7	400	75
	Mo$_2$C	9.18	1660	2517	-	7.8-9.3	540	57
Borides	TiB$_2$	4.50	3000	3225	-	7.8	560	7
	VB$_2$	5.05	2150	2747	-	7.6	510	13
	NbB$_2$	6.98	2600	3036	-	8.0	630	12
	TaB$_2$	12.58	2100	3037	-	8.2	680	14
	CrB$_2$	5.58	2250	2188	-	10.5	540	18
	Mo$_2$B$_5$	7.45	2350	2140	-	8.6	670	18
	W$_2$B$_5$	13.03	2700	2365	-	7.8	770	19
	LaB$_6$	4.73	2530	2770	-	6.4	400	15

ergy of covalent bonding is several electronvolts (e.g., 4.5 eV for H-H). In coatings, such bonds are created by borides, carbides and nitrides of aluminum, silicon and boron, as well as diamond (Table 6.6).

Table 6.6
Properties of hard coating materials with covalent bonds (Data from [5, 88, 90] and various other sources.)

Type of coating		Density [g/cm³]	Hardness [HV]	Melting point [°C]	Thermal conductivity [J/(cm·s·K)]	Coeff. of linear exp. [10⁻⁶/K]	Young's Modulus [kN/mm²]	Resistivity [µW×cm]
Nitrides	BN (cubic)	2.52	3000–5000	2730	-	-	660	10^{18}
	BN (FCC)	-	4700	1200-1500	1.8-2.01	-	-	-
	AlN	3.26	1230	2250	-	4.2-5.7	350	10^{15}
	Si₃N₄	3.19	1720	1900	-	2.5-3.9	210	10^{18}
Carbides	B₄C	2.52	3000–4100	2450	-	4.5-6.1	441	$0.5×10^6$
	C (diamond)	3.52	≈8000	3800	-	1.0	910	10^{20}
	SiC	3.22	2600	2760	-	5.3-6.1	480	10^5
Borides	B	2.34	2700	2100	-	8.3	490	10^{12}
	TiB₆	2.43	2300	1900	-	5.4	330	10^7
	AlB₁₂	2.58	2600	2150	-	-	430	$2×10^{12}$

Table 6.7
Properties of hard coating materials with ionic bonds. (Data from [5, 88, 90] and various other sources.)

Type of coating		Density [g/cm³]	Hardness [HV]	Melting point [°C]	Thermal conductivity [J/(cm·s·K)]	Coeff. of linear exp. [10⁻⁶/K]	Young's Modulus [kN/mm²]	Resistivity [µW×cm]
Oxides	Al₁₂O₃	3.98	1800–2500	2047	0.301	8.4-8.6	400	10^{20}
	Al₁₂TiO₃	3.68	-	1894	-	0.8	13	10^{16}
	TiO₂	4.25	1100	1867	-	9.0	205	-
	ZrO₂	5.76	1200–1550	2677	-	7.6-11.1	190	10^{16}
	HfO₂	10.2	780	2900	-	6.5	-	-
	ThO₂	10.0	950	3300	-	9.3	240	10^{16}
	BeO₂	3.03	1500	2550	-	9.0	390	10^{23}
	MgO	3.77	750	2827	-	13.0	320	10^{12}
High speed steel (for comparison)		7.9	850–1100	1300	0.502	11.0-13.0	205	10^5

– **Ion** or heteropolar, electrovalent, polar (**I**) which occur between atoms of non-metals, caused by electrostatic attraction by ions of opposite sign, formed as the result of complete transition of valence electrons from the less electronegative atom to the more electropositive one. The energy of ion bonding is several electronvolts (e.g., 7.9 for NaCl). In coatings, such

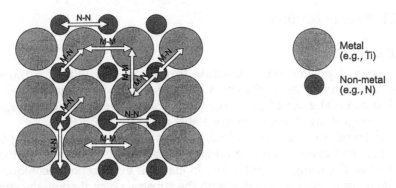

Fig. 6.16 Types of atomic bonds occurring in hard materials of coatings produced by PVD techniques.

bonds are formed by various oxides (Table 6.7). Such bonds are most brittle and are characterized by the greatest coefficient of thermal expansion.

In coating materials, the above-described bonds do not occur in their pure form but rather as mixed bonds, forming complex combinations of interactions of different systems, e.g., metal - metal, non-metal - non-metal, metal - non-metal (Fig. 6.16), with the predominance, however, of a characteristic group: in metallic materials - of metallic bonds, in ion materials - of ionic bonds with a small participation of covalent bonds (e.g., in ceramics).

Table 6.8
Physico-chemical properties of hard coating materials (From *Subramanian, C., and Strafford, K.N.* [97]. With permission from Elsevier Science.)

Value	Hardness	Brittle-ness	Melting point	Stability	Coeff. of linear exp.	Adhesion to metallic substrate	Reactivity	Suitability for multi-layer systems
High level ↓ Low level	C	I	M	I	I	M	M	M
	M	C	C	M	M	I	C	I
	I	M	I	C	C	C	I	C

M - metallic; C - covalent; I - ionic bonds in materials.

Table 6.8 shows the general correlation between physico-chemical properties of coating materials and the type of atomic bonds. It follows that none of the three groups of materials has completely sufficient properties needed for obtaining a coating with good allaround properties. For example, materials which feature high hardness (I) are at the same time brittle, while metallic materials (M) ensure very good adhesion to the substrate but feature high reactivity with the mating material. Properties closest to versatile are featured by materials with metallic bonds and this is what accounts for their practical utilization. Very good chemical resistance and stability are featured by ceramic materials with ionic bonds.

6.4.2 Types of coatings

6.4.2.1 General

A coating deposited on a substrate forms together with it a transition layer of greater or lesser thickness but always playing a major role.

The **external zone of the layer** fulfills protective (enhancing resistance to tribological wear and corrosion) and decorative functions.

The **transition layer**, first of all, ensures adhesion of the coating to the substrate and compensates deformations caused by differences in thermal expansion of coating and substrate. In classical evaporation techniques, the transition layer joins the coating with the substrate by adhesion, the strength of this bond being proportional to the purity of the substrate during the deposition process. In ionic deposition techniques the bond between coating and substrate is stronger because the transition layer formed is of a pseudo-diffusion character [13]. It is formed as the result of ion bombardment of the substrate and of the deposited coating and its sputtering which causes significant point defects to be generated. These facilitate diffusion and favor the formation of the transition layer. Such a layer can also be formed in these conditions by materials which normally cannot diffuse into each other. By sputtering of substrate atoms and with the initially discontinuous layer of the deposited material, the sputtered atoms are reflected back from the gas atoms and recondense on the substrate. The result is a mixing of sputtered substrate atoms and deposited atoms of the coating material until a continuous layer of deposited material is formed. The layer thus formed diminishes residual stresses caused by differences in physico-chemical properties of substrate and coating materials and also reduces the concentration gradient [13]. Coherent transition layers, strongly binding the coating to the substrate, is formed especially by carbides and nitrides of transition metals with transition metals themselves.

The **substrate** carries mainly mechanical loads, while its tribological and chemical resistance is substantially lower than that of the coating.

The appropriate selection of coating material for a given substrate material is of great importance because of the type of transition pair being formed. If a coating with required properties is to be obtained, it is necessary to appropriately design the properties of the external zone of the coating, of the transition layer and of the substrate. The design of the first two consists of an appropriate selection of materials constituting the coating and application of appropriate technological parameters of the deposition process. Design of substrate properties requires earlier assurance of appropriate mechanical properties (e.g., strength and hardness which can be obtained through heat and thermo-chemical treatment) as well as surface smoothness (dependent on machining) and chemical properties (obtained by mechanical, chemical or ion cleaning of the surface).

It happens very often that not all of the required properties may be obtained at the same time. Enhancement of one may cause deterioration of another, e.g., an increase in hardness and strength usually causes a decrease of ductility and adhesion which is required through a broad temperature range. High hardness and high melting point usually do not go hand in hand with a decreased tendency to brittle cracking [93]. Moreover, requirements placed on the external layer of the coating are different from those placed on the intermediate and internal layers (Table 6.9).

Table 6.9
Service requirements placed on coatings and wear mechanisms with tribological wear serving as example (From *Holleck, H.* [93]. With permission.)

Material		Requirements	Mechanism of tribological wear
Coating	external layer	- weak adhesion of coating to material of tribological mating pair	adhesive
		- appropriate hardness	abrasive
	middle layer	- high hardness - high mechanical and fatigue strength	surface fatigue
	internal layer	- appropriate hardness - character of chemical bonds akin to bonds in sub-strate material	abrasive
	transition	- good adhesion to substrate	
Substrate	layer	- good adhesion to coating - good mechanical strength	(in case of wear of coating, part to be scrapped)

When it is not possible to meet all requirement at the same time by a coating made from one material, complex coatings may be deposited [98-102].

6.4.2.2 Classification of coatings

Coatings deposited by PVD techniques can be divided into two groups:

– **simple**, known as **monolayer coatings**, comprising one material - a metal (e.g., Al, Cr, Mo, Cu, Ag, Au) or a phase (e.g., TiN, TiC);

– **complex**, comprising more than one material (metal, phase or compound), with the materials distributed in a varied manner relative to each other.

Five types of complex coatings are distinguished (Fig. 6.17) [14, 97]:

– *alloyed (multi-component) coatings* in which the sub-lattice of one metallic element is partially filled by another metallic element, similar to substitution-type solutions. There are over 600 ternary compounds known involving carbon and nitrogen with transition metals belonging to groups IVb, Vb and VIb of the periodic table. Of these, the best researched compounds are solutions of nitrides: TiN, VN, ZrN, NbN, HfN, TaN, WN, CrN and MoN with TiC, VC, ZrC, NbC, HfC and TaC carbides. Compounds of carbides and nitrides usually together form solid solutions and as ternary and quaternary compounds feature better properties, especially tribologi-

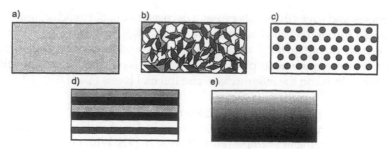

Fig. 6.17 Schematic representations of compound structures [14]: a) alloy coating (multi-component); b) multi-phase coating (multi-component); c) composite coating (multi-component); d) multi-layer coating; e) gradient coating.

cal, than simple coatings. The obtaining of appropriate properties may be controlled, taking advantage of the broad range of mutual solubility of these compounds. Best researched and featuring best tribological properties is the titanium carbonitride Ti(C,N), as well as the titanium and aluminum alloy nitride (Ti,Al)N [98]. Less researched are the following nitrides: (Ti,B)N, (Ti,Zr)N, (Ti,Nb)N, (Ti,Al,V)N and (Ti,Al,V,Cr,Mo)N. Good properties are also exhibited by the (Ti,W)C carbide;

– *multi-phase coatings,* constituting a mixture of two or more divisible phases, e.g. TiN/Ti$_2$N or TiC/TiN [97];

– *composite coatings,* also comprising a mixture of two or more phases and constituting a specific type of multi-phase coating in which one phase is discreetly dispersed in another phase, occurring in a continuous manner, e.g., Ti/Al$_2$O$_3$;

– *multi-layer coatings,* otherwise known as *microlaminates,* comprising consecutive simple layers of different materials, deposited on top of one another, with different properties, and also forming between themselves transition layers, e.g., TiN/AlN. Usually, the internal layer (transition, relative to the substrate) ensures good adhesion to the core, one or more of the middle (intermediate) layers ensures hardness and strength of the coating, while the external layer ensures good tribological properties, e.g., low coefficient of friction, or anti-corrosion properties (resistance to various aggressive environments), and decorative properties, like color and luster. An example of a multi-layer coating, applied on sintered carbide tooling in order to enhance their service life, is the tri-layer coating composed of TiC/Ti(C,N)/TiN [99] or TiC/TiN/Al$_2$O$_3$ [90]. Another example is the six-layer coating, composed of TiN/Ti(C,N)/ZrN/(Ti,Al)N/HfN/ZrN or the octolayer TiN/Ti(C,N)/Al$_2$O$_3$/TiN/Al$_2$O$_3$/TiN/Al$_2$O$_3$/Ti(C,N) [97]. Good bonding is ensured by such layers as TiC/TiN, NiCr/TiN, TiC/TiN/Al$_2$O$_3$, and WC/TiC/TiN [121]. Monolayers constituting the multi-layer coating should be arranged in such a way that the transition layer formed between any of them allows best mutual adhesion, i.e., that the transition layer forms a coherent interphase zone. Coherent connections are formed by coating materials with metallic bonds with metals or other coating materials which also feature metallic bonds, e.g., TiC/TiB$_2$. Weaker connections are formed between materials with metallic bonds and materials with ion bonds (and strongly depend on the chemical composition and structure of the transition layer), e.g., TiC/Al_2O_3. Weakest connections occur between materials with covalent bonds and other materials with the same type of bonds or ionic bonds, e.g., B_4C. Good connections are obtained between those materials which are mutually soluble, forming alloys, e.g., TiC and TiN or Al_2O_3 and AlN [89, 90];

– *gradient coatings*, constituting a modification of multi-layer coatings, in which the change of chemical composition and of properties of the individual layers does not occur in leaps (as in typical multi-layer coatings), but in a continuous manner. An example of a gradient coating is $TiN/Ti(C,N)/TiC$.

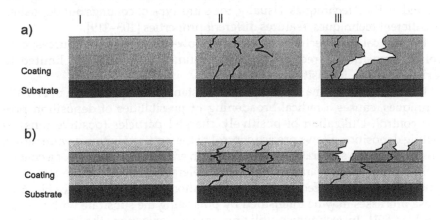

Fig. 6.18 Schematic representations of mechanical destruction of coatings: a) single layer; b) multi-layer; *I* - coating diagram; *II, III* - successive stages of destruction.

The mechanism of mechanical destruction of an alloy coating and its modifications is akin to that of destruction of a simple coating, although its properties are better. Both belong to the monolayer coating group. In a monolayer coating the initiation of microcracks occurs both at the surface and at the interface with the substrate. Propagation and coalescence of microcracks destroy the coating across its entire cross-section (Fig. 6.18a). On the other hand, the mechanism of destruction of a multi-layer coating is different. Initiation of microcracks occurs mainly at the surface of the coating, while interfaces between layers change the direction of microcrack propagation, thus enhancing the mechanical resistance of the coating. This type of coating wears in a laminar manner (Fig. 6.18b) [14, 97].

The earliest coatings and the most broadly applied are simple coatings. Of the complex coatings, most often used are multi-layer modifications, usually three- or four-layer ones. The maximum number of layers may even reach several tens, although currently, these are only laboratory-scale experiments. Gradient coatings appear to be promising for the future, along with alloy coatings and their modifications [100].

6.4.3 Control of structure and properties of coatings

6.4.3.1. General

Service properties of coatings depend predominantly on chemical composition and metallographical microstructure, as well as on adhesion of coating to substrate. Chemical composition depends, of course, on the type and proportion of deposited materials, like structure and adhesion. Not all materials, however, may be deposited in the form of a coating on a substrate, using a freely chosen technique. For that reason, the type of technique, primarily, and its technological parameters in a particular vapour depositor are the essential factors determining service properties of hard coatings deposited by PVD techniques. Usually, the same type of coatings but deposited by different techniques, features different properties [103–119].

In classical PVD techniques, utilizing low-energy vapour sources, control of deposition is reduced to a minimum and, in effect, is limited to variation in the intensity of evaporation.

Utilization of plasma as a means of assistance, i.e., application of PAPVD techniques, causes a radical broadening of possibilities of deposition process control. Utilization of positively charged particles (positive ions) to coating deposition allows precise control of their energy within a wide range. Ions reaching the substrate usually have high energy. As the result of a change of this value, i.e., a change of applied accelerating voltage, these ions may sputter, heat or even affect shallow implantation of the substrate (see Fig. 6.1). These processes may be intensified by polarizing the substrate with a negative bias [14]. In techniques utilizing vapour ionization, the environment from which the coating material is deposited is constituted by plasma, activating chemical reactions between metal vapours and vapours of the reactive gas. In this way it influences the kinetics of coating growth and shapes the morphology and properties of the deposited material [14]. The degree of plasma ionization and activation and, hence, coating properties depend on the chosen technique.

6.4.3.2 Models of coating deposition

In the initial stages of the vapour deposition process, atoms (ions) are deposited on the substrate surface and - unfortunately - on elements of the vacuum chamber of the depositor. This occurs as the result of attraction to the surface by the action of dipole moments of surface atoms of the substrate, as well as other electrical forces (e.g., caused by negative polarization of the substrate). Due to surface diffusion, atoms (ions) migrate across the surface. When they encounter other migrating atoms, they form 2- and 3-dimensional nuclei (clusters of several or more atoms) which grow, expand across the surface and create the coating. The flux density of the atoms (ions) reaching under the surface may be high or low.

When a high density flux of atoms (ions) reaches the cold substrate, many nuclei are formed on the substrate surface. These nuclei form centers of crystallization, expand, transform into fine grains, cover the substrate surface and form the coating. The quickest rate of expansion is exhibited by defected grains with unstable crystallographic orientations, in particular those perpendicular to the surface. The flux of atoms (ions) flows fastest around them and exhibits greatest intensity in reaching their surface which is parallel to that of the substrate. They grow quickly upward in the form of overturned pyramids or columns, at the same time expanding laterally, thereby retarding the growth of smaller grains [77].

When the flux reaching the substrate surface exhibits low density, the growth of the coating is slower and more laminar in nature.

The type of coating growth depends on the material and the geometry of the substrate, on the type of atoms or ions reaching it, on the deposition temperature, the energy of atoms or ions and on other material and technological parameters.

In the 1960s, efforts were undertaken to generalize the effect of deposition parameters of hard coatings by PVD techniques on their structure and properties. The melting point of the deposited metal T_m [K] was assumed as the basic and - to this day - main, although not the only - material-related parameter, while the temperature of deposition (substrate temperature) T was assumed as the main process parameter. In stricter terms, their ratio T/T_m is the value taken into consideration [120-125].

In 1968 W.A. Movchan and A.W. Demchishin, after investigating the metallographic structure of thick (up to 2 mm) coatings obtained by electron beam evaporation of Ni, Ti, W, Al_2O_3, ZrO_2 and deposition, proposed the first model of coating formation for vacuum deposition by evaporation. They distinguished three structural zones, dependent on the ratio of T/T_m (Fig. 6.19a) [120]. With $T/T_m < 0.3$ for metals and $T/T_m < 0.22$ to 0.26 for oxides, type I structure occurs (zone 1) in which fine crystallites, ending in spheroidal surfaces, dominate. With a rise of temperature, the diameter of this convex shape grows and the structure of the crystallites becomes columnar with pores between the crystallites. With $0.3 < T/T_m < 0.45$ to 0.5 for metals and oxides, type II structure occurs (zone 2), characterized by greater surface diffusion, greater columnar grains and significant surface asperities. This is the equilibrium structure of the material during volume crystallization. If $T/T_m > 0.5$, the dense type III structure occurs, similar to a crystallized structure (in zone 3), with coarse equiaxial grains [121]. Its hardness and strength, similarly to those of type II structure, correspond to solid material. This type of structure is desired in barrier layers [125]. In the case of traditional evaporation techniques, the only way of controlling properties of the deposited coating is a change of substrate temperature [14].

In thermally vapour deposited Bi coatings Sanders distinguished 5 structural zones [77].

In PAPVD techniques, by changing pressure and ion energy, it is possible to deposit coatings of predictable properties within a wide range of substrate temperatures and it should be considered as very significant that the substrate temperature may drop substantially [86].

In 1974 and during the ensuing years, the Movchan-Demchishin model was modified by J.A. Thornton [122] who applied it to cathode and magnetron sputtering and introduced a new parameter - pressure in the sputtering zone. With constant pressure of gas mixture, formation of type I structure is the result of weak surface diffusion of atoms; grains assume the character of fibers. Growth of surface diffusion, stimulated by a rise of substrate temperature causes the formation of a transition, type IV structure (T - Thornton zone), with fine, dense fiber crystallites, changing with a rise of temperature to columnar grains [121]. This structure features high strength and hardness, with low ductility, high surface smoothness and compressive stresses. It therefore has the most favorable physical and chemical properties. For the temperature range of T/T_m >0.5, Thornton's model is analogous to that proposed by Movchan and Demchishin. For the entire temperature range, a rise in gas pressure causes a shift in the ranges of occurrence of the particular types of structures I through IV in the direction of higher values of the T/T_m ratio (Fig. 6.19b). In techniques utilizing cathode and magnetron sputtering, Thornton's model is to this day universally used, on account of its simplicity and good agreement with industrial practice [121]. Lardon et al. distinguished six zones in the structure of coatings deposited by plating on a strongly polarized substrate [77].

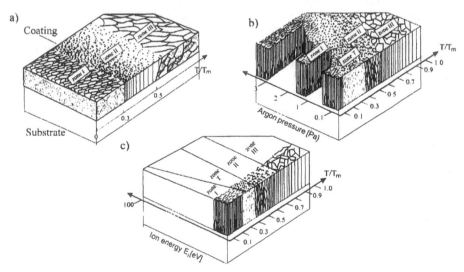

Fig. 6.19 Dependence of models of metallographic structures on technological conditions of coating deposition by PVD techniques: a) Movchan-Demchishin model for vacuum vapor deposition; b) Thornton model for cathode and magnetron sputtering; c) Messier model for ion beam deposition; *T* - substrate temperature; T_m - melting temperature of evaporated or sputtered material. (Fig. a - from *Movchan, V.A., and Demchischin, A.V.* [120]. With permission; Fig. b - from *Thornton, J.A.* [122]. With permission, from the *Annual Review of Materials Science*, Volume 7, © 1977, by Annual Reviews; Fig. c - from *Messier, R., et al.* [125]. With permission.)

For ion beam techniques of coating deposition R. Messier modified in 1984 Thornton's model, by replacing gas pressure in the sputtering zone by energy of ions reaching the substrate surface. This energy is equal to

$e(V_p\text{-}V_b)$, where V_p - plasma potential (positive of several V) and V_b - potential of substrate polarization (several tens to several hundred V) (Fig. 6.19c) [125]. In this model the same structures are distinguished as in Thornton's model. For the temperature ratio range of $0.2<T/T_m<0.5$ the mobility of condensing atoms and the range of existence of T-type structure are shifted in the direction of lower temperatures.

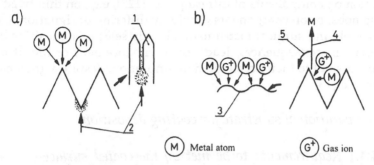

M Metal atom   G⁺ Gas ion

Fig. 6.20 Diagrams of formation of I and T structures in ion sputtering: a) without substrate polarization; b) with negative substrate polarization; *1* - Type I columnar structure; *2* - shaded zones; *3* - Type T intermediate structure; *4* - back sputtering; *5* - back scattering. (From *Zdanowski, J.* [13]. With permission.)

In ion deposition of coatings the structure and smoothness of the coating depend on whether the substrate is polarized or not. During ion sputtering, without polarization, the coating material is deposited mainly on visible roughness peaks and practically does not even reach screened cavities. Consequently, columnar crystallites grow on the peaks while between them remain unfilled gaps (Fig. 6.20a). During ion sputtering with substrate polarization, roughness peaks are sputtered strongest while their atoms, recoiling from collisions with gas atoms, fill roughness cavities, together with some atoms sputtered directly from side walls of these roughness cavities. As the result of atom diffusion into cavities the substrate surface is smoothed, and the structure type formed is T, not the columnar I. The formation of one structure type or of the other depends on whether shadowing or atom diffusion prevails. Limitation of formation of type I open structures and the formation of T-type structures with low values of T/T_m requires intensive ion bombardment in which secondary sputtering covers only 30 to 60% of the deposited material and that rather rough substrate surface [13].

In all PVD techniques of hard coating deposition, the structure, thickness and even stoichiometric composition [40-43] of deposited coatings are significantly influenced by the distance from vapour source (vapour deposition head) and by the distance of radiation from the source axis (perpendicular to the source surface) [123].

Moreover, in many deposition techniques it is possible to utilize specific possibilities of process control, e.g., assisted heating or cooling of substrate, counteracting the formation of droplets of coating material on the substrate [39], ion cleaning of substrate, metering of reactive gases, increasing or decreasing of intensity evaporation, sputtering, ionization, etc.

In all cases, coating deposition by PVD techniques requires high process precision on which, to an extent greater than in other techniques, depend deposition results. It is also of importance to select conditions individually for each load. It is inadmissible to carry out simultaneous deposition on components of differing size [127], e.g., on thin twist drills and big hobs, or on components of widely differing designation, e.g., on cutting tools and surgical instruments. Poorly selected deposition conditions may, in consequence, lead not to enhancement of life of coated components but to their accelerated deterioration, despite a good coating appearance [13].

6.4.4 Preparation of substrate for coating deposition

6.4.4.1 Requirements to be met by the coated surface
In all coating deposition techniques with, perhaps, one exception, i.e., pulsed plasma [40-44], good adhesion of the coating to the substrate and, hence, service properties depend on proper preparation of the substrate surface which, in turn, depends on the coated component and its designation.

In order for the coating to fulfill its task, the surface of the coated component should feature the following characteristics [127]:

– **Hardness**: obtained by heat treatment (e.g., hardening and tempering) or thermo-chemical treatment (e.g., nitriding, chromizing), less often by mechanical treatment.

– **Smoothness**: the surface should be smooth, ground or polished to $R_a < 0.8$ μm and with deburred edges.

– **Cleanliness**: The surface should be free of particles of mechanical contaminants (pollen, dust), organic contaminants (fats, greases, antirust protective media, sweat) and of products of chemical reactions (corrosion products, e.g., oxides and sulfides).

6.4.4.2 Initial cleaning
Initial cleaning is carried out outside of the working chamber of the vapor depositor. It consists of removing mechanical, organic and chemical contaminants from the surface of the components designated for coating. This initial cleaning may be accomplished by the following means:

– **Mechanical**: by removal of scale and permanent discoloration by glass-beading (only in exceptional cases), by rotary finishing and by abrasive techniques.

– **Chemical**: by removal of organic fats, through saponification, in acidic or alkaline baths, in organic solvents (e.g., in trichloroethylene or tertrachloroethylene) or in alkaline aqueous solutions.

– **Physical**: by removal of contaminations in cleaning baths through their dissolution or emulsification.

– **Physico-chemical**: by removal through detachment of the more stable contaminants in washing baths (e.g., chemical solvents) and alkaline aqueous solutions, assisted by ultrasonic vibration.

Not in all deposition techniques and not all components are cleaned by all techniques, as enumerated in the above sequence. In some techniques it is only possible to brush off dust and grease and to degrease chemically (usually in specially prepared baths and organic solvents). Degreasing in different baths is usually separated by single or double rinsing and a possible drying cycle. Quite often, the final stage is ultrasonic washing in trichloroethylene. Ultrasonic washing in freon baths has been eliminated, mainly on account of the detrimental effect of freon on the earth's atmosphere because freon causes depletion of the ozone layer.

Table 6.10 shows an example of the operation of cleaning twist drills of 100 mm diameter, practically used in the Institute for Terotechnology in Radom, Poland [127].

Table 6.10

Initial cleaning of twist drills, prior to coating deposition (From *Bujak, J., et al.* [127]. With permission.)

Type of operation	Washing medium	Time[min.]	Temperature[$^\circ$C]
Degreasing	1) $10g/dm^3$ NaOH $10g/dm^3$ $Na_3PO_4 \cdot 10H_2O$ $10g/dm^3$ Na_2CO_3 $5g/dm^3$ wetting agent or 2) $20g/dm^3$ NaOH $20g/dm^3$ $Na_3PO_4 \cdot 10H_2O$ 0.03 g/dm^3 oleic acid, oxyethylized by seven particles of ethyl oxide	31	5065
Rinsing I	distilled water	1.66 (100 s)	80-95
Rinsing II	distilled water	1.66 (100 s)	20
Drying	pure air	10	110
Rinsing III	trichloroethylene	6	80
Rinsing IV	trichloroethylene + ultrasonic	5	20
Rinsing V	trichloroethylene vapours	10	90
Drying	air	5	110

6.4.4.3 Final (ion) cleaning

Final cleaning is carried out in the vapour depositor's chamber, directly prior to deposition, and it has the following aims [127-129]:

– precise cleaning of component surface,
– activation of component surface,
– heating of component to desired temperature.

Final heating is accomplished by ion beam techniques and is termed **ion etching** or **ion beam etching**. In practice, the process of ion cleaning, followed by coating deposition, may be carried out without any interruption between the two operations during which residual gas atoms could condense on the substrate. This is possible because at low pressures employed in ion beam techniques of deposition, a monolayer of atoms condenses on a substrate that is not bombarded by ions within seconds [13].

During ion etching some structural components of the vapour depositor are also sputtered, especially load fixtures. For this reason, a condition of effective ion beam etching is limiting to a minimum vapour deposition on the substrate of structural materials (not coating material, e.g., titanium), and ensuring that the rate of sputtering is greater than the rate of condensation of residual gases. Meeting the second condition enables heating of the substrate (by ion bombardment or by additional heaters) which causes a rise of desorption of residual gas atoms adsorbed by the surface, as well as of ions of the working gas used to clean the surfaces which are trapped in the substrate [13].

A rise of substrate temperature causes an increase of the force of adhesion of coating to substrate. Overheating the substrate may, however, cause a loss of acquired mechanical properties of substrate, e.g., exceeding of tempering temperature causes a loss of hardness. For this reason, in the case of metallic loads, heating is carried out at a temperature approximately 50 K lower than that of tempering (150 to 350°C). In ion beam heating, temperature depends on shape, mass and size of load surface [127].

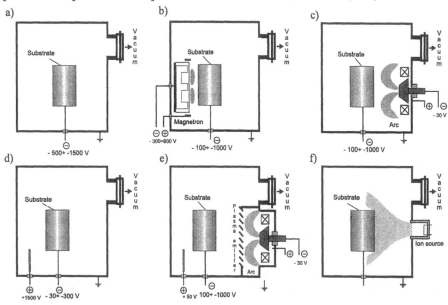

Fig. 6.21 Schematic diagrams of systems for ion etching techniques: a) with load as cathode; b) in metallic plasma magnetron; c) in metallic plasma, assisted by arc discharge; d) in triode system with hollow cathode; e) in triode system with arc source; f) by ion beam.

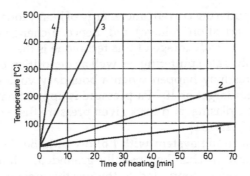

Fig. 6.22 Rate of load heating in different techniques of ion etching: *1* - traditional sputtering with load as cathode; *2* - in triode system with hollow cathode; *3* - in metallic plasma from magnetron discharge; *4* - in metallic plasma from arc discharge. (From *Bujak, J., et al.* [127]. With permission.)

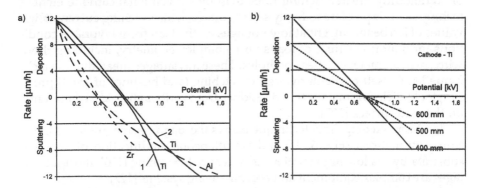

Fig. 6.23 Rate of deposition/sputtering in metallic plasma from arc discharge vs. potential: a) dependence on material of sputtered cathode; b) dependence on distance between load and cathode. (Fig. a - curve Zr and Ti (*1*) - from *Andreev, A.A., et al.* [130]. With permission; Curve Al and Ti (*2*) - from *Vyskocil, J., and Musil, J.* [131]. With permission from Elsevier Science; Fig. b - from [14]. With permission.)

Four techniques of ion beam etching are known and in use, of which two are utilized regularly while the remaining two rather seldom [128].

Ion etching with the load as cathode - *cathodic etching*. This technique, the oldest, simplest and most widely used, may be carried out in the majority of vapor depositors equipped with systems for substrate polarization (Fig. 6.21a) and with systems for rapid extinguishing arc microdischarges. Its disadvantages are low rate of etching (Fig. 6.22), low rate of heating and possibility of occurrence of arc microdischarges, on account of high voltage applied to the load [127].

Ion etching in metallic plasma. This technique consists of utilization of magnetron discharge (Fig. 6.21b) or/and arc discharge (Fig. 6.21c), with low target sputtering currents and with a negative bias of the load. The density of metallic plasma bombarding the load is independent of its size which allows precise control of sputtering current and energy of heavy ions of target metal (by a change of polarization voltage) and, hence, a change of rate of load heating (Fig. 6.22). A disadvantage of this technique is the possibility of occurrence of both substrate sputtering, as well as deposition of a metal layer on it (Fig. 6.23). This may happen when a boundary voltage is exceeded, which for a titanium target at 10^{-3} Pa pressure is approximately 700 V. For each type of metallic plasma and for a given pressure there exists a boundary etching voltage [129-131].

Ion etching in a triode system with hollow cathode. In this technique the cathode is a vacuum chamber, usually grounded, while the anode is the vacuum feed-through, connected to the positive pole of the high voltage generator. This feed-through takes up approximately 0.1% of total cathode surface. The triode system is constituted by: the load, the anode and the plasme source, e.g., hollow cathode (Fig. 6.21d) or arc source (Arc Enhanced Glow Discharge - AEGD method; Fig. 6.21e). This technique allows obtaining of significantly greater etching rates than by conventional cathode etching techniques. It is possible to freely select voltage, hence, etching current. High values of ion beam current at low voltages on the load form favorable conditions for uniform etching of loads with complex geometry, including those with small diameter cavities and holes. These advantages mean that the technique lends itself particularly well to etching (and heating) of small precision tools, dielectric materials and residues from various chemical compounds on the load surface [127].

Ion beam etching. This technique allows thorough cleaning of loads made from any material (Fig. 6.21f) and simultaneous modification of both the substrate, by shallow implantation, as well as of the coating [130]. Its disadvantages are complex equipment and accompanying systems [127].

6.5 Service characteristics of coatings deposited by PVD techniques

6.5.1 General

PVD techniques are characterized by the following features:
   – Possibility of application of raw materials in the form of pure metals and gases, in place of often harmful compounds.
   – Broad possibilities of choice of coating materials, thus broad range of properties of deposited coatings.
   – Relatively high deposition productivity with utilization of specialized vacuum vapor depositors.
   – Quite high deposition costs (high investment costs), although more than compensated by a severalfold increase in life of coated objects or by possibility of their achieving properties impossible to achieve by other techniques.
   – Necessity of maintaining high degree of cleanliness and strict adherence to procedures which call for high operator qualifications.

– Ecological friendliness of deposition processes (no harmful products of chemical reactions and need for their disposal).

Well-deposited coatings have one feature in common, i.e., in principle, they simultaneously embrace all the basic service characteristics, such as very good tribological and anti-corrosion properties, coupled with high decorative properties. Some coating materials additionally feature unique optical and dielectric properties.

It is essential to emphasize two special characteristics:

- the same coating materials deposited by different techniques usually do not feature the same coating characteristics;

- properties of coatings deposited by one technique, with the utilization of the same materials, do not have to be the same because their chemical composition may be different; they may form stoichiometric, substoichiometric and superstoichiometric compounds. For example, coatings of titanium nitride with general chemical formula of TiN_x may contain nitrogen within the range of 28 to 50% (atomic) which corresponds to the value of $0.42 \leq x \leq 1$ [13]. The chemical composition of the layer depends, of course, on the conditions of deposition, in stricter terms, on the proportion of reagents in the plane of the substrate, in accordance with the formulas:

$$Ti + N_2 = TiN + 0.5N_2 \qquad (6.1)$$

$$2Ti + N_2 = 2TiN \qquad (6.2)$$

$$3Ti + N_2 = Ti_2N + TiN \qquad (6.3)$$

$$4Ti + N_2 = Ti_2N + TiN + Ti \qquad (6.4)$$

This proportion depends on the rate of condensation of titanium on the substrate and on partial pressure of nitrogen (uniform in different zones of the working chamber), as well as on the degree of ionization which, similarly to degree of condensation, is not uniform in all zones of the working chamber [13].

Usually, the thickness of coatings does not exceed several microns (most often 2 to 5 μm) for monolayer coatings and 15 μm for multi-layer coatings [121]. In some cases, it may reach 100 μm, as in CrN coatings. The rate of deposition varies and in most cases it is within the range 3 to 18 μm/h. Table 6.11 shows properties of some coatings deposited by the reactive magnetron sputtering technique.

As opposed to CVD techniques, in the development of PVD techniques a striving is observed for departure from low-temperature processes and for reaching into the higher temperature range, up to a substrate temperature of 600°C, in an effort to activate the reaction of compound formation and to enhance adhesion of coating to substrate through a partial diffusion-type connection. In the future, new modifications of techniques

Table 6.11
Characteristics of some coatings deposited by the reactive magnetron sputtering technique
(From *Müntz, W.D.* [132]. With permission.)

Coating	Rate of deposition ×10⁻³ [μm/s]	Deposition temperature [°C]	Microhardness* [HV0.01]	Maximum operating temperature[°C]	Critical load**(adhesion) [N]
TiN	3.5	400–500	2500–2800	550	35–75
TiC	4	450–550	2800–3700	400	
Ti(CₓNᵧ)	4	350–400	3500–4500	450	
(Ti,Al)N(50:50)	4	400–500	2200–2300	700	45–60
Cr	8	100–150	500–700	600	80
CrN	4	200–350	2000–2200	600	35
Cr/CrN	4	200–350	2000–2200	650	50
Me-C	4	150–200	1000–2000	300	20–40
WN	3	280–300	2500–4000	400	20–55

* Coating thickness: 3–4 μm; substrate: high speed steel
** Depending on hardness of substrate

may be expected, all with the aim of improving adhesion of coating to substrate, enhancement of uniformity and rate of deposition, as well as broad application of complex multi-component and multi-layer coatings, enrichment of assortment of deposited compounds and broadening of application possibilities [3, 126].

6.5.2 Decorative characteristics

Unique decorative characteristics, such as shiny or matte surface, as well as a broad spectrum of colors possible to obtain by using various coating materials (Table 6.12), coupled with good wear properties, renders PVD coatings suitable to replace traditional electrodeposited decorative coatings, especially gold and silver. For example, stoichiometric titanium nitride has a golden color and strongly resembles gold. Deposited on a smooth substrate it has an attractive luster and, with its high hardness and low wear, constitutes a coating that is more durable than gold. For this reason, the TiN titanium nitride has been used for quite some time as a good and inexpensive decorative coating for artificial jewelry, wrist watch cases, small metallic objects, like elements of pens, ball-pens, calculators, cutlery, bathroom appliances, dishes (including glass and ceramic), cigarette lighters, spectacle frames, theater glasses, musical instruments, etc.

Table 6.12
Colors of thin, hard coatings (From different sources.)

Coating	Color	Hardness [HV0.015]
TiN	yellow-golden (from bright yellow, through red-yellow to bright brown) to gray-silver	2300-2700
ZrN	pale golden	2300-3200
HfN	yellow-green	2700-3100
Ti(C_xN_{1-x}) (for x = 0.05–50)	reddish-golden-brown	2450-2900
Zr (C_xN1-x) (for x = 0.05–20)	golden	3250-3450
Zr (C_xN_{1-x}) (for x > 0.9)	silver	3300-3600
(Ti_xAl_{1-x})N (for x = 0.1–70)	golden-brown-black	2400-2900
(Ti_xZr_{1-x})N (for x = 20–80)	golden	2400-2900
Cr(C,N)	silver	1500-2000
TaN	yellow-silver metallic	1700-2100
SiC	from gray to yellow	2800-3200

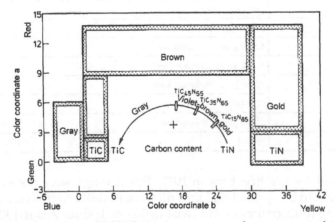

Fig. 6.24 Colors of coatings of Ti(C,N) titanium compounds vs. content of carbon and nitrogen. (From *Knotek, O., et al.* [134]. With permission.)

By utilizing compounds of titanium with carbon (carbides) and with nitrogen (nitrides), it is possible to obtain a quadruple spectrum of colors, gently blending one into another (Fig. 6.24).

6.5.3 Tribological properties

6.5.3.1 Coating of tooling and machine parts
Good tribological properties of PVD coatings (Table 6.13), coupled with low thermal conductivity (see Tables 6.5 to 6.7) and resistance to temperature effects (see Table 6.11), are the basic factors which caused that these coatings

have come to be used for anti-wear purposes, deposited primarily on cutting tools, such as lathe tools, twist drills, milling cutters and sintered carbide inserts. Moreover, PVD coatings are used on cold forming tools, such as blanking punches, dies and stamps, on injection molds, as well as on precision components, bearing races, etc. Especially good effects are obtained by coating accurate tooling, properly machined, heat and diffusion treated.

Table 6.13
Tribological properties of different rubbing pairs
(From *Müntz, W.D.* [132]. With permission.)

	Rubbing pair	Coefficient of friction μ
High speed steels	52100 grade chromium steel	0.73
	stainless steel	0.75
	carburized steel	0.75
TiN	steel	0.06–0.16
	52100 grade chromium steel	0.44
(TiAl)N	52100 grade chromium steel	0.58
(TiAl6V4)CN	carburized steel	0.12
(TiAl6V4) N	52100 grade chromium steel	0.44
WN	52100 grade chromium steel	0.85
TiC	52100 grade chromium steel	0.20
W:C	52100 grade chromium steel	0.10
	(Ti,Al)N	0.10
CrN	52100 grade chromium steel	0.2–0.3
MeC	W:C-H	0.05–0.15

Utilized for the first time in 1977, TiN coatings were the earliest PVD application to be applied and are still most popular to this day as coatings for cutting tools (reamers and twist drills) and, since about 1980, for cold forming tools [90]. These thin, hard layers feature a coefficient of friction on steel approximately twice lower than that of high speed steel on steel in the same combination. They protect tools against overheating and oxidation and reduce the tendency to cold weld the cut material to the cutting edge. Moreover, they allow a reduction of the contact force in cutting operations by 20 to 30%, of temperature in the contact zone by 90 to 120 K and of the force of friction by 25 to 40%, while tool life extension is in the range of 100 to 700%. Usually, cutting speed can be raised by 90%, down-time is reduced by 50%, electrical energy consumption is reduced by 60 to 80% and the tool work cycle between resharpening operations can be extended by a factor of 4 to 5. The cost of tools with a TiN coating is higher by 50 to 100% but life extension by 2 to 3 times is economic justification for using PVD coatings. The high cost of equipment (ca. $500,000) be amortized in cutting tool application within 7 to 9 months, assuming continuous one-shift-per-day operation [121].

Different types of tooling are often coated by layers of: TiN, TiC, WC, Cr_3C_2, CrN, Ti(C,N), (Ti,Al)N, (Ti,Zr)N, TiC/TiN, TiC/TiB$_2$ and TiC/Ti(C,N)/TiN [94].

Performance of such tools, coated with an anti-wear layer, depends primarily on coating material, type of tool and its material, material being machined, as well as type and conditions of the machining operation. Since all these values vary within certain limits, their appropriate selection can and should be carried out empirically, based on results of laboratory testing. It should be, however, taken into consideration that the results of such tests are usually better than those of industrial applications [14].

6.5.3.2 Tool performance

Effect of material on tool life. A lower coefficient of friction of the coating on steel, conducive to lesser heat being dissipated in the zone of contact of tool with machined material, hence a lowering of cutting edge temperature, and the fulfillment by the layer of the role of a heat barrier, cause a reduction of cutting forces (Fig. 6.25), improve tool working conditions and result in an extension of tool life, in comparison with that of uncoated tools (Fig. 6.26) [135-145].

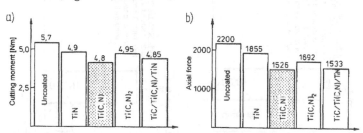

Fig. 6.25 Dependence of drilling conditions in steel drilling by NWKa 10 mm dia. twist drills on type of coating: a) cutting moment; b) axial force. (From *Smolik, J., et al.* [138]. With permission.)

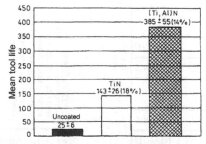

Fig. 6.26 Comparison of results of drilling by twist drills of 6 mm diameter, made of high speed steel, uncoated and coated by TiN and (Ti,Al)N. Drilling depth: 15 mm. Drilling speed: 20 m/min, feed rate: 0.2 mm/rev. Drilled material: steel (hardness: 243 HRC). (From *Münz, W.D.* [132]. With permission.)

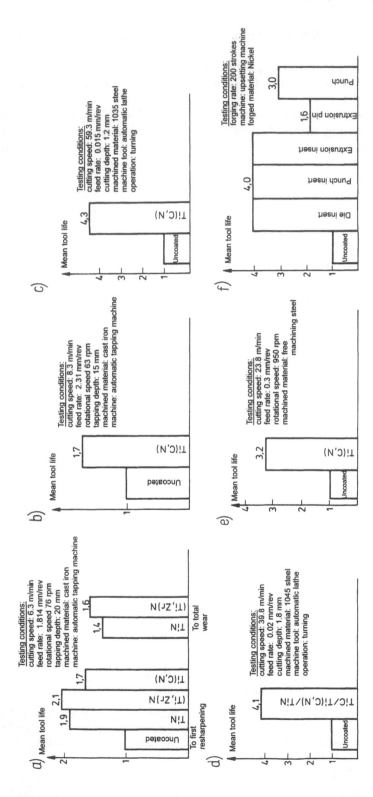

Fig. 6.27 Mean increase of tool life of tools made from high speed steel, coated by anti-wear coating: a) thread taps; b) thread milling cutters; c,d) cutting tools; e) reamers; f) forming tools: inserts, punches, dies, sintered carbide pins. (From *Smolik, J., et al.* [138]. With permission.)

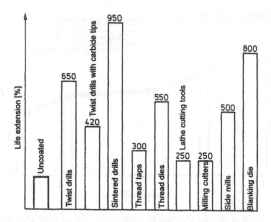

Fig. 6:28 Life of various tools coated by TiN. (From [145]. With permission.)

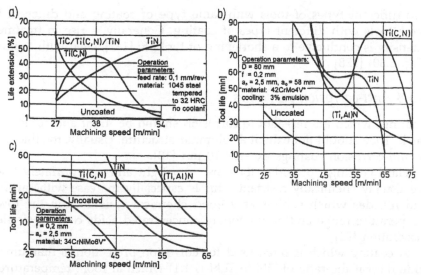

Fig. 6.29 Increase in life of tools coated by different types of anti-wear coatings vs. cutting speed: a) twist drills; b) end mills; c) cutters. (Fig. a - from *Smolik, J., et al.* [138]; Fig. b and c - from *König, W., and Koch, K.F.* [146]. With permission.)

Life extension of different types of tooling. The average life of tooling with hard coatings is 2 to 5 times longer than that of uncoated tools. In particular cases, life extension may even exceed 10 times and more. For tools used in threading operations in industrial conditions this extension is 2-fold; in lathe operations, reaming and extrusion it may be 4 to 5 fold (Fig. 6.27) [138], while in stamping operations it is even higher (Fig. 6.28).

Effect of type of treatment and its parameters on tool life. In all types of cutting operations there is an effect of parameters on the life of tools (Fig. 6.29). For each type of tool and each type of coating there is an optimum intensity of treatment (cutting speed, depth, feed rate), usually greater than the intensity of treatment by an uncoated tool, different for

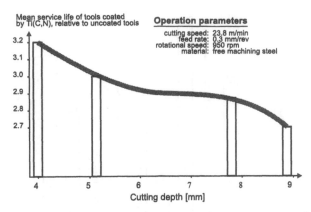

Fig. 6.30 Effect of cutting depth during reaming on life of reamers with Ti(C,N) coating. (From *Smolik, J., et al.* [138]. With permission.)

two different types of tools with same type of coating and different for same tools with different types of coating. Generally, a rise in treatment intensity is conducive to a shortening of tool life [147], even with coatings (Fig. 6.30) [138].

6.5.4 Anti-corrosion properties

Coatings which are resistant to corrosion should be tight and sufficiently thick. They should not exhibit a columnar structure. Usually, the thickness of anti-corrosion coatings exceeds 10 μm [13].

Hard coatings, provided they are very tight and appropriately thick, are usually chemically resistant. This is especially the case with carbides and nitrides which exhibit very good chemical resistance in the lower temperature range and only strongly concentrated acids cause their slow dissolution [92].

A coating which is often used for corrosion protection is titanium nitride (best if the ratio of TiN to Ti$_2$N is 1:1) which at room temperature is chemically inactive and features small chemical affinity to materials with which it mates in service. It is resistant to acids and alkalis, it reacts weakly with concentrated acids and strong oxidizing agents and the more amorphous its structure, the more resistance it exhibits. Oxidation of coatings in dry air begins near the lower limit of glow temperatures, i.e., approximately 650°C, causing a gradual change off coloring from golden to brown after cooling [148]. The (Ti,Al)N coating appears to have even better anti-corrosion properties than TiN.

The titanium nitride as a material is neutral with respect to the human organism, featuring total corrosion resistance in human saliva and does not exhibit any cytotoxic effects in body fluids. Thanks to these properties it is used for coating working surfaces of surgical and dentistry instruments, as well as of dentures. The thickness of coatings utilized in medicine is small - usually about 1 μm [149].

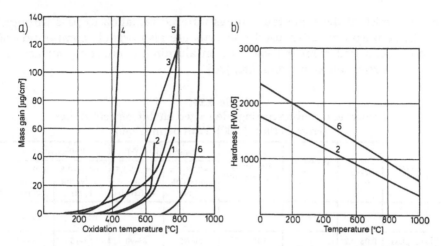

Fig. 6.31 Effect of temperature of hard coating on stainless steel on: a) resistance to oxidation (specimens were held at all temperatures for 1 h); b) hardness drop; 1 - stainless steel; 2 - TiN; 3 - Ti(C$_{0.3}$N$_{0.7}$); 4 - TiC; 5 - (Ti$_{0.75}$Al$_{0.25}$Cr$_{0.25}$)N; 6 - (Ti$_{0.5}$Al$_{0.5}$)N. (Fig. a - from *Münz, W.D.* [132].)

Corrosion resistance at elevated temperatures of the majority of hard coatings is also good - decidedly better than that of carbon and low alloy structural steels. Resistance to high temperature oxidation of carbide coatings is lower than that of stainless steels, while that of coatings containing aluminum is higher (Fig. 6.31a). The hardness of coatings decreases with a rise of temperature (Fig. 6.31b) [150]. Generally, it can be stated that titanium nitride and carbonitride coatings are resistant to corrosion in an oxygen-bearing atmosphere up to 1000°C. For service at higher temperatures and in molten salts, Si$_3$N$_4$ and (Si,Al)N coatings are used.

Hard coatings also exhibit good resistance to erosion at elevated temperatures. For this reason, they are used on steam, gas and water turbine blades and on nozzles of rocket engines. These are mainly coatings like TiC, TiN, TaC, Al$_2$O$_3$ and, more recently, CoCrAl and NiCoCrAlY [151], as well as CoCrAlY, NiCrAlY, CoNiCrAlYHfSi and CoCrAlYSi. Hard coatings are also deposited on sliding surfaces of highly stressed bush bearings, on walls of atomic reactors, etc.

6.5.5 Optical and electrical properties

Very good properties, like hardness comparable with that of diamond, very good optical transparency and very good resistivity are featured by diamond-like amorphous carbon films, plasma deposited from hydrocarbons (Table 6.14). They are used for coating machining tools and tools for cutting non-metals (such as laminations of plastics and wood), aluminum and non-ferrous metals to extend their life. Furthermore, such coatings are used on optical lenses in order to increase their chemical, mechanical and thermal resistance, on microchips, as insulation and mechanical protection which at the same time allows infrared radiation to pass through. For these purposes, often used materials are also organic compounds of silicon, deposited from the vacuum on aluminum reflecting surfaces in autos or street lamps, mirrors, etc. [30].

Table 6.14
Properties of carbon coatings deposited on different substrates
(From *Büttner, J.* [30]. With permission.)

Substrate material	Coating thickness [μm]	Microhardness HV0.5N	Resistivity [μW×cm×10^2]	Transmissivity [%]		Reflectivity [%]	
				λ=0.38–0.8 μm (visible light)	λ=0.82–2.5 μm (infrared)	λ=0.38–0.8 μm (visible light)	λ=0.82–2.5 μm (infrared)
Plates of Si, GaAs	1	25–30	10^5–10^{12}	-	-	5–30	2–40
Quartz glass	0.05–0.2	-	10^4–10^{10}	50–85	85–90	14–15	6–14
Optical glass	0.05–0.2	-	10^4–10^9	40–80	80–85	4–16	6–12
Optical glass with Cr coating	0.07–0.4	-	10^4–10^{10}	0.2–0.3	0.2–1.7	2–15	15–40
Hard alloy HG10*	1.5	40–50	10^4–10^{11}	-	-	-	-
M2 grade steel	0.1–0.5	15–25	10^4–10^9	-	-	-	-

Moreover, the microelectronics industry uses semiconductor epitaxial layers, dielectric layers, diffusion barriers, anti-wear coatings, e.g., SiC, Si_3N_4, SiO_2, BN, AlN [94].

Coatings deposited by PVD techniques onto optical glass, transparent and organic mineral materials, especially multi-layer coatings, serve the role of anti-glare protection, filtering (permitting passage of radiation of certain wavelengths only), coloring (usually decorative or reflecting and absorbing radiation within certain wavelengths) and interferential surfaces. The most common application is for anti-reflection protection for optical glass (single lenses and lens systems, prisms, head plates of measuring instruments, etc.) because they reduce the reflectivity of visible radiation to 0.05%. In normal optical glass this reflectivity is 1 to 9. Anti-reflection coatings (e.g., weakly reflecting red and green light and better reflecting blue light) are deposited on auto rear-view mirrors to protect against night blinding. PVD techniques are also employed to deposit interference filters. By vapor deposition in vacuum hard anti-glare coatings are generated. They may feature different colors, usually from yellow-brown to brown-gray) for eyeglasses, with controlled absorption of solar radiation usually within 10 to 50%. Similar techniques are used for coatings deposited on auto windshields and in the building construction industry, for household foils and reflecting elements, as well as for coatings which do not transmit (or weakly transmit) visible radiation in one direction - (unidirectional vision) [30] and even vary their translucency with changes of illumination intensity. Optical properties of glasses, predominantly the coefficient of refraction, may be changes by the implantation of ions, e.g., of B, including implantation combined with other coatings deposited by PVD techniques.

References

1. **Burakowski, T., Miernik, K., and Walkowicz, J.**: Manufacturing techniques of thin tribological coatings with utilization of plasma (in Polish). *Metaloznawstwo, Obróbka Cieplna, Inżynieria Powierzchni* (*Metallurgy, Heat Treatment, Surface Engineering*), No. 124-126, 1993, pp. 16-25.
2. **Burakowski, T.**: *Development trends in heat treatment* (in Polish). Proc.: *Meeting of Section of Technology Fundamentals*, Polish Academy of Sciences, 17 March 1988, publ. IMP, Warsaw, 1989.
3. **Burakowski, T., Roliński, E., and Wierzchoń, T.**: *Metal surface engineering* (in Polish). Warsaw University of Technology Publications, Warsaw 1992.
4. **Burakowski, T.**: Superficial layer formation - metal surface engineering (in Polish). *Metaloznawstwo, Obróbka Cieplna, Inżynieria Powierzchni* (*Metallurgy, Heat Treatment, Surface Engineering*), No. 106-108, 1990, pp. 2-18.
5. **Frey, H., and Kienel, G.**: *Dünschicht Technologie*. VDI Verlag, Düsseldorf 1987.
6. **Michalski, A.**: PVD techniques used to deposit layers of hard and refractory materials for cutting tools (in Polish). *Metaloznawsto, Obróbka Cieplna* (*Metallury, Heat Treatment*), No. 79, 1986, pp. 18-23.
7. **Sokołowska, A.**: *Non-conventional means of material synthesis* (in Polish). PWN, Warsaw 1991.
8. **Bunshah, R.F., and Deshpandey, C.V.**: Plasma assisted vapour deposition process: A review. *Journal of Vacuum Science Technology*, A3(3), 1985, pp. 553-560.
9. **Zega, B.**: The physical vapour deposition of hard coatings - A complement to thermal treatments of tools. Proc.: *Justom 83*, 1983, Novi Sad, pp. 393-410.
10. **Kloc, R.**: *Methods of formation of thin metallic layers* (in Polish). PWN, Warsaw 1974, pp. 33-39.
11. **Kienel, G.**: PVD-Verfahren und ihre Anwendung zur Herstellung verschleisshemmender Schichten. Proc.: *37 Härterei-Kolloquium*, Wiesbaden 1981.
12. **Burakowski, T.**: Methods of manufacture of superficial layers - metal surface engineering (in Polish). Proc.: Conference on *Methods of Manufacture of Superficial Layers*, Rzeszów (Poland), 9-10 June 1988, pp. 5-27
13. **Zdanowski, J.**: Ion techniques of deposition of titanium nitride layers as hardening, anti-corrosive and decorative coatings (in Polish). *Elektronika*, No. 2, 1988, pp. 3-8.
14. Research project No. 71232 91C: *Technologies of multicomponent and multilayer, anti-wear coatings deposition with the use of PAPVD methods* (in Polish). Institute for Terotechnology, Radom 1994.
15. **Glang, R.**: *Vacuum evaporation*. Handbook of thin film technology. Ed. L.I. Massel, R. Glang, McGraw-Hill Book Co., New York 1970.
16. **Mattox, D.M.**: Recent advances in ion plating. Proc.: *6th International Vacuum Congress*, Japan Journal of Applied Physics, Suppl. 2(1), 1974, pp. 443-445.

17. **Bunshah, R.F.**: Processes of ARE type and their tribological applications. *Thin Solid Films*, No. 107, 1983, pp. 21-24.
18. **Bunshah, R.F., and Raghuram, A.C.**: Activated reactive evaporation process for high rate deposition of compounds. *Journal of Vacuum Science Technology*, No. 9(6), 1972, pp. 1385-1387.
19. **Berghaus, B.**: Deutsches Patent DPR 668 639, 1932.
20. **Auwärter, M.**: US patent No. 2,920,002, 1952.
21. **Randhawa, H.**: Review of plasma-assisted deposition processes. *Thin Solid Films*, No. 196, 1971, pp. 329-349.
22. **Roliński, E.**: Contemporary surface engineering (in Polish). *Przegląd Mechaniczny (Mechanical Review)*, No. 8, 1987, pp. 5-6.
23. **Fabian, D., Mey, E., Ebersbach, G., and Schurer, Ch.**: Formation of superficial layers on tooling and machine components (in Polish). *Przegąd Mechaniczny (Mechanical Review)*, No. 22, 1984, pp. 19-22.
24. **Kuo, Y.S., Bunshah, R.F., and Okrent, D.**: Hot hollow cathode and its applications in vacuum coating: a concise review. *Journal of Vacuum Science Technology*, A, No. 4(3), 1986, pp. 397-402.
25. **Williams, G.**: Vacuum coating with hollow cathode source. *Journal of Vacuum Science Technology*, No. 11(1), 1974, pp. 364-376.
26. **Yamada, I.**: Ionized cluster beam deposition techniques. *Handbook of plasma processing technology*. Ed. S.M. Rossnagel, J.J. Cuomo, W.D. Westwood, Noyes Publ., Park Ridge, New York 1989.
27. **Yamada, I., Takagi, T., and Younger, P.**: Ionized cluster beam deposition. *Handbook of thin film deposition processes and techniques*. Ed. K.K. Schuegraf, Noyes Publ., Park Ridge, New York 1988.
28. **Drwięga, M., and Lipińska, E.**: Application of ion beams to modification of solid superficial layers, with focus on the Ion Beam Assisted Deposition technique (in Polish). Report No. 1583/AP by the Institute of Nuclear Physics, Cracow 1992.
29. **Moll, E., and Bergmann, E.**: Hard coatings by plasma assisted PVD technologies: industrial practice. *Surface and Coatings Technology*, Vol. 37, 1989, p. 483-486.
30. **Büttner, J.**: Modern methods of layer deposition in vacuum (in Polish). *Elektronika*, No. 7-8, 1986, pp. 45-47.
31. **Sobański, J.**: Modern techniques of thin layer deposition (in Polish). Proc.: *First Conference on Electron Technologies*, Wroclaw (Poland) 1982, pp. 503-507.
32. **Hill, J.R.**: *Physical vapour depositions*. Airco Temescal, Berkeley 1976.
33. **Günter, K.G., Feller, H., Hintermann, H.E., and König, W.**: Advanced coatings by vapour phase processes. *Annals of the CIRP*, Vol. 38, No. 2, 1989, pp. 645-648.
34. **Bunshah, R.F., and Deshpandley, C.V.**: Plasma assisted PVD processes. A review. *Journal of Vacuum Science Technology*, A, No. 3(3), 1985, pp. 553-560.
35. **Bunshah, R.F.**: *Deposition technologies for films and coatings*. Noyes Publ., Park Ridge, New York 1982.
36. **Randhawa, H., and Johnson, P.C.**: Technical note: A review of cathodic arc plasma deposition processes and their applications. *Surface and Coatings Technology*, 31, 1987, pp. 303-318.
37. **Randhawa, H.**: Cathodic arc plasma deposition technology. Proc.: *7th International Conference on Thin Films*, New Delhi, 1987; *Thin Solid Films*, 167, 1988, pp. 175-177.
38. **Labunov, V.A., and Resse, G.**: Ion beam sources for surface treatment of hard solids and obtaining of thin films (in Russian). *Zarubezhnaya Elektronnaya Tekhnika*, No. 1, 1983, pp. 3-5.
39. **Betiuk, M.**: PVD-Arc - process control and layer structure (in Polish). Proc.: Conference on *Modern Techniques in Surface Engineering*, Łódź-Spała (Poland), 22-23 Sept. 1994.

40. **Sokołowski, M., Sokołowska, A., Michalski, A., Gokieli, B., Romanowski, Z., and Rusek, A.**: Crystallization from reactive pulse plasma. *Journal of Crystal Growth*, Vol. 42, 1977, pp. 507-509.

41. **Sokołowski, M., Sokołowska, A., Rusek, A., Michalski, A., and Glijer, J.**: Reactive electro-erosion of metals under pulse-electric discharge. *Journal of Materials Science*, Vol. 14, 1979, pp. 841-842.

42. **Michalski, A., Zdunek, K., Sokołowska, A., and Olszyna, A.**: Pulse-plasma technique of deposition of thin TiN layers on tooling at temperatures below 500 K (in Polish). *Przegląd Mechaniczny (Mechanical Review)*, No. 15, 1991, pp. 7-10.

43. **Sokołowska, A., Olszyna, A., Michalski, A., and Zdunek, K.**: Diamond layers deposited from impulse plasma. *Surface and Coatings Technology*, Vol. 47, 1991, pp. 144-147.

44. **Zdunek, K.**: Mechanism of crystallization of multicomponent metallic coatings using impulse plasma method. *Journal of Materials Science*, Vol. 26, 1991, pp. 4433-4434.

45. **Walkowicz, J., Miernik, K., and Mężyk, K.**: Polish Patent No. 149083, 1988.

46. **Walkowicz, J., Miernik, K., Celiński, Z., and Smolik, J.**: Polish Patent No. 152210, 1989.

47. **Zdunek, K.**: Crystallization of metallic coatings obtained by pulsed plasma (in Polish). *Transactions of Warsaw Technical University*, Vol. 149, 1991.

48. **Scheibe, H.J., Pompe, W., Siemroth, P., Buecken, B., Schultze, D., and Wilberg, R.**: Preparation of multi-layered film structures by laser arcs. *Thin Solid Films*, Vol. 193/194, 1990, pp. 788-792.

49. **Matsunawa, A., Katayama, S., Miyazawa, H., Hiramoto, S., Oka, K., and Phmine, M.**: Basic study on laser physical vapour deposition of ceramics. *Surface and Coatings Technology*, No. 43/44, 1990, pp. 176-184.

50. **Chrisey, D.B., and Hubler G.K. (ed.)**: *Pulsed laser deposition of thin films*. J. Wiley and Sons, Inc., New York, Chichester, Brisbane, Toronto, Singapore 1994.

51. **Chapman, B., and Mangano, S.**: Introduction to sputtering. *Handbook of thin film deposition processes and techniques*. Ed. K.K. Schuegraf, Noyes Publ., Park Ridge, New York 1988.

52. **Patz, U.**: The concept and application of equipment for big surface layering in electronics (in Polish). *Elektronika*, XXVI, 1985, pp. 3-6.

53. **Logan, J.S.**: RF diode sputter etching and deposition. *Handbook of plasma processing technology*. Ed. S.M. Rossnagel, J.J. Cuomo, W.D. Westwood, Noyes Publ., Park Ridge, New York 1989.

54. **Orlinov, V.**: Modern electron and ion technological methods for deposition of thin films. Proc.: *1st Polish National Autumn School*, Szczyrk (Poland), 1979, II-4.

55. **Matthews, A., and Teer, D.G.**: Deposition of Ti-N compounds by thermionically assisted triode reactive ion plating. *Thin Solid Films*, No. 72, 1980, pp. 541-549.

56. **Horwitz, C.M.**: Hollow cathode etching and deposition. *Handbook of plasma processing technology*. Ed. S.M. Rossnagel, J.J. Cuomo, W.D. Westwood, Noyes Publ., Park Ridge, New York 1989.

57. **Asmussen, J.**: Electron cyclotron resonance microwave discharges for etching and thin film deposition. *Handbook of plasma processing technology*. Ed. S.M. Rossnagel, J.J. Cuomo, W.D. Westwood, Noyes Publ., Park Ridge, New York 1989.

58. **Matsuo, S.**: Microwave electron cyclotron resonance plasma chemical vapour deposition. *Handbook of thin film deposition processes and techniques*. Ed. K.K. Schuegraf, Noyes Publ., Park Ridge, New York 1988.

59. **Penning, F.M.**: Die Glimmentladung bei niedrigen Druck zwischen Koaxialen Zylindern in einem axialen Magnetfeld. *Physica*, III, No. 9, 1936, pp. 873-875.

60. **Latham, R., King, A.H., and Ruchforth, L.**: *The magnetron*. Chapman and Hall. London 1952.

61. **Waits, R.K.**: The planar magnetron sputtering. *Journal of Vacuum Science and Technology*, Vol. 15(2), March/April 1978, pp. 179-184.

62. **Chapin, J.S.**: USA Patent No. 438482, 1974.

63. **Chapin, J.S.**: The planar magnetron sputtering sources. *Research and Development*, Vol. 25, No. 1, Jan, 1974, pp. 37-40.

64. **Thornton, J.A., and Penfold, A.S.**: Cylindrical magnetron sputtering. *Thin film processes*. Ed. J.L. Vessen and Kern, Academic Press, New York 1978.

65. **Thornton, J.A.**: Magnetron sputtering: basic physics applications to cylindrical magnetrons. *Journal of Vacuum Science and Technology*, Vol. 5, No. 12, 1978, pp. 171-176.

66. **Thornton, J.A.**: Recent advances in sputter deposition. *Surface Engineering*, Vol. 2, No. 4, 1986, pp. 283-287.

67. **Thornton, J.A.**: Plasma-assisted deposition processes: theory, mechanisms and applications. *Thin Solid Films*, Vol. 107, No. 1, 1984, pp. 17-21. Proc.: *International Conference on Metal Coatings*, San Diego, 1983, pp. 372-376.

68. **Thornton, J.A.**: End-effects in cylindrical magnetron sputtering sources. *Journal of Vacuum Science and Technology*, Vol. 16, No. 1, 1979, pp. 79-85.

69. **Class, W.**: Magnetron deposition of conductor metallization. *Solid State Technology*, No. 6, 1983, pp. 103-108.

70. **Class, W.H.**: Performance characteristics of a new high rate magnetron sputtering system. *Thin Solid Films*, Vol. 107, No. 4, 1983, pp. 67-74.

71. **Fraser, D.B., and Cook, H.D.**: Film deposition with the sputter gun. *Journal of Vacuum Science and Technology*, Vol. 14, No. 1, 1977, pp. 147-152.

72. **Leja, E., Horodyński, T., and Budzyńska, K.**: Magnetron technique of thin layer deposition (in Polish). *Elektronika*, No. 9, 1982, pp. 5-6.

73. **Rossnagel, S.M.**: Magnetron plasma deposition processes. *Handbook of plasma processing technology*. Ed. S.M. Rossnagel, J.J. Cuomo, W.D. Westwood, Noyes Publ., Park Ridge, New York 1990.

74. **Danilin, B.S., and Sirchin, V.K.**: *Magnetron sputtering system* (in Russian). Publ. Radio i Sviaz, Moscow, 1982.

75. **Thornton, J.A.**: Coating deposition by sputtering. *Technologies for thin films and coatings*, Ed. R.F. Bunshah, Noyes Publ., Park Ridge, New York 1982.

76. **Posadowski, W.**: The WMT-100 industrial-laboratory magnetron sputtering system (in Polish). *Elektronika*, No. 6, 1989, pp. 33-34.

77. **Miernik, K.**: Operation and design of magnetron sputtering equipment (in Polish). Publ. Institute for Terotechnology, Radom (Poland) 1997.

78. **Zdanowski, J.**: *Ion etching of solid surface and its applications* (in Polish). Publ. Wrocław Technical University, Wrocław (Poland) 1976.

79. **Chapman, B., and Magnano, S.**: Introduction to sputtering. *Handbook of thin film deposition processes and techniques*. Ed. K.K. Schuegraf, Noyes Publ., Park Ridge, New York 1989.

80. **Zdanowski, J.**: Ion technology (in Polish). Proc.: *First Conference on Electron Technologies*, Wrocław (Poland) 1982, pp. 159-162.

81. **Mattox, D.M.**: Film deposition using accelerated ions. *Electrochemical Technology*, Vol. 2, No. 9-10, 1964, pp. 295-298.

82. **Franks, J.**: Ion beam technology applied to thin films deposition. *Thin Solid Films*, Vol. 86, 1981, pp. 219-224.

83. **Gaytherin, G., and Weissmantel, Ch.**: Some trends in preparing film structures by ion beam methods. *Thin Solid Films*, Vol. 50, 1978, pp. 135-140.

84. **Balikoiev, I., Barchenko, A., Zagranichni, S., Miernik, K., and Celiński, Z.**: On the use of ion mixing for improvement of the microhardness of industrial steels. *Proceedings of International Conference on Ion Implantation.* Ion Implement Elenite 1990, pp. 110-115.

85. **Robertson, D.D.**: Advances in ion beam mixing. *Solid State Technology*, Dec. 1978, pp. 57-60.

86. **Michalski, A., and Sokołowska, A.**: Pulsed plasma method of depositing hard layers (in Polish). Proc.: Conference on *Techniques of Formation of Superficial Metallic Layers*, Rzeszów (Poland), June 1988, pp. 94-98.

87. Commercial brochures by: Balzers, Multi-Arc, Vac-Tec, Ulvac, Tecvac, Hoch-Vakuum, Leybold Heraeus, NPO *Avtoprompokritye*, Dowty, Abar, Plasma und Vakuum Technik.

88. **Broszeit, E., and Gabriel, H.M.**: Beschichten nach den PVD Verfahren. *Zeitschrift für Werkstofftechnik*, No. 11, 1980, pp. 31-40.

89. **Bujak, J., Miernik, K., Smolik, J., and Walkowicz, J.**: Properties of materials used for hard coatings (in Polish) Proc.: VII Symposium on *Utilization of Technical Equipment*, Radom (Poland), 1993; *Tribologia* No. 4/5, 1993, pp. 77-83.

90. **Hebda-Dutkiewicz, E.**: Hard layers deposited by PVD methods (in Polish). Center for Utilization of Equipment, Radom (Poland), 1990.

91. **Holleck, H.**: Designing advanced coatings for wear protection. *Surface Engineering*, Vol. 7, No. 2, 1991, pp. 137-144.

92. **Toth, L.E.**: *Transition metal carbides and nitrides.* Academic Press, New York 1991.

93. **Holleck, H.**: Material selection for hard coatings. *Journal of Vacuum Science and Technology*, A 4(6), Nov./Dec. 1986, pp. 2662-2669.

94. **Kocolapova, T.J. (ed.)**: *Properties, obtaining and applications of refractory compounds* (in Russian). Publ. Metallurgia, Moscow 1996.

95. **Werth, Ch. A., and Thomson, R.N.**: *Solid state physics* (Transl. from original English), PWN, Warsaw 1974.

96. **Kittel, C.**: *Introduction to solid state physics.* John Wiley and Sons, New York 1966.

97. **Subramanian, C., and Strafford, K.N.**: Review of multicomponent and multi-layer coatings for tribological applications. *Wear*, Vol. 165, 1993, pp. 85-95.

98. **Müntz, W.D.**: Reactive sputtering of nitrides and carbides (in Polish). *Elektronika*, No. 5, 1986, pp. 3-5.

99. **Bujak, J., Miernik, K., Smolik, J., and Walkowicz, J.**: Tool life increasing by deposition of Ti-C-N coatings. Proc.: *VIII International Tool Conference*, Miskolc 1993.

100. **Lunk, A.**: Trends in plasma activated deposition of hard coatings. *Contr. of Plasma Physics*, Vol. 31, No. 2, 1991, pp. 231-246.

101. **Holleck, H.**: Basic principles of specific application of ceramic materials as protective layers. *Surface and Coating Technology*, No. 43/44, 1990, pp. 245-258.

102. **Betz, H.T., Olsen, O.H., Schurin, B.D., and Morris, J.C.**: *Thermophysical properties of high temperature solid materials*, Vol. I, II. Mac Millan Co., New York 1967.

103. **Lunk, A., and Schmidt, M.**: Plasma processes in activated thin film depositions. *Contr. of Plasma Physics*, Vol. 28, No. 3, 1988, pp. 275-279.

104. **Fleischer, W., Schultze, D., Wilberg, R., Lunk, A., and Schrade, F.**: Reactive ion plating (RIP) with auxiliary discharge and the influence of the deposition conditions on the formation and properties of TiN films. *Thin Solid Films*, 63, 1979, pp. 347-356.

105. **Freller, H., Günter, K.G., and Hasser, H.**: Progress in physical vapour deposited wear resistant coatings on tools and components. *Annals of the CIRP*, Vol. 37, No. 1, 1988, pp. 165-168.

106. **Machet, J., Lory, C., and Weissmantel, C.**: Ion plating deposition of hard coatings at T<500°C. Proc.: *E.M.R.S.*, Strassbourg, June 1987.
107. **Hollahan, J.L., and Bell, A.T.**: *Techniques and applications of plasma chemistry.* John Wiley and Sons, New York 1974.
108. **Zdanowski, J.**: Ion modification of the properties of a solid (in Polish). *Elektronika*, No. 4, 1993, pp. 3-6.
109. **Lardon, M., Buhl, R., Singer, H., Moll, E., and Pulker, H.K.**: Influence of the substrate temperature and the discharge voltage on the structure of titanium films produced by ion-plating. *Vacuum*, Vol. 30, No. 7, pp. 255-260.
110. **Zdanowicz, L.**: Influence of deposition conditions on the structure and properties of thin films. Proc.: *1st National Autumn School*, Szczyrk (Poland), Oct. 1979.
111. **Wolin, Z.M.**: Ion-plasma techniques of obtaining wear-resistant coatings (in Russian), *Technologia Lyogkikh Splavov (Technology of Light Alloys)*, No. 10, 1984, pp. 55-58.
112. *Handbook of thin film technology*, Ed. L.I. Massel, R. Glang, McGraw-Hill Book Co., New York 1982.
113. *Deposition technologies for films and coatings*, Ed. R.F. Bunshah, Noyes Publ., New York 1982.
114. **Holland, L.**: *Vacuum deposition of thin films.* Chapman and Hall, London 1963.
115. **Enjouji, K., Murata, K., and Nishikawa, S.**: The analysis and automatic control of a relative d.c. magnetron sputtering process. *Thin Solid Films*, Vol. 108, 1983, pp. 118-121.
116. **Kramer, B.M.**: Requirements for wear resistant coatings. *Thin Solid Films*, Vol. 108, No. 2, 1983, pp. 117-120.
117. **Hardwick, D.A.**: The mechanical properties of thin films; a review. *Thin Solid Films*, Vol. 154, 1987, pp. 109-124.
118. **Dieter, G.E.**: *Mechanical metallurgy.* McGraw-Hill Book Co., New York 1976.
119. **Posti, E., and Nieminen, I.**: Influence of coating thickness on the life of TiN-coated high speed steel cutting tools. *Wear*, 129, 1989, pp. 273-283.
120. **Movchan, V.A., and Demchishin, A.V.**: Investigation of structure and properties of thick vacuum condensates of nickel, titanium, tungsten, aluminium oxide and zirconium dioxide (in Russian). *Fizika Metallov i Metallovedenye*, Vol. 28, No. 4. 1969, pp. 83-86.
121. **Celiński, Z., and Miernik, K.**: Plasma-chemical methods of formation of wear resistant layers (in Polish). *Trybologia*, No. 6, 1991, pp. 6-11.
122. **Thornton, J.A.**: *Annual Review of Materials Science*, Vol. 7, 1977, pp. 239-260.
123. **Robinson, P., and Matthews, A.**: Further developments of the multiple mode ionized vapour source system. *Material Science Forum*, Vol. 102-104, 1992, pp. 581-590.
124. **Michalski, A.**: *Crystallization of multi-phase layers from pulsed plasma* (in Polish). Warsaw University of Technology, Warsaw 1987.
125. **Messier, R., Giri, A.P., and Roy, R.A.**: Revised structure zone model for thin film physical structure. *Journal of Vacuum Science and Technology*, A2(2), 1984, pp. 500-511.
126. **Burakowski, T.**: Status quo and development trends of surface engineering (in Polish). Part III: Classification and general characteristic of new generation methods and parameters used for layer formation. *Przegląd Mechaniczny (Mechanical Review)*, No. 15, 1989, pp. 12-16 and 25-29.
127. **Bujak, J., Miernik, K., Rogowska, R., Smolik, J., and Walkowicz, J.**: Preparation of tool surface for deposition of anti-wear coatings by PVD techniques (in Polish). *Przegląd Mechaniczny (Mechanical Review)*, No. 44, 1994, pp. 12-15.
128. **Rogowska, R.**: Preparation of cutting tool surface for deposition of TiN coatings (in Polish). Seminar on *Selected Problems of Surface Engineering*, Rzeszów-Bystre (Poland), 1992, p.30.
129. **Bujak, J., Miernik, K., Smolik, J., Walkowicz, J., and Barczenko, W.**: Preparation of component surface for deposition of layers by PAPVD techniques with

utilization of ion etching (in Polish). Seminar on *Selected Problems of Surface Engineering*, Rzeszów-Bystre (Poland), 1992, p.32.

130. **Andreev, A.A., Gavrilko, I.V., Kunchenko, V.V., and Sopyrkin, L.I.**: Investigation of some properties of Ti-N$_2$, Zr-N$_2$ condensates, obtained by deposition of plasma streams in vacuum (KIB technique) (in Russian). *Fizika i Khimia Obrabotki Materialov (Physics and Chemistry of Material Treatment)*, No. 3, 1980, pp. 64-68.

131. **Vyskocil, J., and Musil, J.**: Arc evaporation of hard coatings: Process and film properties. *Surface and Coatings Technology*, No. 43-44, 1990, pp. 299-304.

132. **Müntz, W.D.**: Continuous hard coating. *Metal Progress*, No. 8, 1987, pp. 65-68.

133. **Randhawa, H., and Johnson, Ph.C.**: New developments in decorative vacuum coatings. *Metal Finishing*, Sept. 1988, pp. 19-22.

134. **Knotek, O., Prengel, H.G., and Brand, J.**: Deposition of Ti(C,N) coatings by arc evaporation. *Materials Science Forum*, Vol. 102-104, 1992, pp. 591-598.

135. **Ciurapiński, A.,Waliś, L., and Kominek, Z.**: An analysis of possibilities of utilization of radioisotope techniques to investigate migration of elements in materials treated by PVD and low temperature CVD (in Polish). Internal report No. 73/I/86. Institute of Nuclear Chemistry and Technology, Warsaw 1986.

136. **Staśkiewicz, J., and Czyżniewski, A.**: TiN layers obtained by the modified method of reactive direct current magnetron sputtering (in Polish). Proc.: Conference on: *Methods of Formation of Metallic Superficial Layers*, Rzeszów (Poland), June 1988, pp. 99-103.

137. **Madera, B.**: Service life of tools coated by TiN (in Polish). Proc.: Conference on *Methods of Formation of Metallic Superficial Layers*, Rzeszów (Poland), June 1988, pp. 104-105.

138. **Smolik, J., Miernik, K., Walkowicz, J., and Bujak, J.**: Service properties of TiN, TiC and Ti(C,N) coatings (in Polish). Proc.: *II Polish Conference on Surface Treatments*. Czestochowa-Kule, 1993, pp. 159-163.

139. **Miernik, K., Walkowicz, J., and Smolik, J.**: Deposition of AlN layers by collimation magnetron sputtering. *Surface and Coatings Technology*, No. 98, 1988, pp. 1298-1303.

140. **Walkowicz, J., Smolik, J., Miernik, K., and Bujak, J.**: Anit-wear properties of Ti(C,N) layers deposited by the vacuum arc method. *Surface and Coatings Technology*, No. 81, 1996, pp. 201-208.

141. **Smolik, J.**: Interface role in forming anti-wear properties of multilayer TiC/Ti(C,N)/TiN coating obtained by vacuum arc method. Ph. D. thesis (in Polish). Warsaw University of Technology, Warsaw 1998.

142. **Bujak, J., Miernik, K., Smolik, J., and Walkowicz, J.**: Obtaining of TiN and TiAlN layers by the magnetron and vacuum-arc methods (in Polish). *Problemy Eksploatacji*, No. 3, 1992, pp. 157-161.

143. **Bujak, J., Miernik, K., Smolik, J., Walkowicz, J., and Rogowska, R.**: Techniques of deposition of multi-component and multi-layer anti-wear coatings on cutting tools, utilizing plasma-chemical methods (in Polish). *Problemy Eksploatacji*, No. 4, 1993, pp. 39-46.

144. **Miernik, K., Balikojew, I., Barczenko, W., Zagraniczny, S., and Lysenko, W.**: Enhancement of wear resistance by ion treatment (in Polish). *Przegląd Mechaniczny (Mechanical Review)*, No. 15-16, 1993, pp. 39-40.

145. Brochure by Technical University in Koszalin: TiN layers, with hardness almost that of diamond, on cutting tools, forming tools and machine components (in Polish).

146. **König, W., and Koch, K.F.**: Tendenzen in der Werkzeugentwicklung. Proc.: *VIIIth International Conference on Tools (25th Jubilee Conference)*, Miskolc (Hungary), 30 Aug. - 1 Sept., 1993, pp. 17-33.

147. **Bałamucki, J., and Żebrowski, H.**: Effectiveness of reaming small bore holes, using reamers coated with a TiN layer (in Polish). *Przegląd Mechaniczny (Mechanical Review)*, No. 15-16, 1993, pp. 33-35.
148. **Müntz, W.D.**: Titanium aluminium nitride films - a new alternative of TiN coatings. *Journal of Vacuum Science and Technology*, A4(6), Nov/Dec 1986, pp. 2717-2725.
149. **Bujak, J., Miernik, K., Smolik, J., Rogowska, R., and Walkowicz, J.**: Application of TiN coatings for dentistry tools (in Polish). *Przegląd Mechaniczny (Mechanical Review)*, No. 10, 1994, pp. 24-26.
150. Commercial brochure from Ceme-Coat company.
151. **Danielewski, N., Roszczynialska, E., Zdunek, K., and Sokołowska, A.**: Application of the pulsed-plasma technique to obtain wear resistant coatings (in Polish). *Archiwum Nauki o Materialach (Material Science Archives)*, Vol. 14, No. 4, 1993, pp. 295-313.

Subject Index

Milton Keynes UK
Ingram Content Group UK Ltd.
UKHW021931071024
449327UK00022B/1769